Treasure Chest of Six Sigma Growth Methods, Tools, and Best Practices

A Desk Reference Book for Innovation and Growth

Lynne Hambleton

PRENTICE HALL

PRENTICE HALL

UPPER SADDLE RIVER, NJ • BOSTON • INDIANAPOLIS • SAN FRANCISCO

NEW YORK • TORONTO • MONTREAL • LONDON • MUNICH • PARIS • MADRID

CAPETOWN • SYDNEY • TOKYO • SINGAPORE • MEXICO CITY

The publisher offers excellent discounts on this book when ordered in quantity for bulk purchases or special sales, which may include electronic versions and/or custom covers and content particular to your business, training goals, marketing focus, and branding interests. For more information, please contact:

U. S. Corporate and Government Sales
(800) 382-3419
corpsales@pearsontechgroup.com

For sales outside the U. S., please contact:

International Sales
international@pearsoned.com

This Book Is Safari Enabled

The Safari® Enabled icon on the cover of your favorite technology book means the book is available through Safari Bookshelf. When you buy this book, you get free access to the online edition for 45 days.

Safari Bookshelf is an electronic reference library that lets you easily search thousands of technical books, find code samples, download chapters, and access technical information whenever and wherever you need it.

To gain 45-day Safari Enabled access to this book:

- Go to http://www.prenhallprofessional.com/safarienabled
- Complete the brief registration form
- Enter the coupon code JUKU-BLUI-EID7-3NJU-CV7J

If you have difficulty registering on Safari Bookshelf or accessing the online edition, please e-mail customer-service@safaribooksonline.com.

Visit us on the Web: www.phptr.com

Publisher
John Wait

Editor in Chief
Karen Gettman

Acquisitions Editor
Bernard Goodwin

Editorial Assistant
Michelle Housley

Development Editor
George E. Nedeff

Marketing Manager
Curt Johnson

Publicist
Andrea Bledsoe

Cover Designer
Louisa Adair

Managing Editor
Gina Kanouse

Copy Editor
Language Logistics, LLC

Graphic Artist
Laura Robbins

Senior Indexer
Cheryl Lenser

Compositor
Eric S. Miller

Manufacturing Buyer
Dan Uhrig

Library of Congress Cataloging-in-Publication Data:

Hambleton, Lynne.

Treasure chest of six sigma growth methods, tools & best practices : a desk reference book for innovation and growth / Lynne Hambleton.

p. cm.

Includes bibliographical references and index.

ISBN 978-0-13-230021-6 (pbk. : alk. paper) 1. Six sigma (Quality control standard) 2. Strategic planning. 3. Business planning. 4. Management. I. Title.

HD62.15.H354 2008

658.4'013—dc22

2007016916

10-Digit ISBN 0-132-30021-4
13-Digit ISBN 978-0-132-30021-6

Text printed in the United States on recycled paper at R.R. Donneley & Sons in Crawfordsville, Indiana.
First printing, *July 2007*

I dedicate this book first and foremost to my loving husband, Bill, and our two wonderful sons, Corbin and Garrett. I also dedicate this book to Skip and Kathy Creveling, whose friendship and support are invaluable gifts; and to Janet Nelson, a fellow consultant and CSSBB, who is courageously and gracefully battling cancer.

Part III Best Practices Articles

Acknowledgments

Thank you to my friends and professional colleagues who contributed to this book. They took precious time out of their hectic schedules to share the wisdom they have gained through their business experiences. Some contributed inadvertently by brainstorming concepts with me, namely Dan Rose, Joe Szostek, and Chris Tsai. Thank you to the peer reviewers for reading the early drafts to test for understanding. This invaluable manuscript input came from colleagues such as Eric Maass and Scott Wise. Other colleagues authored *Best Practices* articles, featured in Part III (listed alphabetically):

- Thank you to **Donna Burnette** and **David Hutchens**, for agreeing to share some of their professional insights on the critical components of learning that have earned them firm national recognition. Their article on discover-based learning programs adds an invaluable perspective on how best to digest and utilize new knowledge and skills. This approach transforms otherwise dry, dense content into a fun and memorable experience.

- Thank you to **Mike Cook**, whose witty, provocative article on the *importance of collaboration and designing relationships* adds color and refreshing change of pace to this book and reminds us of the important human element involved in the work.

- Thank you to **Clyde ("Skip") Creveling** for not only sharing his creativity and thought-leadership in writing an article on how to *"fast track" a product development process*, but also his unending professional support and guidance.

- Thank you to both **Larry Goldman** and **Karl Luce** for sharing best practices experience on *project selection and the portfolio management process using Monte Carlo simulation and optimization techniques*. Their insights help give a competitive advantage to any reader.

- Thank you to **Bill Jewett** for his sage and practical approach to writing not one, but four articles. His depth and breadth of experience as a practitioner, manager, and consultant were shared in two articles about robust design—specifically on *quality loss* and on *performance variation*. In addition, the best practices of collecting and leveraging the *Voice of the Customer* article, as well as a benchmark *product commercialization process* article round out an engineering snapshot of product development for a non-technical business person.

- Thank you to **Sy Zivan**, one of the benchmarking pioneers in the 1980s from Xerox Corporation, for sharing his knowledge on the *best practices of the benchmarking process.* His article reflects his latest thinking on how benchmarking has evolved over the years.

Thank you to each of these professionals and any others I inadvertently have missed mentioning for adding your unique and invaluable perspectives, all of which enhanced this book.

Thank you to my family and friends who put up with me during the writing process. Those who wove a supportive, energizing network around me include L. Berkowitz, K. Creveling, D. Croessmann, L. Judson, L. Markt, M. McCandless, and especially my husband, W. Magee.

Most importantly, thank you also to the professionals at Prentice Hall for their support and hard work to make this book a reality—Heather Fox, publicist; Bernard Goodwin, editor; Michelle Housley, editorial assistant; and George Nedeff, development editor (listed alphabetically).

About the Author

Lynne Hambleton is a business consultant with special focus on strategy development and execution and change management to improve operational processes and expand commercial opportunities. She has held several management positions in Xerox Corporation where she worked for almost 25 years. She also has worked in education, healthcare, and energy public sectors and start-ups. Hambleton's experience spans general management, marketing, field operations, strategic planning, alliance development, and sales/channel management. She also has served as an adjunct professor of strategic planning at Rochester Institute of Technology's School of Business.

Ms. Hambleton received a Master's degree in Business Administration, with an emphasis in industrial marketing; a Master's degree in Adult & Higher Education/Organizational Development; and a Bachelor of Science degree in psychophysiology, all from University of North Carolina—Chapel Hill. Ms. Hambleton is also an active PMI-certified Project Management Professional (PMP) (1998); a Master Black Belt; and Certified Six Sigma Black Belt (CSSBB) from Villanova University (2006).

Hambleton's additional publications include *Six Sigma for Marketing Processes, An Overview for Marketing Executives, Leaders, and Managers* (co-authors C.M. Creveling and B. McCarthy), Prentice Hall, 2006; the chapter titled, "Supporting a Metamorphosis through Communities of Practice," in *Leading Knowledge Management and Learning*, by Dede Bonner, 2000; and the article, "How Does a Company the Size of Xerox Design a Curriculum in Project Management for the Entire Organization?" printed in *In Search of Excellence in Project Management*, Volume 2, by Harold Kerzner, 1999. Ms. Hambleton lives in Rochester, New York, and can be reached best via email at lynne@mageemanagement.com or visiting www.mageemanagement.com.

Preface

The *Treasure Chest* is part of Prentice-Hall's *Six Sigma for Innovation and Growth Series*. This book serves as a consolidated "how to" reference book of Lean Six Sigma, covers growth and innovation tools, provides an overview of methods and the tools to which they align, and offers an overview of additional best practices used to manage a successful Six Sigma growth initiative. The *Treasure Chest of Six Sigma Growth Methods, Tools, and Best Practices* guides you in selecting the *right tool* to answer *the right question* at the *right time*. The *right question* drives the requirement or need to be addressed—regardless whether the requirement comes from your customer or a business need. Understanding the overall objective, or requirement, helps determine which Six Sigma methodology to use. The *right time* dictates what else you know given where you are in your approach. Understanding whether you are in the planning, designing, implementing, or maintaining phases of an overall process determines which process step, thereby the context of the question being asked. Finally, the *right tool* should be the last question asked, as it is based on first knowing what is required and at what point you are in the process.

The *Treasure Chest* is a desk-reference book for people interested in growth, operations excellence, and business-process improvement. This book speaks to the general business practitioner, business analyst, manager, and leader, regardless of the business context. It is for profit or non-profit enterprises; large or small firms; whether in headquarters-function, plant, or field location, regardless of functional discipline.

The book covers a range of applications from strategic planning aspects of business (offering portfolio renewal) to presenting development and launch preparation, from post-launch operations management to offering discontinuance. Whether contributing a new design (product and/or service), proposing a new process, evaluating a portfolio of offerings, or managing a current portfolio of offerings, this book compiles the resources that help drive growth proactively and presents them in a quick-reference format for easy navigation. This book takes the hassle out of researching the methodology and tools so you can immediately begin to find solutions for your discipline.

The *Treasure Chest* speaks primarily to business people who need practical "hands-on" guidance and answers to the following questions:

1. How do you select the appropriate tool based on the business need (or question being asked) and the required deliverables?
2. How do you use the tool, what inputs or data are required, and what comprises a step-by-step procedure for each tool or technique?
3. How do you analyze the tool's output and decide on the next course of action?

This book was specifically written for general business disciplines, such as marketing, strategic planning, pricing, finance, customer administration, sales, services, support, maintenance, and parts and supplies distribution. This book also is targeted to the technical engineering and research community searching for candidate tools that support communication, project management, risk mitigation planning, and change management requirements.

Common Language

Communication presents a challenge when a collaborative team speaks a different language, different filter, different perspective, different interpretation. *Successful* innovation and growth rely on the integration and collaboration of multiple disciplines, often represented in a *cross-functional team.* Such teams may be comprised of internal, functionally distinct professionals or any combination of external partners, clients, and sometimes even competition. Regardless if the goal requires creating something from scratch or fine-tuning the management of current offerings, the combination of multiple disciplines, capabilities, and perspectives greatly enhance the end results of this work. However, this collaborative work requires a common language to understand and integrate the diversity.

The methods and tools presented in this book assist with interpretation of different perspectives and provide a common platform, foundation, and language from which multiple views can work in harmony. Interestingly, *each discipline has its own unique language, different thought-processes, and/or different tools that characterizes its work.* Sometimes, perhaps with minor adaptations, a tool considered commonplace by one discipline can be viewed as an "a-ha" eureka discovery to a second group when properly applied. Tools commonly used by marketing groups can excite a technical team if introduced at the appropriate time. For example, a communication summary tool or prioritization tool may be overlooked because it is a "soft tool," but it actually can fit perfectly when communicating "big picture" thinking. Similarly, if the time is right, marketing teams enjoy the rigor of techniques from the technical counterparts to provide a fresh perspective.

Treasure Chest embraces tools used by both the technical and non-technical communities and describes when and how to use them. It contains the business tools and methods for innovation and growth to facilitate best practices sharing and a "common language" across multi-disciplined teams. It also integrates some key technical tools appropriate for general business use (or understanding).

INTRODUCTION

Different Methods for Different Purposes

The Evolution of Six Sigma

Six Sigma (or "Lean Six Sigma," as some refer to it) evolved into a rich set of different standardized methods, tools, and best practices. Six Sigma started as a problem-solving approach to reduce variation in a product and manufacturing environment. That application has expanded to process improvement and other areas of the business, including product or process redesign, research, and technology design, offering portfolio renewal, product development, and post-launch operations management. Six Sigma application stretches beyond the manufacturing enterprises into the services industry and non-profit organizations. Regardless of the application, businesses search for simplicity without jeopardizing the need for robust data. Six Sigma offers a set of methods and tools from which to choose. Six Sigma methods build from a common core foundation yet allow flexibility to adapt to changing environment needs. Part of the flexibility stems from a plethora of candidate tools available, depending on the situation. The tool library ranges from rigorous statistical and quantitative tools to "soft" qualitative ones. The purpose of this book is to help in selecting the most appropriate method and the most appropriate tool within the suite of available candidate tools.

Common Approach to Leverage Everyone's Contribution, Regardless of Business Model

A *method* establishes the foundation for how work gets accomplished. It defines the who, what, when, where, why, and how of a *process*; wherein a process describes a series of logically sequenced tasks to complete work. It answers the questions such as, "what gets done;" "who does the work;"

"when the work starts and stops;" "where the work is done;" "why the work is being done;" and "how the work is to be completed." A well-deployed method orchestrates and integrates the people working in a process in an efficient and effective (streamlined) set of activities. It organizes the work as defined by a set of customer requirements. The work produces the agreed-to deliverables according to their "acceptance" criteria. A well-constructed method defines a set of tasks that circumvent redundancies and gaps. Activities done beyond the prescribed work to produce the required deliverables arguably could be called unnecessary or a "waste." A method defines which tool best supports a task and will produce the desired results, providing a common language of terms and tools and a common way of working for those involved in the process.

Overview of Six Sigma Method

The methods used in Six Sigma (including Lean Six Sigma) contain several common principles, such as data-driven decision-making and project management fundamentals. Part I, "Six Sigma Methodology Overview—Choosing the Right Approach to Address the Requirements," uses these principles to organize its content.

Tool-Task-Deliverables Linkage

Six Sigma methods represent a structured thought process that starts with thoroughly understanding the *requirements* (or key business questions) before proceeding. The requirements, in turn, define the *deliverables* to be produced and the *tasks* needed to produce those deliverables and, last, the supporting *tools* to be used to complete the tasks and produce the deliverables. This structure is often called the *Tools-Tasks-Deliverables* combination to indicate the interdependencies. The *Tools-Tasks-Deliverables* linkage is executed in "reverse" or from right-to-left, starting with Deliverables. Hence, a tool is selected only *after* the requirements, deliverables, and tasks are well understood to ensure that the appropriate tool is used for a given task and to avoid the "rut" of treating everything as if it were a nail when the only tool you use is a hammer. The various Six Sigma methods suggest a variety of applicable tools to choose from, but rarely does a given project require the utilization of every tool. No tool fits every situation. Determining which tool fits best depends on the situation. Thus, tool selection is done only after the requirements, resulting deliverables, and tasks are completely understood. [Part II of this book provides not only an inventory of potential tools, but also information on how to apply and interpret results to help you in tool selection.] Remember: ***Use the right tool at the right time to help ask and answer the right questions.***

Result-metrics

The *result-metrics* focus is a distinguishing principle of Six Sigma methods. These fact-based metrics determine whether (internal or external) customer requirements are achieved. Performance typically is evaluated via a statistical metric of the process or offering (e.g. product, services, or information).

High-level process and performance metrics define what critical-to-quality is and encompass the critical parameters necessary to meet requirements. Eventually, these metrics should be translated into a language that is "meaningful" to a process worker involved in providing either the inputs or process deliverables (outputs). Depending on the requirements, the result-metrics may be "hard" or "soft" measurements—quantitative or qualitative; continuous or attribute data. A good litmus test for translated critical-to-quality metrics is whether a "new hire" understands clearly what is expected of him/her to meet requirements with no "fuzzy" or nebulous evaluation of what characterizes "good" or "poor" performance.

Process-centric

Another principle employs a *process-centric* view. Understanding *how* inputs to a process are integrated and how value is added to a product, information, or services offering is as important as *what* is being added. Understanding the combination of *what* and *how* inputs and other key variables come together to produce the final outputs (or deliverables) enables a more accurate forecast of whether customer requirements (or targets) will be satisfied. Prior "results" alone are poor predictors of future outcomes, and without knowledge of the process, any forecast is blinded; any successful forecast would be by chance. Because business prefers accurate forecasts of performance, a process-centric view becomes an integral ingredient.

Adaptive and Iterative

Methods used in Six Sigma are *adaptive* and *iterative*. Adaptive implies the fact that it can be tailored to a variety of situations and business contexts. Moreover, any given Six Sigma method can be integrated with another process or methodology as an underpinning to identify, gather, analyze, and report on critical parameters in a proactive or reactive manner. For example, if your firm has an existing standard product development process or customer account selling approach, Six Sigma can supplement it and make it more robust. The adaptive nature of these methods also speaks to the wide array of industries and situations in which they can be applied. The breadth of industries includes military, government, automotive, aerospace, high-tech, manufacturing, office products, financial services, e-commerce, logistics and supply chain, healthcare, and pharmaceutical industries. Within companies, multiple disciplines have embraced Lean Six Sigma: manufacturing, engineering, finance, administration, customer operations, maintenance, services deployment, and marketing and sales. The approach can even be applied to personal and social situations. Of the Top 100 companies in the 2005 Fortune 500 list, 70 of them have been in the top 100 for five or more years. Interestingly, of those 70 companies, 63% of them publicly acknowledge implementing Six Sigma to some degree. Through further analysis, we have found that these same 44 Six Sigma users also reported 49% higher profits (compounded annually) on average than their peers.

The iterative nature of the Six Sigma methods stems from the fact that more information on a variable or potential root cause gets revealed as the project progresses. Hence, one path of inquiry based on one assumption may prove to be a dead-end or altered, as more data on the current state becomes known. Although Six Sigma methods use a project-structure with phase-gates, a fundamental principle across the various approaches encourages informed updates to prior step deliverables, as appropriate, and promotes proper communication. Use the best information available at the time but continue to ask questions and keep an open mind. Six Sigma projects involve a discovery process wherein an individual serves as a "sleuth," investigating, exploring, hypothesizing, and testing assumptions.

Data-driven Decision-making

Given the uncertain nature of projects, when seeking facts that answer key business questions, revisions to earlier project work reflect the evolving discovery of fact-based results. For example, a business operations review may focus on a set of key metrics to manage a process. If a chronically missed target evokes a Six Sigma project, an interim project deliverable could identify that some of the metrics associated with the key variables driving the desired business outcomes are missing, hence the "dashboard" requirements become refined to reflect the vital few parameters, comprised of both leading and lagging metrics. Next, the project could focus on establishing baseline data for the "new" critical metrics to re-evaluate performance and better understand any cause-and-effect relationship(s).

Project-based Methods

As previously referenced, Six Sigma methods tend to use a *project structure*. A project structure has a distinct beginning and end to the work performed. The requirements phase determines the boundaries of this definitive timeframe. A project team, with defined roles, forms only for the duration of the project's timeframe. A project structure adds the rigor of requiring completed deliverables approvals (often gained in a phase-gate review meeting with the project and key stakeholders) before exiting a given phase-gate or step and starting another one. Project structure borrows heavily from the project management discipline and its nine knowledge areas to manage the lifecycle of the project: scope, time, budget/cost, risk, quality, communications, human resources, procurement, and integration.

The *project* context of Six Sigma methods incorporates a rather short-term perspective (averaging a three, six, or twelve-month project scope). A project may involve an improvement or enhancement to something [often focused on reducing defects, minimizing variance from a target, or improving velocity (speed)], clean-sheet innovation, or design and creation (such as product or services development or portfolio assessment). The technical community (for example, engineering or manufacturing) has embraced a category of Six Sigma methods called Design for Six Sigma (DFSS). A newly emerging field is Six Sigma for Marketing (SSFM). SSFM may be a misnomer, because the various methods apply to the remaining ("non-engineering") business disciplines, such as marketing, sales, strategic planning, services, and customer operations. Examples of project-based methods include

- DMAIC (Define-Measure-Analyze-Improve-Control, and its variants DMAIIC (with "II" representing Improve-Innovate) and Lean Six Sigma)
- Lean (and its variants PDCA (Plan-Do-Check-Act) / PDSA (Plan-Do-Study-Act) and Lean Six Sigma)
- DFSS category with DMADV (Define-Measure-Analyze-Design-Verify), CDOV (Concept-Design-Optimize-Verify) (and their variants DMEDI (Define-Measure-Explore-Develop-Implement), PIDOV (Plan-Identify-Design-Optimize-Validate), ICOV (Identify-Characterize-Optimize-Verify), and IIDOV (Invent-Innovate-Develop-Optimize-Verify))
- SSFM category with UAPL (Understand-Analyze-Plan-Launch) and (sometimes) IDEA (Identify-Define-Evaluate-Activate)

Operational-based Methods

Managing an ongoing operation, however, is emerging as a new application area. Hence, the application of the Six Sigma method and tools to an operational process may last for years, rather than months as with a short-term project. The objective of operational-based Six Sigma is to manage or sustain an improvement of a launched product and/or services offering, for example, or to adapt and respond to environmental changes. This operational focus of Six Sigma is being applied to business areas such as customer operations (for sales, services, support, administration, financing, and related business disciplines) and strategic planning (for offerings portfolio management). Some might argue that portfolio management can be handled as a project defined by an annual planning cycle, but others view it as an ongoing area.

Nonetheless, Six Sigma discipline has added a competitive advantage to those firms that have begun to apply its method and tools to this process area. The operations-based methods currently fall within the SSFM (Six Sigma for Marketing) category with LMAD (Launch-Manage-Adapt-Discontinue, for customer operations) and sometimes IDEA (Identify-Define-Evaluate-Activate, for strategic planning), which can be considered operational given that the management and revitalization process of a firm's offerings portfolio is cyclical and can span multiple years.

How Do the Various Six Sigma Methods Fit Together?

In summary, this Introduction overviews the major Six Sigma approaches being used currently. Each Six Sigma method has a valid purpose in today's business world, the selection of which approach best fits a need depends on the key business question being asked at the time. They all fit together and inform one another. The integrated view is as follows:

An enterprise's strategic platform defines its business and offerings, so typically a process flow starts with the business strategy process of portfolio definition and renewal (IDEA — Identify-Define-Evaluate-Activate).

From there, funding gets earmarked for research, tactical, and operational activities. Research and Technical Development (R&TD) efforts are funded to develop forward-looking capabilities that eventually feed product development and commercialization. The approach used to guide Research's activities is called I^2DOV (Invent-Innovate-Develop-Optimize-Verify).

The specific offering's design, development, and commercialization efforts split into two branches: 1) the technical team that uses CDOV (Concept-Design-Optimize-Verify) to guide its activities and 2) the marketing and business areas that use the UAPL (Understand-Analyze-Plan-Launch) approach.

Finally, the operational and supporting infrastructure and business areas of a post-launch customer value chain environment use the LMAD (Launch-Manage-Adapt-Discontinue) approach to guide and direct their activities.

If any of these areas of an enterprise encounter a trouble spot in an existing process or offering, the Lean Six Sigma DMAIC (Define-Measure-Analyze-Improve-Control) method and any of its variants (that is, DMAIIC (Define-Measure-Analyze-Improve-Innovate-Control), DMADV (Define-Measure-Analyze-Design-Verify), DMEDI (Define-Measure-Explore-Develop-Implement)) focus on the problem and its root cause-and-effect to determine the best correction.

Figure 1 depicts how each method integrates with one another.

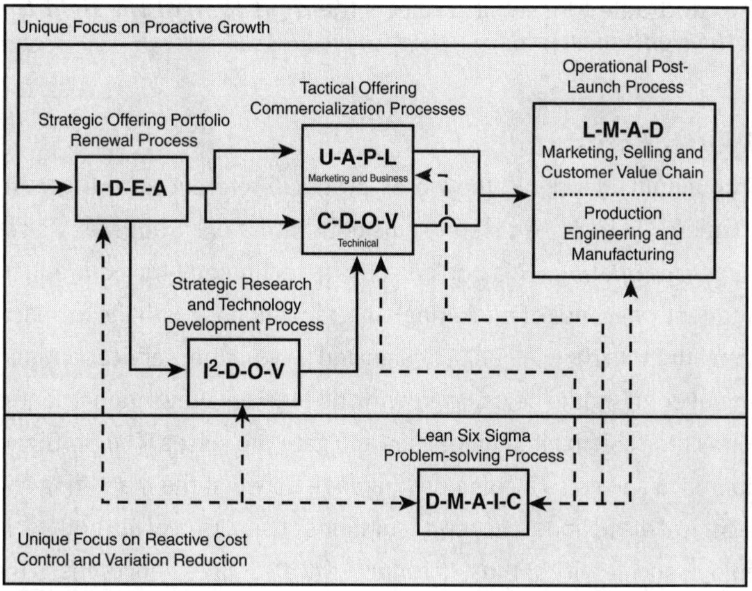

FIGURE 1: Integrated Portfolio of Six Sigma Methods

The various Six Sigma approaches have their appropriate applications. Moreover, their candidate tools and methods sometimes overlap or feature tool variants of another's, as they all build from the core Six Sigma fundamentals.

Right Tool at the Right Time

Part II of this book explores how to select *the most appropriate tool to answer the right question at the right time.* It organizes the tools in alphabetical order and, using an encyclopedia-style article format, describes what question each tool tries to answer and how to use the tool.

Take a closer look at the structure of the various Six Sigma (or Lean Six Sigma) methods to understand their similarities and differences and when best to apply each of them. Start with Part I to understand the various methods and their structure, requirements, deliverables, and list of candidate tools. Afterward examine Part II, the heart of this book, to decide which of the candidate tools might be appropriate for your project. Part II is structured as a desk reference that inventories the different tools and techniques. Each tool "article" describes the main purpose of the tool, how best to use it, how to interpret the results of the tool, and any variations on how to apply the tool or technique. Most importantly, Part I lists the candidate tools aligned with particular requirements and task-deliverable combinations, but Part II identifies in more detail which key question each tool tries to answer. Remember—an effective Six Sigma practitioner scrutinizes the candidate tool set and selects the ***right tool, at the right time, to answer the right question.***

Special Note

Throughout this book, the word "product" references to a generic company "offering" and represents both a *tangible product* as well as a *services offering.* This book discusses technology-based products frequently because of marketing's interdependency with the technical community. In parallel, R&D, design and production/services support engineering should be using growth and problem prevention-oriented forms of Six Sigma in their phases and gate processes. The Six Sigma approach serves as a common language between the marketing and technical disciplines. The term "solutions" usually involves both technology and services; thus, "product" and "services" encompass the scope of a given solution. Regardless of the offering, the Six Sigma approach we are outlining is the same and can be applied to either a tangible product or a services offering.

In addition, the term "Six Sigma" refers to the generic field or discipline and encompasses the many different approaches. People may distinguish Six Sigma (SS), Lean Six Sigma (LSS), Design for Six Sigma (DFSS) and Six Sigma for Marketing (SSFM) from one another. However, this book uses "Six Sigma" categorically unless otherwise noted.

What this Book Covers

The *Treasure Chest* is organized into three parts: 1) Six Sigma methodology, 2) tools and techniques, and 3) best practices applicable to Six Sigma deployment. The book uses an encyclopedia-like format made up of over 60 topics.

Part I Six Sigma Methodology Overview—Choosing the Right Approach to Address the Requirements

This segment of the book presents an overview the various Six Sigma approaches and describes the purpose of each. With the knowledge of the different methods, this book connects the requirements of a method to the appropriate candidate tools and techniques. This Part serves as the foundation for selecting the right tool for a given purpose.

It discusses the various technical, business and marketing Six Sigma methods, including DMAIC, Lean Six Sigma, Design for Six Sigma (DFSS), and Six Sigma for Marketing (SSFM). Within each approach, a general description, common applications, and key requirements provide an overview of its structure and purpose. In addition, the key requirements determine each method's unique *tools-tasks-deliverables* combination. The Method Section provides the foundation for comparing and contrasting the different approaches. Once a method is selected, it establishes a team's common language, regardless of whether or not people are part of a hetero- or homogeneous team. The method defines the requirements for its work activities. It sets expectations, describes required deliverables and their due dates, and identifies who does what when. A common method is the single most critical unifying theme for a team's work. If it is well understood and followed by each team member, then collaboration, coordination, and communication can occur fluidly.

Part II Six Sigma Encyclopedia of Business Tools and Techniques—Choosing the Right Tool to Answer the Right Question at the Right Time

This Part features an in-depth look at a robust library of tools, organized in alphabetical order for easy reference. By design, it represents the bulk of this book, with an exhaustive review of the Growth and Lean Six Sigma tools and techniques. Each tool (tangible item) or technique (for example, brainstorming) features a "how to" description explaining how to utilize the tool and interpret typical outcomes. Each description identifies the deliverable the tool or technique supports. Part II features the following topics for each of the 60+ tools included in this book:

- The question the tool helps to answer
- Any alternative names or variants associated with the tool
- When best to use the tool or technique
- A brief description and useful real-world examples
- How to implement the tool for your application
- How to analyze and apply its output
- Helpful hints and tips that encourage you to think outside of the box
- Supporting candidate tools that link to the featured tool, depending on the question needing to be answered, by providing either input to or using the output from this featured tool

Part II begins with a useful summary table of tools organized by the type of question it helps to answer. The *Treasure Chest* organizes the candidate tools and techniques alphabetically for easy reference. The encyclopedia includes an array of soft tools and techniques, graphical methods, and statistical tools. The statistical tool descriptions leverage some of the appropriate software tools such as MINITAB, Minitab's new Quality Champion, Decisioneering Crystal Ball, Visio, and other Microsoft applications, such as Excel.

Part III Best Practices Articles

This portion of the book contains a series of articles written by well-reputed professionals that complements and extends beyond the world of Six Sigma to provide that competitive advantage in growth and operational excellence. It features a collection of technical and humanistic topics ranging from the latest thinking on benchmarking strategy to determining and offering the best portfolio. Articles describe how best to accelerate the development of an offering, how to ensure design robustness, and how to govern projects. Part III includes a discussion on best practices to introduce, deploy, and sustain a major culture change, such as deploying Six Sigma thinking, by featuring a set of articles to support a change initiative including communities of practice and project management; simulation approach to training of new skills, knowledge, and attitudes; and designing collaborative work relationships.

A final note, the "Six Sigma for Marketing" (or SSFM) terminology in the marketplace may mislead prospective practitioners wishing to use Six Sigma to drive innovation and growth. While the primary application of SSFM involves processes typically associated with marketing, the respective work often encompasses additional functional disciplines within a company. Depending on the size of the firm and its business model, the professionals involved in 1) portfolio renewal, 2) offering development and commercialization preparation, and finally 3) post-launch operations management throughout an offering's lifecycle reach beyond just marketing. General business professionals involved in these three processes also represent disciplines such as strategic planning, pricing, finance, customer administration, customer service and support, professional services, and logistics and supply chain. In fact, the third process involving post-launch operations spans the entire customer value chain. Hence, when this book references "SSFM," it follows the marketplace terminology of the emerging Six Sigma focus. In the context of SSFM, the identification of "marketing" (and sometimes sales) distinguishes the new Six Sigma application as different from the classic variation reduction, problem-solving, and cost cutting approach, and as different from the technical DFSS (Design for Six Sigma). However, the *Treasure Chest* intends the reference to marketing as a "loose" association and prefers the broader reference of "*general business*" to better articulate its broader scope and applicability. This book works nicely as a sequel to the *Six Sigma for Marketing Processes*, co-authored by CM Creveling, L. Hambleton, and B. McCarthy.

Page numbers preceded by PDF: indicate pages available for download at www.prenhallprofessional.com/title/0132300214.

Part IV Appendixes

The Appendixes contain a set of references such as charts and statistical tables for hypothesis testing, a glossary, and a list of references.

PART I

Six Sigma Methodology Overview: Choosing the Right Approach to Address the Requirements

1

Define-Measure-Analyze-Improve-Control (DMAIC)

Six Sigma's most common and well-known methodology is its problem-solving DMAIC approach. This section overviews the methodology and its high-level requirements, given that the requirements define the appropriate deliverables, which dictate the tasks and the tool selection to aid in the task. This section also outlines the DMAIC standard toolset, through the understanding of the tool-task-deliverables linkage, to facilitate appropriate selection of a tool when referencing the "how to" tool articles in Part 2 of this book.

What Is the Main Objective of this Approach?

The DMAIC (Define-Measure-Analyze-Improve-Control) is the classic Six Sigma *problem-solving process*. Traditionally, the approach is to be applied to a problem with an existing, steady-state process or product and/or service offering. **Variation is the enemy**—variation from customer specifications in either a product or process is the primary problem.

Variation can take on many forms. DMAIC resolves issues of defects or failures, deviation from a target, excess cost or time, and deterioration. Six Sigma reduces variation within and across the value-adding steps in a process. DMAIC identifies key requirements, deliverables, tasks, and standard tools for a project team to utilize when tackling a problem.

Brief Description of DMAIC Applications

This classic or traditional Six Sigma methodology was designed to solve a problematic process or product and/or service offering **to regain control**. It addresses improvements in productivity (how many), financial (how much money), quality (how well) and time (how fast)—PFQT. Originally costs dominated the financial aspects, but lately project focus has shifted to revenues and growth. The 5-step DMAIC [pronounced *"duh-**MAY**-ick"*] method often is called the *process improvement* methodology.

The classic strategy reduces process variance (in total, across the activities and within-step) to bring it back on target—the customer specification or requirement. Knowing that **Six Sigma resolves more issues than just cycle time,** Figure 1-1 highlights its impact on cycle time by contrasting a problematic process versus its post-Six Sigma improved state.

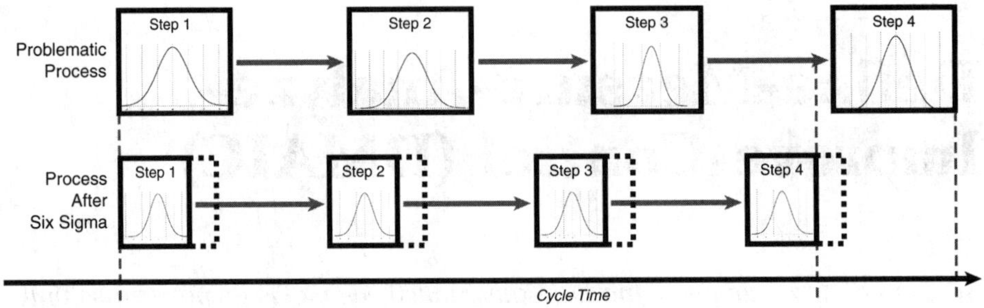

FIGURE 1-1: Six Sigma's Impact on Cycle Time

The DMAIC approach is designed to allow for flexibility and iterative-work, if necessary. As more is learned through the 5-step process, assumptions or hypotheses as to the root cause of the problem may be disproved, requiring the project team to revisit them and modify or to explore alternative possibilities. For example, root cause to a sales force effectiveness issue may have been hypothesized as a sales training problem in a specific geographic region. Rather than jumping to conclusions without facts by implementing a new sales training program, the Six Sigma project team wisely decides to gather facts about the problem first. After some investigation and analysis, the team discovers that the root cause points to an issue with sales management direction, not lack of sales representatives' knowledge and skills. If the project team acted upon the original assumption, time and money would have been wasted on developing a mismatched solution that would have produced poor results; the team's hard work would have gone to waste. Instead, the team did a mid-course correction based on facts, adjusted its hypothesis, and developed a solution directly aimed at the true root cause—hence favorable results ensued.

DMAIC builds on three fundamental principles:

- Results-focused; driven by data, facts, and metrics.
- Work is project-based (short-term in nature, with length depending on scope and complexity) and **project-structured**, versus an ongoing process.

- Inherent **combination of tools-tasks-deliverables** linkage that varies by step in the method.

The DMAIC methodology uses a process-step structure. Steps generally are sequential; however, some activities from various steps may occur concurrently or may be iterative. Deliverables for a given step must be completed prior to formal gate review approval. Step Reviews do occur sequentially. The DMAIC five steps are

Step 1. **DEFINE** ***the problem and scope the work effort of the project team.*** The description of the problem should include the pain felt by the customer and/or business as well as how long the issue has existed. Hence, identify the customer(s), the project goals, and timeframe for completion.

The appropriate types of problems have unlimited scope and scale, from employee problems to issues with the production process or advertising. Regardless of the type of problem, it should be systemic—part of an existing, steady-state process wherein the problem is not a one-time event, but has caused pain for a couple of cycles.

Step 2. **MEASURE** ***the current process or performance.*** Identify what data is available and from what source. Develop a plan to gather it. Gather the data and summarize it, telling a story to describe the problem. This usually involves utilization of graphical tools.

Step 3. **ANALYZE** ***the current performance to isolate the problem.*** Through analysis (both statistical and qualitatively), begin to formulate and test hypotheses about the root cause of the problem.

Step 4. **IMPROVE** ***the problem by selecting a solution.*** Based on the identified root cause(s) in the prior step, directly address the cause with an improvement. Brainstorm potential solutions, prioritize them based on customer requirements, make a selection, and test to see if the solution resolves the problem.

Step 5. **CONTROL** ***the improved process or product performance to ensure the target(s) are met.*** Once the solution has resolved the problem, the improvements must be standardized and sustained over time. The standard-operating-procedures may require revision, and a control plan should be put in place to monitor ongoing performance. The project team transitions the standardized improvements and sustaining control plan to the process players and closes out the project.

A DMAIC project typically runs for a relatively short duration (three to nine months), versus product development projects (using UAPL or DFSS) and operational line management (using LMAD), which can run years. Given the relatively shorter duration to other types of Six Sigma methodologies, we distinguish the DMAIC as having five *steps*, rather than phases.

The DMAIC method is primarily based on the application of statistical process control, quality tools, and process capability analysis; it is **not** a product development methodology. It can be used to help redesign a process—any process, given that the redesign fixes the initial process problem. To be implemented, the method requires four components:

- A **measurement system** (a gauge) of the process or product/service offering in trouble.
- **Standard toolset** that supports tasks to produce deliverables (including statistical, graphical, and qualitative tools and techniques).
- An ability to define an **adjustment factor(s)** to correct the process or product/service offering back on target.
- **A control scheme** to maintain the improvement or correction over time by implementing a control plan with a monitoring system to audit the response performance against statistical control limits and defined action plans if needed.

What Key Overall Requirements Define this Approach?

Requirements come from the customer and the business, depending on the problem scenario. The (internal and external) customer requirements get translated into what is *critical-to-quality* (CTQ). These CTQs define the criteria to evaluate what good looks like—how well the project scope and deliverables meet requirements. Hence, the project team must meet the requirements of a phase before declaring completion and closing it out.

The DMAIC method was designed and structured to answer the following overall business questions:

- What does the customer define as the problem?

  (Secondarily, is the problem sustained over time, is it chronic, or is it a one-time occurrence?)

- What characterizes the current problem (e.g., process and performance metrics), and how has it changed over time?

  (Secondarily, is the process in control, and does it have a good measurement system? Is the process capable of producing the customer requirements?)

- What are the root causes, and what improvement actions correct them to meet customer requirements again?

  (Secondarily, is the process capable of producing the customer requirements?)

- What controls should be implemented to sustain this improvement, including a warning system, action plan, and communication plan needed in case requirements fail to be met?

  (Secondarily, can the improvements be sustained over time?)

What Requirement Determines the Key Activities in this Approach?

The preceding key business questions determine the DMAIC architecture. Figure 1-2 depicts a high-level process flow of the DMAIC method through its five steps.

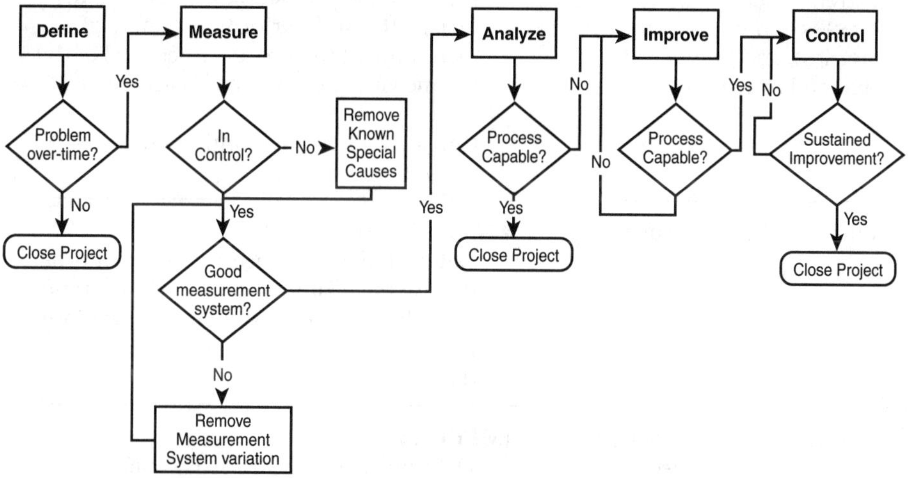

FIGURE 1-2: High-Level DMAIC Process Flow

Table 1-1 shows the linkage between the high-level business requirements and the five-step DMAIC method.

Table 1-1: DMAIC Requirements-Step Linkage

Requirements	Resulting High Level Task (Step)
What does the customer define as the problem?	1. **DEFINE** • Describe in the words of the external or internal customer—Voice of Customer (VOC). • Define the boundary conditions set forth by the business, including regulatory environment—Voice of Business (VOB). • Understand the current process. What has happened over time, examine process control charts to identify incidents of common and special cause variation—Voice of the Process (VOP).
What characterizes the current problem (that is, process and performance metrics), and how has it changed over-time?	2. **MEASURE** • Measure the problem; describe it with facts, data, and performance metrics. Determine if the process in control and if the measurement system is accurate. • Considered iterative until metrics are gathered over time.
What are the root causes?	3. **ANALYZE** Determine if the process capable of producing the customer requirements. If not, consider it iterative until root causes are identified and verified with facts and data.
What improvement actions correct the root causes to meet customer requirements again?	4. **IMPROVE** Determing if the process is capable of producing the customer requirements. If not, consider it iterative until improvements are identified and verified with facts, data, and performance metrics.
What controls should be implemented to sustain this improvement, including a warning system, action plan, and communication plan needed in case requirements fail to be met?	5. **CONTROL** • Demonstrate how the improvements and/or changes can be sustained. • Manage Risks

Figure 1-3 provides a DMAIC icon that reinforces both the overall flow of a method and the purpose of each step and respective interrelationships. It summarizes the five-step DMAIC process and its notable iterative nature. Throughout the remainder of this text, Figure 1-3 will symbolize the DMAIC approach and indicate a particular step within it if appropriate.

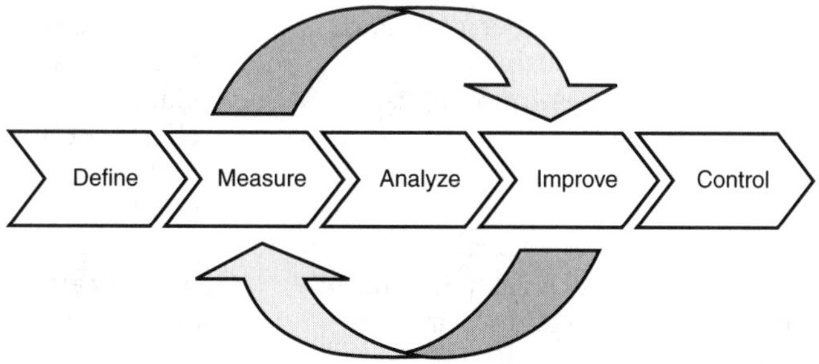

FIGURE 1-3: DMAIC Icon

What Tools Are Aligned to Each Step of the Process?

Given the preceding High Level Task Step(s), the following series of tables summarize the subsequent tool-task-deliverables combination associated with each individual step within the five-step approach.

The detail behind how to use each tool can be found in Part II, "Six Sigma Encyclopedia of Business Tools and Techniques: Choosing the Right Tool to Answer the Right Question at the Right Time."

TABLE 1-2: Define Tools-Tasks-Deliverables

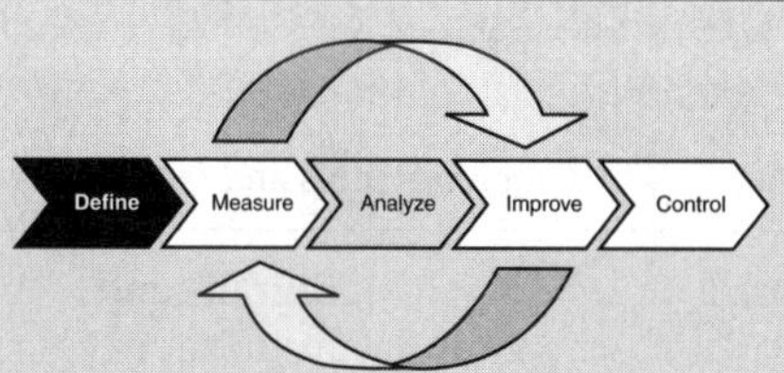

Step 1: DEFINE
What does the customer define as the problem?

Deliverables	Tasks	Candidate Tools and Techniques
Project Charter Approved (contract with the Sponsor regarding problem, project scope, project goal(s), key deliverables, timeframe, and budget)	Identify Problem statement/Opportunity and Goal statement.	• SMART • Project Charter Form containing: Problem Statement (As-Is), Desired State (To Be), and the Business Reasons • Big "Y" over time

continues

Table 1-2: Continued

Deliverables	Tasks	Candidate Tools and Techniques
High-level Process Map Constructed	Develop High-level Process Map	• Process Map • RACI Matrix
Critical Parameters Hypothesized	Gather VOC and Business Requirements	• VOC/VOB Gathering techniques • Current process control charts (VOP) • Stakeholder Analysis • CTQ
Project Charter Published and Communicated	Develop Communication Plan	Communication Plan template
High-Level Project Plan Defined and Approved	Finalize Project Charter	• Project Charter Form • High-level Process Map • SIPOC • Project RACI Matrix

Table 1-3: Measure Tools-Tasks-Deliverables

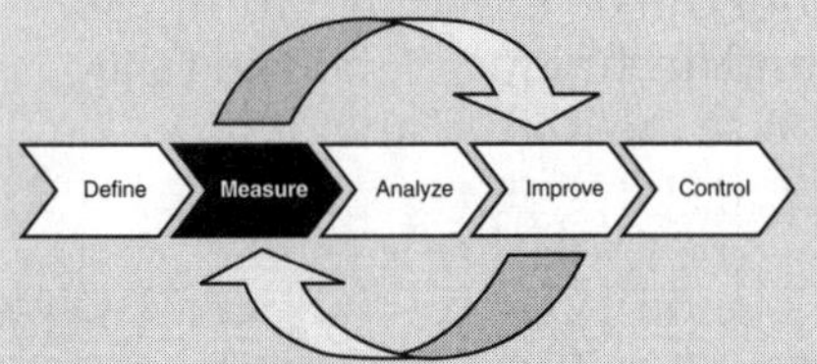

Step 2: MEASURE
What characterizes the current problem, and how has it changed over time?

Deliverables	Tasks	Candidate Tools and Techniques
Data Collected	• Identify Sources of Data • Collect Baseline Data from existing process • Determine current Process Performance; is it in control? • Remove any known special causes; verify if process is in control	• Y = f(X); Big "Y" and little "Ys" • Data Gathering Plan template • Control Charts • Statistical Sampling • Graphical Methods • QFD (Quality Function Deployment)

Deliverables	Tasks	Candidate Tools and Techniques
Process Map Defined In-depth With Current Measures	Develop Detailed Process Map	• Detailed Process Map • RACI Matrix, revised
Current Measurement System Capability Evaluated	• Validate measurements and collection system • Is the process capable of meeting requirements?	• Measurement System Analysis (MSA) • Process Capability Analysis
Project Charter and Plan updated, as necessary	• Revise Problem and Goal statements as needed • Update Project Plan, as needed	• Project Charter; its plan and milestones • Project RACI Matrix

TABLE 1-4: Analyze Tools-Tasks-Deliverables

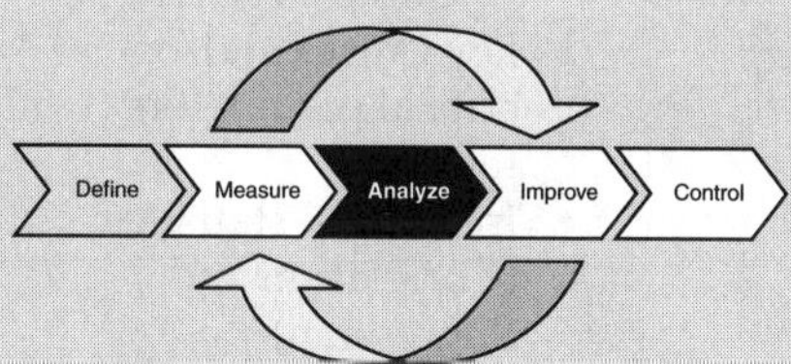

Step 3: ANALYZE
What are the root causes of the current problem?

Deliverables	Tasks	Candidate Tools and Techniques
Data Analyzed	• Validate gaps in requirements vs. current metrics • Establish Y=f(X) • Quantify Opportunity to close gaps	• Y = f(X); Big "Y"; little "Ys" and the "Xs" • Critical Gap/Step Analysis • Pareto Charts • Statistical Analysis: Normal Distribution, Variation • Correlation and Regression

continues

TABLE 1-4: Continued

Deliverables	Tasks	Candidate Tools and Techniques
Process Analyzed	• Develop Detailed Process Map • Establish Y=f(X) • Quantify Opportunity to close gaps	• Detailed Process Map (inputs, outputs, metrics, process step owners) • RACI Matrix, revised • Process Mapping of Critical Parameters • Y = f(X) • Pareto Charts • Process Capability Analysis
Root Cause Analyzed	• Conduct Root Cause Analysis • Prioritize Root Causes • Quantify Opportunity to close gaps	• Brainstorming Technique • Cause and Effect Diagrams • Five Whys • Affinity Diagram (KJ) • Hypothesis Testing of key causes and/or critical parameters (vital few Xs) • Inferential statistics (Correlation and Regression) • DOE • FMEA
Project Charter and Plan updated, as necessary	• Revise Problem and Goal statements as needed • Update Project Plan, as needed	• Project Charter; its plan and milestones • Project RACI Matrix

TABLE 1-5: Improve Tools-Tasks-Deliverables

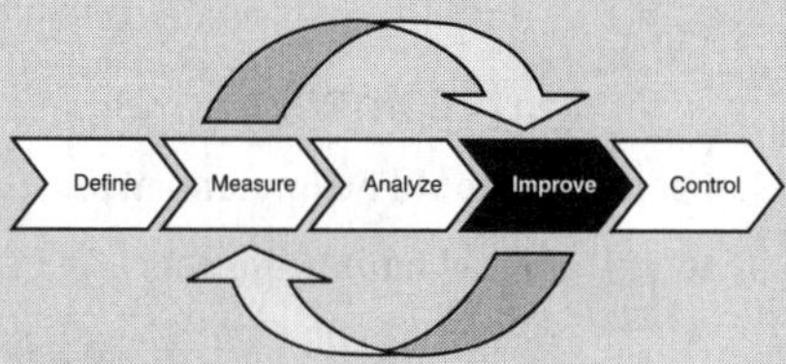

Step 4: IMPROVE
What improvement actions correct the root causes to meet customer requirements again?

Deliverables	Tasks	Candidate Tools and Techniques
Potential Solution Generated	Develop Potential Improvements or solutions for root causes	• Brainstorming Technique • Positive Deviance • TRIZ
Potential Solution Evaluated	• Develop Evaluation Criteria • Measure results • Evaluate improvements meet targets • Evaluate for Risk	• Basic DOE • Pilots/Tests • FMEA • Cost/Benefit Analysis
Solution Selected	Select and Implement improved process and metrics	• Pugh Concept Evaluation • Solution Selection matrix • Force Field diagram • QFD • Measurement System Analysis (MSA) • Process Capability Analysis
Improved Path Forward Implemented	Develop Detailed Future Process Map of improvement	• Detailed Process Map • RACI Matrix, future • Procedure manual (standard operating procedure) • Implementation and Transition Plan
Project Charter and Plan updated, as necessary	• Revise Problem and Goal statements as needed • Update Project Plan, as needed	• Project Charter; its plan and milestones • Project RACI Matrix

TABLE 1-6: Control Tools-Tasks-Deliverables

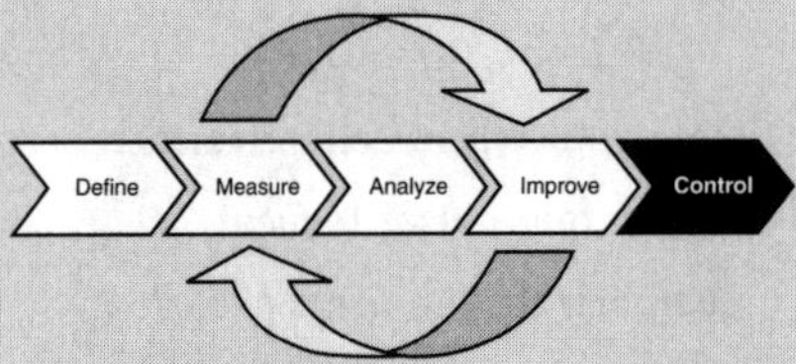

Step 5: CONTROL
What controls should be implemented to sustain this improvement?

Deliverables	Tasks	Candidate Tools and Techniques
Control Plan Defined	• Document New Measurement Process • Define control plan	• Control Plan Design • Control Charts (SPC) • FMEA/Risk Analysis • Communication Plan • Stakeholders Analysis
Improvements/ Innovation Implemented	Validate metrics and collection systems	• Measurement System Analysis (MSA) • Process Capability Analysis • Cost/Benefit Analysis
Training Conducted	Train	• Training/Transition plan
Process Documented	Document recommendation or improvement summary and highlight changes from As-Is to Improved	• Process Map • RACI • Procedure manuals
Tracking System Deployed	Establish Tracking Procedure	• Scorecard or Dashboard • Data Mining (MINITAB graphical data analysis)
Lessons Learned Documented and Project Closed	• Revise Problem and Goal statements to reflect actual • Update Project Plan to reflect actual • Record lessons learned and file along with final project documentation	• Project Charter; its plan and milestones • Project RACI Matrix • New SIPOC

What Are Some of the Key Concepts that Characterize this Approach?

There are some key characteristics that distinguish DMAIC from other Six Sigma methods. The following overview wraps up the DMAIC highlights and introduces some of its variants.

How Is the Problem Defined?

The problem statement in a Project Charter typically speaks to defects or variance from a target *over time* with an *existing, steady* state, process, or product. (The charter is part of a standard Six Sigma toolset used to document the project scope. **See Also** "SMART," in Part II, **p. 665**) Typically, the customer should determine the target; however, at times the business, industry standard, or regulatory agency may set it. Time-based problem statements indicate the problem may be chronic (has persisted for a period of time), which helps create a case for change (versus a one-time occurrence) to incite interest in and resources to tackle the issue.

Common metrics include DPMO (Defects per Million Opportunities (or units)), PPM (Parts per Million), Mean Time between Failures (MTBF), Cost, Percent Variance, or Errors.

What Is Commonly Measured?

Typically, three key items are measured:

- **Output (or Outcome)**—The end result of the process (or product) requiring improvement
- **Process**—The workflow (of activities and items) that produces the output
- **Inputs**—The raw materials and information used by the process to produce the output

The relationship of these three key items often is described as an equation: Y = f(x), which reads, *"Y is a function of X."* The "Y" refers to the output(s); the "X" refers to the key measures from the process variables (inputs and/or the process itself). **See Also** "Y=f(x)," in Part II, **p. 758**.

The DMAIC project goal is to identify the critical (or vital few) Xs—the *root cause* of the problem—and select their optimal level(s) to best drive the desired improvement in the output performance (sometimes called the "Big Y"). This language sounds foreign to many people not comfortable with mathematically-structured sentences; however, it is readily used in most Six Sigma texts. A simpler articulation is the **goal of a DMAIC project is to improve *PFQT—Productivity (how many), Financial (how much money), Quality (how well), and Time (how fast).***

Are There any DMAIC Variations?

There are two prevalent variations to the traditional DMAIC method. Both build on the DMAIC fundamentals but add new dimensions to extend its applications. The first is DMAIIC, wherein **innovation** is added for situations where a simple improvement modification is inadequate and a new design may be required. (Note that in the technical, engineering arena that an innovation adaptation typically aligns with the DMADV method. DMADV distinguishes itself from DMAIIC by not only its often unique environment scenario, but also it usually calls for a requirement of building a new process (or product design) from scratch at the start of the project; whereas, DMAIIC often is unaware of the redesign requirement until much later into the project lifecycle. **See Also** "DFSS," in Section 3, **p. 45**) The second is Lean Six Sigma, which adds concepts of *velocity*, *value-add*, and *flow* to the DMAIC concepts.

1. **DMAIIC—Adding an "I" for Innovation**—Many organizations have found that improving a current process or product may not be enough to deliver the desired results, and at times *innovation* is needed. Since the project teams have just completed the Define-Measure-Analyze stages of the process and are in the midst of Improve, rather than starting over from scratch, project teams have found that the work done to this point is a good foundation for innovation work. Hence, some companies have built on the DMAIC framework already in-place and added a second "I" for innovation to keep the project team progressing. Therefore, the variation is Define-Measure-Analyze-Improve/Innovate-Control.

Figure 1-4 shows the DMAIIC flow diagram and depicts how the Innovate tasks integrate into the classic DMAIC model.

2. **Lean Six Sigma—Adding Lean Concepts**—By incorporating **lean** concepts into DMAIC, the project adds a dimension of velocity (i.e. improved cycle time), value-add, and flow to what Six Sigma already offers.

Both concepts share similar views on customer-focus, process-centric work, and appropriate tools. Lean simply adds a deeper set of tools to eliminate waste between process steps handoffs. Often DMAIC provides a project the big picture view (what the customer values balanced by business values) and process stabilization and capability—while Lean introduces speed and flow concepts at a more detailed level. The Define-Measure-Analyze-Improve-Control structure still holds true for Lean Six Sigma projects. **See Also** "Lean and Lean Six Sigma," Section 2, **p. 29**, for more details on Lean.

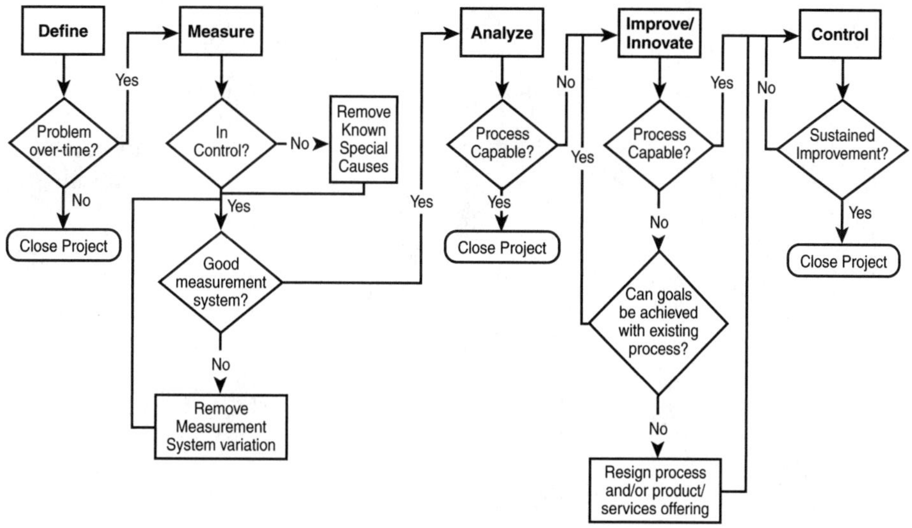

FIGURE 1-4: High-level DMAIC Process Flow

Summary

Many view DMAIC as the foundation of Six Sigma. DMAIC is best used as an iterative problem-solving method to combat variation in an existing, steady-state process. Students of the quality and process improvement often start by learning the DMAIC approach because most other methodologies derive from its fundamental structure and concepts. Components of the standard DMAIC toolset also can be found in the tool suite of other Six Sigma approaches.

SECTION 2

Lean and Lean Six Sigma

The Lean and Lean Six Sigma approach expands Six Sigma concepts into a philosophy focused on velocity, value-add, and process flow. This section overviews Lean concepts and its high-level requirements, given the requirements define the appropriate deliverables, which dictate the tasks and the tool selection to aid the task. This section outlines the Lean standard toolset through the understanding of the tool-task-deliverables linkage to facilitate appropriate selection of a tool when referencing the "how to" tool articles in Part II of this book.

What Is the Main Objective of this Approach?

The Lean method streamlines a process to its essential value-adding activities from the perspective of the customer. It drives efficient process performance by focusing on doing things right the first time. Lean practitioners scrutinize the flow of value-adding tasks by selecting those critical tasks to execute and those to eliminate. Paying customers define what is classified as *value-add* by indicating that they are willing to pay for it. A *value-added activity* is the *process of physically changing* an object per the customers' requirements, and the task is done right the first time. In Lean terms, *work is a value-added activity; waste (or non-value-added activities) is the opposite.*

The purpose is to improve speed of a continuously flowing process that operates on-demand (from customer-pull perspective) and with minimum to no waste. The pull-perspective refers to just-in-time production in that nothing gets produced until the customer orders it—providing what the customer wants, when they want it, and how many they want without any lead-time. The goal is to balance resources to customer demand—defined as "flow." The Lean method reduces or eliminates wasteful activities, such as wait-time, unnecessary hand-offs and transportation, excess inventory, and overproduction.

Brief Description of Typical Applications

Lean methods traditionally are applied to factory operations, production, and supply chain processes to trim out their non-value activities. However, the approach works well with administrative and other non-production areas. Lean Production or *Just-in-Time* (JIT) was popularized in Japan, post-World War II, to improve factory efficiency.

The Japanese marketplace sought a great variety of automotive offerings, and Toyota developed and applied Lean concepts to improve production flexibility and efficiency to serve market demands better. The *Toyota Production System* challenged the idea of large lot production from Ford and introduced *flexibility through diversified small lot production*. Both Taiichi Ohno and Shigeo Shingo pioneered most of these concepts at Toyota in the 1970s. Toyota focused on systematically and continuously reducing waste (or "muda" in Japanese). Muda included product defects, overproduction, excess inventories, unnecessary processing, wasted movement of people, excess transporting of goods, and waiting. MIT coined the term "lean manufacturing" in the 1990s while studying Toyota's manufacturing successes.

Figure 2-1 shows how Lean improves the speed of a problematic process by eliminating waste, minimizing hand-offs, and reducing unnecessary activities. In this case, Step 3 is problematic (too close to the lower specification limit). Lean is able to remove Step 3 without jeopardizing quality and improve cycle time—the (former) Step 4 becomes the (new) Step 3 in the improved process.

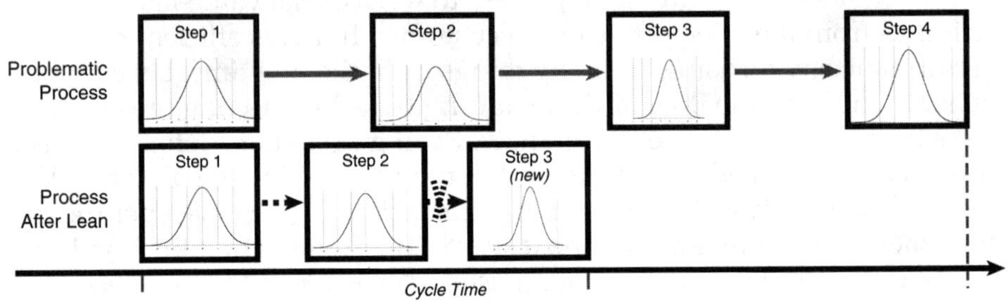

FIGURE 2-1: Lean's Impact on Cycle Time

Some compare Lean and Six Sigma differences as a matter of a degree of complexity, size, and scale of the initiative. Lean is the simpler of the two. The optimal approach integrates Lean with Six Sigma to add not only more tools to the arsenal, but also a different perspective. Six Sigma offers a big picture context, balancing the customer and business requirements and criteria to evaluate overall impact of leaning, such that Lean initiatives no longer stay at the local level and potentially out of synch with the rest of the organization. For example, a common manufacturing mistake is to allow lean initiatives to outrank any other process voice, such that the process loses its ability to design and produce robust products.

See Also "The Anatomy of Variations in Product Performance" and "Lean Six Sigma for Fast Track Commercialization High Risk-High Reward, Rapid Commercialization: PROCEED WITH CAUTION!" articles, in Part III, "Best Practices Articles," **p. PDF:777 and PDF:835**, respectively, for further elaboration on this point.

What Key Overall Requirements Define this Approach?

The Lean approach was designed and structured to answer the following overall business questions:

- Is the process delivering the most optimal value to the customer?
- Is the process operating as efficiently as possible, absent of any unnecessary waste?
- Is the process output produced only at the pull of the customer?
- Does the process strive for continuous improvement such that the process players (internal and partners) are empowered to adjust to changes or opportunities for improvement?

What Requirement Determines the Key Activities in this Approach?

The preceding key business questions determine the Lean architecture. Figure 2-2 depicts a high-level Lean workflow:

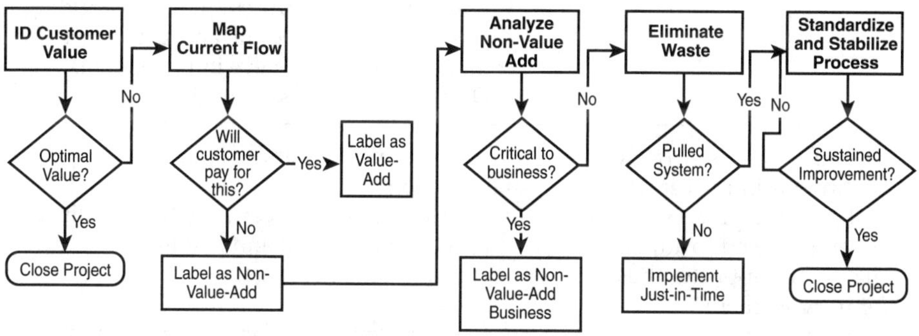

FIGURE 2-2: High-Level Lean Process Flow

Lean often is not associated with a rigorous method, but rather its concepts are often integrated in with other methodologies, such as DMAIC or DFSS. However, a pure Lean approach generally follows the following high-level approach.

Table 2-1 shows the link between the high-level business requirements with the Lean approach.

TABLE 2-1: Lean Requirements-Step Linkage

Requirements	Resulting Key Activities
Is the process delivering the most optimal value to the customer?	1. Identify customer value. • Collect and summarize exactly what they would be willing to pay for. 2. Map the current process workflow. • Diagram the current process. • Group activities into either value-add or non-value-add categories. • Identify value currently delivered.
Is the process operating as efficiently as possible, absent of any unnecessary waste? Is the process output produced only at the pull of the customer?	3. Analyze the non-value-add activities and determine problem(s). • Look for bottlenecks (i.e., Theory of Constraints). • Group into non-value categories and problem(s) identified. • Apply appropriate waste elimination strategies and tools. 4. Eliminate the appropriate waste. • Apply appropriate Lean Six Sigma strategies for improvement. • Eliminate just-in-case activities and resources. • Map and document the future state process. 5. Standardize the process and develop a transition plan. • Work with all the process players, including suppliers, to collaborate and gain agreement on how to achieve improvements. • Document changes.
Does the process strive for continuous improvement such that the process players are empowered to adjust to changes or opportunities for improvement?	6. Validate the process and inspect for stabilization. • Create a system that automatically adjusts to relevant internal and external environmental changes.

What Tools Are Aligned to Each Step of the Process?

Given the preceding high-level activities, the following series of tables summarize the subsequent tool-task-deliverables combination associated with each individual step within the Lean approach.

The detail behind how to use each tool can be found in Part II, "Six Sigma Tools and Techniques: Choosing the Right Tool to Answer the

Right Question at the Right Time," of this book. Table 2-2 summarizes the tool-task-deliverable linkage for Lean's first requirement.

TABLE 2-2: Lean Tools-Tasks-Deliverables for Requirement 1

Requirement: Is the process delivering the most optimal value to the customer?		
Step 1: Identify customer value.		
Deliverables	**Tasks**	**Candidate Tools and Techniques**
Customer Value documented	• Define exactly what the customer would be willing to buy (both current and potential customers); required volume • Determine the desired average completion rate	• VOC Gathering techniques • Takt time (customer demand rate)

Table 2-3 summarizes the tool-task-deliverable linkage for Lean's Requirement 2.

TABLE 2-3: Lean Tools-Tasks-Deliverables for Requirement 2

Step 2: Map the current process workflow.		
Deliverables	**Tasks**	**Candidate Tools and Techniques**
Value Stream process flow documented	Diagram the current customer value stream from two perspectives: goods/services and the production process (and its players)	• Detailed process map • Value Stream mapping • Spaghetti map
Value and Non-value-add activities documented	• Measure and collect time associated with each activity in the process • Identify and categorize activities	• Value Stream matrix • Detailed process map or swim-lane process map
Identify value currently delivered	Calculate process capacity and Takt time	• Total Lead Time • Process Cycle Efficiency • Workstation Turnover Times

LSS

Table 2-4 outlines the tool-task-deliverable linkage for Lean's third requirement.

TABLE 2-4: Lean Tools-Tasks-Deliverables for Requirement 3

Requirements: • **Does the process operate as efficiently as possible, absent of any unnecessary waste?** • **Is the process output produced only at the pull of the customer?**		
Step 3: Analyze the non-value-add activities and determine problem(s).		
Deliverables	**Tasks**	**Candidate Tools and Techniques**
Non-value-adds categorized and problem(s) identified	• Categorize non-value-add activities into business and pure waste • Determine the type of problems that exist	• Categories of waste • Theory of Constraints
Lean strategy developed	• Develop and apply appropriate flow, balance, capacity, waste elimination, and process pull strategies • Validate to ensure that the faster processes, less waste, continuous flow, on-demand improvements are achieved. [that is, reduced cycle time, reduced product defects (reworks and maintenance), reduced changeover time, reduced equipment failure, reduced accidents, and so on.]	• Value stream matrix • Detailed process map • Spaghetti map • Flow Manufacturing • Multi-process Operation • Visual Control; 5S • Constraint management • Level loading; Leveled Production • Changeover • Manpower Reduction • Pull systems; Kanban; Just-in-Time • Flexible Process • Lot size reduction • Autonomation (Jidoka) • Quality Assurance (QA) • Standard Operation • Maintenance and Safety • Cost/Benefit analysis

Step 4: Eliminate the appropriate waste.		
Deliverables	**Tasks**	**Candidate Tools and Techniques**
Lean strategies implemented	Select and implement Lean strategies and supporting implementation plans	• Prioritization matrix, or Solution Selection matrix • Implementation plan • Communication plan • Training plan
Improved state documented	Document standard process and procedures [this should be state and highlight improvements (or changes to current state)]	• Standard Operating Procedures, revised • Should be value stream matrix • Should be process map • Should be Standard Operating Procedures • Control plan • Risk management plan

Step 5: Standardize the process and develop a transition plan.		
Deliverables	**Tasks**	**Candidate Tools and Techniques**
Improved process stabilized	• Work with all the process players to collaborate and gain agreement on how to achieve improvements, including suppliers • Document any changes or modifications to drafted plans	• Implementation plan • Communication plan • Should be value stream matrix • Should be process map • Should be Standard Operating Procedures • Control plan • Risk management plan

Table 2-5 outlines the tool-task-deliverable linkage that address Lean's Requirement 4.

TABLE 2-5: Lean Tools-Tasks-Deliverables for Requirement 4

Requirement: Does the process strive for continuous improvement such that the process players (internal and partners) are empowered to adjust to changes or opportunities for improvement?		
Step 6: Validate the process and inspect for stabilization.		
Deliverables	**Tasks**	**Candidate Tools and Techniques**
Process improvements validated and stabilized	Create and/or ensure a system that automatically adjusts to relevant internal and external environmental changes	• Control plan • Risk management plan • Kaizen (See "What Does Kaizen Mean?" later in this chapter.)

What Are Some of the Key Concepts that Characterize this Approach?

This next section provides an overview of some key Lean concepts that distinguish it from the other Six Sigma approaches. It covers topics such as velocity (or speed), value-add, wastes, Kaizen (or continuous improvement), and variations on the Lean philosophy.

How Is Value-add Defined?

Value is defined as what a customer wants or needs and is willing to pay for. *Value-added* describes those activities that add value (as defined by the customer) to the product (or service). These activities often are thought of as in-process activities done within the context of producing or manufacturing the final customer offering.

How Do You Know if the Process Activities Are Value-added?

Value-added activities should withstand the 3C litmus test:

1. **Change**—The activity changes the product or service.
2. **Customer**—The customer actually cares about the activity's outcome (its impact on the overall offering).
3. **Correct**—The customer's perspective determines if an activity is executed, and its outcome is produced correctly the first time—no rework.

What Is a Non-value-add Activity?

Lean defines *work* as *valued-added activities* (designing and building goods and services) for which a *customer is willing to pay*. Work is made up of those activities that create value from the customer's perspective.

The opposite of work is *non-value-added* activities, or waste. Waste is often a symptom, rather than a root cause, of a problem. Waste simply leads to the root causes within the system (process, goods, and/or services).

Are There Different Types of Non-value-added Activities?

Non-value-added activities fall into two categories:

1. **Business Non-value-add**—Those items required for the business to operate efficiently, such as legal requirements, regulatory requirements, recording financials, and maintaining an intellectual capital management system.
2. **Non-value-add**—Items that lack value to both the external customer and the business; hence the customer is unwilling to pay for it, and it does not add value to the final product (or outcome).

Their various types of waste include

Overproduction*—Producing parts ahead of schedule (before a customer requests or needs it) while other items in the process wait; working on unneeded parts (the wrong parts at the wrong time); evidenced by producing too many parts or finished goods or producing them too early

Inventory Space*—Excess space consumed by shelving, floor space, excessively wide aisles, bins, filing cabinets, or files that house accumulated in-process or finished goods, including parts waiting for rework or scrap storage

Waiting Time*—People unnecessarily waiting, stalled in the workflow due to either shared equipment, unbalanced work activities (operations), or waiting for decisions, approvals, or inspections

Transportation*—Moving parts or objects unnecessarily (including papers or files); excess travel distance, which also can waste time and equipment

Motion*—Excess, unnecessary, or non-valued people activities such as searching, walking, sitting, choosing, copying, stapling, sorting, climbing, bending over, and lying down

Unnecessary Processing*—Unnecessary operations, steps (including inspection or approvals) and complexity (at times excessive documentation)

Defects*—Repairs and rework needed to get something to function properly, which slows the process flow and reduces first-pass throughput yield

Materials—Scrap or excess ordering of raw materials

Idle Material—Material that waits or sits in inventory

Energy—Wasted power or people energy (closely related to motion)

Excessive Labor—Too many workers in the process causing inefficient operations.

Safety hazards—Unsafe work environments

Management Wastes—Failure to orchestrate and unleash the inherent capabilities of the organization (people, process, infrastructure, and technology)

A memory-jogger acronym often used in the United States to remind people of the different types of waste is known as **DOWNTIME**, which stands for:

- **D**efects
- **O**verproduction
- **W**aiting
- **N**on-utilized talent
- **T**ransportation
- **I**nventory
- **M**otion
- **E**xtra processing

Purists of the Japanese Lean concepts recognize only seven key wastes (defects, overproduction, waiting, transportation, inventory, motion, and excess processing) and drop non-utilized talent. Utilizing talent to its fullest is an ingrained part of the Japanese culture, so they do not need to be reminded of this potential waste. Other cultures, however, benefit from the reminder to appropriately utilize or leverage the people talent.

* The first seven categories listed here that are labeled with the asterisk (*) represent the traditional Lean *Seven Wastes* from Japan.

Moreover, Japanese often group things in sets of seven, and DOWNTIME is an eight-letter acronym.

Regardless of which set of wastes is used, one can observe these wastes, seeing evidence of them in the workplace. Lean practitioners constantly look for waste to see if it exists. As a result, practitioners have expanded the list of wastes to include managerial wastes, which include

- **Ownership Waste**—Lack of accountability, authority, or empowerment to perform responsibilities in full
- **Focus Waste**—Dispersed energy working on different sets of critical issues due to inconsistent alignment of management and employees
- **Structure Waste**—Lack of infrastructure support (processes, reporting relationship, culture, rewards, and recognition) needed to focus on continuous improvement (Kaizen) or to gain ownership
- **Discipline Waste**—Failure to maintain the process, performance, and behaviors needed to achieve standards (or specifications)

What Are the Key Lean Strategies to Eliminate Waste?

The appropriate strategy for waste elimination depends on the scenario in which the waste exists. Customer requirements, the data depicting the waste, the root cause, the work environment, the process players, and the culture represent some of the factors to consider when developing a waste elimination strategy. Lean offers some generally accepted generic waste elimination strategies including

- **Waste elimination**—Rework and loop-backs
- **Value Stream Mapping**—Aids in visualizing the end-to-end flow and identifies potential sources of waste. (**See Also** "Value Stream Analysis," **p. 727** in Part II)
- **Value Stream Analysis**—Is a tool used to dissect the process flow and analyze any waste by categorizing, defining its impact and identifying the countermeasure to it. (**See Also** "Value Stream Analysis," **p. 727** in Part II)
- **Work standardization**—Tasks organized in the optimal sequence with the most effective conditions (the 5Ms): manpower, materials, methods (motion sequences, technical procedures, administrative procedures), machinery (equipment, tools) and measurements (quality requirements); plus uniform work conditions and workplace arrangements

- **Inventory reduction or elimination**—Including Work In Process (WIP) (using a **pull system**—potentially with a kanbans (Japanese for place card or sign)
- **Workload balancing**—For example, constraint management, level loading, excess resources elimination)
- **Batch size reduction**—Or Lot
- **Flexible process reconfiguration**
- **Quality control**
- **Complexity reduction**—And segregation into similar levels
- **Linked value-add process steps**—Contiguously placed within the workplace
- **5S established**—For good housekeeping, which are defined as
  - Sort (or tidiness), known as "Seiri"
  - Set in order (or orderliness, simplify, sequence), known as "Seiton"
  - Shine (or sweep, cleanliness), known as "Seiso"
  - Standardized clean-up, known as "Seiketsu"
  - Sustain (or self-discipline), known as "Shitsuke"
    - Visual workplace established
    - Flat organization structure
    - Empowered employees
    - Teamwork established
    - Employee morale improved

What Is the Theory of Constraints?

Theory of Constraints (TOC) describes a process for continuous improvement by removing bottlenecks in a process with limited production or throughput. The concept focuses on the identification of the system constraint. A process chart or a value stream map are useful tools to assist in finding bottlenecks. (**See Also** "Process Map (or Flowchart)—7QC Tool," **p. 522** and "Value Stream Analysis" **p. 727** in Part II.) The concept was introduced by E. Goldratt in two books: *The Goal* (1986) and *Theory of Constraints* (1990). He proposed three basic measures to be used in the evaluation of a system:

- **Throughput**—The rate which money is generated through sales (measured as *incoming money*)
- **Inventory**—The money locked up in the system as investments (or purchases) that are intended to be sold in the future (measured as *money stuck inside* the company)
- **Operational expenses**—The money spent to convert inventory into throughput (measured as *money going out* of the company). This includes depreciation, scrap, carrying costs

These measures are said to be more reflective of the true system impact than machine efficiency, equipment utilization, downtime, or balanced plants.

What Does Kaizen Mean?

Kaizen is a Japanese philosophy in which one strives for continuous improvement in all aspects of life. Literally, "kai" means *change* and "zen" means *to become good.*

With respect to the workplace, Kaizen improvements are often thought of as constant, small, incremental improvements that occur as work happens. As you work, if you see something that would improve the process (or product) or make work easier, you are empowered to implement the modification. Kaizen improvements are viewed as a constant initiative to seek an optimal level of operation, rather than a radical change or innovation that would require the efforts of a Six Sigma project.

Given the incremental adjustment approach of Kaizen, it may involve short-term projects (one to six-day timeframe), involving only those people involved in working in the process (or product). Often a Kaizen initiative starts with an impromptu conversation among empowered colleagues, who simply brainstorm about the improvements they could make. This conversation evolves into a self-directed work team that focuses on fixing the problem.

They may continue to meet during breaks, before or after work, or they could take time out of work to take a more formal examination of how best to optimize the system. An important perspective of Kaizen is that the improvements are documented. The documentation often is in the form of updating Standard Operating Procedures (SOPs).

Kaizen thrives in a workplace environment that encourages and supports empowered employees to maintain SOPs as part of an evergreen process that continuously reviews, refreshes, and renews.

The approach, tools, and techniques the empowered employees follow could be the DMAIC approach or the classic E. Deming approach of Plan-Do-Check-Act (PDCA). Common tools used by Kaizen teams include:

Affinity diagrams, Tree Diagrams, Process Decision Program Charts, Matrix Diagrams, Interrelationship Diagraphs, Prioritization Matrices, and Activity Network Diagrams.

Are There Any Lean Variations?

There are two prevalent variations to the Lean method:

1. **PDSA (Plan-Do-Study-Act) or PDCA (Plan-Do-Check-Act)**—A mainstay of the Total Quality Management discipline, Edwards Deming popularized the PDSA model (sometimes called the PDCA cycle) to improve a process flow. This often is referred to a simplified foundation of other problem-solving methods.

   Original writings about this approach originated from Shewhart in the late 1930s when he was describing the power of prediction as an essential ingredient to using experimental data to build knowledge. Hence, Shewhart starts with original data, predicts an outcome, and gains knowledge from observing experiments. Shewhart proposed that since the new findings were based on experimental data, that the knowledge was *probable* and that the non-experimental circumstances only could be inferred.

   The approach starts with *planning* as the important groundwork that establishes the objectives and timeframe for a study (Plan). Next, the approach executes the plan (Do). Third, the study results are collected, checked, and analyzed, evaluating the hypothesis and drawing new conclusions (Study or Check). Last, based on the experimental data that provided the new knowledge, follow-on improvement actions are implemented (Act). Because this is a cyclical model, the learnings from Act feed into another iterative Plan phase, as part of continuous improvement.

   Figure 2-3 depicts the cyclical nature of the PDSA model.

   Plan-Do-Study-Act also inspired early development of the DMAIC approach. Figure 2-4 shows the alignment of the two structures by step.

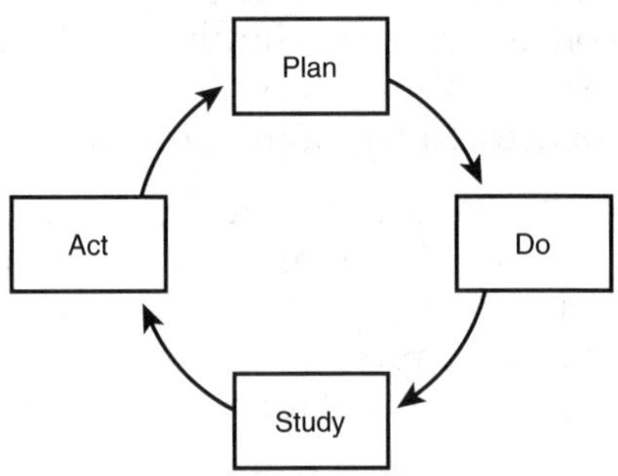

FIGURE 2-3: PDSA Model

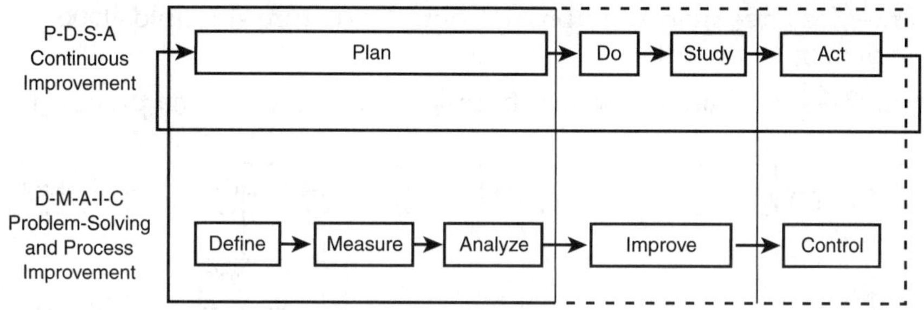

FIGURE 2-4: Comparison of PDSA and DMAIC Models

2. **Lean Six Sigma—Adding Lean concepts**—By incorporating Lean concepts into DMAIC, the project adds dimensions such as velocity, value-add, and flow to what the Six Sigma problem-solving method already offers.

   Six Sigma examines a process to improve quality and reduce variation—variation from target or customer requirements, including variation reduction within a problematic step. Lean concentrates on process timing—overall cycle time, including the timing between process steps by removing non-value-added activities. Combining both Lean and Six Sigma produces a compounding impact on the improved state. Figure 2-5 depicts the impact of integrating the two disciplines on a given process.

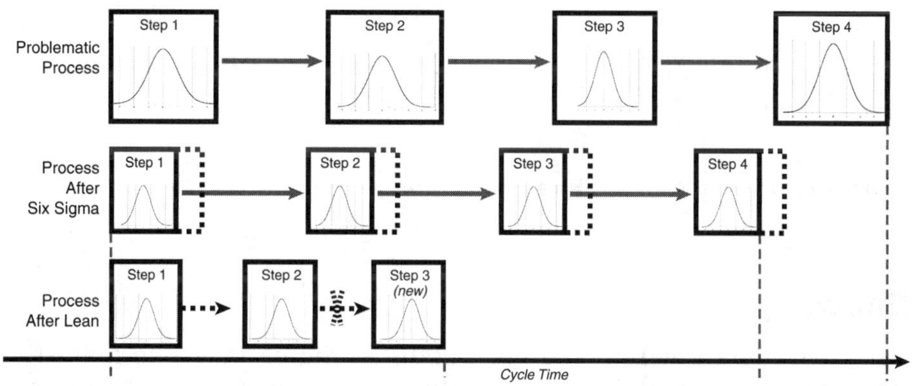

FIGURE 2-5: Lean and Six Sigma's Impact on Cycle Time

Both concepts share similar views on customer-focus, process-centric work, and appropriate tools. Lean simply adds a deeper set of tools to eliminate waste between process steps handoffs. Often DMAIC provides a project the big picture view (what the customer values balanced by business values), process stabilization, and capability, while Lean introduces speed and flow concepts at a more detailed level. The

LSS

Define-Measure-Analyze-Improve-Control structure still holds true for Lean Six Sigma projects.

Figure 2-6 shows an integrated, high-level, Lean Six Sigma process flow.

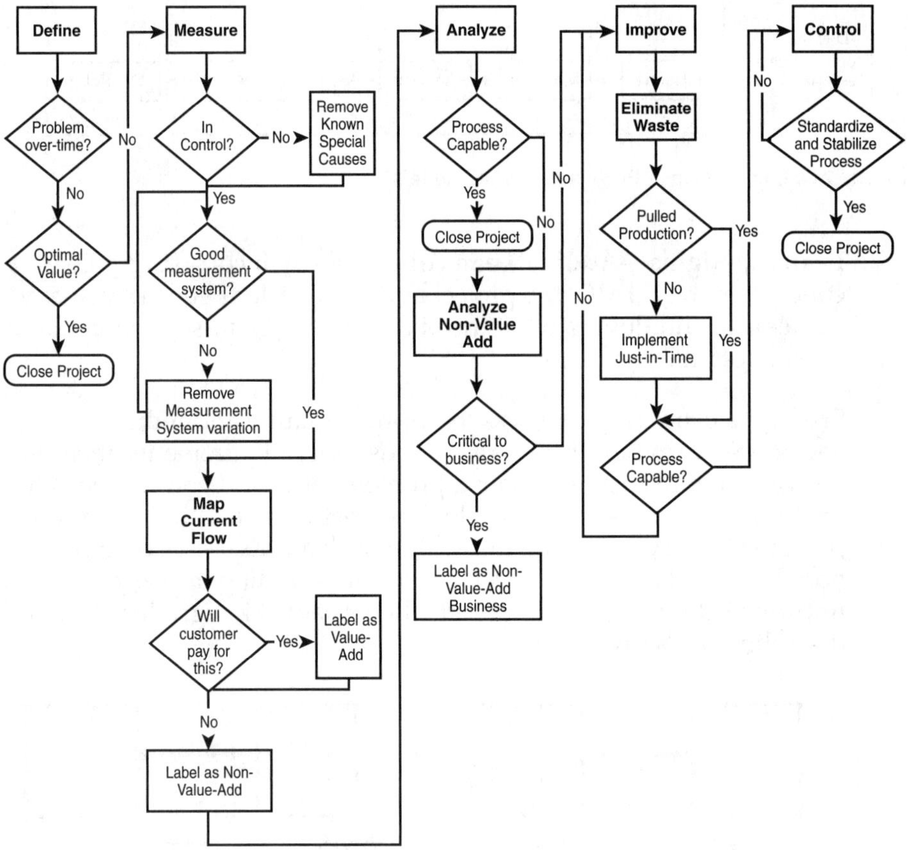

FIGURE 2-6: High-level Lean Six Sigma Process Flow

Summary

Lean concepts are an important part of the Six Sigma arsenal to combat variation. The Lean philosophy introduced a simple but powerful focus on waste elimination and continuous improvement to better meet customer requirements on-demand, sans any lead-time. It is an important cornerstone of the Six Sigma approach and has contributed invaluable tools and techniques to create flow, eliminate waste, and strive for continuous improvement.

SECTION 3
Design for Six Sigma (DFSS)

Design for Six Sigma (DFSS) represents a portfolio of methods and tools that expands Six Sigma concepts to take a preventative approach by designing quality into a product (or process). This section overviews the DFSS concepts and its key methods. Next, the high-level requirements are discussed and the appropriate deliverables are defined, which dictate the tasks and the tool selection to aid the task. While the DFSS category applies to technical design applications, it also features a unique subset of tools applicable to general business applications. This section outlines the DFSS standard toolset, to build an understanding of its tool-task-deliverables linkage and facilitate in the selection of the appropriate business-oriented tools borrowed from DFSS when referencing the "how to" tool articles in Part II of this book.

DFSS Evolution and Approach Overview

Design for Six Sigma (DFSS) defines a category of Six Sigma methods that evolved from the DMAIC problem-solving method. Businesses would realize great success in fixing problematic processes and/or product offerings when they first implemented DMAIC. Initial successes often prevail in the manufacturing area and across the board for cost cutting. However, after awhile, they find it increasingly difficult to maintain quality while continuing to reduce defects or variations and take costs out of the business, particularly if the process was poorly designed. This experience of diminishing success occurs more quickly in situations where the process or product is poorly designed. Eventually, modifying the existing process through the DMAIC method alone will fail to meet ever-demanding, often changing customer requirements. For example, additional cost reductions could rob the (process or product) design of its robustness, thereby jeopardizing the quality. Alternatively, a poorly designed process may demand some level of rework or scrap to uphold quality. Such defective processes drain resources of time and money attempting to deliver expected quality levels. In response to this phenomenon, DFSS emerged to design in quality.

DFSS evolved to address the need for a redesign or new design—an innovation in response to a problem. If the process is incapable of meeting the desired customer specifications, it requires a redesign or an all together new design. The new design reduces sensitivity to variation and noise, which yields improved reliability, durability, and quality—a more robust design. The need for a redesign versus new design may be a matter of degree. Note that a redesign modifies an existing design, and a new design starts with a clean sheet. However, a new design also could apply only to a module or sub-component of a current design. The degree of difference between a redesign and new depends on the portion that the sub-assembly represents relative to the entire architecture.

DMAIC Versus DFSS Methods

Relative to the classic problem-solving DMAIC method, the DFSS arena of Six Sigma is still maturing. As a result, two consequences have materialized:

1. A fuzzy delineation between DMADV (for reactive redesign) and CDOV (for clean-sheet, proactive) methods, causing the deployment and application to vary from company to company
2. The emergence of other DFSS variants

The DMADV (Define-Measure-Analyze-Design-Validate) method aims to *redesign* a problematic process or product. The approach initially follows the first three steps of DMAIC and then deviates in the last two-steps by introducing *Design*/Redesign and *Validate* steps to gain the improvements needed. This approach *prevents* problems from happening through quality and robust design concepts.

Figure 3-1 delineates the alignment of the DMADV and DMAIC steps:

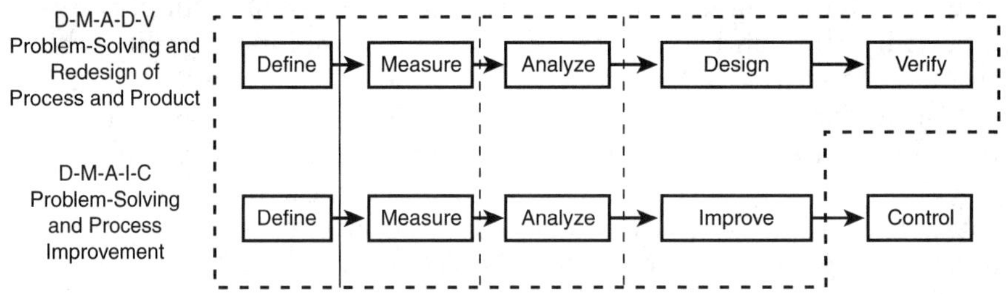

FIGURE 3-1: Comparison of DMADV and DMAIC Methods

If the situation calls for a clean-sheet design and commercialization, wherein the project team is starting from scratch, then the CDOV (Concept-Design-Optimize-Verify) method is preferred. CDOV takes on more of a *proactive* approach by creating a robust design to prevent the need for a redesign. Given its forward-thinking purpose, CDOV is part of a portfolio of Six Sigma methods that drive growth—top-line business growth through new revenue initiatives.

Business people and academicians alike have proven repeatedly that fixing issues early on in a process or lifecycle saves money. DFSS, especially CDOV, builds on that notion and starts with a clean-sheet approach, addressing the early stages of an offering's lifecycle—research, design, and development. DFSS differs from the classic problem-solving (DMAIC) methods that activate once the process moves into production and responds to problems that arise. Hence, the proactive-based DFSS methods cost a business less to deploy in the long-run based on an offering's lifecycle expenses, as shown in Figure 3-2.

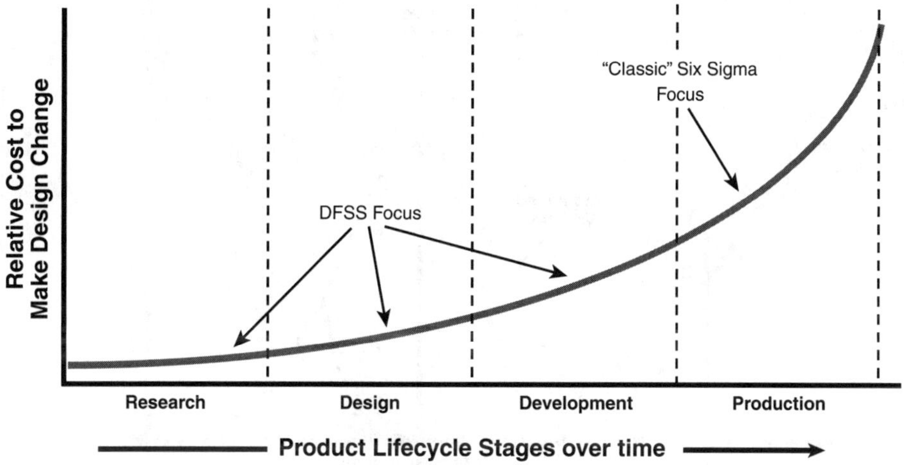

FIGURE 3-2: Cost Comparison of DFSS and Classic Six Sigma Over Product Lifecycle

Key DFSS Concepts

DFSS focuses primarily on growth through *product development*; hence, it relates typically to the *technical, engineering, and manufacturing environments*. Its main application addresses two main processes 1) the technical product/services development and commercialization and 2) the packaging design and assembly. In the product design world, George Box, Douglas Montgomery, and Genichi Taguchi have contributed crucial concepts to *proactively* improve both quality and ultimately customer satisfaction. Two such concepts are the Loss Function Model and Robust Design Principle.

Taguchi's Loss Function Model

First, the concept behind the *Loss Function* model is to *avoid economic loss* by *minimizing deviation of a product's function* (from its desired target) and *optimizing its tolerance* (often down to the parts-level). The amount of *deviation* from the target represented the *tolerance* in the system (or the delta). Calculating the amount that a product's (or part's) *quality characteristic* deviates from its target is called the *quality loss*. When the quality loss is zero (that is, exactly on target), then the quality characteristic is at the target value, indicating the absence of economic loss to society.

Figure 3-3 illustrates how the model's quality loss increases as the product (or output) moves off the (customer) target and/or as variation increases. As a result of Taguchi's work, quality metrics aim to prevent potential *cost of poor quality* that may occur later in the product's lifecycle.

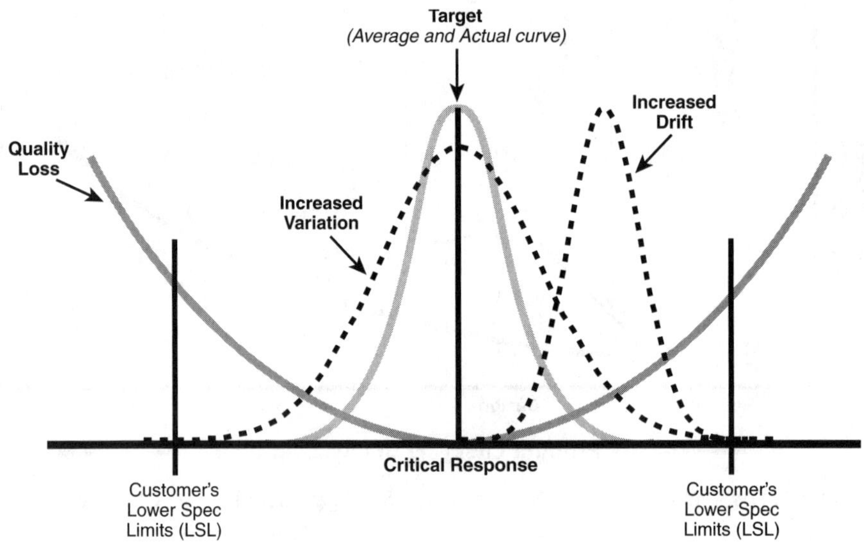

FIGURE 3-3: Taguchi Loss Function Model

Robust Design Principle

The second contribution, the *robust design* principle, promotes stability by minimizing both the impact of noise factors and the cost of failure on a product design. Taguchi incorporated the noise factors into the experimental design to determine the optimal design. This class of experimentation is referred to as Design of Experiments (DOE) techniques, used to select the critical parameters and their optimal levels such that they are insensitive to noise (or stresses). Noise sources fall into three categories: unit-to-unit variation, internal (such as deterioration, contamination, durability, and so on) and external environment. The noise factors can be applied to systems other than tangible manufactured products and apply to "everyday life." By using the DOE techniques, Taguchi called attention to the importance of actually "manipulating" variables to witness (or observe) the effects in the critical response factors to demonstrate the cause-and-effect relationships. G. Box and D. Montgomery recommended the use of response surface methodology for robust design and optimization, wherein the experimental design identifies the system response to one or more variables and the response is graphed using geometric concepts.

DMADV: Define-Measure-Analyze-Design-Validate

Let us explore in more depth the DMADV method, which is probably the most popular of the DFSS approaches.

What Is the Main Objective of the Approach?

The DMADV (Define-Measure-Analyze-Design-Validate) builds off the DMAIC method as a data-driven Six Sigma method fundamentally to **redesign a problematic process or product offering**.

The technical community (that is, engineering, manufacturing) was the first discipline to embrace this approach, but it is applicable in other areas of the business. This approach, traditionally, is to be applied to a problem with an existing, steady-state process or product and/or service offering. Sometimes this approach may be also used to design a new process or product when new requirements emerge.

As part of the Design for Six Sigma (DFSS) branch of Six Sigma, DMADV is the most prevalent method. The objective of the DFSS category of methods is to develop a new or radically redesigned product, process, or service.

Brief Description of Typical Application

The 5-step **DMADV** [pronounced *"duh-MAD-vee"*] method is often called the *process innovation* methodology. It was designed to solve a problematic process or product and/or service offering through an innovative redesign when the current design proves to be incapable of meeting customer requirements (or targets).

Similar to the DMAIC approach, **DMADV** is designed to allow for flexibility and iterative-work, if necessary. As more is learned through the 5-step process, assumptions or hypotheses may be disproved, requiring the project team to revisit them and modify or to explore alternative possibilities. It builds on three fundamental principles:

- Work is **results-focused**; driven by data, facts and metrics.

- Work is **project-based** (short-term in nature, with length depending on scope and complexity) and **project-structured**, versus an on-going process.
- Inherent **combination of *tools-tasks-deliverables*** linkage that varies by step in the method.

The DMADV 5-Phase-Gate Structure

The DMADV methodology uses a phase-gate structure. Phases generally are sequential; however, some activities may occur concurrently or may be iterative. A formal phase-gate review requires the completed set of respective deliverables prior to approval. Phase-Gate Reviews do occur sequentially. The DMADV five steps are

Phase 1. **DEFINE** the problem and/or the new requirements (or goals) of the design activity, including external and internal customer requirements. This involves a significant amount of VOC gathering activities to shape the design requirements.

Phase 2. **MEASURE** the current process or product performance, gather appropriate data, and determine the customer's needs and related specifications—what is critical-to-quality. This may involve a comparison between the current process and the new requirements.

Phase 3. **ANALYZE** the current performance data to isolate the problem and identify the potential improvement options to better meet the external and internal customer requirements. Identify the cause-and-effect relationships between the problem(s) and key variables.

Phase 4. **DESIGN** a new process so the problem is eliminated or new requirements are met. This may involve revamping (or developing a new) process, product, or service offering using predictive models, simulations, experimentation, and pilot testing to identify the optimal design to meet the goals.

Phase 5. **VALIDATE** that the new process is capable of meeting the new process requirements. This may involve validating that the process or product design meets (little Y) targets for performance, robustness, and stability to ensure the overall requirement(s) can be met over time in a non-test (real-world) environment.

DMADV Duration

A DMADV project may run from a relatively short duration (nine to twelve months) to several years depending on the complexity, size, scale, and scope of the product or process. Given the relatively long duration of a Design for Six Sigma (DFSS) project, the method uses a phase- structure to distinguish one step from another. At the completion of each phase, the project sponsor conducts a *phase gate review* with key project members and other key stakeholders to verify and approve the phase requirements and deliverables and to set (or clarify) expectations for the next phase. Phase deliverables include both "administrative" and "design" outputs. Administrative outputs involve items often associated with project management-type deliverables such as timeline and project life-cycle management, budget (resources and funding), scope (for example, scope creep; what is in and out of scope), communication, risk management, procurement, quality, people resources, and overall integration. Design outputs address the actual process, product, service offering, or information being developed. Albeit the approach may be iterative in nature, the project team should withhold starting a new phase until the sponsor approves the current phase deliverables.

Four Components of the DMADV Method

Similar to the DMAIC method, DMADV is based on the application of *statistical process control*, *quality tools*, and *process capability analysis* wherein the process or product exists and requires a redesign to fix the current problem. Some apply DMADV to a pure clean-sheet design effort; however, the CDOV method is better constructed for this purpose. To effectively implement DMADV, the method requires four components:

- A **measurement system** (a gauge) of the process or product/service offering in trouble.
- **Statistical analysis tools** to assess samples of data.
- An ability to define an **adjustment factor(s)** to redesign the process or product/service offering back on target.
- **A control scheme** to maintain the improvement or correction over time by implementing a control plan with a monitoring system to audit the response performance against statistical control limits and defined action plans if needed.

CDOV: Concept-Design-Optimize-Verify

Another DFSS approach, CDOV, utilizes the prevention approach of DMADV but promotes a unique structure that presumes a cross-functional design team.

The structure addresses the design team's technical community and incorporates a parallel integration with a sister-method uniquely focused on the marketing and business partners. CDOV introduces new concepts that expand the proactive-nature of DFSS to fuel the Six Sigma for Growth momentum.

What Is the Main Objective of this Approach?

The CDOV (Concept-Design-Optimize-Verify) structure applies a flexible, adaptive approach to help a technical or engineering project team to design and commercialize a new product or services offering and/or new manufacturing system using a clean sheet design. CDOV can be used to design a new platform, extend product lines, and integrate new modular designs within an existing platform or product family. In addition, CDOV can be applied to a product improvement project.

The CDOV structure compliments and integrates with a Six Sigma for Marketing (SSFM) method targeted at the other cross-functional team members, namely marketing and business professionals. This sister SSFM method is called UAPL (pronounced "***YOU**-apple*")—Understand-Analyze-Plan-Launch. *While some companies successfully deploy CDOV as a standalone method, its power is best realized when integrated with the non-technical commercialization method.* (Commercialization encompasses preparation for the marketplace and launch.) Together, CDOV and UAPL provide a common structure and language for multiple disciplines to work toward a common goal—the design, development, and commercialization of an offering.

The Commercialization Model

Over and above the actual technical design and development of an offering, the technical community should contribute to the entire *market-readiness* process. The product commercialization process features requirements and deliverables from multiple disciplines of the cross-functional team. Too often companies fail to involve the technical arm of product development team in this portion of work. Commercialization success is defined at minimum as *market acceptance* and achieving a *smooth launch throughout the customer value chain*. Successful commercialization needs the technical community to partner with marketing and other functional disciplines to provide critical deliverables and complete the process.

The different disciplines within a cross-functional team share some joint requirements and deliverables and also own complimentary requirements and deliverables. CDOV and UAPL orchestrate and integrate the cross-functional team's work. Each method identifies the mutual dependencies in terms of primary roles, key deliverables, and timing. The deployment of a Stage-Gate process reinforces cross-functional integration and validation of requirements and deliverables. **See Also** "SSFM," **p. 67** for more details.

Brief Description of Typical Applications

The 4-phase **CDOV** [pronounced *"see-**DOVE**"*] method, a *product design, development*, and *commercialization* method, integrates the Voice of the Customer with the Product and Manufacturing Process Design to generate top-line growth through new product (or services) launches. This approach relates to not only the technical/design development environment, but also the manufacturing, assembly, materials management, and supply chain development areas.

CDOV was constructed as a *tactical* product design engineering process that parallels and integrates with the activities of the marketing and business areas also involved in the offering commercialization process. While the technical team follows the CDOV method, their counterparts in other areas of the business follow the UAPL (Understand-Analyze-Plan-Launch) approach. In concert, the multi-functional effort creates a robust offering ready for successful release to the marketplace. **See Also** "SSFM," **p. 67** for more details.

The Kano and New, Unique, and Difficult (NUD) Principles

The Kano principles inspires CDOV in that it pursues the *New, Unique, and Difficult* (NUD) elements and incorporates them into the design to create *customer-pull*. Noriaki Kano, a Japanese engineer, developed a model at the product offering level, showing the relationship between customer satisfaction and a company's ability to fulfill its customer requirements with the offering. If a firm's offering includes NUD features and functionality, it would tend to "delight" its customers, particularly for those latent requirements. If the offering only included the bare minimum "must have" features and functionality, the customers would be dissatisfied. Finally, if a firm's offering only included the "expected" features and functionality, declaring it "fit for use," then the customer satisfaction would depend on the number of items included in the offering (where more equates to better) and probably would remain at a "neutral" level. Figure 3-4 is a diagram of this three-prong Kano Model.

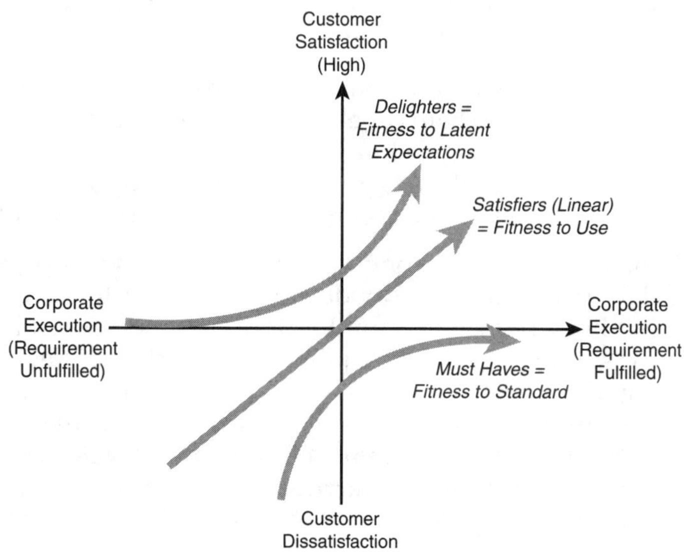

FIGURE 3-4: The Kano Model

Using the NUD technique, the "New" provides the customer "delighters"—the features, functionality, and services that neither your firm nor competition offers. The "Unique" describes a requirement that the marketplace offers (from either competition or a substitute product) but your firm does not. The "Difficult" may represent something complex in size, scale or scope, or technically challenging, but if provided, creates a competitive advantage and perhaps even a barrier to entry.

In addition to the Kano model, CDOV leverages from other principles to be discussed in a subsequent section: robust design, critical parameter management, and systems engineering to prevent problems. CDOV has been successfully deployed as a means to design and improve integrated hardware, firmware, and software for optical-mechanical products, production systems, pharmaceutical, chemical, materials-based systems, and software products.

The CDOV 4-Phase-Gate Structure

Similar to DMADV, the CDOV methodology uses a phase-gate structure. The phases generally are sequential; however, some activities may be iterative or may occur concurrently. Formal phase-gate reviews occur sequentially and require the completed set of respective deliverables to gain approval. The four-phase CDOV represents

Phase 1. **CONCEPT** (at a system-level) is developed to meet specific business goals and customer market segment requirements.

Phase 2. **DESIGN** the concept's subsystem, subassembly, and part-level elements.

Phase 3. **OPTIMIZE** the design for robustness at both a sub-component and integrated system level.

Phase 4. **VERIFY** the final design, production processes, and customer value chain capabilities are prepared for launch. Verify that 1) the design's functional performance remains robust and stable and 2) the design, supporting processes, and capabilities all meet requirement(s) in a non-test (real-world) environment.

If the new concept is complex in size, scale, and scope, each of the four CDOV phases may subdivide into sub-phases to correspond with subassemblies, particularly for the Optimize and Verify Phases. As a result, these sub-phases would require "sub-phase-gate" reviews before assembling into an integrated system. Regardless of the phase, each phase-gate review examines the following five topics: 1) the **business case** (financials and assumptions), 2) **technical design and development**, 3) **manufacturing and distribution preparation** (including assembly, materials management, and supply chain development), 4) **business and operating**

constraints (including regulatory, health, safety, environmental and legal), and 5) **post-launch readiness** (including shipping, sales, service and support). The CDOV structure supports the PACE® Commercialization Process from PRTM.

Figure 3-5 flowcharts the CDOV approach at a high-level view, along with its key requirements.

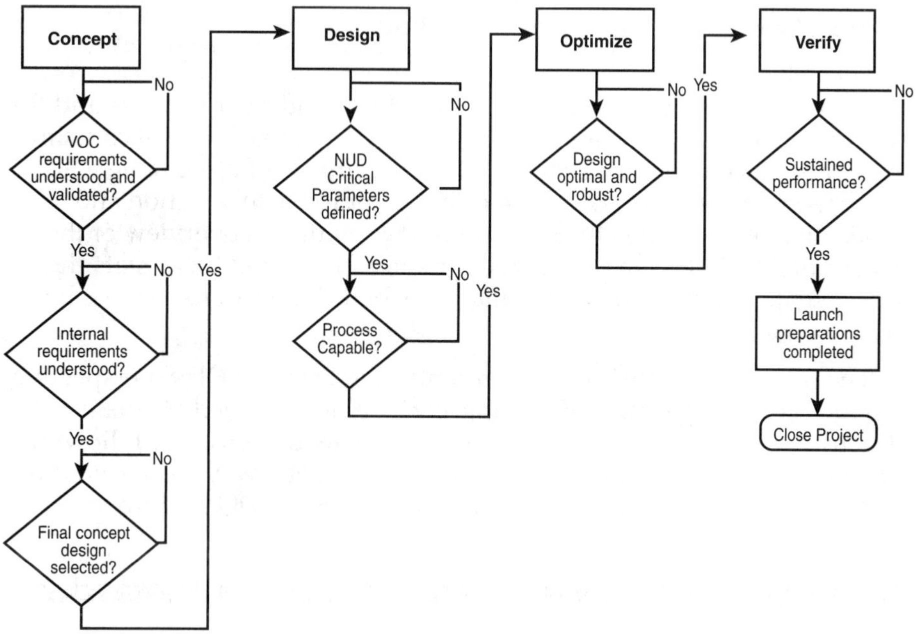

FIGURE 3-5: CDOV Model Process Flow

Comparison Between DMADV and CDOV

As part of the DFSS family of methods, Six Sigma experts might argue that DMADV and CDOV share similar objectives but differ in emphasis. As practitioners become exposed to the various DFSS models, the promoted toolset with any one given approach borrows from another. As long as the practitioners comprehend that tool selection follows the understanding of requirements, deliverables, and tasks linkage, then the sharing and changeability of candidate tools only strengthens the execution. The key is that the right tool is selected at the right time to answer the right question.

Figure 3-6 compares the DMADV and CDOV methods.

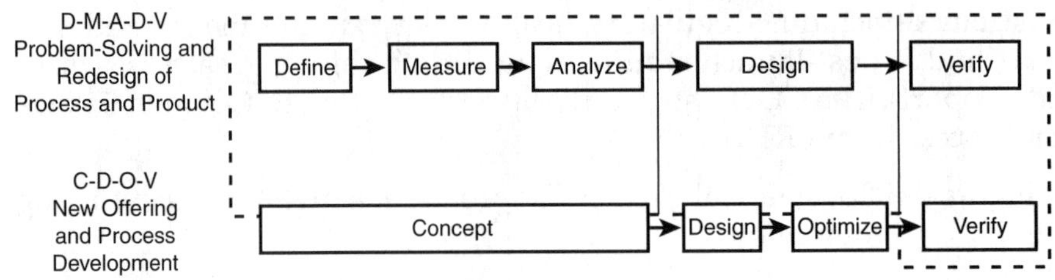

FIGURE 3-6: Comparison of DMADV and CDOV Methods

Given the fluidity between DMADV, CDOV and their variants and the fact that the toolset associated with each is similar, this book takes a simplistic approach to describe them. With the purpose of organizing a Six Sigma desk reference of tools, to aid in the appropriate selection and application of tools, the DFSS section next provides an overview of the structure of DMADV and CDOV and their variants and then treats them in aggregate to outline the candidate tools by key questions the tool helps to answer.

The remainder of this section presents an integrated DFSS perspective, presuming that regardless of the approach being deployed, its main objective will involve a new design to some extent. Hence, the following information applies to all DFSS methods in a generic way; however, this information most closely aligns with the proactive CDOV approach.

What Key Overall Requirements Define this Approach?

Recall that the public delineation among the various DFSS methods has blurred as this evolving field stabilizes. All DFSS methods focus on "design" at some level, either a proactive clean-sheet approach or reactive redesign. For purposes of simplicity, the remainder of this section will generalize and treat DFSS as one category focused on concept design and with a common set of requirements and corresponding *tool-task-deliverable* combination. This book's appendix lists several resources to reference if more detail is needed on the various approaches.

CDOV Prerequisites

Before getting started on a Product design, development, and commercialization initiative, the technical team will need require several inputs. According to Clyde M. Creveling, contributing author of several books, including most recently *Six Sigma in Technical Processes* and *Design for Six Sigma in Technology and Product Development*, the list of CDOV prerequisites entails the following:

- Business goals defined and a strategy on how to meet them.
- Target market segments identified.
- Customer requirements (VOC) and "Voice of Technology."
- Product/Offering portfolio plan and how the specific market segment requirements are fulfilled by the product-line (or product family).
- Technical strategy defined to address the product-line strategy.
- New platform and subsystem technology developed and certified.

The first four inputs probably come from the Business strategy and/or marketing and Offering Portfolio Renewal process (using perhaps the IDEA method). Research & Technology Development (R&TD), working in part with the Business strategy in the Offering Portfolio Renewal process, typically would provide the last two inputs (using perhaps the I^2DOV method).

The overall requirements that drive a concept design initiative entail the following:

- What is the best concept design based on the complete set of requirements (both market segment and business)?
- What are the design's New, Unique, and Difficult critical parameters?
- Are the design and its component-levels robust?
- Can the performance be sustained in non-test environments (including not only the system design, but also all the supporting post-launch functions)?

What Requirement Determines the Key Activities in this Approach?

Again, for purposes of simplicity, the remainder of this section overviews DFSS as one category using a high-level common set of requirements and corresponding *tool-task-deliverable* combination.

Table 3-1 uses the CDOV phases to tie to the overall requirements and lists the corresponding DMADV phases (for redesign) in the parentheses when they differ.

Table 3-1: CDOV Requirements-Phase Linkage (with Corresponding DMADV Phases)

Requirements	Resulting High-Level Phase (Corresponding DMADV Phase)
What is the best concept design based on the complete set of requirements (both market segment and business)?	1. CONCEPT (*and Define, Measure, Analyze*) • Define concept requirements based on the current VOC and VOB requirements. • Identify certified technologies ready for market use. • Design concept as a high-level system.
What are the design's New, Unique, and Difficult critical parameters?	2. DESIGN • Design subsystem, subassembly, and part-level elements of system.
Are the design and its component-levels robust?	3. OPTIMIZE *(and Design)* • Develop robust subsystems and subassemblies and evaluate performance. • Integrate the robust subsystems and subassemblies as a complete system and evaluate performance. • Conduct initial system reliability assessment.
Can the performance be sustained in non-test environments (including not only the system design, but also all the supporting post-launch functions)?	4. VERIFY • Verify the capability of the product design functional performance. • Verify the capability of production assembly and manufacturing processes and the extended supply chain, customer value chain, service, and support organizations.

Which Tools Are Aligned to Each Step of the Process?

Given the previous high-level phases, the following set of tables summarizes the subsequent tool-task-deliverables combination associated with each of the four phases in the CDOV approach, which is transferable to the DMADV approach. Table 3-2 summarizes the Concept Phase.

Part II of this book overviews the DFSS tools that are commonly applied to the marketing, sales, customer value chain, and supporting infrastructure and business areas. For a more in-depth understanding of the DFSS tools applied to the technical environments, please refer to Part IV, "Appendixes," for additional DFSS resources.

TABLE 3-2: Concept Phase Tools-Tasks-Deliverables (Including Corresponding DMADV Elements)

Phase 1: CONCEPT (and Define, Measure, Analyze)—What is the best concept design based on the complete set of requirements (both market segment and business including the subsystem, subassembly, and part-level elements of system design)?

High-level Deliverables	High-level Tasks	Candidate Tools & Techniques
• Voice of Business requirements understood • VOC data documented as New, Unique, Difficult (NUD) requirements	Work with marketing and other business colleagues to categorize and prioritize NUD, easy, common, and old requirements.	• Business case • VOC Gathering techniques • KJ Analysis • Market segmentation analysis • NUD • CTQs
System functional concept models designed	• Create product or system-level HOQ. • Conduct competitive product benchmarking. • Generate product or system-level requirements. • Generate system-level concepts. • Test and analyze various concepts.	• Concept generation techniques (e.g. TRIZ, Brainstorming, Brain writing) • Design techniques—architecture, platform, modular. • House of Quality (HOQ) • Functional Flow Diagramming and Critical Parameter Mapping • Competitive benchmarking • Taguchi Noise Diagramming and System Noise Mapping • Taguchi Stress Testing • Highly Accelerated Life and Stress Testing • Reliability modeling • Monte Carlo simulation • Multi-vari studies

continues

TABLE 3-2: Continued

High-level Deliverables	High-level Tasks	Candidate Tools & Techniques
Final concept selected	Evaluate test data and select concept model.	Pugh's Concept Evaluation Process
Technology transfer control plans documented	• Develop reliability requirements and initial model. • Develop system FMEA.	• Design and Process FMEA • Statistical Process Control and Capabilities Studies (for both design functions and process outputs)
Next steps planned	Create DESIGN Phase project plan and risk analysis.	• Project Management tools, including Monte Carlo simulation of project cycle time • FMEA

Table 3-3 summarizes the tool-task-deliverables linkage for the Define Phase of CDOV.

TABLE 3-3: Design Phase Tools-Tasks-Deliverables (Including Corresponding DMADV Elements)

Phase 2: DESIGN—What are the design's New, Unique, and Difficult critical parameters? (Including 1) Concept requirements based on requirements, 2) Certified market-ready technologies, and 3) high-level concept design.)

High-level Deliverables	High-level Tasks	Candidate Tools and Techniques
VOC data validated as New, Unique, Difficult (NUD) details at a sublevel	Work with marketing and other business colleagues to categorize and prioritize NUD, easy, common, and old sub-level requirements.	• VOC Gathering techniques • KJ Analysis • Market segmentation analysis • Business case • NUD
Sub-level models for functionality and reliability designed	• Gather Certified Sub-level technologies from Research and Technology Development process.	• Concept generation techniques (e.g. TRIZ, Brainstorming, Brain writing) • Design techniques—architecture, platform, modular

High-level Deliverables	High-level Tasks	Candidate Tools and Techniques
	• Model sub-level alternatives, test and analyze.	• House of Quality (HOQ) • Functional Flow Diagramming and Critical Parameter Mapping • Competitive bench-marking • Taguchi Noise Diagramming and System Noise Mapping • Taguchi Stress Testing • Highly Accelerated Life and Stress Testing • Reliability modeling • Monte Carlo simulation
Final sub-level design selected	Evaluate results and select sub-level design.	• Pugh's Concept Evaluation Process
Baseline design documented under nominal conditions, including Critical Adjustment Parameters (CAPs) and reliability plans	• Develop sub-level reliability plan. • Develop sub-level FMEA.	• Design and Process FMEA • Statistical Process Control and Capabilities Studies (for both design functions and process outputs) • Design for manufacture and assembly • Value engineering and analysis • MSA • CPM • Engineering methods and math modeling • DOE • Descriptive and inferential statistical analysis • ANOVA data analysis • Regression and empirical modeling • Design capability studies
Next steps planned	Create OPTIMIZE Phase project plan and risk analysis.	• Project Management tools, including Monte Carlo simulation of project cycle time • FMEA

Table 3-4 summarizes the *tool-task-deliverables* linkage for the Optimize Phase of CDOV.

Table 3-4: Optimize Phase Tools-Tasks-Deliverables (Including Corresponding DMADV Elements)

Phase 3: OPTIMIZE (and Design)—Are the design and its component-levels robust? (Including 1) Robust subsystems and subassemblies, 2) Integrates the robust system, 3) Initial system reliability assessment.) C D O V		
High-level Deliverables	**High-level Tasks**	**Candidate Tools and Techniques**
Robustness, adjustability, and reliability documented for each sub-level and integrated system	• Review and finalize the critical functional responses. • Develop noise diagram and map. • Conduct robustness experiments. • Conduct and analyze reliability/ capability evaluations.	• Noise diagramming and mapping • Measurement System Analysis (MSA) • Design of Experiments (screening and modeling) • Analysis of Means (ANOM) • ANOVA • Baseline signal-to-noise characterizations • Taguchi methods for Robust Design • Analysis and Empirical Tolerance Analysis • Additive experiments and modeling • Regression • Capability studies
Risk profile and assessment documented for each sub-level and integrated system	Analyze data, build predictive model and run verification experiments.	• Statistical process control (SPC) • Reliability analysis life testing • Design FMEA
Critical parameters documented at the sub- and integrated system level	Document critical functional parameter nominal set points.	• Design capability studies • Critical Parameter Management (CPM)

High-level Deliverables	High-level Tasks	Candidate Tools and Techniques
Next steps planned	Create VERIFY Phase project plan and risk analysis.	• Project Management tools, including Monte Carlo simulation of project cycle time • FMEA

Table 3-5 summarizes the *tool-task-deliverables* linkage for the last CDOV phase, Verify.

TABLE 3-5: Verify Phase Tools-Tasks-Deliverables (Including Corresponding DMADV Elements)

Phase 4: VERIFY—Can the performance be sustained in non-test environments (including not only the system design, but also all the supporting post-launch functions)? This includes 1) the capability of the product design functional performance and 2) the capability of production assembly and manufacturing processes, as well as the extended supply chain, customer value chain, and service and support organizations.

C D O V

High-level Deliverables	High-level Tasks	Candidate Tools and Techniques
Sub-level and system capabilities documented	• Conduct final tolerance design and document. • Verify product design meets all requirements.	• MSA • Analysis and Empirical Tolerance Design • Worst case analysis • Root Sum of Squares analysis • Monte Carlo simulation • DOE • ANOVA • Regression • Multi-vari studies • Design capability studies • System noise mapping • System-level sensitivity testing • Nominal system CFR • Stress case system
Sub-level and system design reliability growth	Evaluate system performance and reliability under nominal conditions.	Reliability assessments

continues

Table 3-5: Continued

High-level Deliverables	High-level Tasks	Candidate Tools and Techniques
Sub-level and system design risk profiles and assessment	Complete corrective actions on problems.	Design FMEA
Manufacturing, assembly, and supply chain capability	• Establish statistical process control (SPC) for critical-to-function components and critical functional responses. • Build product design verification using production parts	• SPC • CPM • MSA • Analysis and Empirical Tolerance Design • Worst case analysis • Root Sum of Squares analysis • Monte Carlo simulation • DOE • ANOVA • Regression • Multi-vari studies • Design capability studies • System noise mapping • System-level sensitivity testing • Nominal system CFR • Stress case system • Kaizen • DMAIC and Lean
Critical parameter management database	Develop transfer plan for the critical parameter database for production, supply chain, and service organizations.	• CPM • Competitive benchmarking studies • SPC
Next steps planned	Create launch plan and risk analysis.	• SPC • CPM • Project Management tools • FMEA

What Are the DFSS Variations?

Design for Six Sigma encompasses a portfolio of design methods. The origin of each approach stems from the key business requirement that it addresses and the timing of its evolution. Some grew out of

DMAIC-based projects, some from a deep understanding of engineering techniques and principles. What they all have in common is the objective of **designing quality into the design of a product or process**.

Other DFSS variations of note over and above DMADV and CDOV include the following four models:

- **DMEDI (Define-Measure-Explore-Develop-Implement)**

  This method closely aligns with the DMADV approach as a redesign technique and incorporates elements of Lean Six Sigma. It uses a slightly different vocabulary from DMADV and adds tools from the Lean methodology to ensure efficiency or "speed." DMEDI phase-gates include

  Phase 1. **DEFINE** the problem or new requirements.

  Phase 2. **MEASURE** the process and gather the data that is associated with the problem or new requirements.

  Phase 3. **EXPLORE** the data to identify cause-and-effect relationships between key variables.

  Phase 4. **DEVELOP (or DESIGN)** a new process so the problem is eliminated; measured results meet new requirements.

  Phase 5. **IMPLEMENT** the new process under a control plan.

- **PIDOV (Plan-Identify-Design-Optimize-Validate)**

  [Reference: *Design for Six Sigma Statistics*, by Andrew Sleeper, McGraw-Hill, New York, 2006; ISBN 0-07-145162-5.]

- **ICOV (Identify-Characterize-Optimize-Verify)**
- **IIDOV or I^2DOV (Invent/Innovate-Develop-Optimize-Verify)**

  A technology development process used in a Research and Development environment, wherein the goal is to prepare technology for commercialization.

Summary

Design for Six Sigma (DFSS) refers to a category of methods and tools that designs quality into a product (or process). While DFSS generally applies to technical design applications, it also features a unique subset of tools applicable to general business applications. The DFSS standard toolset contains invaluable tools applicable to non-technical scenarios as

DFSS

well. General Six Sigma practitioners should include the business-oriented DFSS tools and techniques in their portfolio of candidate tools; hence, many of the applicable tools can be found in Part II of this book, the "How To" utilize tools section.

SECTION 4
Six Sigma for Marketing (SSFM)

Six Sigma for Marketing (SSFM) represents a portfolio of methods and tools that expands Six Sigma concepts to take a preventative approach primarily for the non-technical community. It complements the DFSS design methods, but also encompasses the business processes before and after product design and development. SSFM spans across an organization's processes, starting with its strategic planning front-end to deciding what to offer, through its offering design and development, through its marketing, selling, and supporting of offerings and ending up with the discontinuance of an offering. SSFM methods benefit an organization taken individually or in its entirety. This section overviews the three SSFM methods and each of their high-level requirements and appropriate deliverables, which dictate the tasks and the tool selection to aid the task. This section outlines the SSFM standard toolset to build an understanding of its tool-task-deliverables linkage and facilitate in the selection of the appropriate tools when referencing the "How To" tool articles in Part II of this book.

What Is the Main Objective of the Approach?

Six Sigma for Marketing (SSFM) defines a category of Six Sigma methods applied to marketing and sales to drive growth. In addition, SSFM concepts can be applied to other general business functions (such as finance and customer service and support).

SSFM is the newest category of Six Sigma. It builds off the fundamental principles of both *Lean Six Sigma* and *Design for Six Sigma* (DFSS) and their respective tools. Its focus is proactive in nature, to drive growth in a balanced way, rather than simply cutting costs. SSFM concepts concentrate on lead indicators of issues, to incorporate preventative mechanisms as part of ongoing operations.

SSFM fills a Six Sigma void that both Lean Six Sigma and DFSS have had difficulty filling. The marketing and sales portion of a business traditionally tend to have unique work approaches and language that did not adapt well to the conventional cost cutting and revenue growth approaches. SSFM, albeit in its infancy, has begun to yield benefits for companies such as 3M. At a recent Six Sigma conference, 3M claimed that its SSFM efforts have taken longer to produce the benefits but have positively impacted the bottom-line profit far greater than the classic DMAIC initiatives. In fact, its benefits continue given the ongoing nature of the revenue annuities.

Brief Description of Typical Application

Recall that SSFM is the newest category of Six Sigma. Initially, SSFM addresses the three core business processes that marketing, as a functional discipline, typically supports:

- **Strategic Planning**—to develop, manage, and refresh a portfolio of offerings. This activity defines the initial product concept, position in the portfolio, and high-level pricing targets.
- **Offering Development**—to design, develop, commercialize and launch a new offering (including products, services, and information) often as part of a cross-functional team. This activity refines the product, its position and pricing, and begins to define launch promotions.
- **Operations Management**—to manage a launched offering throughout the customer value chain, involving the *go-to-market*, customer-facing functional partners such as marketing, sales, telemarketing, services (for example, consulting, field engineering, or customer training), service and maintenance, customer administration, customer support centers, and customer financing services. This set of activities includes continued refinement of product positioning and pricing, as well as new product promotions.

SSFM, when applied to each of these three areas, applies a unique method for each process—IDEA ((Identify-Define-Evaluate-Activate) for strategic planning, UAPL (Understand-Analyze-Plan-Launch) for offering development, and LMAD (Launch-Manage-Adapt-Discontinue) for operational management. The three SSFM methods define the work that marketing performs in each area and define how that work links together in a progressive closed-loop fashion, as shown in Figure 4-1.

Figure 4-1: The Strategic-Tactical-Operational Triangle

The Strategic Planning and Portfolio Renewal Process

The *Strategic Planning and Portfolio Renewal* process defines a business' set of marketplace offerings. This fundamental process refreshes an enterprise's offerings to sustain its existence over time. Multiple functional disciplines may be involved in this process, or the enterprise may limit this work to a small set of corporate officers depending on the size of the enterprise and scope of its offerings. This process generally calls for a cross-functional team comprised of finance, strategic planning, marketing, and sometimes research, engineering, sales, service and customer support. Some businesses, with a unique strategic planning department, may serve as a surrogate for the other various functional areas. If this is the case, the strategy office typically staffs people with various backgrounds (research, finance, and marketing). This process can span a year and should get refreshed on a regular basis. Portfolio planning and management are the foundation from which to build and grow a business. Our experience tells us that successful businesses have marketing play a key role in the Strategic Planning and Portfolio process.

The Tactical Planning and Portfolio Renewal Process

A business' tactical practices entail its Product and/or Services Commercialization process. This process defines, develops, and "readies" a business' offering for the marketplace. The industry, market segment, and size/scale/complexity of the offering dictate the number of the functional disciplines involved in this process and the amount of time it spans. The timeframe ranges from several months to several years. A business usually manages this process by establishing a unique *project* team to develop a single product or service from the portfolio of opportunities. At a minimum, two types of disciplines are needed—the "technical" functions to drive content and the "customer-facing" functions. The "technical" experts develop the offering and may include engineering, research, and manufacturing. The "customer-facing" disciplines represent those roles along the value chain that interface with a business' customer or client, such as marketing, sales, services, and customer support. In the *Commercialization* process, marketing may represent the "customer-facing" touchpoints throughout the process, and may bring in the other functional areas toward the conclusion of the process in preparation for "hand-off" to ongoing operations.

The Post-Launch Operational Management Process

The *Post-Launch Operational Management* process unifies the operational dimensions of a business along its customer supply chain. This process manages and coordinates a business' offerings of products and services and all its supporting activities. The offering and go-to-market strategy dictates the variety of functional disciplines involved across the value chain. The various functions may include sales, service, support, customer financing, customer administration, customer training, other services, and third party partners. Again, marketing may serve a representative role, integrating the multiple functional areas as it manages the product-line (or offering) throughout its lifecycle. This process represents long timeframes (often years) depending on the lifecycle of a given offering (product or service).

The natural flow of marketing work starts with strategic renewal of the offering portfolios to the tactical work of commercializing new offerings and finally to the operational work of managing the product and services lines in the post-launch sales, support, and service environment. Marketing professionals frequently overlook the fact that their contributions are part of a process (or set of related processes). They view their work as part of a program or project. However, marketing work is repeatable. Though the timeframe for repetitiveness may extend over a year or more, the work is nonetheless procedural in nature.

Note

The American Society for Quality (ASQ) defines a *process* as "a set of interrelated work activities characterized by a set of specific inputs and value added tasks that make up a procedure for a set of specific outputs." Most marketers would agree that strategic planning and launching a product meet this definition of process. The Six Sigma approach embraces a process view to communicate its structure and flow of interrelated tasks. Although it may seem unnatural to marketing professionals, the best way to describe *Six Sigma for Growth* is through a process lens.

The Strategic and Tactical process areas are internally focused; hence, we refer to them as *inbound* marketing areas. Yes, external data are critical to a successful portfolio definition and development and product commercialization. However, those processes' outputs are earmarked for internal use. These processes' outputs are not yet ready for external consumption. The outputs ready for prime-time market exposure are part of *outbound* marketing. The operational processes involving post-launch product marketing, sales, services, and support are customer-facing activities. Given the different customers of inbound and outbound marketing, the requirements for each differ. These requirements ultimately define the success (or failure) of the deliverables. Figure 4-2 depicts how these three marketing areas interrelate. The inner-circle shows the circular connection, and the outer-circle depicts the demarcation between inbound and outbound marketing.

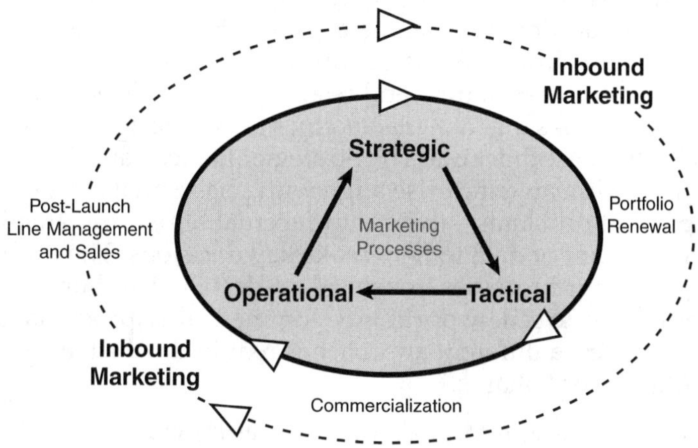

Figure 4-2: Interconnectivity of Marketing and the Process Triangle

Inbound Marketing
How does SSFM apply problem-prevention to marketing processes? *Inbound marketing* focuses on strategic product portfolio definition and development and tactical product commercialization. Inbound marketing can cause problems by under-developing the right data needed to renew product portfolios. The data are needed to define specific, new product requirements, thereby directing commercialization activities. Moreover, inbound marketing data define launch plans, which determine downstream operational success. Marketing can design and launch an inappropriate mix of products. As a result, a business could miss the growth targets promised in the business cases that were supposed to support the long-term financial goals of the company.

Outbound Marketing
Outbound marketing is focused on customer-facing operations. It encompasses post-launch product line management across the value chain (sales and services—including customer support). Outbound marketing can create problems and waste by failing to develop the right data to make key decisions about managing, adapting, and discontinuing the various elements of the existing product and service lines. Outbound marketing

also could fail to get the right information back upstream to the product portfolio renewal teams. They need to renew the portfolio based upon real, up-to-date data and lessons learned from customer feedback and the marketing and sales experts in the field.

The importance of the comprehensive, *closed-loop Strategic-Tactical-Operational* scope provided the structural underpinnings used to create the unique Six Sigma methods for marketing. Each of these arenas has a flow of repeatable work—a process context that is quite different from the steps found in the traditional Six Sigma methods. However, the Six Sigma fundamental elements from the classic approaches have been maintained: tool-task linkage, project structure, and result metrics. This new work is made up of specific tasks that are enabled by flexible, designable sets of tools, methods, and best practices. The Strategic, Tactical, and Operational processes within an enterprise align with phases that can be designed to prevent problems—to limit the accrual of risk and enable the right kind and amount of data to help make key decisions. The traditional methods help get your processes improved, re-designed and under control. If the objective is to renew portfolios, commercialize products, or manage product lines, a different approach is required, which employs a different set of steps or "phases."

Each of these processes feature distinct phases in which sets of tasks are completed. Each task can be enabled by one or more tools, methods, or best practices that give high confidence that the marketing team develops the right data to meet the task requirements for each phase of work. A *gate review* at the end of a phase is commonly used to assess the results and define potential risks. Marketing executives and professionals find phase-gate reviews an important part of risk management and decision-making. In the post-launch environment, "gates" are replaced by *key milestone reviews* because you are in an ongoing process arena unlike portfolio renewal or commercialization processes that have a very strictly defined end date.

Tackling the job of adding process discipline and metrics to the functions within a portfolio renewal process is by far the most difficult task when integrating Six Sigma into the three marketing processes. Adding process metrics and discipline to marketing functions in any environment is not a trivial matter. It must be done carefully, taking into account the unique culture that exists in marketing organizations. There is art and science here. Attention must be paid to how marketing teams interact with other teams within adjacent business processes—such as Research and Development (R&D), engineering, and the customer-facing partners.

SSFM attempts the balance between marketing creativity with the necessary process discipline and metrics to add rigor and robustness to the deliverables, thereby improving the likelihood for marketplace success.

A unique Six Sigma method for marketing was created for each of the three areas: Strategic, Tactical, and Operational. The method to guide marketing's strategic work is called IDEA. A second approach was developed for its tactical work, called UAPL. A third method to direct marketing's operational work is called LMAD. A brief description of each can be found in the following sections.

An organization gains the biggest benefit from implementing all SSFM methods because they interrelate. However, depending on the organization's maturity, size, complexity, needs, and resources, it could be quite an undertaking to deploy all three approaches at once. A *staged deployment* may start with the greatest need or the one easiest to establish. **Any one SSFM method can be used successfully by itself, but the true power is in the aggregate of the three-part suite.** An overall *change management plan* is helpful to define which method is appropriate and the execution approach that will ensure success.

IDEA (Identify-Define-Evaluate-Activate)

Brief Description of Typical Applications

In the Strategic marketing process environment, the unique SSFM method is known as the IDEA Process to renew and refresh the portfolio of offerings. This approach [pronounced *"Idea" or "**I**-'d**E**"*] works well at either a corporate, division, or department level to identify, develop, and fund an *offering concept*.

It may trigger an initiative that lasts a few months to a year or more before funding an offering development and commercialization team to make it real. Hence, a portfolio renewal process can be viewed as either a project-based or an operationally-based process. Regardless of the time it takes to complete the renewal and refresh process, the IDEA method contains four distinct phases:

Phase 1. **IDENTIFY**—markets, their segments, and the opportunities they offer

Phase 2. **DEFINE**—portfolio requirements and product portfolio architectural alternatives

Phase 3. **EVALUATE**—portfolio alternatives against competitive portfolios, by offering

Phase 4. **ACTIVATE**—ranked and resourced individual commercialization projects

Figure 4-3 diagrams the high-level flow of the IDEA approach and its key requirements.

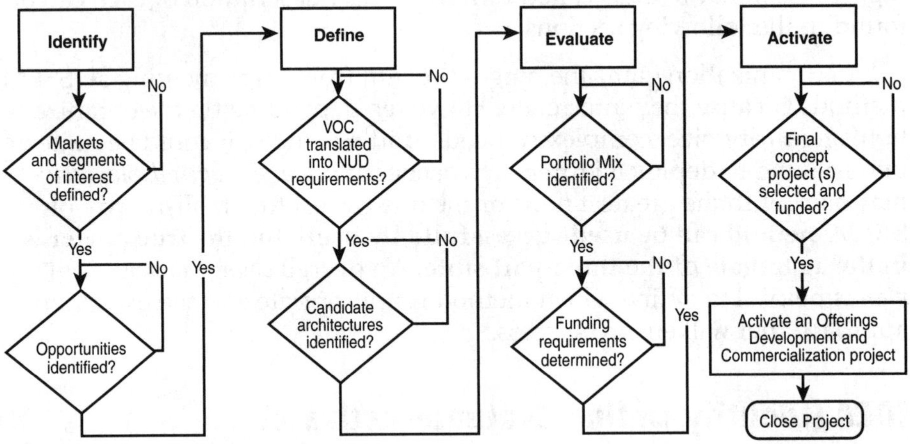

Figure 4-3: High-level IDEA Model Process Flow

The design of IDEA is to be flexible, such that it can be tailored to serve as additional foundational structure to any existing strategic portfolio renewal and refresh process at a company-wide or division level. Its *tools-tasks-deliverables* combination can be incorporated to make the current approach more structured and robust. Its structure provides clear and concise deliverables that link strategic direction to organizational capabilities. Moreover, the IDEA approach reinforces an evergreen process that continually renews and fuels the tactical development and commercialization process, which flows through the post-launch operations environment back into the strategic portfolio renewal planning process.

What Key Overall Requirements Define this Approach?

Prior to entering the IDENTIFY Phase; the *business strategy* must be clearly defined and documented. The business strategy breaks down into several general areas:

- Financial Growth Goals
- Core Competencies and Capabilities
- Innovation Strategy (including both marketing and technology)

These goals and capabilities serve as critical criteria to evaluate, prioritize, and select potential opportunities.

As with all projects, the appropriate team members must participate in the project. The right mix of disciplines are needed to understand not only what the customer wants and what the business goals are, but also what is possible and what is feasible. What is possible and feasible refers not necessarily to technological hurdles, although that is an important ingredient for the concept under consideration and any related infrastructure support, but for the technology acceptability in the marketplace. Services and information concepts may require as careful a consideration around technological feasibility as a tangible product-offering concept when comprehending the activities along the full customer value chain. For example, customer support requirements or a system to manage intellectual capital or complex projects all require some level of technological consideration, often beyond the given offering.

The appropriate amount of funding (resources) is required for the team to adequately complete their deliverables. Recall that though this Portfolio Renewal represents an inbound marketing function, it requires the proper funding for solid benchmarking and customer requirements gathering.

The appropriate mix of renewal and commercialization projects partly determines whether the right people are available to work on the critical projects and if the available funding exists. Companies (and/or divisions) often take on too many projects in the hopes that enough of them will deliver financial results that will meet the growth requirements of the business strategy. They overload the workforce—particularly understaffed marketing organizations. Applying Six Sigma thinking can minimize this dilemma.

What Requirement Determines the Key Activities in this Approach?

The requirements that define the method's respective key activities (or phases) are found in Table 4-1.

TABLE 4-1: IDEA Requirements-Phase Linkage

Requirements	Resulting High-Level Phase
What market holds what opportunity?	1. IDENTIFY • Define the key markets of interest and describe their key characteristics. • Determine if they contain any segments and describe their key characteristics. • Identify the opportunities within each.
What are the customer requirements, and how do they translate into an offering?	2. DEFINE • Translate and document the opportunities as NUD requirements (e.g. New, Unique, and Difficult) • Document the requirements by segment. • Identify candidate portfolio architectures or platforms.
What is the recommended mix of portfolio offerings and funding requirements?	3. EVALUATE • Summarize the candidate offering concepts (product and services) and position them against the current offerings. • Document Growth Potential of the balanced offering portfolio architecture (including financial targets and tolerances). • Select the best portfolio architecture from among the candidates based on market potential and business case analysis. • Document product portfolio architecture. • Summarize the financial requirements and potential revenue and entitlement from the selected portfolio architecture as a comparison to the overall growth goals for the - business. • Document Risk Assessment for the Portfolio that shows a balance across multiple dimensions of risk, with a Portfolio FMEA (Failure Modes and Effects Analysis) to identify possible failures, respective frequencies, and potential impact of those failures. • Document Real-Win-Worth (RWW) Analysis across the elements of the portfolio.

Requirements	Resulting High-Level Phase
Which offering concept projects will be activated as a development and commercialization project(s)?	4. ACTIVATE • Rank the potential "activation" projects, requesting an offering development and commercialization team. • Define the timing and headcount requirements for activation of projects based on a risk-balanced commercialization project control plan.

What Tools Are Aligned to Each Phase of the Process?

Given the preceding high-level phases, the following series of tables summarize the subsequent *tool-task-deliverables* combination associated with each individual phase within the four-phase IDEA approach. The detail behind how to use each tool can be found in Part II of this book, "The Six Sigma Encyclopedia of Business Tools and Techniques: Choosing the Right Tool to Answer the Right Question at the Right Time." Table 4-2 outlines the Identify Phase into its tool-task-deliverables linkages.

TABLE 4-2: Identify Phase Tools-Tasks-Deliverables

Phase 1: IDENTIFY—What market holds what opportunity (including 1) General market defined, 2) Specific segments identified, and 3) Specific opportunities determined)?		SSFM Portfolio Renewal I D E A
Deliverables	**Tasks**	**Candidate Tools and Techniques**
Documented Growth Goals and Core Competencies	Define Business Strategy and Financial Goals.	Project Charter
Documented Innovation Strategy	Define Innovation Strategy.	Idea capture and database development tools
Documented Markets of interest and respective market segmentation, if any • Market and Segment Behavioral Dynamics Map • Porter's 5 Forces Chart	Define Markets and Market Segments.	• Market Identification and Segmentation Analysis • Porter's 5 Forces Analysis and segmented risk profile • Market Behavioral Dynamics Map Methods • Value chain diagrams • Key event timelines

continues

TABLE 4-2: Continued

Deliverables	Tasks	Candidate Tools and Techniques
Opportunities Documented—Supporting deliverables include: • Competitive Benchmarking Data and Trend Charts • Market Perceived Quality Profile and Gap Matrix • SWOT Matrix	• Design and conduct VOC validation surveys. • Document VOC findings. • Conduct competitive benchmarking (for marketing, sales channel, and technical disciplines). • Define Opportunities across Markets and within segments. • Create a database of internal ideas based on opportunity categories.	• VOC Gathering Methods • Competitive Benchmarking and Best Practice Analysis • SWOT analysis matrix • Market Perceived Quality Profile Method • Idea (or Concept) capture database
Next steps planned	Create DEFINE Phase project plan and risk analysis.	• Project Management tools • FMEA

Table 4-3 organizes the Define Phase of IDEA into its tool-task-deliverables linkages.

TABLE 4-3: Define Phase Tools-Tasks-Deliverables

Phase 2: DEFINE—What are the customer requirements, and how do they translate into an offering (including 1) Documented NUD portfolio requirements, 2) Documented requirements by segment, and 3) Identified candidate portfolio architectures)?		
Deliverables	**Tasks**	**Candidate Tools and Techniques**
Opportunities translated and documented as NUD (**N**ew, **U**nique, **D**ifficult) requirements • Segmentation statistics summary • VOC-based requirements data	• Analyze VOC data to find common versus differentiated requirements for use in time and circumstance. • Conduct statistical analysis on VOC data to refine segment identification.	• Statistical survey design and analysis • Statistical data analysis

Deliverables	Tasks	Candidate Tools and Techniques
• Common requirements across segments • Differentiated requirements across segments • Customer Survey Results	• Translate NUDs into Portfolio Requirements (including metrics). • Document NUD customer needs across segments.	
Requirements by segment documented	Construct and/or refine market and segment Behavioral Dynamics Map.	Quality Function Deployment (QFD)
Candidate portfolio architectures or platforms identified	Define candidate port-. folio architectures	• Portfolio House of Quality (HOQ) • Product Portfolio Architecting methods
Next steps planned	Create EVALUATE Phase project plan and risk analysis.	• Project Management tools • FMEA

Table 4-4 presents the Evaluate Phase of IDEA into its tool-task-deliverables linkages.

TABLE 4-4: Evaluate Phase Tools-Tasks-Deliverables

Phase 3: EVALUATE—What is the recommended mix of portfolio offerings and funding requirements (including 1) Summary candidate offering portfolio architectures, 2) Best candidates selected based on market potential and business case analysis, and 3) Summary financials)?		
Deliverables	**Tasks**	**Candidate Tools and Techniques**
Documented Growth Potential	• Create Preliminary Business Case. • Develop Portfolio Financials [including NPV (Net Present Value), ECV (Expected Commercial Value), ROI (Return on Investment)].	• Financial Modeling and Forecasting tools such as NPV, ECV, ROI analysis and Monte Carlo simulation • Business case development and valuation methods

continues

TABLE 4-4: Continued

Deliverables	Tasks	Candidate Tools and Techniques
Documented Product Portfolio Architecture	• Develop a Portfolio Evaluation Criteria and benchmark portfolio architecture • Evaluate and Select the best portfolio architecture from the candidates. (It may be a hybrid of the candidates.)	Portfolio balancing methods
• Documented Risk Assessment • Documented Real-Win-Worth (RWW) Analysis	• Assess market dynamics and fit with trend lines. • Assess technical risk profiles. • Assess Portfolio Financials.	• Real-Win-Worth analysis. • Pugh Concept Evaluation and Selection Method
Next steps planned	• Evaluate concept risk.	• Create ACTIVATE Phase project plan. • Project Management tools • FMEA

Table 4-5 organizes the last phase of IDEA, the Activate Phase, into its tool-task-deliverables linkages.

TABLE 4-5: Activate Phase Tools-Tasks-Deliverables

Phase 4: ACTIVATE—Which offering concept projects will be activated as a development and commercialization project(s) (including the 1) Rank order of activation project and 2) Timing for activation of projects, based upon a risk balanced commercialization project control plan)?

Deliverables	Tasks	Candidate Tools and Techniques
Documented availability, readiness, and deployment of Core Competencies and Resources	• Rank projects for activation priority and strategic value. • Determine required core competencies and resources for top projects.	• Dynamic Rank Ordering methods • Pareto process • Resource planning • RACI

Deliverables	Tasks	Candidate Tools and Techniques
	• Document availability of resources.	
Documented Project Activation Plan and Control Plan	Create project activation timing schedule and control plan.	Control Plan
Documented risk response plan	Conduct a risk analysis on the activation plan and the portfolio for financial performance against growth goals.	FMEA
Documented enabling technologies maturity and readiness	Define stability of enabling technologies and document and balance resources across the project activation plan.	Project Management tools for resource planning and budgeting
Next steps planned	• Create high-level Development and Commercialization project plan. • Document Lessons Learned. • Close Portfolio Renewal project.	Project Management tools

U-A-P-L (Understand-Analyze-Plan-Launch)

Brief Description of Typical Applications

The successful completion of an offering concept passing the IDEA Activation Phase triggers the need to initiate a project team to design, develop, and commercialize the concept. The new project team takes the strategic planning processes deliverables as key inputs to start the new process.

The project team may be comprised of multiple functional teams or one integrated team made up of the required technical and business resources to convert the strategic concept into a viable, market-ready offering. This team bridges the strategic initiative with the operational management of a launched offering; hence, this team performs the *tactical* activities.

The marketing and business project team members develop the inbound marketing deliverables in parallel with their technical counterparts, who design the offering. Using a Six Sigma approach, the technical discipline follows a DFSS method (DMADV (Define-Measure-Analyze-Design-Verify) or CDOV (Concept-Design-Optimize-Verify)) to develop and stabilize the offering design. In turn, the marketing and business professionals should follow a complimentary Six Sigma approach, namely UAPL [pronounced "***YOU**-apple*"]. This tactical marketing process recognizes the interdisciplinary dependency for design and commercialization deliverables, regardless if the offering is a product, service, or combination. The UAPL method reinforces the bi-directional communication between the two core disciplines and aids in the translation of functional requirements and specifications.

The integrated flow between CDOV and UAPL with their complimentary approach is depicted in Figure 4-4.

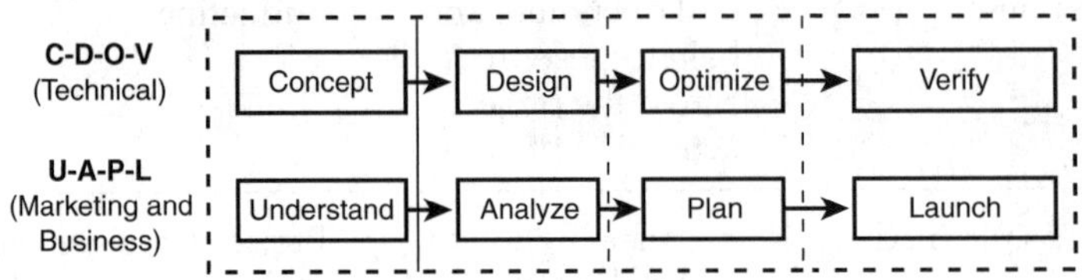

FIGURE 4-4: Comparison of CDOV and UAPL Methods

The design of UAPL is to be flexible, such that it can be tailored to serve as additional foundational structure to any existing tactical offering development and commercialization method. Its tools-tasks-deliverables combination can be incorporated to make the current approach more robust. For example, if a Product Development Process (PDP) process is well established and successful, critical elements of UAPL can strengthen the design and development in two key ways. First, UAPL reinforces a proactive perspective to the process. Second, UAPL reinforces a bi-directional communication through a structured translation approach between the customer-supplier relationships for both external and internal process players across any functional language or other perspective barriers.

The standard UAPL approach is segmented into four distinct phases, defined as follows:

Phase 1. **UNDERSTAND**—the market opportunity and specific customer requirements translated into product (or service) requirements.

Phase 2. **ANALYZE**—customer preferences against the value proposition.

Phase 3. **PLAN**—the linkage between the value chain process details (including marketing and sales) to successfully communicate and launch the product (or service) concept as defined in a maturing business case.

Phase 4. **LAUNCH**—preparation of the new product (or service) under a rigorously defined launch control plan.

Figure 4-5 provides a diagram of the high-level flow of the UAPL approach and its key requirements:

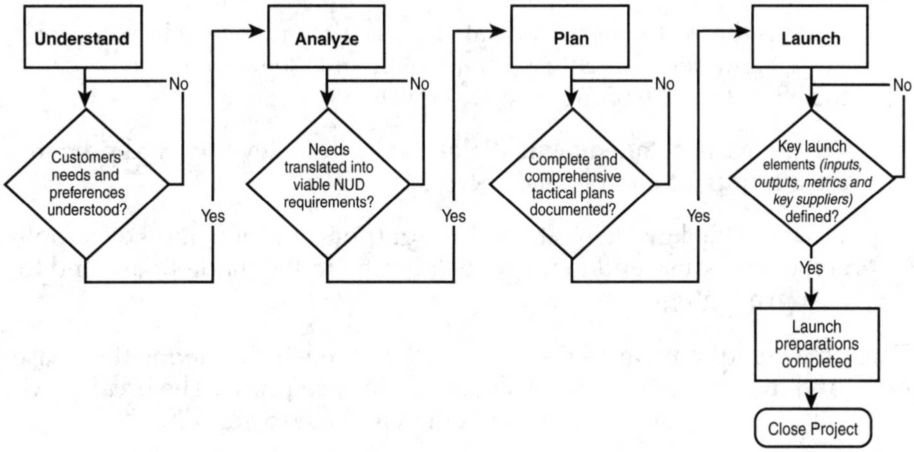

FIGURE 4-5: High-level UAPL Model Process Flow

What Key Overall Requirements Define this Approach?

Prior to entering the UNDERSTAND Phase of UAPL; the offering development/commercialization team must fully comprehend the *concept strategy* from the offerings portfolio and renewal (IDEA) process. The concept strategy inputs should include

- Offering positioning within the portfolio mix
- Offering's value proposition to its target market segments (including environmental influences, such as marketplace, political, regulatory, competitive, and technological assumptions)
- Financial growth goals and assumptions for the offering.
- Core competencies and capabilities required for the offering
- Innovation strategy and assumptions (impacting technology, marketing, and business across the customer value chain and any infrastructure)

When the tactical project team has these inputs in hand, it can embark on the specific design, development, and commercialization efforts of the specific offering concept.

The overall requirements that direct the activities within UAPL incorporate the efforts of marketing, general business, and the customer value chain players. These UAPL activities compliment any required technical activities needed to design, develop, and commercialize the offering concept. The UAPL approach answers the follow overarching questions:

- What are the customers' specific offering needs and preferences?
- What is the best way to translate those customer needs into concept design and deployment requirements that take into consideration the NUD (New, Unique, and Difficult)?
- What are the detailed tactical plans to design, develop, and prepare the offering for commercial success?
- What are the key elements (inputs, outputs, metrics, and key suppliers) to introduce and manage the offering in the marketplace and to ensure a high-quality launch?

These four requirements define the UAPL approach and define the phase-gates and the respective key activities within each phase. The breakdown of the *requirement-phase* linkage is detailed in the next section.

What Requirement Determines the Key Activities in this Approach?

The requirements that shape the UAPL method into its four phases can be found in Table 4-6.

TABLE 4-6: UAPL Requirements-Phase Linkage

SSFM Commercialization U A P L

Requirements	Resulting High-Level Phase
What are the customers' specific offering needs and preferences?	1. UNDERSTAND • Document customer requirements. • Document offering requirements and superior offering concepts. • Refine and update Real-Win-Worth (RWW) Analysis. • Refine and update business case.

Requirements	Resulting High-Level Phase
What is the best way to translate those needs into concept design and deployment requirements that take into consideration the NUD (New, Unique, and Difficult)?	2. ANALYZE • Finalize offering requirements from customer testing. • Document value proposition linked to brand. • Refine price model and sales forecasts. • Update critical parameters for marketing, sales channel, customer value chain participants, and supporting infrastructure and business areas. • Refine and update Real-Win-Worth (RWW) Analysis. • Refine and update business case. • Draft a high-level marketing and business plan, including a preliminary risk mitigation plan (defined in concert with the technical development plan, as appropriate).
What are the detailed tactical plans to design, develop and prepare the offering for commercial success?	3. PLAN • Develop a robust marketing and business plan, including a refined risk mitigation plan (defined in concert with the technical development plan, as appropriate). • Develop and/or coordinate the development of a customer value chain plan (including the sales channel(s), customer administration and financing, customer services and training (pre- and post-purchase), customer support and maintenance, supplies and parts, and other supporting infrastructure and business). • Develop marketing collaterals, advertising, promotional, and other customer support materials. • Document channel management plan for marketing, sales channel, and other customer value chain participants and include both risk and Critical Parameter Management. • Update critical parameters appropriate for marketing, sales channel, customer value chain participants (including any third-party partners), and supporting business and infrastructure. • Refine and update Real-Win-Worth (RWW) Analysis. • Refine and update business case. • Develop project management plan for launch.

continues

TABLE 4-6: Continued

Requirements	Resulting High-Level Phase
What are the key elements (inputs, outputs, metrics, and key suppliers) to introduce and manage the offering in the marketplace and to ensure a high-quality launch?	4. LAUNCH • Document the launch plan, including critical inputs, respective suppliers, key outputs, and performance metrics. • Document post-launch customer value chain processes and control plans including marketing, business, selling, support, and services. • Document customer relationship management process and control plan. • Finalize marketing, sales, services, and support collaterals, advertising, and promotional materials. • Update critical parameters appropriate for marketing, sales channel, customer value chain participants (including any third-party partners), and supporting business and infrastructure. • Update Real-Win-Worth (RWW) Analysis, as appropriate. • Update business case and sales forecast, as appropriate.

Which Tools Are Aligned to Each Step of the Process?

Given these high-level phases, the series of tables that follow summarize the subsequent tool-task-deliverables combination associated with each individual phase within the four-phase UAPL approach.

The detail behind how to use each tool can be found in Part II of this book. Table 4-7 organizes the Understand Phase into its tool-task-deliverables linkages.

TABLE 4-7: Understand Phase Tools-Tasks-Deliverables

Phase 1: UNDERSTAND—What are the customer's specific offering needs and preferences (including 1) Customer requirements, 2) Offering requirements and superior offering concepts, 3) Real-Win-Worth (RWW) Analysis, and 4) The offering's business case)?

Deliverables	Tasks	Candidate Tools and Techniques
Business case goals documented	• Gather offering financial goals, positioning, and assumptions from the strategic plan. • Conduct a Goals, Objectives, Strategies, Plans, and Actions (GOSPA) analysis.	GOSPA Analysis
Segmented markets identified	Verify markets and their segments and specific customer types and characteristics for the offering concept.	• Market Perceived Quality Profile • Customer Behavioral Dynamics Map
Customer requirements documented	• Create a customer interview guide. • Gather and translate specific VOC data. • Document the specific NUD offering requirements. • Construct specific customer behavioral dynamics maps. • Conduct competitive benchmarking against the NUD customer requirements.	• VOC gathering methods • Customer Interview Guide • Customer Value Map • KJ diagrams (images and requirements) • Competitive benchmarking data and trend analysis. • NUD: *What is New, which no one fulfills today; What is Unique to what you offer today but what competitors fulfill; What is Difficult, that may give you a competitive advantage and compliment current competencies and business model?* • Porter's 5 Forces Analysis • Offering category SWOT Analysis

continues

Table 4-7: Continued

Deliverables	Tasks	Candidate Tools and Techniques
Offering requirements documented	Translate the VOC into the value proposition and help translate them into technical requirements.	• Value chain analysis • CTQ tree or matrix
Superior offering concept documented	• Help generate and evaluate concepts that are candidates to fulfill NUD. • Generate and/or revise competitive position analysis.	• Product Category Data Mining • Offering category SWOT • Concept generation tools • Pugh concept • QFD/HOQ methods
Innovation strategy documented	• Define the innovation strategy (s) and portfolio fit for marketing, business, customer value chain and infrastructure. • Generate the qualitative and quantitative elements of value that the concept provides to substantiate the specific offering opportunity.	
Core competencies documented	• Conduct customer value chain analysis. • Determine availability and readiness at time of launch.	Value chain mapping
Critical parameters defined for marketing, sales channel, customer value chain, and supporting business and infrastructure	• Update market perceived quality profiles. • Develop offering category specific key event timeline.	• Market Perceived Quality Profile • Value Chain analysis • Key events timeline • CTQ tree or matrix

Deliverables	Tasks	Candidate Tools and Techniques
Real-Win-Worth Analysis updated	• Conduct Won-Lost analysis. • Revise the offering Real-Win-Worth Analysis.	• Real-Win-Worth Analysis • Won/Lost Analysis
Business case updated	Develop the offering business case and document assumptions.	• Business case modeling
Next steps planned	Create ANALYZE Phase project plan and risk analysis.	• Project Management tools • FMEA

The Analyze Phase of UAPL can be characterized as a robust phase due to the many requirements directing this phase's activities. Several crucial elements are needed to answer the critical question, *"What is the best way to translate those needs into concept design and deployment requirements that take into consideration the NUD (New, Unique, and Difficult)?"* For purposes of emphasis, these additional phase requirements are listed here as follows:

1. Finalized offering requirements from customer testing.
2. Documented value proposition linked to brand.
3. Refined price model with Monte Carlo simulations.
4. Refined sales forecasts with Monte Carlo simulations.
5. Updated critical parameters (and respective database(s)) for marketing, sales channels, customer value chain, and supporting infrastructure and businesses.
6. Refined and updated Real-Win-Worth (RWW) Analysis.
7. Refined and updated business case.
8. PLAN Phase project plan, including a preliminary risk mitigation plan for marketing and supporting business areas.

Table 4-8 presents the Analyze Phase of UAPL into its tool-task-deliverables linkages.

Table 4-8: Analyze Phase Tools-Tasks-Deliverables

Phase 2: ANALYZE—What is the best way to translate those needs into concept design and deployment requirements that take into consideration the NUD (including 1) Finalized offering requirements, 2) Documented value proposition, 3) Refined price model, 4) Refined sales forecast, 5) Updated critical parameters, 6) RWW Analysis, 7) Refined business case, and 8) Project plan for the PLAN Phase)?

Deliverables	Tasks	Candidate Tools and Techniques
Critical Parameter Management defined	Assess strategic fit of evolving customer value chain support and services.	• Critical Parameter Management (CPM) • Multi-vari studies
Organizational capability assessment completed	Examine offering and/or Product Development Process (PDP) capability to meet launch date.	Statistical Process Control (SPC)
Customer preferences defined	• Define customer preferences. • Conduct a Conjoint Analysis if offering's complexity supports potential "bundling" of options.	• Customer-based concept testing • DOE (Design of Experiment) • Descriptive and inferential statistical data analysis. • ANOVA (Analysis of Variance) • Regression and empirical modeling methods • Conjoint Analysis • Demand Elasticity
Price model finalized	Simulate price model and select optimal configuration.	Monte Carlo simulation
Marketing, selling, and customer value chain support strategy defined	Evaluate capabilities of marketing, sales channels, customer value chain, and supporting infrastructure and business areas and develop plan to close any gaps and meet customer needs.	• Sales channel analysis • Marketing plan template

Deliverables	Tasks	Candidate Tools and Techniques
Sales forecasts finalized	Simulate sales forecasts by channel, and select optimal configuration.	Monte Carlo simulation
Marketing plan drafted	• Develop a marketing plan. • Refine RWW Analysis	• RWW Analysis • Marketing Plan template
Business case refined	Update the business case and financials.	Business case analysis
Next steps planned	Create PLAN Phase project plan and risk analysis.	• Project Management tools • FMEA

Similar to the UAPL Analyze Phase, the Plan Phase has many requirements directing its activities. Several critical elements are needed to answer the critical question, *"What are the detailed tactical plans to design, develop, and prepare the offering for commercial success?"* For purposes of emphasis, these additional phase requirements are listed here as follows:

1. Finalized robust plans for both marketing and the sales channels.
2. Developed marketing collateral, advertising, promotional, training, and customer support materials.
3. Documented management plan for the post-launch operations, including the sales channels, customer value chain components, and supporting infrastructure and business areas.
4. Updated critical parameters (and respective database) for marketing, the sales channels, customer value chain components, and supporting infrastructure and business areas.
5. Refined and updated Real-Win-Worth (RWW) analysis.
6. Refined and updated business case.
7. LAUNCH Phase project plan, including a preliminary risk mitigation plan, for marketing and supporting business areas.

Table 4-9 arranges the Plan Phase of UAPL into its tool-task-deliverables linkages.

TABLE 4-9: Plan Phase Tools-Tasks-Deliverables

Phase 3: PLAN—What are the detailed tactical plans to design, develop and prepare the offering for commercial success (including 1) Marketing and the sales channels plans, 2) Marketing collateral, advertising, promotional, training, and customer support materials, 3) Post-launch operations management plans, 4) Post-launch operations critical parameters, 5) Real-Win-Worth (RWW) analysis updated, 6) Business case updated, and 7) LAUNCH Phase project plan)?

Deliverables	Tasks	Candidate Tools and Techniques
Updated Critical Parameter Management documentation and database	• Update, revise and implement the Critical Parameter Management plan for the functional areas involved in post-launch operations. • Develop Who/What matrices to map responsibility for delivery of all CTQs.	• CPM • CTQs • RACI
Post-Launch Process Maps	• Develop detailed process maps for marketing and sales processes. • Develop high-level (or appropriate level) process maps for customer value chain organizations and supporting infrastructure and business areas. • Document process maps depicting how to fulfill the CTQ requirements.	• Process Mapping • Kaizen on current processes • Lean on current processes

Deliverables	Tasks	Candidate Tools and Techniques
Plans: Preliminary Marketing, Sales Channel Management, Market Communications, Customer Relationship Management, and Brand Positioning	• Audit and update the marketing plan. • Establish Market Accountability Plan. • Create and implement internal communications plan to the company employees and across the value chain including the sales force, customer support, services, and suppliers. • Develop external communications strategy (comprehending multiple venues and format options such as collaterals, the Internet/ Web). • Develop sales support plan.	• Plan Development Methods • Marketing and Sales Process Noise Diagramming • Customer Relationship Management methods • Marketing Communications Planning • Brand Positioning and Management • RACI
Preliminary designs for marketing and sales collateral, advertising, and promotional materials	Develop preliminary designs for marketing and sales collateral, advertising, and promotional materials.	• Marketing Collateral and Promotional Material Planning and Development
Refined and updated RWW Analysis	Test effectiveness of offering positioning.	• Real-Win-Worth (RWW) Analysis
Refined and updated business case	• Develop Pricing Plan by segment. • Update Business Case and Sales Forecast.	• Business case analysis • The Hybrid Grid • Monte Carlo simulation
Next steps planned	• Create LAUNCH Phase project plan and risk analysis. • Document critical risks, problems, and assumptions.	• Project Management tools • FMEA

The UAPL Launch Phase contains a robust set of requirements (the key phase-gate question plus seven supporting requirements), like its two preceding phases. Its list of deliverables reaches 11 items. The tasks to produce those 11 outputs can be consolidated into only three core categories: 1) Conducting an audit of the marketing plan, 2) Assessing the plan completeness and readiness of both internal and external communications, and 3) Assessing the plan completeness and readiness of post-launch operations. The brevity of the task list can deceive project teams into a false sense of control and excess available time. However, this concise three-task list represents a great deal of required work effort to produce the 11 Launch Phase-Gate deliverables.

Table 4-10 organizes the last UAPL phase, Launch, into its tool-task-deliverables linkages.

TABLE 4-10: Launch Phase Tools-Tasks-Deliverables

Phase 4: LAUNCH—What are the key elements (inputs, outputs, metrics and key suppliers) to introduce and manage the offering in the marketplace and to ensure a high-quality launch (including 1) Launch plan, 2) Post-launch process maps and control plans with Discontinuance criteria, 3) CRM process plan, 4) Final marketing, sales, customer support collaterals, advertising and promotional materials, 5) Post-launch CPM, 6) Refined RWW Analysis, and 7) Final Business case and sales forecast)?

SSFM Commercialization U A P L

Deliverables	Tasks	Candidate Tools and Techniques
Integrated Value Chain Plan	• Develop an integrated customer value chain plan that encompasses the marketing coordination, services, support, supply chain, customer financing and administration, and supporting infrastructure and business areas. • Assess Implement ation Plan readiness of the External Communications Strategy Implementation.	• Value chain development methods • Process mapping • RACI • Kaisen of currently operating processes • Lean and DMAIC of currently operating processes

Deliverables	Tasks	Candidate Tools and Techniques
Risk management plan	• Assessment of post-launch operations risk (sales, support, and so on) • Post-launch operations Process Noise Diagrams and Failure Modes and Effects Analysis. • Conduct final marketing plan audit.	• Process Noise Diagrams • FMEA • Market plan audit
Launch Plan	• Finalize Launch Plan. • Assess Implementation Plan readiness of the Communications Plan for both external and internal (employees and value chain players such as the sales force, customer support and service, partners, suppliers).	• Launch planning methods • Kaizen
Control plans and process maps of post-launch processes	• Finalize the post-launch operations process maps and control plans for marketing, sales, entire value chain, and supporting infrastructure and business areas (including marketing, advertising, promotion, public relations, selling, CRM, customer training and support, supporting alliances and partnerships, supplies management, customer financing and administration, and so on). • Establish discontinuance criteria to determine appropriate conditions wherein the offering needs to be discontinued.	• Control plan development methods • Process Mapping and analysis • RACI • Customer relationship management methods

continues

Table 4-10: Continued

Deliverables	Tasks	Candidate Tools and Techniques
Statistical Process Control Charts on critical data	• Finalize Statistical Process Control Charts for marketing, sales, and post-launch operations.	• SPC • Statistical data mining and analysis tools • Market plan audit
Critical Parameter Management databases	• Finalize the post-launch operations CPM database with requirements, metrics, and controls.	• CPM
Final launch materials	• Finalize marketing and sales collaterals, advertising, promotional, training, and customer support materials.	• Marketing communications planning
Final documentation for brand alignment with value proposition	• Conduct final marketing plan audit.	• Brand positioning and management • Marketing plan audit
Updated and refined Real-Win-Worth (RWW) Analysis	• Conduct final marketing plan audit.	• RWW Analysis • Market plan audit
Updated and refined business case and sales forecast	• Conduct final marketing plan audit.	• Business case analysis • Market plan audit
Next steps planned	• Prepare to implement Launch Plan. • Document Lessons Learned. • Close Development and Commercialization project.	• Project Management tools • FMEA

L-M-A-D (Launch-Manage-Adapt-Discontinue)

Brief Description of Typical Applications

In the Operational marketing process environment, the *LMAD* method [pronounced *"Elle-mad"*] is used for managing the portfolio of launched products and/or services across the customer value chain. This approach solely applies to ongoing field operations, with the middle two phases specifically describing the cyclical lifecycle management process until the offering reaches end-of-life, triggering a withdrawal of commercial support. This approach has the following four distinct phases:

Phase 1. **LAUNCH**—the offering through its introductory period into the market according to the Launch Control Plan of the prior process.

Phase 2. **MANAGE**—the offering in the steady-state marketing and sales processes, while integrating the partner go-to-market value chain functions.

Phase 3. **ADAPT**—the marketing and sales tasks and tools, integrating the cross-functional go-to-market partners when marketplace events and noises require a course correction, modification, or any other kind of change.

Phase 4. **DISCONTINUE**—the offering with discipline to sustain brand loyalty.

Recall that as with the other SSFM methods, the design of LMAD supports any existing operational method such that its tools-tasks-deliverables can be incorporated to help make the current approach more robust. For example, within sales, standard processes for account management and selling may be well established and successful. If this is the case, LMAD should be tailored to fit as an underpinning to these current account management and selling methods.

The four-phased LMAD approach ebbs and flows in and out of the MANAGE and ADAPT phases to stay "on plan" throughout the life of an offering. Figure 4-6 shows the need for a course correction, across a generic LMAD lifecycle.

The Phase-Gate Review approach applied to the strategic IDEA and tactical UAPL processes transform with LMAD because of the post-launch environment. The *post-launch operational processes* behave differently than portfolio renewal or commercialization processes. Like steady-state manufacturing, operational environments *lack formal phase gates*. Hence, operations require timely, periodic reviews of progress against a plan. (Recall that the post-launch operations plans developed during the tactical UAPL commercialization project cover marketing, sales channels, customer service and support, customer financing and administration, supplies, and supporting infrastructure and business areas.) These periodic operations reviews should be thought of as key milestones in the continuum of executing the appropriate plan—not as gate reviews. Data and results can be gathered and summarized on an hourly, daily, weekly, monthly, quarterly, and yearly basis. A balance between proactive and reactive performance measures should be maintained—with an emphasis on leading indicators when possible to stay on plan and under control.

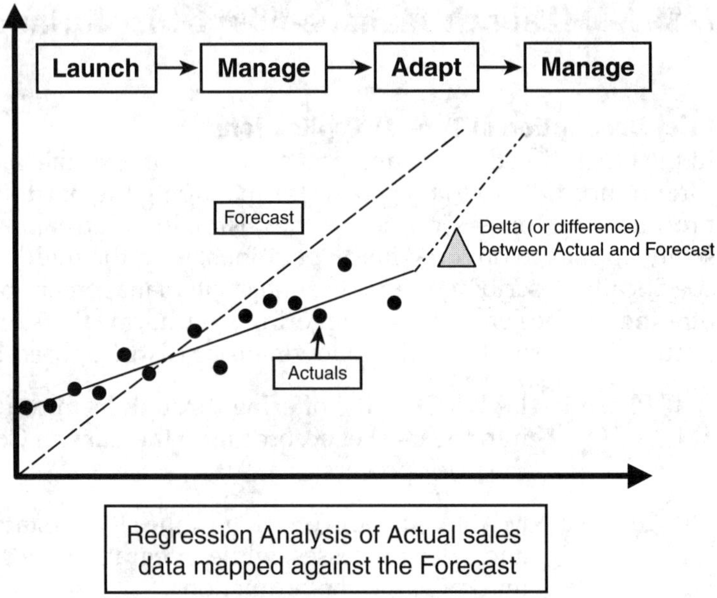

FIGURE 4-6: Generic Revenue Versus Forecast Sales Data

Most of the tools, methods, and best practices of outbound marketing are the same as inbound marketing. LMAD focuses on the tools and techniques that add value to staying on a given operational plan. The marketing, sales, and customer value chain professionals in an operational environment use these tools to refine forecasts, adjust estimates and to analyze data streams. They respond to change—change in the environment, customer needs, competition, technology, regulation, and so on. They adapt particularly when the change jeopardizes meeting their goals.

Given that in a post-launch steady-state environment, a business will fluidly cycle in and out of the MANAGE and ADAPT phases as necessary.

The LMAD approach also applies to the production engineering and manufacturing arena with relatively minor adaptations; however, this book focuses on the customer-facing value chain processes and activities.

Figure 4-7 depicts the high-level flow of the LMAD approach and its key requirements for the post-launch operations of the customer value chain environment.

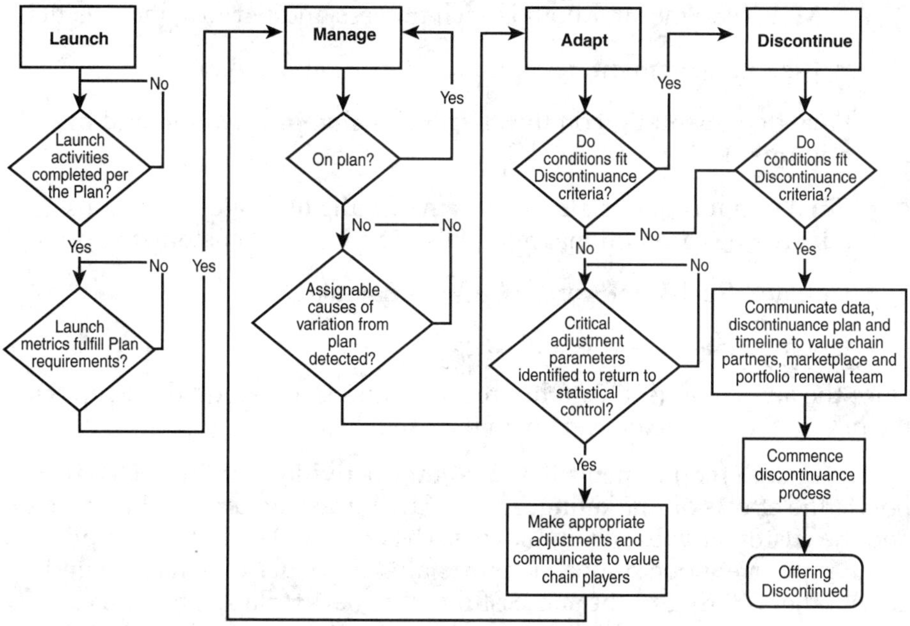

FIGURE 4-7: High-level LMAD Model Process Flow

What Key Overall Requirements Define this Approach?

Prior to entering the LAUNCH Phase of LMAD, the tactical offering development/commercialization project team must provide the following inputs:

- **Integrated value chain plan** across each of the various post-launch processes areas, including the following components:
  - Critical Parameter Management (CPM) requirements, metrics, and controls
  - Readiness assessment
  - Control plan and Discontinuance criteria
  - Risk management plans
  - Process maps
- **Launch plan** addressing market readiness activities, communication, and deployment to the external marketplace (customers, industry analysts, media/public relations, investors), employees, and each of the various post-launch processes areas.
  - **Launch control and change management plan**, which has built-in robustness to handle nonrandom sources of variation and sensitivities to change issues.

    - **Metrics** to monitor both launch process and performance results.
    - **Risk management, response, and mitigation plan**.
- Final documentation on the offering's **value proposition and brand alignment**.
- **Final support materials**, such as marketing and sales collaterals, advertising, promotional materials, training, and customer support.
- Updated **Real-Win-Worth (RWW) Analysis**.
- Updated business case and sales forecast.

Once the launch project team has received and understood these inputs, the offering launch team can begin its activities.

The overall requirements that direct the activities within LMAD incorporate the efforts of marketing, supporting infrastructure and business and the customer value chain players. These LMAD activities compliment any current operational activities already in place to manage and adapt other offerings to be successful in the marketplace. The LMAD approach answers the follow overarching questions:

- Has the Launch activities completed per the Launch Plan and fully satisfy the Launch Plan metrics? Including:
    - Communication and training activities completed to internal and external parties.
    - Launch deliverables fully distributed and available.
    - The post-launch customer value chain process players understand their respective plans and how they affect one another (marketing, sales, services, customer financing and administration, support and maintenance, and supporting infrastructure and businesses).
- Can the post-launch marketing, sales, customer value chain, and supporting infrastructure and businesses detect assignable causes of variation that indicate that adaptive actions are required to stay on plan?
- What are the critical adjustment parameters needed to return to a state of statistical control? In addition, what are the leading indicators (assignable causes of variation) that signal the post-launch customer value chain players to adjust their critical adjustment parameters?
- Do the current conditions fit the discontinuance criteria?

These four high-level requirements define the LMAD approach and define the key milestones and the respective key activities within each phase. The breakdown of the *requirement-milestone linkage* is detailed in the next section.

What Requirement Determines the Key Activities in this Approach?

The requirements that shape the LMAD method into its four milestones can be found in Table 4-11.

TABLE 4-11: LMAD Requirements-Milestone Linkage

Requirements	Resulting High-Level Milestones
Have the launch activities completed per the Launch Plan and fully satisfied the Launch Plan metrics?	1. LAUNCH • Deploy and stabilize offering's launch activities, per the Launch Plan, across the post-launch customer value chain processes. • Communicate and train internal and external parties. • Distribute launch deliverables. • Ensure that the post-launch customer value chain process players understand their respective plans and how they affect one another.
Can the post-launch operations players detect assignable causes of variation that indicate that adaptive actions are required to stay on plan?	2. MANAGE • Manage the ongoing, steady-state commercialization process throughout the lifecycle of the offering. • Depending on the offering's lifecycle, both the Manage and Adapt phases can repeat several times and last years, or the timeframe could be very brief, before the final Discontinuance phase is reached.

continues

TABLE 4-11: Continued

Requirements	Resulting High-Level Milestones
What are the critical adjustment parameters needed to return to a state of statistical control? In addition, what are the leading indicators (assignable causes of variation) that signal the post-launch customer value chain players to adjust their critical adjustment parameters?	3. ADAPT • Determine when to react or be proactive to events to stay on plan. (Can be in response to changes in the marketplace, economy, technology, regulatory, or simply the performance of the value chain processes.)
Do the current conditions fit the discontinuance criteria?	4. DISCONTINUE • Discontinue an offering based on its planned or forced end-of-life criteria. • Develop a preplanned set of deliverables and tasks to increase the likelihood of an efficient, more cost-effective transition to a replacement offering that renews the portfolio.

Which Tools Are Aligned to Each Step of the Process?

Given these high-level milestones, the following series of tables summarize the subsequent tool-task-deliverables combination associated with each individual phase within the four-phase LMAD approach.

Because of the operational environment within which LMAD fits, a common set of tools repeatedly gets utilized across the suite of LMAD phases. Hence, a single LMAD scorecard for performance evaluation featuring a common tool set the can be adapted on an "as needed" basis as part of the annual and/or operations review process. The tool set aligns within three categories: 1) **Process Definition**, 2) **Risk Management**, and 3) **Operational Models, Data Analysis and Controls**. In total, the tool set enables the required post-launch operational tasks throughout the lifecycle of a commercialized offering for marketing, sales, and customer value chain teams and supporting businesses.

Process definition: There are five major types of tools, methods, and best practices used to define the Outbound Marketing and Sales Process.

1. **Process Requirements Development**—Including Interviewing and Requirements Data Gathering; Requirements Structuring, Ranking and Prioritization (for example, KJ Analysis); and Quality Function Deployment (to define detailed metrics).

2. **Process Mapping**—This captures the marketing and post-launch functions Inputs, Outputs, and Constrains in a process map, along with Process Noise Mapping and Failure Modes and Effects Analysis (FMEA).

3. **Concept Generation**—For marketing and post-launch operations professionals (including any third-party partners).

4. **Pugh Concept Evaluation and Selection Process** and the **Concept Innovation Process**.

5. **Critical Parameter Management**—This entails Critical Parameter Identification and Metrics for Product Promotion; Advertising; Channel and Distribution; Customer Relationship and Support; and Marketing Communications.

Process Risk Management: This includes three types of major types of tools, methods, and best practices used to define the outbound marketing, sales, customer value chain, and supporting infrastructure and business processes.

1. **Offering Management Scorecard** to drive risk analysis, risk management, and decision-making

2. **Failure Modes and Effects Analysis (FMEA)** to delve further into Risk Analysis, Risk Management, and decision-making

3. **SWOT Analysis** to evaluate Strengths, Weaknesses, Opportunities, and Threats

Operational Models, Data Analysis, and Controls: These have several appropriate Six Sigma for Growth tools, methods, and best practices available, including the following:

1. Line Management Control Planning for the offering

2. Project Management Methods, including two key Six Sigma tools: 1) Cycle-time Monte Carlo Simulation and 2) Critical Path Task FMEA

3. Cost Modeling for outbound marketing, sales, customer value chain, and supporting infrastructure and business processes

4. Price Modeling for the offering, services, support, and supplies using the Monte Carlo Simulation

5. Sales Forecasting models using the Monte Carlo Simulation

6. Market Perceived Quality Profiles (MPQP)

7. Customer Value Chain Mapping

8. Surveys and Questionnaires design
9. Post-launch data structures, analysis, and management utilizing four types of tools:
   a. **Descriptive and Inferential Statistics**—such as Graphical Data Mining, Multi-vari Studies, Hypothesis Testing, Confidence Intervals, t-Tests, Data Sample Sizing, and last Regression and Model Building
   b. **Capability Studies**
   c. **Statistical Process Control**
   d. **Design of Experiments (DOE)**[md]on the critical parameters and Conjoint Analysis for the outbound marketing, sales, customer value chain, and supporting infrastructure and business processes
10. Data feedback structures for strategic portfolio renewal process and "advanced" or variant offering (product/services) planning.
11. Offering discontinuance planning.

Recall that the candidate LMAD tool set is common across the four phases. The detail behind how to use each tool can be found in Part II of this book.

Table 4-12 organizes the Launch Milestone into its tool-task-deliverables linkages.

TABLE 4-12: Launch Milestone Tools-Tasks-Deliverables

Milestone 1: LAUNCH—Have the launch activities been completed per the Launch Plan and fully satisfied the Launch Plan metrics?		SSFM Post-Launch L M A D
Deliverables	**Tasks**	**Candidate Tools and Techniques**
• Capability Studies • Customer Identification and Qualification Metrics • Customer Purchase Experience and Behavioral Analysis • Customer Satisfaction Assessment	• Gather critical data. • Analyze the performance data: the actual versus forecast, the initial trend assessment, and growth rate versus the plan. • Generate statistical process control charts and capability indices for key metrics.	• Process Requirements Development • Process Mapping • RACI • Concept Generation • Pugh Concept Evaluation and Selection Process and the Concept Innovation Process • CPM (Critical Parameter Management)

Deliverables	Tasks	Candidate Tools and Techniques
	• Refine applicable forecast models.	• Offering Management Scorecard • Line Management Control Planning • Market Perceived Quality Profiles (MPQP) • Customer Value Chain Mapping • Surveys and Questionnaires • Post-launch data • Data feedback structures • Kaizen
Risk assessment and management	Conduct a Failure Modes and Effects (FMEA) Analysis, take appropriate action, and document Lessons Learned.	• FMEA • Offering Management Scorecard
Competitive Assessment	Conduct a competitive assessment.	• Offering Management Scorecard • SWOT
Advertising and Promo tion Effectiveness Evaluation	Evaluate any adver. tising and promotion effectiveness	• Offering Management Scorecard • SWOT
Critical parameters updated	Update CPM and SPC charts.	• CPM • SPC • Offering Management Scorecard • Offering discontinuance planning
Business Case Fulfillment Assessment	Assess performance against the offering business case.	• Offering Management Scorecard • Cost modeling • Price modeling • Forecast modeling
Next steps planned	Create MANAGE Phase project plan and risk analysis.	• Project Management tools • FMEA • Offering discontinuance planning

To govern over the ongoing operational marketing, sales, and customer value chain processes, the *same set of tasks and resulting deliverables found in the Launch Milestone also are used in the Manage portion of LMAD*. The difference is that they continue to get updated and modified as the environment and information change over time. Table 4-13 portrays the LMAD Manage Milestone into its tool-task-deliverables linkages.

Table 4-13: Manage Milestone Tools-Tasks-Deliverables

Milestone 2: MANAGE—Can the post-launch operations players detect assignable causes of variation that indicate that adaptive actions are required to stay on plan?

Deliverables	Tasks	Candidate Tools and Techniques
• Capability Studies • Customer Identification and Qualification Metrics • Customer Purchase Experience and Behavioral Analysis • Customer Satisfaction Assessment	• Gather critical data. • Analyze the performance data: the actual versus forecast, the initial trend assessment, and growth rate versus the plan. • Generate statistical process control charts and capability indices for key metrics. • Refine applicable forecast models.	• Process Requirements Development • Process Mapping • RACI • Concept Generation • Pugh Concept Evaluation and Selection Process and the Concept Innovation Process • CPM (Critical Parameter Management) • Offering Management Scorecard • Line Management Control Planning • Market Perceived Quality Profiles (MPQP) • Customer Value Chain Mapping • Surveys and Questionnaires • Post-Launch data • Data feedback structures • Kaizen
Risk assessment and management	Conduct a Failure Modes and Effects (FMEA) Analysis, take appropriate action, and document Lessons Learned.	• FMEA • Offering Management Scorecard

Deliverables	Tasks	Candidate Tools and Techniques
Competitive Assessment	Conduct a competitive assessment.	• Offering Management Scorecard • SWOT
Advertising and Promotion Effectiveness Evaluation	Evaluate any advertising and promotion effectiveness	• Offering Management Scorecard • SWOT
Critical parameters updated	Update CPM and SPC charts.	• CPM • SPC • Offering Management Scorecard • Offering discontinuance planning
Business Case Fulfillment Assessment	Assess performance against the offering business case.	• Offering Management Scorecard • Cost modeling • Price modeling • Forecast modeling
Next steps planned	Create ADAPT Phase project plan and risk analysis.	• Project Management tools • FMEA • Offering discontinuance planning

Table 4-14 arranges the Adapt Milestone of UAPL into its tool-task-deliverables linkages.

TABLE 4-14: Adapt Milestone Tools-Tasks-Deliverables

Milestone 3: ADAPT—What are the critical adjustment parameters needed to return to a state of statistical control? In addition, what are the leading indicators (assignable causes of variation) that signal the post-launch customer value chain players to adjust their critical adjustment parameters?

SSFM Post-Launch L M A D

Deliverables	Tasks	Candidate Tools and Techniques
• Updated critical parameters • Updated Process Noise Maps	• Apply critical adjustment parameters to get results back on target.	• Process Requirements Development • Process Mapping • RACI

continues

TABLE 4-14: Continued

Deliverables	Tasks	Candidate Tools and Techniques
	• Conduct Designed Experiments as necessary to improve effectiveness of Critical Adjustment Parameters for current conditions. • Generate Statistical Process Control Charts and Capability Indices after adjustments. • Refine advertising, promotion, customer support services, and channel management plans. • Refine ADAPT Phase Control Plan. • Develop Noise Maps.	• Concept Generation • Pugh Concept Evaluation and Selection Process and the Concept Innovation Process • CPM (Critical Parameter Management) • Offering Management Scorecard • Line Management Control Planning • Market Perceived Quality Profiles (MPQP) • Customer Value Chain Mapping • Surveys and Questionnaires • Post-launch data • Data feedback structures • Offering Management Scorecard • SPC • Offering discontinuance planning • Kaizen • Lean and DMAIC
Updated Process FMEA	Analyze FMEAs, respond accordingly, and document actions.	• FMEA • Competitive situation analysis • SWOT • Market Perceived Quality Profile and Gap Analysis • Offering Management Scorecard
Update Annual Operating Plan	• Refine price models. • Refine sales forecast models. • Assess Business Case against current performance.	• Offering Management Scorecard • Cost modeling • Price modeling • Forecast modeling

Deliverables	Tasks	Candidate Tools and Techniques
Documented DISCONTINUE Phase Plan approved and ready for use	Refine DISCONTINUE Phase Control Plan.	• Offering Management Scorecard • RACI
Next steps planned	• Create LAUNCH Phase project plan and risk analysis. • Document Critical Risks, Problems, and Assumptions.	• Project Management tools • FMEA • Offering Management Scorecard

Table 4-15 organizes the last LMAD Milestone, Discontinue, into its tool-task-deliverables linkages.

TABLE 4-15: Discontinue Milestone Tools-Tasks-Deliverables

Milestone 4: DISCONTINUE—Do the current conditions fit the discontinuance criteria?

SSFM Post-Launch L M A D

Deliverables	Tasks	Candidate Tools and Techniques
Refined discontinuance criteria	• Apply CPM to adjust marketing, sales, and customer value chain functions and results to control the discontinuance, protect the brand, and maximize the business case. • Document discontinuance criteria based on CPM data to assess the business case versus current performance. • May conduct Designed Experiments as necessary to improve the effectiveness of critical adjustment parameters for discontinuance conditions.	• CPM • SPC charts • Capability studies • Trend analysis

continues

TABLE 4-15: Discontinue Milestone Tools-Tasks-Deliverables

Deliverables	Tasks	Candidate Tools and Techniques
Discontinuance Phase Project Plan	• Develop Discontinuance Plan and risk assessment. • Update FMEA. • Update Noise Maps. • Update competitive assessment. • Refine price models for discontinuance. • Refine advertising, promotion, and channel management plans for discontinuance. • Generate discontinuance forecast models for sales, supplies, support. • Document final business case assessment against goals.	• Project Plan template • FMEA • Noise Maps • Competitive assessment • Project Management tools • RACI • Pricing models • Forecast models • Business case template
Documented discontinuance data sent to portfolio renewal team	• Gather and summarize discontinuance data. • Send performance report to portfolio renewal team, including recommendations for sustaining the brand through next-generation offerings.	• VOC • Gathering survey data from marketing, sales, and customer value chain players • Market Perceived Quality Profile (MPQP) and Gaps • SWOT • Porter's 5 Forces Analysis • FMEA • Lessons Learned

Do DMAIC and Lean Six Sigma Apply to Marketing?

Yes; if a marketing process is broken, ineffective, or out of control, then use one of the traditional Six Sigma approaches to improve or re-design it. SSFM presumes that the strategic, tactical, and operational marketing processes have been designed to function properly. The three SSFM methods define *what to do* and *when to do it* within structured marketing processes.

Where to Get More Information on SSFM

For more in-depth understanding of the three Six Sigma for marketing and business processes (IDEA, UAPL, and LMAD), please refer to *Six Sigma for Marketing Processes, An Overview for Marketing Executives, Leaders, and Managers*, by Clyde M. Creveling, Lynne Hambleton, and Burke McCarthy, published by Prentice-Hall, New Jersey, 2006; ISBN 0-13-199008-X.

Summary

Six Sigma for Marketing (SSFM), though a bit of a misnomer, refers to a category of proactive growth methods and tools for the marketing, sales, and other business professionals involved in strategic planning, offerings development, and ongoing operations management of the launched offering portfolio. The SSFM standard toolset contains invaluable candidate tools applicable to several business processes throughout an organization and can be found in Part II of this book, the "How To" utilize tools section.

PART II

Six Sigma Tools and Techniques: Choosing the Right Tool to Answer the Right Question at the Right Time

Flowchart—7QC Tool

G Gantt Chart

GOSPA (Goals, Objectives, Strategies, Plans and Actions)

Graphical Methods

H Histogram—7QC Tool

House of Quality (HOQ)

Hypothesis Testing

I Interrelationship Diagram—7M Tool

K KJ Analysis

M Market Perceived Quality Profile (MPQP)

Matrix Diagrams -7M Tool

Measurement System Analysis (MSA)

Monte Carlo Simulation

Multi-vari Chart

N Normal Probability Plot

P Pareto Chart—7QC Tool

PERT (Program Evaluation and Review Technique) Chart

Poka-Yoke

Porter's 5 Forces

Prioritization Matrices— 7M Tool

Process Capability Analysis

Process Decision Program Charts (PDPC) - 7M Tool

Process Map (or Flowchart)—7QC Tool

Pugh Concept Evaluation

Q Quality Function Deployment (QFD)

R RACI Matrix (Responsible, Accountable, Consulted, Informed)

Real-Win-Worth (RWW) Analysis

Regression Analysis

Risk Mitigation Plan

Rolled Throughput Yield

Run Chart - 7QC Tool

S 7M—Seven Management Tool

7QC—Seven Quality Control Tool

Sampling

Scatter Diagram—7QC Tool

Scorecards

SIPOC (Supplier-Input-Process-Output-Customer)

SMART Problem & Goal Statements for a Project Charter

Solution Selection Matrix

Stakeholder Analysis

Statistical Tools

Stratification—7QC Tool

SWOT (Strengths-Weaknesses-Opportunities-Threats)

T Tree Diagram— 7M Tool

TRIZ

V Value Stream Analysis

Voice of Customer Gathering Techniques

W Work Breakdown Structure (WBS)

Y Y = f (X)

ENCYCLOPEDIA

The Six Sigma Encyclopedia of Business Tools and Techniques

Summary Tool Matrix

For easy navigation, use the Jump Table that follows, which lists the different tools and techniques featured in this book and organizes them by purpose and the key questions they answer.

The statistical and graphical tools include additional descriptors about their primary applications—describe, compare, and predict. In addition, the project management tools also include appropriate descriptors—scheduling, planning, scope, human resources (HR), deliverables, and tasks.

The Encyclopedia Summary Tool Matrix Jump Table

Generic Hypothesis Testing Schematic			
Purpose	**Key Question**	**Tool Name**	**Page**
Competitive Marketplace and Positioning			
	What are the experts doing, and how do they do it?	Benchmarking	**p. 160**
	How does the market perceive the quality of the product and/or services offerings versus competition?	Market Perceived Quality Profile (MPQP)	**p. 390**

continues

Continued

Generic Hypothesis Testing Schematic			
Purpose	**Key Question**	**Tool Name**	**Page**
	How are we performing relative to the competitive business threats?	Porter's 5 Forces	**p. 464**
	How would a potential offering be valued in the marketplace, and positioned well against competition? Is it worth the investment to develop the idea? Would it be successful? Would the concept outpace competition?	Real-Win-Worth (RWW) Analysis	**p. 560**
	How do the organization's strengths and weaknesses compare with the competitive opportunities and threats?	SWOT (Strengths-Weaknesses-Opportunities-Threats)	**p. 699**
Financial			
	What is the payback time period for an investment?	Cost/Benefit Analysis	**p. 238**
Predict	What are the probabilities and risk associated with several financial possibilities? (that is, pricing, forecasting)	Monte Carlo Simulation	**p. 435**
Graphing			
Describe	What does the distribution of a set of data look like? Alternatively, what do the distributions look like and compare for multiple sample groups?	Boxplot—Graphical Tool	**p. 164**

Generic Hypothesis Testing Schematic			
Purpose	**Key Question**	**Tool Name**	**Page**
Describe	How is the data of a process distributed; what does the data set look like?	Dotplot	**p. 280**
Describe	How is the data of a process distributed; what does the data set look like? Are the data normally distributed?	Histogram—7QC Tool	**p. 330**
Describe & Compare	Across multiple sources of variability, which one contributes the most?	Multi-vari Chart	**p. 439**
Compare & Predict	Are the data distributed normally?	Normal Probability Plot (**See Also** "Control Charts"—Normal Versus Non-normal Data section **p. 227**)	**p. 444**
Describe & Compare	What are the vital few items with the biggest impact? Which 20% of items produce 80% of the impact (the 80/20 rule)?	Pareto Chart—7QC Tool	**p. 445**
Describe	How does the data look over time? Are the data randomly distributed over time? Does the process look stable and random over time?	Run Chart—7QC Tool	**p. 610**
Describe & Compare	Are these two factors correlated? What is the relationship between these two factors?	Scatter Diagram—7QC Tool	**p. 640**
Describe & Compare	Are there any patterns in the data? Is the data set homogeneous?	Stratification—7QC Tool	**p. 697**

continues

Continued

Generic Hypothesis Testing Schematic			
Purpose	**Key Question**	**Tool Name**	**Page**
Governance			
	What is the progress of the project and project team or the process and process players?	Scorecards	**p. 653**
Idea/Solution Generation and Selection			
	How can you elicit new ideas in a short period of time?	Brainstorming Technique	**p. 168**
	Which technical specifications for a product or services offering best meet a specific set of customer requirements?	House of Quality (HOQ) (**See Also** Quality Functional Deployment (QFD)," **p. 543**	**p. 335**
	Which design or potential solution option is best?	Pugh Concept Evaluation	**p. 534**
	Which technical specifications for a product or services offering best meet a specific set of customer requirements?	Quality Function Deployment (QFD)	**p. 543**
	Which solution option best meets requirements?	Solution Selection Matrix	**p. 672**
	What is the best solution to address this problem to create a competitive advantage?	TRIZ	**p. 715**

Generic Hypothesis Testing Schematic			
Purpose	**Kcy Question**	**Tool Name**	**Page**
Organization & Planning			
	What activities or deliverables are required to meet requirements? Alternatively, what was observed?	Checklists—7QC Tool	**p. 204**
	What information is available about the current process, product, or services offering?	Data Collection Matrix	**p. 248**
	How do the activities of a function, program, or project align with the overall organizational strategy?	GOSPA (Goals, Objectives, Strategies, Plans, and Actions)	**p. 320**
	How do these two (or more) groups relate to one another?	Matrix Diagrams—7M Tool	**p. 399**
	What are the details behind this general topic breakdown into smaller components? How does it break down into its piece-parts?	Tree Diagram—7M Tool	**p. 712**
Prioritization			
	Which key variables (process steps or process inputs) best meet customer requirements (or the key process output variables)?	Cause-and-Effect Prioritization Matrix	**p. 188**
	What is the best option among the several possibilities for a crucial (often mission-critical) decision that carries risk of significant consequences if wrong?	Prioritization Matrices—7M Tool	**p. 470**

continues

Continued

Generic Hypothesis Testing Schematic			
Purpose	**Key Question**	**Tool Name**	**Page**
Process Evaluation (Also see Variation Analysys)			
Describe & Compare	Is the process able to to meet customer requirements?	Capability Analysis (**See Also** "Process Capability Analysis," **p. 486**)	**p. 173**
Describe & Compare	How is the process behaving; is it in control?	Control Charts—7QC Tool	**p. 217**
Describe	What are the components of the process; what is involved?	Flowchart —7QC Tool (**See Also** "Process Map," **p. 522**)	**p. 316**
Describe & Compare	Is the process able to meet customer requirements?	Process Capability Analysis	**p. 486**
Describe	What are the components of the process; what is involved?	Process Map—7QC Tool	**p. 522**
Describe	What is the yield of a process?	Rolled Throughput Yield (**See Also** "Process Capability Analysis," **p. 486**)	**p. 610**
Describe	Where is value being added in the process? Conversely, where does waste exist in the process?	Value Stream Analysis	**p. 727**
Describe	Where is value being added in the process? Conversely, where does waste exist in the process?	Value Stream Mapping (**See Also** "Value Stream Analysis," **p. 727**)	

Generic Hypothesis Testing Schematic			
Purpose	**Key Question**	**Tool Name**	**Page**
Project Management			
Schedule	What is the most efficient way to complete this process or project?	Activity Network Diagram (AND)—7M Tool	**p. 136**
Schedule	What is the most efficient way to complete this process or project?	Arrow Diagrams (**See** "AND")	**p. 159**
Schedule	What is the most efficient way to complete this process or project?	Critical Path Method (CPM) (**See** "AND")	**p. 242**
Schedule	How long will the project take? What are the key milestones, and when should you expect them?	Gantt Chart	**p. 317**
Schedule	What are the probabilities and risks associated with several schedule possibilities?	Monte Carlo Simulation	**p. 435**
Schedule	What is the most efficient way to complete this process or project?	PERT (Program Evaluation and Review Technique) Chart	**p. 453**
Planning	What might go wrong during the planning of this complex, significant project?	Process Decision Program Charts (PDPC)—7QC Tool	**p. 515**
Scope	What are the project scope, boundary conditions, deliverables, budget and schedule?	Project Charter (**See** SMART Problem and Goal Statemants for Project Charter **p. 665**)	

continues

Continued

Generic Hypothesis Testing Schematic			
Purpose	**Key Question**	**Tool Name**	**Page**
HR, Deliverables & Tasks	Who is responsible for what?	RACI Matrix (Responsible, Accountable, Consulted, Informed)	**p. 554**
Scope	What is the scope of the project or process?	SIPOC (Supplier-Input-Process-Output-Customer)	**p. 663**
Scope	What is the most succinct description of the project's goal and problem statements?	SMART Problem & Goal Statement for a Project Charter	**p. 665**
Risk	Who supports the project, who doesn't, and why?	Stakeholder Analysis	**p. 681**
Deliverables & Tasks	What are the project deliverables, and what activities will produce them?	Work Breakdown Structure (WBS)	**p. 753**
Risk Management			
	What are the potential risk response strategies to a potential risk or failure?	Cause and Prevention Diagram	
	What can go wrong? What can be done to prevent or minimize it? What are the potential failures that could occur, and what is the best response (action) plan to minimize its impact if it does happen?	FMEA: Failure Modes and Effects Analysis	**p. 198**

Generic Hypothesis Testing Schematic			
Purpose	**Key Question**	**Tool Name**	**Page**
	What are the potential root causes of a single problem or problematic outcome and how can they be prevented or mitigated? Typically applied to a design of a system, product, or process that involves human interactions.	Fault Tree Analysis (FTA)	**p. 309**
	What are the probabilities and risk associated with several financial possibilities? (pricing and forecasting, for example)	Monte Carlo Simulation	**p. 435**
	How best to prevent or correct in-process errors (often human mistakes)?	Poka-Yoke	**p. 462**
	What might go wrong during the planning of this complex, significant project?	Process Decision Program Charts (PDPC)—7M Tools	**p. 515**
	How best to plan for, manage, and mitigate unforeseen risk?	Risk Mitigation Plan	**p. 601**
	Who supports the project and who doesn't, and why?	Stakeholder Analysis	**p. 681**
Root Cause			
Compare & Predict	What differences exist between two groups, if any?	Analysis of Variance (ANOVA)—7M Tool	**p. 142**

continues

Continued

Generic Hypothesis Testing Schematic			
Purpose	**Key Question**	**Tool Name**	**Page**
	What are the potential causals of a problem or problematic outcome?	Cause-and-Effect Diagram—7QC Tool (a.k.a. Fishbone, Ishikawa Diagram)	**p. 173**
Compare & Predict	Which variables or combination of variables proves as the best in producing the desired results, based on experimental data?	Design of Experiment (DOE)	**p. 250**
	What is the root cause of this problem or problematic outcome?	5-Whys	**p. 305**
	What are the potential root causes of a single problem or problematic outcome and how can they be prevented or mitigated? Typically applied to a design of a system, product, or process that involves human interactions.	Fault Tree Analysis (FTA)	**p. 309**
	What are the potential causals of a problem or problematic outcome?	Fishbone Diagram—7QC Tool (**See also** Cause-and-Effect Diagram)	**p. 316**
Compare & Predict	Is this population (or sample) of data different from another by chance alone, or because of an outside influence?	Hypothesis Testing	**p. 335**
	How do the various cause-and-effect ideas relate to one another in a complex situation?	Interrelationship Diagram—7M Tool	**p. 369**

Generic Hypothesis Testing Schematic			
Purpose	**Key Question**	**Tool Name**	**Page**
Describe & Predict	What is the cause-and-effect model that describes the process and its critical variables?	Regression Analysis	**p. 571**
Describe & Compare	Are these two factors correlated? What is the relationship between these two factors?	Correlation Analysis (**See also** Scatter Diagram—7QC Tool, **p. 640**)	
Describe	What are the critical parameters in the process?	Y = f(x)	**p. 758**
Sampling			
Predict	How best to collect a representative sample of the population?	Sampling	**p. 618**
Variation Analysis and Process Evaluation			
Describe & Compare	Is the process able to to meet customer requirements?	Capability Analysis (**See also** Process Capability Analysis, **p. 486**)	**p. 173**
Describe & Compare	How is the process behaving; is it in control?	Control Charts—7QC Tool	**p. 217**
Describe & Compare	How accurate is the measurement system? Is the process truly performing the way the data seems to be reporting, or is the measurement system inaccurate?	Measurement System Analysis (MSA)	**p. 412**

continues

Continued

Generic Hypothesis Testing Schematic			
Purpose	**Key Question**	**Tool Name**	**Page**
Describe, Compare & Predict	Are the data distributed normally?	Normal Probability Plot (**See also** Control Charts—Normal Versus Non-normal Data section, **p. 227**)	**p. 444**
Describe & Compare	Is the process able to meet customer requirements?	Process Capability Analysis	**p. 486**
Voice of the Customer (VOC); Voice of the Business (VOB)			
	What are the major themes of ideas, opinions, issues, and so on found in large amounts of language data?	Affinity Diagram—7M Tool	**p. 136**
	What product and/or services features and functionality do customers prefer and are willing to buy?	Conjoint Analysis	**p. 207**
	How does my work relate to the customer requirements and how do I know when I have fulfilled them?	Critical to Quality (CTQ) Matrix	**p. 242**
	What matters to your customers? What are the natural groupings, categories, or affinities of a large number of topics, ideas, and quotations, and how best to translate the verbal input into requirements?	KJ Analysis	**p. 375**
	How best to capture the customer requirements?	Voice of Customer Gathering Techniques	**p. 737**

A

Activity Network Diagram (AND)—7M Tool

What Question(s) Does the Tool or Technique Answer?

What is the most efficient way to complete this process or project?

An AND helps you to

- Manage the timing of activities and overall completion of a project or process
- Graphically organize process steps into the most efficient sequence
- Show the most critical path and any parallel paths
- Evaluate and reorganize the step sequence. It identifies any simultaneous tasks and tasks that will take the longest to complete
- Identify any slack time—that is, the amount of time a non-critical path task can be delayed without delaying the project
- Manage resources and understand upstream and downstream dependencies

Alternative Names and Variations

This tool is also known as

- Activity chart
- Activity-on-Arrow (AOA)
- Arrow diagram or Arrow Diagramming Method (ADM)
- Activity on Node (AON)
- Critical Path Method (CPM) or CPM chart
- Network diagram
- Node diagram
- Precedence Diagram (PDM)

Variations on the tool include

- Program Evaluation and Review Technique (PERT) chart. (**See Also** "PERT (Program Evaluation and Review Technique) Chart," **p. 453**)

When Best to Use the Tool or Technique

The Activity Network diagram (AND) is an important Project Management tool. It helps to sequence the steps of the project into the most efficient order such that some steps are completed simultaneously. Most project managers will use the AND technique with the project team prior to developing the project schedule that commits the team to a target project completion date. The AND technique is helpful when managing a project whose deadline is critical and for which any delay has significant consequences—and conversely whose acceleration could yield great benefits.

Another application of the AND tool is from the perspective of the process of interest. Similar to a process map, the AND technique evaluates the existing steps in a process to identify the critical path and any parallel activities.

The *critical path* is comprised of a set of tasks that if delayed will delay the outcome of a process or the completion of the project. Even if one critical path activity is delayed, the timing of the overall process or project is negatively impacted. As a result, the critical path identifies the least flexible activities critical to maintaining the schedule and the *float activities*. The critical path also calculates *slack time* associated with the non-critical path activities to understand the flexibility of their start and finish without impacting the overall schedule determined by the critical path.

Brief Description

The Activity Network diagram (AND) is a simple graphical representation of the process or project steps. It depicts which steps must be completed when to complete a project and in what sequence. It can be done either manually or aided by computer software.

A typical AND includes the following symbols, shown in Figure A-1:

- Rectangles, called *nodes*, represent the process or project's activities (hence, the technique is sometimes called *Activity on Node*).
- Arrows connect the activities and show dependencies between two activities.

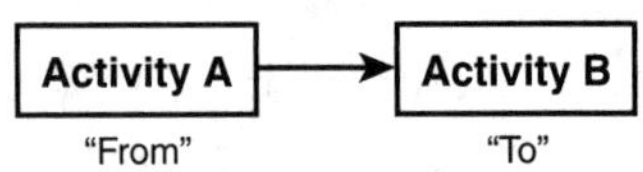

FIGURE A-1: AND Using Activity on Node Symbols

There are four types of dependencies. These dependencies describe the logical relationship between two activities. Sometimes they also are referred to as *precedence relationships*; hence, the technique is sometimes called the *Precedence Diagramming Method* (PDM). Figure A-2 illustrates the various types of dependencies.

1. **Finish-to-Start**—Where the "from" activity must **finish** before the "to" activity can **start**. This is the most common dependency.
2. **Finish-to-Finish**—Where the "from" activity must **finish** before the "to" activity can **finish**.
3. **Start-to-Start**—Where the "from" activity must **start** before the "to" activity can **start**.
4. **Start-to-Finish**—Where the "from" activity must **start** before the "to" activity can **finish**. This type of dependency is rarely used and is usually only used in the engineering discipline.

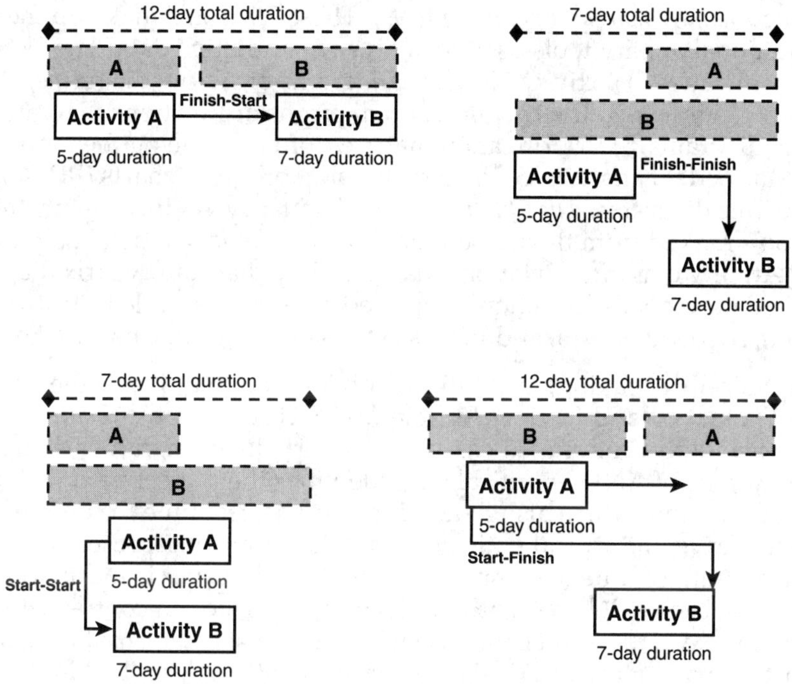

FIGURE A-2: AND Four Different Dependencies

A less common set of symbols includes circles rather than rectangles, as illustrated in Figure A-3. However, this less common approach has the added nuance of using the arrows to represent the activities, rather than the shape, which is reversed from the more traditional approach. Both methods work equally well, and it is a matter of preference as to which technique to use.

- Arrows represent activities (hence, the technique is sometimes called *Activity-on-Arrow*). The tail of the arrow represents the beginning of the activity, and the head of the arrow represents its completion.

- Connected at nodes showing the dependencies between two activities. The nodes often contain lettering in the center of the circle to identify it. As shown in Figure A-3, the "Prepare Spec" task also may be referred to as the AB activity.

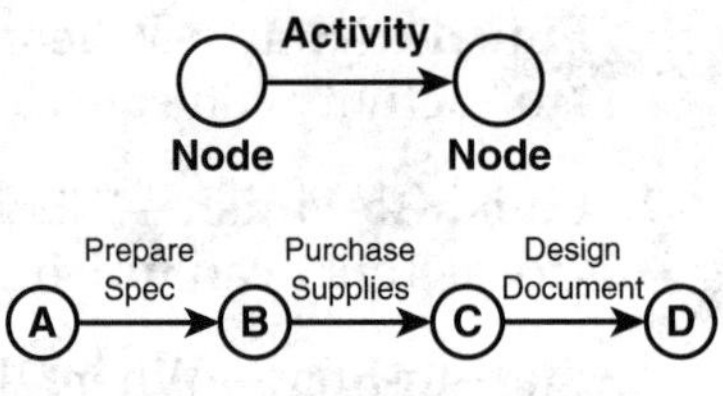

FIGURE A-3: AND Using Activity-on-Arrow Symbols

The Activity on Arrow technique, by design, uses the Finish-to-Start dependency.

The AND technique is a member of the 7M Tools, attributed in part to Dr. Shewhart, as seven management tools, sometimes referred to as the 7MP (or *seven management and planning tools*). These 7M Tools make up the set of traditional quality tools used to analyze quantitative data. The 7M Toolset includes 1) Activity Network diagrams or Arrow diagrams, 2) Affinity diagrams, 3) Interrelationship digraphs or Relations diagrams, 4) Matrix diagrams, 5) Prioritization matrices, often replacing the more complex Matrix data analysis, 6) Process decision program charts (PDPC), and 7) Tree diagrams. *The Quality Toolbox* by Nancy Tague presents the 7M Tools ranked from those used for abstract analysis to detailed planning: Affinity diagram, Relations diagram, Tree diagram, Matrix diagram, Matrix data analysis (commonly replaced by a more simple Prioritization matrix), Arrow diagram, and Process Decision Program Chart (PDPC).

Technically speaking, the Critical Path Method (CPM) and the Program Evaluation and Review Technique (PERT) chart are *estimating techniques applied* to a completed Activity Network diagram, rather than a different type of AND. A critical path identifies the project path with the longest duration, which also determines the shortest time to complete the project. Hence, the critical path lacks any slack time; it represents the minimum amount of time to complete the project. *Slack time* is the amount of time a task can be delayed without delaying a project. In contrast, if parallel paths exist, they will contain slack time or a shorter duration than the critical path. How to calculate the critical path is described in this encyclopedia, in the PERT technique section under the Ps. (**See Also** "PERT (Program Evaluation and Review Technique) Chart," **p. 453**)

How to Use the Tool or Technique

This technique can be done individually or in a small group. When using the simple approach to develop an AND, the procedure involves the following steps:

Step 1. **Identify the topic** and then collect any data about the task. Examples include

a. Brainstorming

b. Process maps

c. Procedure manuals

Step 2. **List the necessary tasks.** One technique used is to create a master list and then to put one activity on either index cards or sticky notes to make it easy to sequence the tasks.

> **Warning**
>
> Be sure you only assign one activity step per index card or sticky note!

Step 3. **Place the tasks into a logical order sequence.** Identify which tasks must precede others, and arrange them accordingly. If some tasks can occur simultaneously, put them on the same vertical plane (parallel to one another). If using sticky notes, follow the same procedure and place the activities on a flip-chart sheet or section of the wall.

Continue this step until all the individual steps are placed in a logical flow. Be sure to leave space in-between each step to allow additional notes to be added.

Step 4. **Add times to the steps** (usually above the activity card or note).

Step 5. **Determine the critical path.**

a. Identify all the possible path combinations from start to finish.

b. Sum the times of each activity in each of the possible path combinations and record the path and its respective duration.

c. Identify the longest path and highlight by bolding (highlighting or darkening) the path arrows—this is the *critical path*.

d. Calculate the *Earliest Start* (ES) times and *Earliest Finish* (EF) times based on how long the preceding task(s) take. Hence, the first activity has zero ES time, and its EF time equals the total duration for the task. Preceding tasks have a compounding (or cumulative) effect on subsequent tasks' ES and EF times.

i. ES = the latest EF from the preceding task

ii. EF = the ES time minus the actual task time

e. Calculate the *Latest Start* (LS) times and *Latest Finish* (LF) times by starting at the end of the network (project or process completion) and working backward toward the start activity.

   i. LF = the shortest LS of the subsequent task

   ii. LS = the LF time minus the actual task time

f. Calculate the slack times for each task and for the project. *Total slack* is the time an activity can be delayed without affecting the project schedule. *Free slack* is the time an activity can be delayed without affecting the early start of any subsequent task.

   i. Total slack = LS minus ES = LF minus EF

   ii. Free slack = the earliest ES of all subsequent tasks minus EF

Step 6. **Examine and adjust the diagram as needed.**

How to Analyze and Apply the Tool's Output

After the Activity Network diagram is complete, use the output to

- Reveal the interdependencies of activities.
- Determine what elements are critical to meeting the completion date (the critical path), including necessary resources and time requirements.
- Determine projected completion date.
- Facilitate "what if" discussions (or alternatives) as to what could be shortened or rearranged; consider any trade-offs.
- Highlight critical activities that will require monitoring and provide a baseline to evaluate the actual process or project performance. Acquire a tool for planning and predicting.
- Help to manage uncertainties, particularly where slack time exists.

Examples

Next is an example of an AND depicting six activities with a Finish-Start dependency. The list of activities includes the following:

- Activity A @ 3-day duration
- Activity B @ 5-day duration

- Activity C @ 7-day duration
- Activity D @ 4-day duration
- Activity E @ 9-day duration
- Activity F @ 3-day duration

This 6-step scenario with its critical path and slack time calculations can be found in Figure A-4.

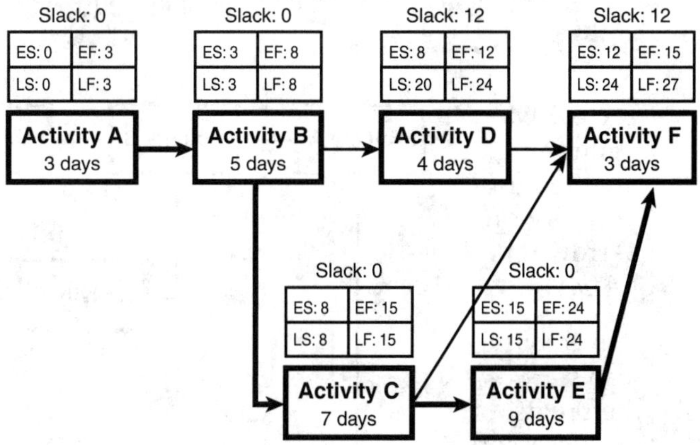

Figure A-4: 6-Step AND Example

This example's path calculations include

- A-B-D-F = 15 days
- A-B-C-F = 18 days
- A-B-C-E-F = 27 days

The longest path is A-B-C-E-F, at 27 days; therefore, it is the critical path.

The calculations for the Earliest and Latest times are found in Figure A-4, and the formulas are as follows:

- ES = the latest EF from the preceding task
- EF = the ES time—the actual task time
- LF = the shortest LS of the subsequent task
- LS = the LF time—the actual task time

The calculations for the slack times are

- Total slack = LS—ES = LF—EF;
- Free slack = the earliest ES of all subsequent tasks—EF

> **Note**
>
> There is zero slack time along the critical path.

A B C D E F G H I J K L M N O P Q R S T U V W X Y Z

Hints and Tips

A summary activity, called a *hammock*, represents a group of related activities as one to show the lapsed time for the collection of tasks. The hammock spans between two activities that may be separated by several other activities, as illustrated in Figure A-5.

A B C D E F G H I J K L M N O P Q R S T U V W X Y Z

- Arrow Diagramming method [or Activity-on-Arrow (AOA)]:
- No activity can start before its predecessors have been completed.

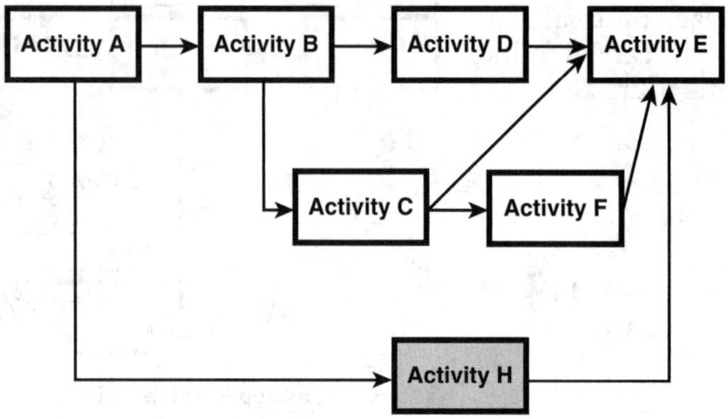

FIGURE A-5: AND Hammock (as Activity H)

- The arrow length has no meaning; it simply represents logical precedence.
- Each activity must have a unique identifier (a number or *alpha*); duplicates are not permitted.
- If an Arrow Diagramming method is used, only finish-to-start dependencies are appropriate.
- AOA may use *dummy* activities (with zero time durations) to help complete the logic in the network diagram by showing the flow of activities. Dummy activities are identified with hashed or dotted arrows, as shown in Figure A-6. They typically show relationships between activities that have more than one predecessor by keeping the sequence correct.

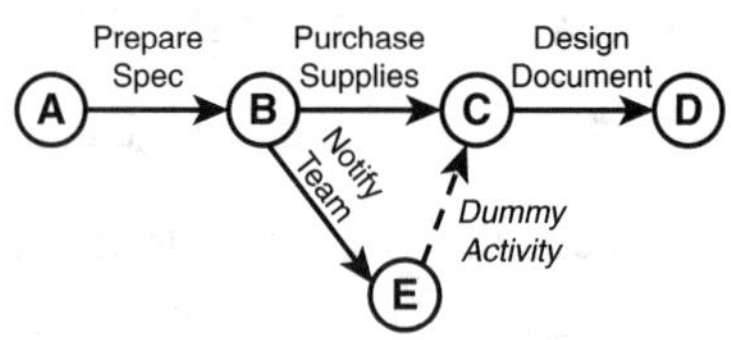

FIGURE A-6: Activity-on-Arrow with Dummy Activity

- Non-sequential steps, such as *loops* or *conditional branches* using decision-diamonds are not used in any of the AND techniques. These special conditions are found in process mapping and more specialized diagramming techniques called Graphical Evaluation and Review Technique (GERT) and Systems Dynamics modeling.

Supporting or Linked Tools

Supporting tools that might provide input when developing an AND include

- VOC Data Gathering Tools and Techniques, such as surveys and interviews (**See Also** "Voice of Customer Gathering Techniques," **p. 737)**
- Written reports
- Brainstorming sessions (**See Also** "Brainstorming Technique," **p. 168)**

A completed AND provides input to tools such as

- Tree diagram and matrix (**See Also** "Tree Diagrams," **p. 712)**
- Fishbone (**See Also** "Cause-and-Effect Diagrams," **p. 173)**
- Simple matrix (**See Also** "Matrix Diagrams," **p. 399)**

Figure A-7 illustrates the link between the AND and its related tools and techniques.

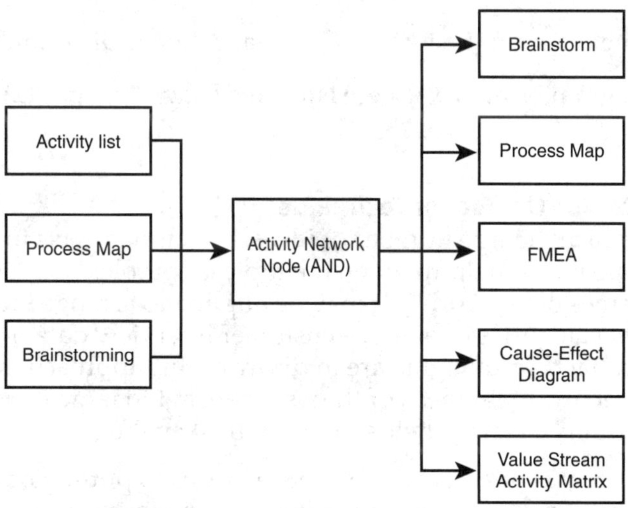

Figure A-7: Activity Network Diagram Tool Linkage

A B C D E F G H I J K L M N O P Q R S T U V W X Y Z

A B C D E F G H I J K L M N O P Q R S T U V W X Y Z

Affinity Diagram—7M Tool

What Question(s) Does the Tool or Technique Answer?

What are the major themes of ideas, opinions, issues, and so on found in large amounts of language data?

An Affinity diagram helps you to

- Organize large volumes of language-based data (for example, text or verbal input) into related categories (natural groupings, categories, or affinities of a large number of topics, ideas, and quotations).
- Show a parent-child relationship between more detailed ideas to their higher-order themes, also called *affinity groupings*.

Alternative Names and Variations

This tool is also known as

- Affinity chart
- KJ Method or analysis (**See Also** "KJ Analysis," **p. 375)**
- Idea Map

Variations on the tool include

- Fishbone diagram (**See Also** "Cause-and-Effect Diagram," **p. 173**)
- Tree diagram or matrix (**See Also** "Tree Diagram," **p. 712**)

When Best to Use the Tool or Technique

The tool organizes a large number of ideas, quotations, facts into themes. This technique is also helpful when the topic is complex and lacking apparent themes or categories. It may be applied to arrange brainstorming ideas into categories or analyze customer interview data or survey results. When facts or thoughts are in disarray, this approach organizes the data. When the main ideas of the issues seem too large or complex to grasp, the Affinity diagram helps to reveal the key themes.

Often this tool is used as an interim step prior to putting data into a Fishbone or Tree diagram. This tool often accompanies a brainstorming session to organize and sort through the ideas generated.

Brief Description

A Japanese anthropologist, Jiro Kawakita, developed the Affinity diagram in the early 1960s; hence, it is also referred to as the KJ Method. [Recall: A Japanese name reverses the order of the given and sir name relative to an English name.] Dr. Kawakita was the founder of the Kawayoshida Research Center.

Today's use of Affinity diagrams is less complex than the original KJ Analysis approach developed by Jiro Kawakita. The reference to this tool often is used hand-in-hand with brainstorming as a means to organize the brainstormed ideas. Hence, this entry discusses the common application of the Affinity diagram. In a separate entry, the more in-depth KJ approach can be found in the section under the Ks. (**See Also**, "KJ Analysis," **p. 375**)

The Affinity diagram tool is a member of the 7M Tools, attributed in part to Dr. Shewhart, as seven management tools, or sometimes referred to as the 7MP, for *seven management and planning tools*. These 7M Tools make up the set of traditional quality tools used to analyze quantitative data. The 7M Toolset includes 1) Activity Network diagrams or Arrow diagrams, 2) Affinity diagrams, 3) Interrelationship digraphs or Relations diagrams, 4) Matrix diagrams, 5) Prioritization matrices, often replacing the more complex Matrix data analysis, 6) Process decision program charts (PDPC), and 7) Tree diagrams. *The Quality Toolbox*, by Nancy Tague, presents the 7M Tools ranked from those used for abstract analysis to detailed planning: Affinity diagram, Relations diagram, Tree diagram, Matrix diagram, Matrix data analysis (commonly replaced by a more simple Prioritization matrix), Arrow diagram, and Process Decision Program Chart (PDPC).

How to Use the Tool or Technique

This technique can be done individually or in a small group. The procedure to develop an Affinity diagram involves the following steps:

Step 1. **Identify the topic** and then collect and/or capture the text or verbal data. Examples of different means of gathering such input include

a. Interviews

b. Brainstorming

c. Surveys

Step 2. **Prepare the data** by putting one idea, thought, or suggestion (either written or verbal input) on either

a. Single piece of paper (or index cards) or

b. Sticky note

(Only one idea per piece of paper (index card or sticky note).

> **Note**
>
> If the data already exists in a text document, a short cut may be to simply cut the document into strips, such that one strip of paper contains only one written idea.

Step 3. **Sort the ideas.** Review each individual piece of paper and begin to place them in piles whose subjects relate to one another (those that have a similarity or an "affinity" for one another).

If using sticky notes, follow the same procedure, but place the ideas with a common theme on a single flip chart sheet or section of the wall. Continue this step until all the individual ideas are placed in a group.

a. A few ideas may not fit well into the groupings or themes; they may be independent ideas or "loners." If so, place them in a "Miscellaneous" grouping.

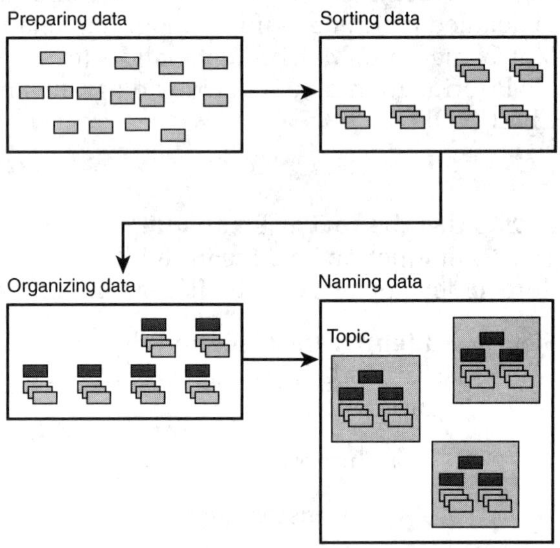

FIGURE A-8: Affinity Technique Illustration of Preparing, Sorting, Organizing, and Naming Steps

Step 4. **Organize the ideas** by reviewing each within a grouping or affinity and then sort them again into further subsets. Identify those that are similar and group them.

a. Look across groups and determine if any groups are related to one another.

Step 5. **Name the themes** by first identifying a category heading or the highest-level of commonality in a given group. Next, name any of the subsets within a group, if possible, until all the groupings have a label.

a. Reexamine if the "Miscellaneous" ideas relate to any of the named groupings. If so, move them to that category.

Figure A-8 illustrates at a high level the flow of the Preparing, Sorting, Organizing, and Naming steps of the Affinity technique.

How to Analyze and Apply the Tool's Output

After the Affinity diagram is complete, its contents (categories and their respective detail) can be transferred to a format that better communicates the themes and respective linkages, such as one of the following formats:

- **Matrix**—Either one category or group per matrix or one per column, depending on how complex the topic or diverse the data
- **Tree diagram**—Either one category or group per tree or one major grouping per major branch, depending on how complex the topic or diverse the data
- **Fishbone diagram**—Either one category or group per tree or one major grouping per major branch, depending on how complex the topic or diverse the data

Examples

An example of an Affinity diagram depicting the selection criteria for choosing a restaurant can be found in Figure A-9.

Figure A-10 illustrates the same "Criteria for Selecting a Restaurant" example but as a Tree diagram, using Quality Companion by MINITAB.

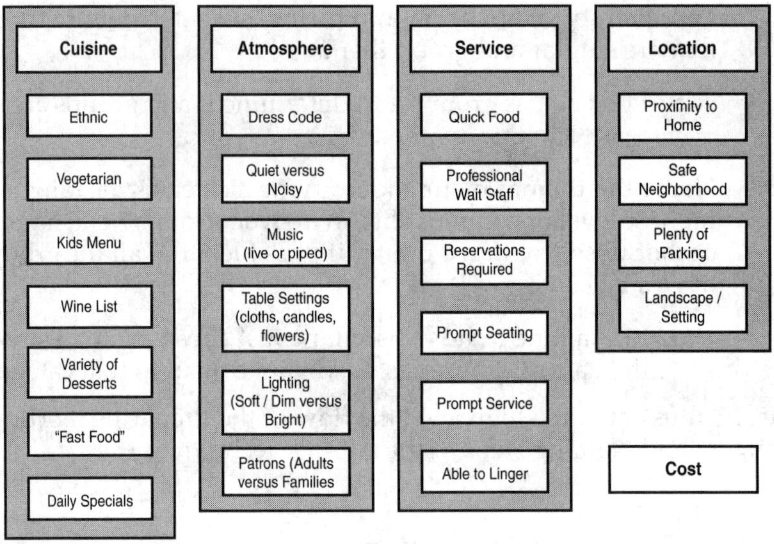

FIGURE A-9: Example Affinity Diagram—Criteria for Selecting a Restaurant

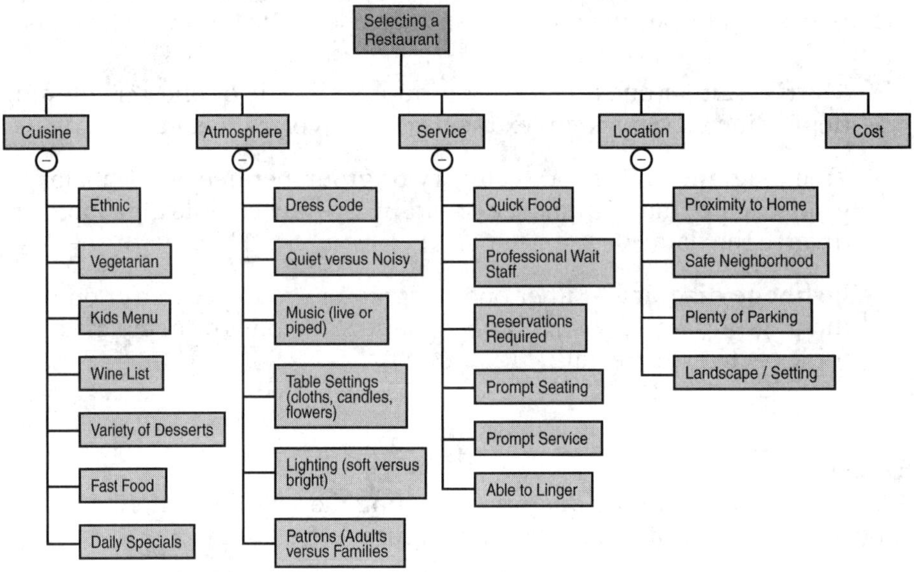

FIGURE A-10: Example Affinity Tree Diagram—Criteria for Selecting a Restaurant

Hints and Tips

- Take time to complete the diagram. Initially start it in a concentrated team session but allow the team members to reflect on it and edit it with any additional ideas.
- Multiple perspectives are important. Ensure that individuals with different perspectives (that is, different roles, different levels, different functional responsibilities) have an opportunity to contribute ideas.
- If a perspective is missing from the initial creation of the Affinity diagram, have the individuals contribute as "editors."
- The naming of the Affinity categories is often best done at the end of the building process, rather than at the beginning.
- If an idea can fit in more than one grouping, create a duplicate note(s) and place it in all of the applicable places. This may trigger additional ideas or show unique relationships that should not be lost.

During the organizing step, a stronger affinity with one group versus the others may become evident and outweigh the other placements. If so, the team may choose to reduce the idea to one representation.

Supporting or Linked Tools

Supporting tools that might provide input when developing an Affinity diagram include

- VOC Data Gathering Tools and Techniques, such as surveys and interviews (**See Also** "Voice of the Customer Gathering Techniques," **p. 737**)
- Written reports
- Brainstorming sessions (**See Also** "Brainstorming Technique," **p. 168**)

A completed Affinity diagram provides input to tools such as

- Tree diagram and matrix (**See Also** "Tree Diagram," **p. 712**)
- Fishbone (**See Also** "Cause-and-Effect Diagram," **p. 173**)
- Simple matrix (**See Also** "Matrix Diagrams," **p. 399**)

Figure A-11 illustrates the link between the Affinity diagram and its related tools and techniques.

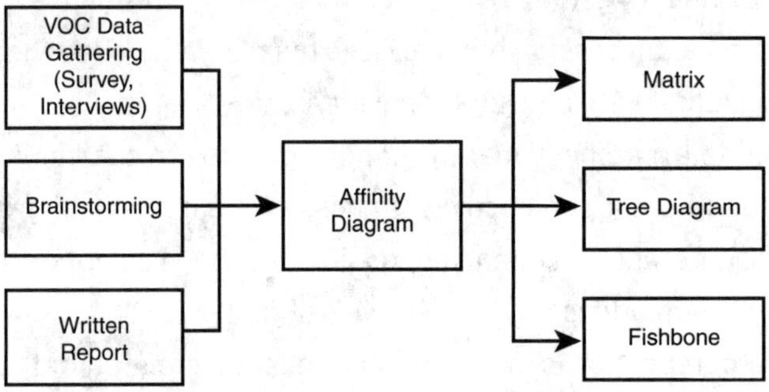

FIGURE A-11: Affinity Diagram Tool Linkage

Variations

KJ Analysis

The KJ Analysis involves more detail and evaluation than what is described in this section. **(See Also** "KJ Analysis," **p. 375)**

Analysis of Variance (ANOVA)—7M Tool

What Question(s) Does the Tool or Technique Answer?

What differences exist between two groups, if any?

The ANOVA helps you to

- Make comparisons among many items. Are their means different? Are their variances different?
- Determine the relationship between groups, within some degree of confidence, using statistical analysis to analyze and compare variables.
- Make inferences as to whether an "improvement initiative" shows statistically significant differences compared with a "non-intervention" group.

Alternative Names and Variations

This tool is also known as

- A type of hypothesis testing
- F-test
- One-way ANOVA
- Two-way ANOVA without replicates
- Two-way ANOVA with replicates
- Analysis of Means (ANOM), a type of ANOVA when comparing three or more means
- Test for Equal Variances, a type of ANOVA when comparing three or more variances

When Best to Use the Tool or Technique

Use ANOVA when comparing among many items. It determines if the average of a group of data is statistically different from the average of other (multiple) groups of data. It helps draw conclusions about a hypothesis that determines if two or more populations are different. This technique can be used when analyzing either of the following:

- The current environment to determine if the variation is due to natural (common) variation or a special cause variation
- An intervention (or improvement) that was newly implemented

ANOVA **tests for a difference between two or more means** and compares and analyzes the data from group to group and the variation within a group. For example, it could analyze the output of multiple machines, examining the output from each machine and from machine to machine.

A similar analysis could be achieved by running multiple two-sample t-Tests. However, ANOVA is more efficient by essentially running the required two-way comparisons **simultaneously**. ANOVA is a statistical technique sometimes called a "poor man's" *Design of Experiment* (DOE). A DOE is a structured, organized method for determining the relationship between factors (Xs) affecting a process and the output of that process.

Brief Description

Ronald Fisher developed the ANOVA tool in 1920. He designed this statistical method to conduct group tests that determined if differences were

A B C D E F G H I J K L M N O P Q R S T U V W X Y Z

statistically significant. Group tests involve multiple data sets and multiple independent and dependent variables, similar to Regression Analysis. Fisher's work was important because it examined the impact of interventions and whether any variance was because of the intervention or random chance.

Part of the technique uses a statistical distribution to which the calculations are compared, called an F-statistic, named for its inventor, Fisher. The F-statistic is the *ratio of the explained variation to the unexplained variation* and is compared with the F-distribution. The technique relies on the assumption that the data comes from a normal distribution. The formula for F-statistic is Factor Variance divided by the Error Variance [F=Factor Variance/Error Variance].

The null hypothesis (H_0) would be that the F-distribution would not be a good fit for the model; therefore, the p-value would be greater than the significance level. The p-value is the *probability* that the calculated F-statistic would be large if the group means were equal. The hypothesis would be written as the following mathematical sentences:

- Null hypothesis: H_0: means of all tested factors are equal
- Alternative hypothesis: H_a: at least two tested factors have different means

Typically, the default significance level is 5% [also called "alpha" (α)]. If the calculated ratio (called *F-calculated*) is **greater than** the F-critical for the specified significance level found in the F-distribution table, then the null hypothesis is rejected. The F-statistic has a negative correlation to the p-value. As the F-statistic increases, the explained variance grows, and the distance between the two means increases as the p-value decreases to show that there is greater confidence in two means being statistically different.

If the p-value is less than or equal to the significance level (α), then at least one source of explained variation is significant in helping the distribution fit the F-distribution. That is if the p-value is less than 0.05, for example, then at least one group of data is different from at least one other group. This paragraph may become clearer after reviewing the subsequent procedure section, "How to Use this Tool or Technique." For more discussion on p-value, reference the Glossary and "Hypothesis Testing," **p. 335**.

ANOVA is a type of hypothesis testing. It determines the source of variation, whether due to a technician (or operator), the part (or process), the interaction, and/or the error. The following terms describe some of the sources of variance:

- **Repeatability**—When the variation comes from repeated measurements of the same item, by the same person, with the same measuring device. This is variation within a factor and is also called *Error* or *Error Variance*.
- **Reproducibility**—When the variation source is the averages from repeated measurements (or factors) made by different people (or machines) on the same item. This is variation between factors and is also called *Operator* or *Technician Error*.

ANOVA is a flexible technique. It is used to identify a statistically significant difference between means of three or more levels of a factor. It is used when two or more means (a single factor with three or more levels) must be compared with each other. The technique falls into three different categorizes. Why and when to use each different type of ANOVA is outlined as follows.

One-Way ANOVA (sometimes referred to as the "fixed effects" ANOVA)

When to Use—Compares two components or sources of variance, treatments, and experimental error.

Example—Analyzes one factor from one source of variation and one measurer.

Two-Way ANOVA without Replicates

When to Use—Compares three components of variance (Factor A treatments, Factor B treatments, and experimental error).

Example—Analyzes two factors and only one technician (or measurer) unless the measurer is one of the factors being analyzed.

Two-Way ANOVA with Replicates

When to Use—Compares three components of variance (Factor A treatments, Factor B treatments, and experimental error), with multiple repetitions.

Example—Analyzes two factors *with multiple repetitions* of each combination and only one technician (or measurer) unless the measurer is one of the factors being analyzed.

In preparation for conducting an ANOVA, collect the data to be analyzed. This technique can use any number of data sets, which can be of unequal size. Within each data set, the sample size can be as small as three or four measurements or as large as necessary and economically feasible.

The structure of an ANOVA test has an independent variable (a category of interest) and a dependent variable(s), which is hypothesized to be affected by differing levels of treatments (or interventions). For example, if the dissolution time of different types of tablets were of interest, tablets would be the category (independent variable), and the dependent variable would be the time it takes for the different tablets to dissolve. Typically, the sample size is 20 or more per group.

How to Use the Tool or Technique

This technique can be done manually, computing the statistical calculations by hand and referencing an F-statistic table. First run an ANOVA by hand before using the computer programs, to help understand where the mechanics and source of the calculations. Afterward, it is wise to use a computer for assistance. Excel (using statistical add-ins) can run an ANOVA. The current software packages that specialize in statistical analysis simplify the technique procedures and reduce any calculation errors.

Two procedures are provided: 1) the manual calculation to understand the mechanics behind generating the ANOVA output and 2) the computer aided approach. While there are several good statistical application software packages available, this book uses MINITAB.

ANOVA tests are best done with a minimum sample size of 20 per group. To simplify the explanation of the process, the following examples use less than 20.

One-Way ANOVA Manual Process

Step 1. State the Null (H_0) and Alternative (H_a) hypotheses.

a. Null hypothesis: H0: means ($\bar{X}$) of all 3 factors are equal

i. $\bar{X}A = \bar{X}B = \bar{X}C$; where ($\bar{X}$) "x-bar" is the mean

b. Alternative hypothesis: H_a: at least two factors have different means

Step 2. Choose the significance level or risk level (α)

a. Default is $\alpha = 0.5$

Step 3. Randomly select the parts or processes.

Step 4. Identify the factors (that is, parts, processes, and technicians).

Step 5. Collect the data.

Step 6. Calculate the F-statistic (or F-calculated) using the sample data.

a. Sum up the values for each factor.

b. Calculate the average value (sum divided by the number of trials).

c. Calculate the variance (s^2) (sum of the squared difference of each value x) from the average ($\bar{X}$), then divide by the sample size minus 1.

$$s^2 = (\Sigma(x - \bar{x})^2) \div (n - 1)$$

A B C D E F G H I J K L M N O P Q R S T U V W X Y Z

d. Calculate the various Sum of Squares (SS):

i. Total Sum of Squares (Total SS):

$$\text{Total SS} = \Sigma(x)^2 - \text{CM}$$

Where CM stands for the "Correction for the Mean" and "N" is the total population, determined by

$$\text{CM} = (\Sigma x)^2 \div \text{N}$$

ii. Part Sum of Squares (Part SS), sometimes called the Machine SS, and "n" is the sample size:

$$\text{Part SS} = [(\Sigma x_A)^2 \div n_A] + [(\Sigma x_B)^2 \div n_B] + [(\Sigma x_C)^2 \div n_C] - \text{CM}$$

iii. Error Sum of Squares (Error SS)

$$\text{Error SS} = \text{Total SS} - \text{Part SS}$$

Where the "Part Sum of Squares" represents all "Other" Sum of Squares that are calculated.

e. Determine the degrees of freedom (df) = *(n—1)*.

f. Calculate the Mean Squares (MS) = [*Sum of Squares/df*].

g. Calculate the F-statistic = [*MS for Part/MS for Error*].

Step 7. Find the F-critical (or F-α) in the F-distribution table.

Step 8. Compare the F-statistic to the F-critical.

One-Way ANOVA Manual Example—Dissolving Rate of Different Types of Tablet

Step 1. State the Null (H_0) and Alternative (H_a) hypotheses.

a. Null hypothesis: H_0: means of all three factors are equal—Average time of A = Average time of B = Average time of C.

b. Alternative hypothesis: H_a: at least one average time is different from the other means.

Step 2. Choose the significance level or risk level (α)— α *=0.5, or 95% confidence.*

Step 3. Randomly select the parts or processes—*Dissolving three different types of tablet (Brand A, Brand B, and a generic C).*

Step 4. Identify the factors (that is, parts, processes, and technicians)—*Dissolution rate in seconds; one person stirring and measuring the time (in seconds) for the different types of tablet to dissolve in room temperature water.*

Step 5. Collect the data *in seconds.*

Tablet Type	Trial 1	Trial 2	Trial 3	Trial 4	Trial 5
Brand A	34	39	42	47	40
Brand B	55	48	61	49	54
Generic C	45	41	47	51	49

Step 6. Calculate the F-statistic (or F-calculated) using the sample data.

Tablet Type	Trial 1	Trial 2	Trial 3	Trial 4	Trial 5	n	Sum	Avg	Variance (S2)	Sum of each X2	Sum of Squares
Brand A	34	39	42	47	40	5	202	40.4	4.72	8250	
Brand B	55	48	61	49	54	5	267	53.4	5.22	14367	
Generic C	45	41	47	51	49	5	233	46.6	3.84	10917	
Totals						15	702	234	32.51	33534	680.40

$CM = (702)^2/15 = 32853.6$

Total SS = Sum of each X^2–CM = 33534–32853.6 = 680.4

Part SS = $(202)^2/5 + (267)^2/5 + (233)^2/5 - 32853.6 = 422.8$

Error SS = 680.4–422.8 = 257.6

Mean Square = Sum of Square/Degrees of Freedom, per row.

Mean Square for Part = 422.8/2 = 211.4

Mean Square for Error = 257.6/12 = 21.5

F-statistic = Mean Square for Part/Mean Square for Error

F-statistic = 211.4/21.5

Source	Sum of Squares	Degrees of Freedom	Mean Square	F-statistic	F-critical
Part	422.8	2	211.4	9.8	3.89
Error	257.6	12	21.5		
Total	680.4	14			

A B C D E F G H I J K L M N O P Q R S T U V W X Y Z

Degrees of Freedom:

- Total df = 15 total samples–1 = 14
- Parts df = 3 tablet types–1 = 2
- Error df = Total df–Parts df = 14–2 = 12

Step 7. Find the F-critical (or F-α) in the F-distribution table.

Look up F-critical for 5% significance, and the numerator df of 2 (Parts) and denominator df of 12 (Error) = 3.89

Step 8. Compare the F-statistic to the F-critical.

F-statistic of 9.8 > F-critical of 3.89; therefore Reject the Null Hypothesis.

This means that the dissolving time for at least one type of tablet differs from another.

One-Way ANOVA Automated Process Using MINITAB for the Same Tablet Example

Step 1. State the Null (H_0) and Alternative (H_a) hypotheses.

a. Null hypothesis: H_0: means ($\bar{X}$) of all 3 factors are equal

i. $\bar{X}A = \bar{X}B = \bar{X}C$; where ($\bar{X}$) "x-bar" is the mean.

b. Alternative hypothesis: H_a: at least two factors have different means.

Step 2. Choose the significance level or risk level (α); select $\alpha = 0.5$.

Step 3. Randomly select the parts or processes.

Step 4. Identify the factors (that is, parts, processes, and technicians).

Step 5. Collect the data and record in MINITAB Worksheet, as illustrated in Figure A-12.

A	B	C
34	55	45
39	48	41
42	61	47
47	49	51
40	54	49

FIGURE A-12: Data Recorded in MINITAB Worksheet

Step 6. Calculate the F-statistic (or F-calculated) using the sample data.

a. Select the following commands: ***Stat > ANOVA > One-Way (Unstacked)***...... (if the recorded data is in columns).

b. Put data into the Responses field and select acceptable confidence level in the ANOVA dialog box, as illustrated in Figure A-13.

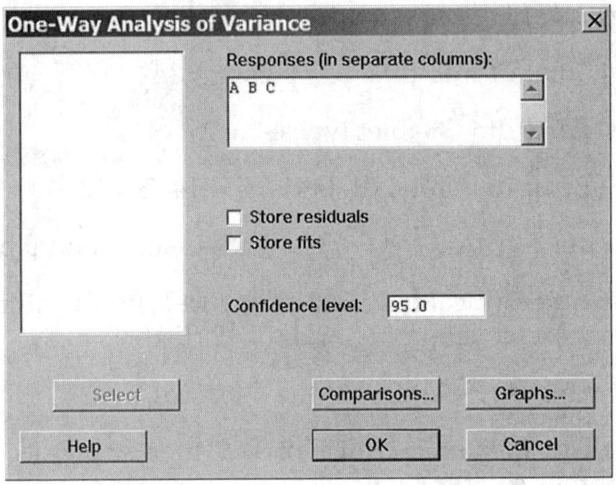

FIGURE A-13: Tablet Example in MINITAB ANOVA Dialog Box

c. Select OK, and the following output is displayed in the session window, as illustrated in Figure A-14.

```
One-way ANOVA: A, B, C

Source  DF     SS     MS     F      P
Factor   2  422.8  211.4  9.85  0.003
Error   12  257.6   21.5
Total   14  680.4

S = 4.633   R-Sq = 62.14%   R-Sq(adj) = 55.83%

                             Individual 95% CIs For Mean Based on Pooled
                             StDev
Level  N    Mean  StDev      +---------+---------+---------+---------
A      5  40.400  4.722      (------*-------)
B      5  53.400  5.225                          (-------*-------)
C      5  46.600  3.847               (-------*------)
                             +---------+---------+---------+---------
                             36.0      42.0      48.0      54.0

Pooled StDev = 4.633
```

FIGURE A-14: Tablet Example—ANOVA Output in MINITAB Session Window

Two-Way ANOVA Without Replicates

Recall: Two-way ANOVA has three components of variance (two factors and experimental error) but no replicates.

Because this is similar to the Two-Way ANOVA With Replicates, this book discusses the more complex of the two approaches. Either way, Two-Way ANOVA adds a level of complexity that is best addressed using application software for statistical hypothesis testing.

Two-Way ANOVA With Replicates—Manual Process

Recall: Two-way ANOVA has three components of variance (two factors and experimental error) and multiple replicates of the factor combinations.

Setup: The procedure is essentially the same as the One-Way ANOVA; however, two factors are being studied. In the case of the tablet test, there are the same three types of tablet used in the One-Way ANOVA, but this test involves varying the temperature of the water in which the different tablets will be dissolved by five degrees. Per the water set-up procedure, the lab analyst dissolved each type of tablet five times (replicates). The sample data is described next.

Example Dissolving Rate of Different Types of Tablet in Different Temperatures of Water

Step 1. State the Null (H_0) and Alternative (H_a) hypotheses.

a. Null hypothesis: H_0: there is no difference between the means of one type of tablet in the different water temperatures (the interaction) and no difference between the means of the different tablets; all are equal.

b. Alternative hypothesis: H_a: at least one factor causes the tablet to have different means, either the variation among the different types or the interaction between the tablet and the different water setups.

Step 2. Choose the significance level or risk level (α)— *α =0.5, or 95% confidence.*

Step 3. Randomly select the parts or processes—*Dissolving three different types of tablet (Brand A, Brand B, and a generic C).*

Step 4. Identify the factors (that is, parts, processes, and technicians)—*Dissolution rate in seconds; one person stirring and measuring the time (in seconds) for the different types of tablet to dissolve in the different water setups, varying by five degrees.*

Step 5. Collect the data—*five runs per tablet-type, per water setup, in seconds.*

Tablet Type	Hot	Warmer	Room Temp	Cooler	Cold
Brand A	34	39	42	47	40
	33	35	39	48	48
	33	37	38	47	47
	33	35	42	45	45
	32	37	40	47	48

continues

A B C D E F G H I J K L M N O P Q R S T U V W X Y Z

Tablet Type	Hot	Warmer	Room Temp	Cooler	Cold
Brand B	55	48	61	49	54
	50	52	60	50	50
	55	53	55	52	54
	54	52	59	49	50
	53	52	60	50	55
Generic C	45	41	47	51	49
	44	43	49	49	53
	42	40	49	53	49
	45	43	49	51	53
	43	44	46	51	55

Step 6. Calculate the F-statistic (or F-calculated) using the sample data.

Tablet Type	Summary	Hot	Warmer	Room Temp	Cooler	Cold	Total
Brand A	Count	5	5	5	5	5	25
	Sum	165	183	201	234	228	1011
	Average	33	36.6	40.2	46.8	45.6	40.44
	Variance	0.50	2.79	3.20	1.20	11.89	31.75
Brand B	Count	5	5	5	5	5	25
	Sum	267	257	295	250	263	1332
	Average	53.4	51.4	59	50	52.6	53.28
	Variance	4.28	3.80	5.48	1.49	5.81	13.37
Generic C	Count	5	5	5	5	5	25
	Sum	219	211	240	255	259	1184
	Average	43.8	42.2	48	51	51.8	47.36
	Variance	1.69	2.69	1.99	1.99	7.18	17.72

Tablet Type	Summary	Hot	Warmer	Room Temp	Cooler	Cold	Total
Total	Count	15	15	15	15	15	75
	Sum	651	651	763	739	750	
	Average	43.4	43.4	49.1	49.27	50	
	Variance	76.21	42.5	66.75	4.80	17.39	

$CM = (3527)^2/75 = 165{,}863$

Total SS = Sum of each X^2–CM = 169,437–165,863 = 3574

Tablet SS = (Sum of Row Square–CM) = $(1011)^2/25 + (1332)^2/25 + (1184)^2/25 – 165{,}863 = 2065$

Water SS = (Sum of Column Square–CM) = 665

Interaction SS = (Interaction Square–CM–SS Column–SS Row) = 622; where the **Interaction Square = Sum of all the Cell Square** = $(\text{Sum Cell})^2/k$; where k = #replicates per cell.

Error SS = Total SS–All other SS = 222

Mean Square = Sum of Square/Degrees of Freedom, per row.

Mean Square for Tablets = 2065/2 = 1032.5

Mean Square for Water Temp = 665/4 = 166.2

Mean Square for Interaction = 622/8 = 77.8

Mean Square for Error = 222/60 = 3.7

F-statistic = Mean Square/Mean Square for Error

F-statistic for Tablets = 1032.5/3.7

F-statistic for Water Temp = 166.2/3.7

F-statistic for Interaction = 77.8/3.7

Source	Sum of Squares	Degrees of Freedom	Mean Square	F-statistic	F-critical
Tablets	2065	2	1032.5	279.05	3.15
Water Temp	665	4	166.2	44.92	2.53
Interaction	622	8	77.8	21.02	2.10
Error (or Within)	222	60	3.7		
Total	3574	74			

Degrees of Freedom:

- Total df = 75 total samples–1 = 74
- Tablets (rows) df = 3 tablet types–1 = 2
- Water Temp (columns) df = 5–1 = 4
- Interaction (column df times row df) = 2 x 4 = 8
- Error df = (Total df–Rows df–Column df–Interaction df)= 14–2 = 12

Step 7. Find the F-critical (or F-α) in the F-distribution table.

Look up F-critical for 5% significance.

Tablets: numerator df of 2 and denominator df of 60 (Error) = 3.15

Water: numerator df of 4 and denominator df of 60 (Error) = 2.53

Interaction: numerator df of 8 and denominator df of 60 (Error) = 2.10

Step 8. Compare the F-statistic to the F-critical.

F-statistic of each factor > F-critical; therefore Reject the Null Hypothesis.

This means that each of the factors has different means, including the interaction between the tablet and the different water setups.

Two-Way ANOVA With Replicates—Automated Process

Compare with the ANOVA Two-Way output for the same example. Select the following commands: ***Stat > ANOVA > One-Way*** to bring up the dialog box as illustrated in Figure A-15.

Select ***OK***, and the following output will be displayed in the session window, as illustrated in Figure A-16.

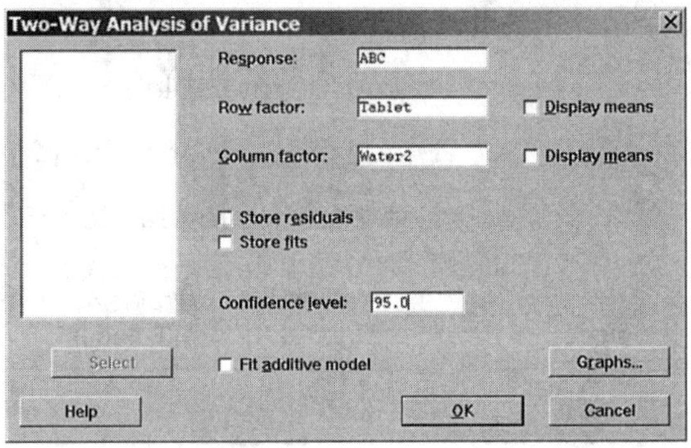

FIGURE A-15: Tablet Example in MINITAB ANOVA Two-way Dialog Box

Two-way ANOVA: ABC versus C15, Water2

Source	DF	SS	MS	F	P
C15	2	2064.99	1032.49	279.05	0.000
Water2	4	664.88	166.22	44.92	0.000
Interaction	8	622.08	77.76	21.02	0.000
Error	60	222.00	3.70		
Total	74	3573.95			

S = 1.924 R-Sq = 93.79% R-Sq(adj) = 92.34%

FIGURE A-16: Tablet Example—ANOVA Two-Way Output in MINITAB Session Window

Note

All three p-values are less than 0.05. Hence, there is sufficient data to reject the Null Hypothesis, meaning that at each of the factors have different means, including the interaction between the tablet and the different water setups.

At this level of complexity, clearly the application software approach is more efficient and minimizes errors than calculating the results manually.

How to Analyze and Apply the Tool's Output

For the manual process, compare the calculated F-statistic to the F-critical to determine if the Null hypothesis was correct or not.

- If the F-statistic is greater than the F-critical, then the Null Hypothesis of all factors' averages being equal can be rejected. This means that at least one factor's mean was different from another's.
- In the tablet example, the calculated F-statistic was 9.8, which was greater than the F-critical of 3.89; therefore Reject the Null Hypothesis. [Note that the F-critical requires an F-Distribution table for the selected significance (for example, 5%). Look up the F-critical at the intersection of the column with the numerator df of 2 (Parts) and row with the denominator df of 12 (Error). This intersection is 3.89.]

For the automated process, there are two pieces of output used to indicate if the Null hypothesis was correct or not. Examine the output in the session window (shown in Figure A-14). The following data determines the results:

- The "P" column in the ANOVA table connotes the p-value. If the p-value ≤ 0.05 (for a 5% significance level), then at least one factor is significant and differs from another; therefore, reject the Null hypothesis. The converse is true if the p-value > 0.05, wherein there would be insufficient evidence to reject the Null.

- In the tablet example, the p-value = 0.003, which is less than 0.05 (for a 5% significance level), so the Null hypothesis is rejected.
- The graph in the lower portion of the session window shows the confidence interval as a range for each tablet type. Notice that Brand A and Brand B's ranges **do not** overlap. Hence, with 95% confidence, Brand A can be said to be statistically faster (on average 40.4 seconds) than Brand B (with an average of 53.4 seconds).

Figure A-17 highlights the points of interest in the ANOVA session window:

One-way ANOVA: A, B, C

```
Source  DF      SS     MS     F      P
Factor   2   422.8  211.4  9.85  0.003
Error   12   257.6   21.5
Total   14   680.4

S = 4.633   R-Sq = 62.14%   R-Sq(adj) = 55.83%

                              Individual 95% CIs For Mean Based on Pooled
                              StDev
Level  N    Mean  StDev      +---------+---------+---------+---------
A      5  40.400  4.722      (-------*--------)
B      5  53.400  5.225                          (--------*--------)
C      5  46.600  3.847              (-------*------)
                             +---------+---------+---------+---------
                            36.0      42.0      48.0      54.0

Pooled StDev = 4.633
```

FIGURE A-17: Highlighted Tablet Example—ANOVA Output in MINITAB Session Window

Examples

If the questions of interest were to study and compare the variances among multiple items (or the standard deviation), then the ANOVA analysis would follow the same procedure but substitute the test data means (seconds) with variances (s^2). The manual procedure would follow the same steps. The automated procedure also would follow the same steps, except utilizing a different MINITAB ANOVA test called "*Test for Equal Variances*."

The procedure steps for the same tablet example would include

Step 1. State the Null (H_0) and Alternative (H_a) hypotheses.

a. Null hypothesis: H_0: the variances (s^2) of all three factors are equal

i. $s^2_A = s^2_B = s^2_C$; where s^2 is the variance

b. Alternative hypothesis: H_a: at least two factors have different variances

Step 2. Choose the significance level or risk level (α); α = *0.5*.

Step 3. Randomly select the parts or processes.

Step 4. Identify the factors (that is, parts, processes, and technicians).

Step 5. Collect the data and record in MINITAB Worksheet. For this test, the data is best represented in a stacked fashion, which lists the trial times (called responses) in sequence for each factor (that is, type of tablet), and in a second column lists the corresponding factor alongside the response time, as illustrated in Figure A-18.

Time	Apririn Type
42	A
47	A
40	A
55	B
48	B
61	B
49	B
54	B
45	C
41	C
47	C
51	C
49	C

FIGURE A-18: Stacked Tablet Data Recorded in MINITAB Worksheet

Step 6. Calculate the F-statistic (or F-calculated) using the sample data.

a. Select the following commands: ***Stat > ANOVA > Test for Equal Variances***.

b. Put the response time data into the "Responses" field and put the corresponding Tablet Type in the factors field. Select acceptable confidence level in the ANOVA Dialog box, as illustrated in Figure A-19.

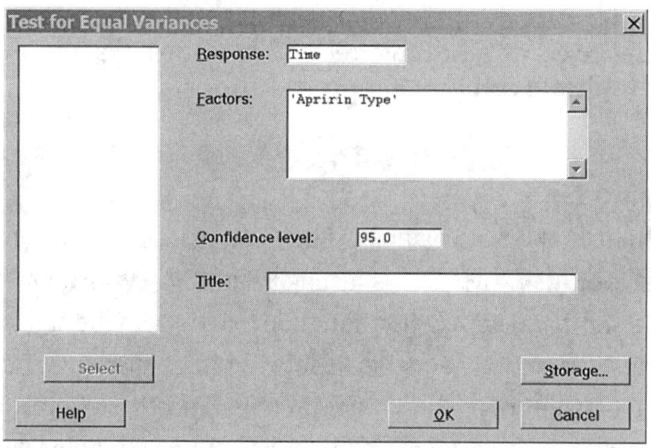

FIGURE A-19: Tablet Example in MINITAB ANOVA Dialog Box

c. Select *OK*, and the following output will be displayed in the session window, as illustrated in Figure A-20.

A B C D E F G H I J K L M N O P Q R S T U V W X Y Z

```
Results for: HTests.MTW

Test for Equal Variances: Time versus Apririn Type

95% Bonferroni confidence intervals for standard deviations

Apririn
   Type  N    Lower    StDev    Upper
      A  5  2.55208  4.72229  18.1756
      B  5  2.82373  5.22494  20.1102
      C  5  2.07909  3.84708  14.8070


Bartlett's Test (normal distribution)
Test statistic = 0.34, p-value = 0.845


Levene's Test (any continuous distribution)
Test statistic = 0.15, p-value = 0.859


Test for Equal Variances: Time versus Apririn Type
```

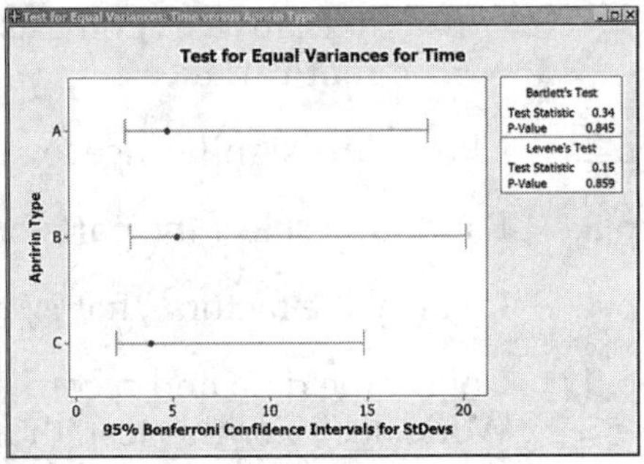

FIGURE A-20: Tablet Example—ANOVA Output in MINITAB Session Window and Graph

The graph in Figure A-20 provides the essential output information. Notice that the confidence intervals overlap, showing that it is difficult to distinguish one range from another, thereby supporting the Null hypothesis that the variances in dissolution time for the three different types of tablet are the same. The session window information on the left of Figure A-20 supports the graph. The p-value for both the Bartlett's and Levene's tests are greater than 0.05 (the selected significance level); hence, there is insufficient evidence to reject the Null hypothesis. Therefore, the variability in the time to dissolve the different types of tablet is the same.

The Figure A-20 session window output mentions two different tests, Bartlett and Levene. The Bartlett test applies to normal distributions, while the Levene test applies to skewed (non-normal) distributions.

This ANOVA test shows that although at least two of the tablet means are different with a 95% certainty, they come from populations and have the *same variance*. The MINITAB ANOVA Help states, "The effect of unequal variances upon inferences depends in part upon whether your model includes fixed or random effects, disparities in sample sizes, and the choice of multiple comparison procedure."

Hints and Tips

- Understand how to establish and analyze a statistical hypothesis test (**See Also** "Hypothesis Testing", Part II, **p. 335** for more details).
- Use statistical application software to help with the calculations.
- If an "interaction" effect is suspected to be present, a replicate of the experiment should be run to estimate its effect. [Interaction is defined as when the response of one variable is dependent on another. Hence, an Interaction Plot would depict perpendicular lines for two factors where the output of one (input) factor depends on the level of a second (input) factor.]

Supporting or Linked Tools

Supporting tools that might provide input when developing an ANOVA study include

- Data Gathering Plan (**See Also** "Data Collection Matrix," **p. 248**)
- Cause-and-Effect diagram or Fishbone (**See Also** "Cause-and-Effect Diagram," **p. 173**)
- Process map (**See Also** "Process Map," **p. 522**
- Hypothesis Testing (**See Also** "Hypothesis Testing," **p. 335**)

A completed ANOVA provides input to tools such as the following:

- Potential solution brainstorming (**See Also** "Brainstorming Technique," **p. 168**)
- Solution Selection Matrix (**See Also** "Solution Selection Matrix," **p. 672**)
- QFD (**See Also** "Quality Function Deployment (QFD)," **p. 543**)
- FMEA (**See Also** "FMEA," **p. 287**)
- Control plan (**See Also** "Matrix Diagrams," **p. 399** for a brief discussion on control plans.)

Figure A-21 illustrates the link between ANOVA and its related tools and techniques.

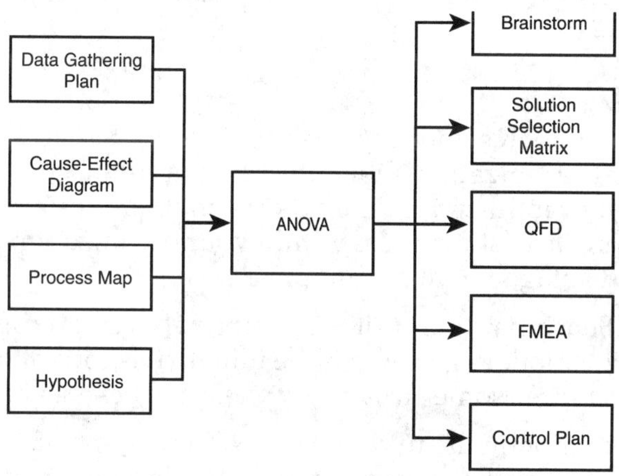

Figure A-21: ANOVA Tool Linkage

Arrow Diagram

See Also "Activity Network Diagram (AND)," **p. 127.**

B

Benchmarking

A B C D E F G H I J K L M N O P Q R S T U V W X Y Z

What Question(s) Does the Tool or Technique Answer?

What are the experts doing, and how do they do it?

Benchmarking helps you to

- Identify "best practices" in the marketplace for both what organizations are doing and how they are doing it.
- Innovate, think creatively, about solving a problem or doing something new to establish a competitive advantage.
- Focus people on continuous improvement.

When Best to Use the Tool or Technique

Benchmarking is best used when new ideas are needed to improve or solve a situation. It can be conducted formally or informally; however, having an organizational impact should involve the company's leaders and/or decision-makers. Benchmarking results that have been selected to act on should be integrated into the strategic direction and goals of an organization. That takes the commitment of the organization's leadership team.

Brief Description

Benchmarking compares your organization's processes and practices with an external organization. Often the best marketplace "expert" may not be part of your industry. Benchmarking another industry and figuring out how to apply its best practices within your organization creates a competitive advantage for your enterprise.

Even if the Benchmarking visit is an informal event, it can spawn creativity and innovation. Regardless if the initiative is formal or informal, the technique requires some structure to yield the best benefits.

Xerox Corporation in the mid- to late-1970s was one of the first companies to document and popularize Benchmarking under the leadership of David Kearns, then CEO and President. Xerox chartered Sy Zivan to improve its parts distribution process. In turn, Sy and his team (including Bob Camp) stepped outside the office products industry and sought to benchmark retailer L.L. Bean. L.L. Bean was leading the way and enjoying great success with e-commerce as measured by customer satisfaction and overall financial growth. [**See** S. Zivan's article, "Benchmarking—Avoid Arrogance and Lethargy," in Part III, **p. PDF:789**, of this book.]

How to Use the Tool or Technique

This technique can be done individually or in a small group but requires the support and active commitment from the organization's leadership team. Regardless if the Benchmarking activity is formal or informal, the technique requires some structure. Successful Benchmarking often is conducted as a project in and of itself and involves these steps at a high level:

Step 1. Plan the benchmark activity. All successful projects require good planning.

a. Identify the **key topic of interest** (what).

b. **List organizations** thought to be "experts" (learn from the best practices) or "not-experts" (learn from their mistakes) (why).

Experience shows that business people tend to skimp on the appropriate time it takes to research the best organizations to benchmark. Some clients assign outside reading to the Benchmarking and/or strategy team, only to reconvene and do book reports on key points learned. Other clients hire market research and Benchmarking firms for advice. Thinking outside of the box often reveals industry leaders that otherwise would be overlooked. The most successful mental model often entails unrelated industries at first glance.

c. Define the data collection plan.

i. **Who** will collect the data, and who is the benchmark firm's key contact?

ii. **Where and how** will the data collection take place? Will it take place at the firm's location, or will it be by phone, written survey, or a combination thereof?

iii. **When** will this data collection take place, and what is the target schedule? Make certain to have a defined beginning and end and avoid scope creep and schedule slippage.

iv. **How** will this Benchmarking occur? What is the budget, and what resources are needed? Is the appropriate level of organization involved? Are the resources properly trained in what to do and how?

Step 2. **Design the data collection approach.** This step answers the questions, "What specific data will be collected?" and "How will the data be collected and communicated?"

A B C D E F G H I J K L M N O P Q R S T U V W X Y Z

a. Who is the **target audience** (consumer) of the benchmark data?

b. What is **purpose**—the gap or opportunity to be addressed?

c. How will the data be collected? Develop an **interview guide** to collect the data.

   i. **Identify the process (or practice) and performance metrics** to be studied. What does success look like? Anticipate how success might be measured by the benchmark organizations, compared with how your organization does today. ***Note:*** *Make certain that* ***process*** *and* ***cultural*** *dimensions are studied; avoid limiting it to* ***critical parameters*** *that are* ***easy to measure****.*

   ii. Develop the **key questions** to be asked (limit it to a small set of questions, that is, no more than about 10).

   iii. **Test** the questions with the Benchmarking Stakeholders.

   iv. Determine the **format** or combination of approaches to best collect the desired data.

   If possible, allow for time to **observe** the benchmark firm. Often seeing the benchmark practice in operation triggers additional questions, helps put the success in context, and identifies the implicit elements to success.

   v. Develop the final document.

Step 3. **Implement the plan** by conducting the Benchmarking activities, gathering the data, and analyzing it. Analysis tools (which are detailed in their own entry within the encyclopedia section of the book) include

a. Affinity diagrams

b. KJ Analysis

c. Cause-and-Effect diagrams

d. QFD

Step 4. **Summarize and recommend a *Go Forward* strategy.** Determine the key messages, the lessons learned from the Benchmarking initiative, and then develop and document a Go Forward plan. The documentation should include the goals, a recommended action plan to achieve these goals, and a control plan to monitor performance.

Hints and Tips

- Confirm proper sponsorship within the context of the organization's strategic initiative. Benchmarking produces creative ideas for change and improvement. To ensure that the changes will be supported (properly funded and resourced), the leadership needs to buy into the Go Forward strategy and lead its implementation, thereby making it part of the organization's strategy and goals.
- Take time to research and develop a list of potential Benchmarking candidates. Research who is an industry leader outside of your own industry.
- If possible, conduct Benchmarking at the benchmarked organization's location to observe the work in process.
- Encompass the full context of the benchmark's success beyond the simple metrics and include process and cultural elements that are probably main contributors and critical parameters.
- Offer reciprocal Benchmarking opportunities to establish a win-win scenario for the benchmarked organizations.

Supporting or Linked Tools

Supporting tools that might provide input when developing Benchmarking include

- VOC Data Gathering Tools and Techniques, such as surveys and interviews (**See Also**, "Voice of Customer Gathering Techniques," **p. 737**)
- Written reports
- Brainstorming sessions (**See Also** "Brainstorming Techniques," **p. 168**)

A completed benchmark imitative provides input to tools such as

- Affinity diagram (**See Also** "Affinity diagrams," **p. 136**)
- KJ Analysis (**See Also** "KJ Analysis," **p. 375**)
- Cause-and-Effect diagram or Fishbone (**See Also** "Cause-and-Effect Diagram," **p. 173**)

- Conjoint Analysis (**See Also** "Conjoint Analysis," **p. 207**)
- QFD (**See Also** "Quality Function Deployment (QFD)," **p. 543**)
- Implementation (**See Also** "Matrix Diagrams," **p. 399** for a brief discussion on transition and launch plans)
- Control plan (**See Also** "Matrix Diagrams," **p. 399** for a brief discussion on control plans)

Figure B-1 illustrates the link between Benchmarking and its related tools and techniques.

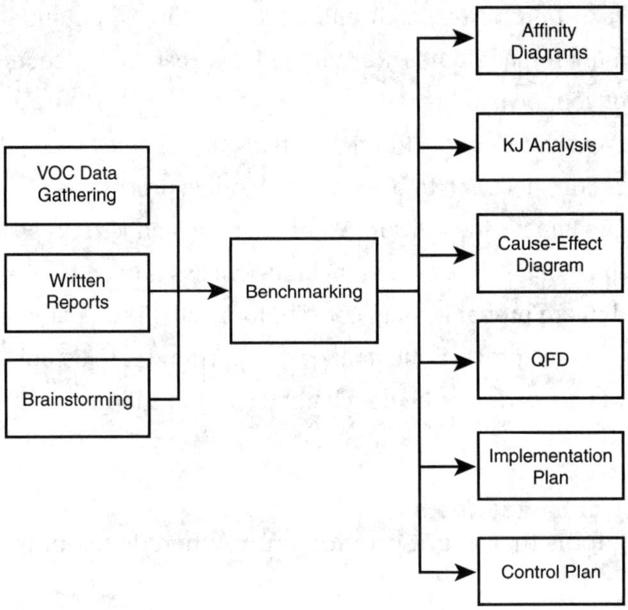

FIGURE B-1: Benchmarking Tool Linkage

Boxplots—Graphical Tool

What Question(s) Does the Tool or Technique Answer?

What does the distribution of a set of data look like? Alternatively, what do the distributions look like, and how do they compare with multiple sample groups?

Boxplots help you to

- Visualize what the distribution looks like for a sample set of data.

- Display the data distribution over its range (the spread) and indicate its quartiles and median.
- Graphically summarize and compare frequency distribution information about one or more sets of data.

Alternative Names and Variations

This tool is also known as

- Box-and-whisker plot

Variations on the Tool

- Dotplot—displaying the data detail versus a summary, therefore used for a small sample size of about 15. (**See Also** "Dotplot," **p. 280**)

When Best to Use the Tool or Technique

The old adage holds true, "A picture is worth a thousand words." The graphical method is the picture that visually displays the distribution of a set of data or that compares multiple sets of data. A Boxplot is best used when the *individual data points are **not** of interest*. It specifically is a good summary graphical tool when the sample size is too small to build a meaningful histogram (less than 50 pieces of data). A Boxplot works well when a Dotplot is too busy with over 15 data points.

Brief Description

The Boxplot is a simple graphical tool containing a great deal of information about a continuous, quantitative data set's distribution, its range (indicating the minimum and maximum), its median, and its quartiles. This tool was developed by John W. Tukey to display five summary data characteristics:

1. the data median
2. the upper quartile of data
3. the lower quartile of data
4. the minimum data point
5. the maximum data point

Figure B-2 illustrates the basic shape and features of a Boxplot graph.

Notice how the diagram divides the data into quarters (25% segments), where the first quartile (Q1) demarcation is the bottom of the box, the second quartile (identifying 50% of the data on either side) is marked by the median line dissecting the box, and the third quartile demarcation is the top of the box. The lines on either end of the box, referred to as "whiskers," depict data on either end of the range within a limit that are calculated as

- Lower Limit = Q1–1.5 (Q3–Q1)
- Upper Limit = Q3 + 1.5 (Q3–Q1)

If data exceeds the whisker limits, an asterisks (*) represents the unusual observation as an "outlier" data point. *Outliers* are defined as data beyond 1.5 times the inter-quartile distance from each quartile. In Figure B-2, the outlier indicates an unusually large value.

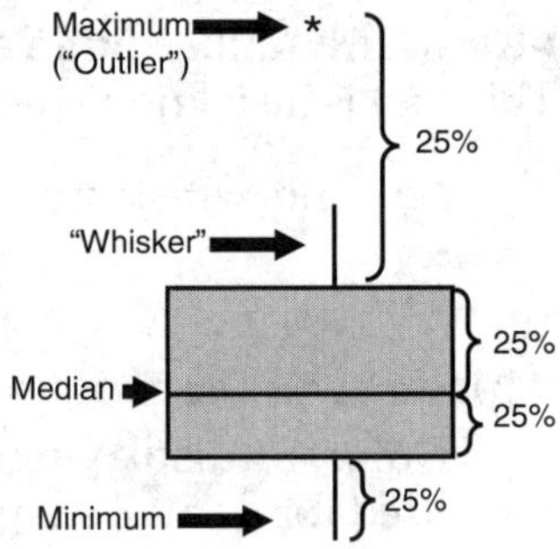

Figure B-2: Basic Boxplot Illustration

How to Use the Tool or Technique

This graphical technique can be used on a single data set or to compare multiple data sets. A Boxplot can be created by hand or by using statistical software. The procedural steps and illustrations are based on the MINITAB software for a simple Boxplot for one variable.

Step 1. Collect the data and enter the variable in a single column in the MINITAB Worksheet.

Step 2. From the tool bar, select ***Graph > Boxplot....***

Step 3. Select One Y ***Simple***; click the ***OK*** button as illustrated in the left portion of Figure B-3.

Step 4. Enter the column of data in the Graph variables window of the dialog box, and click the ***OK*** button, as illustrated in the left portion of Figure B-3.

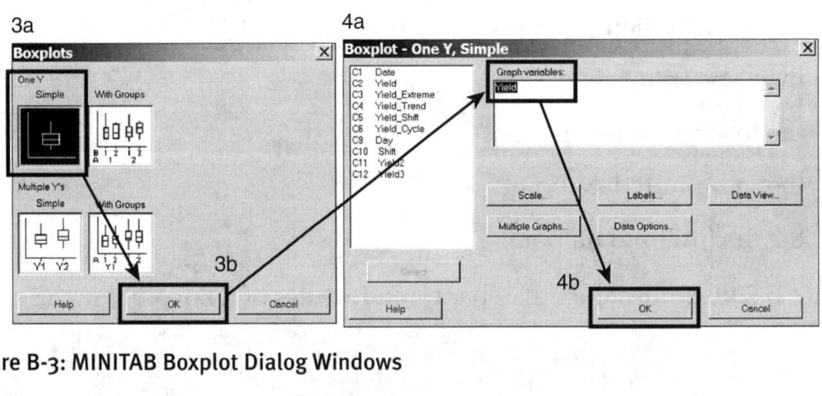

Figure B-3: MINITAB Boxplot Dialog Windows

The output will resemble the illustration in Figure B-2. To examine two groups of data, follow the same procedure, but in the first dialog box, select the appropriate graph type depending on how many Y variables there are. Figure B-4 depicts a two-group "simple" Boxplot example with multiple "Ys" comparing "Yield2" and "Yield3" variables.

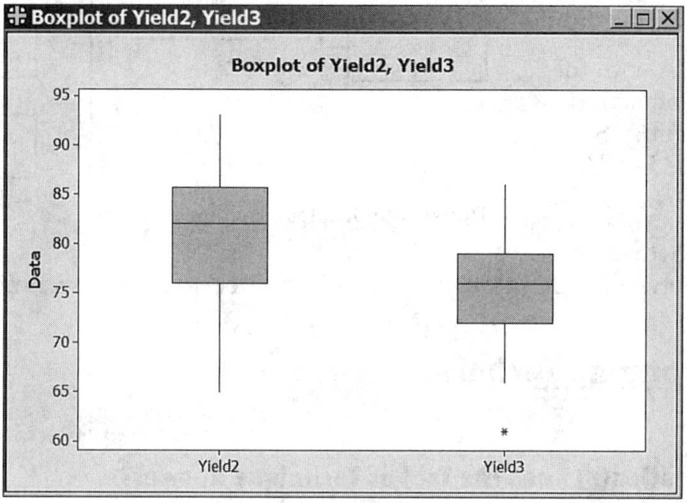

FIGURE B-4: Simple Boxplot with Multiple Ys Illustration

Supporting or Linked Tools

Supporting tools that might provide input when creating a Boxplot graph include

- Data gathering plan to collect the appropriate metrics (**See Also** "Data Collection Matrix," **p. 248)**
- Performance charts and dashboards

A Boxplot graph can provide input to tools such as

- Cause-and-Effect diagram (**See Also** "Cause-and-Effect Diagram," **p. 173**)
- QFD (**See Also** "Quality Function Deployment (QFD)," **p. 543**)

A B C D E F G H I J K L M N O P Q R S T U V W X Y Z

- Statistical analysis tools and other root cause analysis techniques (**See Also** "Statistical Tools," **p. 684**)
- FMEA with follow-on action planning (**See Also** "FMEA," **p. 287**)

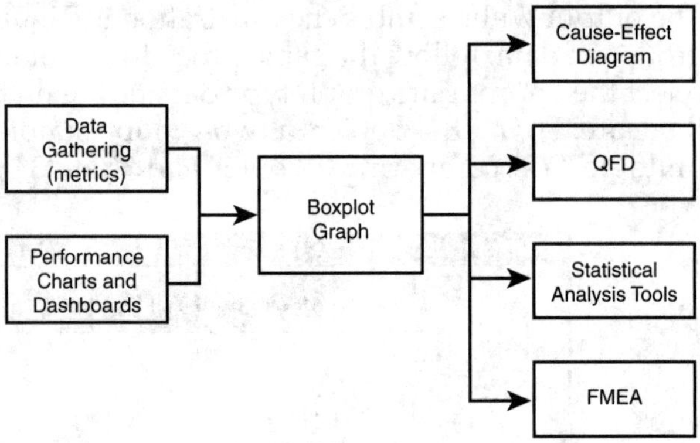

Figure B-5: Boxplot Graph Tool Linkage

Figure B-5 illustrates the link between a Boxplot graph and its related tools and techniques.

Brainstorming Technique

What Question(s) Does the Tool or Technique Answer?

How can you elicit new ideas in a short period of time?

Brainstorming helps you to

- Generate and document a large volume of ideas or concepts in one meeting.
- Create out-of-the-box thinking to develop new ideas.

Alternative Names and Variations

This tool is also known as

- Round-robin brainstorming

Techniques that incorporate brainstorming include

- Nominal group technique
- Mind-mapping
- Affinity diagramming (**See Also** "Affinity Diagram," **p. 136)**
- Fishbone diagramming (**See Also** "Cause-and-Effect Diagram," **p. 316)**
- Tree diagram (**See Also** "Tree Diagram," **p. 712)**

A B C D E F G H I J K L M N O P Q R S T U V W X Y Z

- Matrix diagram (**See Also** "Matrix Diagrams," **p. 399**)
- QFD (**See Also** "Quality Function Deployment," **p. 543**)
- FMEA (**See Also** "FMEA," **p. 287**)

When Best to Use the Tool or Technique

When breakthrough ideas are needed, to expand the team's thinking beyond the traditional concepts, the brainstorming technique can help generate new ideas or perspectives. Use brainstorming when each team member's participation is desired.

Brief Description

Brainstorming features a team of people generating ideas in an intentionally **uninhibited** fashion. An ideal group size consists of about eight people (from six to ten is acceptable). The best results are produced by a diversity of perspectives, such as cross-functional disciplines, different perspectives, tenure, roles, and so on. Often a mix of written (or silent) and verbal generation techniques help to draw people out and allow for 100% participation. Brainstorming often is accomplished in a single meeting that produces as many ideas as possible.

The purpose of brainstorming is to generate several ideas, not necessarily to solve a problem. At the conclusion of the idea generation process, techniques such as multi-voting can help to prioritize or select the top "candidates" for appropriate actions.

How to Use the Tool or Technique

This technique should be used in a small group and should follow these general guidelines:

Step 1. **Prepare for the meeting.**

Identify the topic of interest and assemble a small group that represents a cross-section of perspectives. Invite a "neutral" person to serve as the meeting Recorder and facilitator, who will usually not be participating in the process itself. Choose the documentation vehicle (list on paper or individual ideas per sticky note or index card).

Step 2. **Initiate the meeting.**

Review the topic of interest and discuss the brainstorming approach and guidelines to be used in the meeting.

Step 3. **Individual idea generation.**

Ask each member to write down as many ideas as possible. Silence should be observed while people are writing down their ideas and until everyone has finished. There are two approaches for documentation—people can either

a. List their ideas on pieces of paper, which will be shared later verbally

b. Take a stack of sticky notes or index cards and write **one idea per note** or card

Step 4. **Share ideas with the group.**

Ideas are shared one at a time until all individual ideas are shared with the group. Best results tend to occur when one person shares only one idea at a time and then defers to the next person. This cycle is repeated until all ideas are shared.

a. If a list of paper was used in a round-robin fashion, have each person share one idea at a time and then move on to the next person. The meeting Recorder documents the ideas as a list on a flip chart.

b. If the sticky notes or index cards are used, again in a round-robin fashion, each person shares one idea at a time and then posts the one idea on the wall or flip chart paper and then gives the floor to the next person. (Post index cards with masking tape.)

This approach is preferred if the facilitator's ideas are important to capture as part of the group.

Step 5. **Review and build on ideas.**

Ask the team to reflect on the shared ideas and explore whether an idea spawns an additional idea, expands one, is combined with others, or needs to be modified. Ask the team to reflect on situations outside of the current market segment or change perspectives (such as age, country, and so on) to see if new ideas evolve. Continue to generate additional ideas and record each one as it is shared using the selected documentation approach.

How to Analyze and Apply the Tool's Output

After the brainstorming session is completed, the team can decide how best to use the ideas. The list can be prioritized using multi-voting or a nominal group technique. The ideas can serve as input to an Affinity diagram, Cause-and-Effect diagram (or Fishbone), Mind map, or FMEA.

Hints and Tips

- Select a comfortable environment for the meeting place. Often a casual place away from the workplace (perhaps outdoors) elicits more creativity.
- All ideas are recorded.
- Include everyone and avoid inhibiting anyone. Encourage all to participate by utilizing written and verbal idea generation.
- Allow each person to generate ideas without discussing, evaluating, analyzing, or criticizing any concept at the time of generation/development.
- Build on one another's ideas. Sometimes a courteous "free-wheeling" verbal approach is used.
- Expanding on someone's idea or piggybacking should be encouraged.
- Initially strive for a large quantity of ideas, regardless of how far-fetched they may seem. The process later clarifies, consolidates, and prioritizes the concepts.

Supporting or Linked Tools

Supporting tools that might provide input when conducting a brainstorming session include

- Data Gathered (metrics), performance charts, dashboards (**See Also** "Data Collection Matrix," **p. 248**)
- Documentation or written reports

A completed brainstorming session provides input to tools such as

- Affinity diagram (**See Also** "Affinity Diagram—7M Tool," **p. 136**)
- Cause-and-Effect (or Fishbone) diagram (**See Also** "Cause-and-Effect Diagram—7QC Tool," **p. 173**)
- FMEA (**See Also** "Failure Modes and Effects Analysis (FMEA)," **p. 287**)
- Matrix diagram (**See Also** "Matrix Diagrams—7M Tool," **p. 399**)
- Mind-mapping

- QFD (**See Also** "Quality Function Deployment (QFD)," **p. 543**)
- Tree diagram (**See Also** "Tree Diagram—7M Tool," **p. 712**)

Figure B-6 illustrates the link between a brainstorming session and its related tools and techniques.

A B C D E F G H I J K L M N O P Q R S T U V W X Y Z

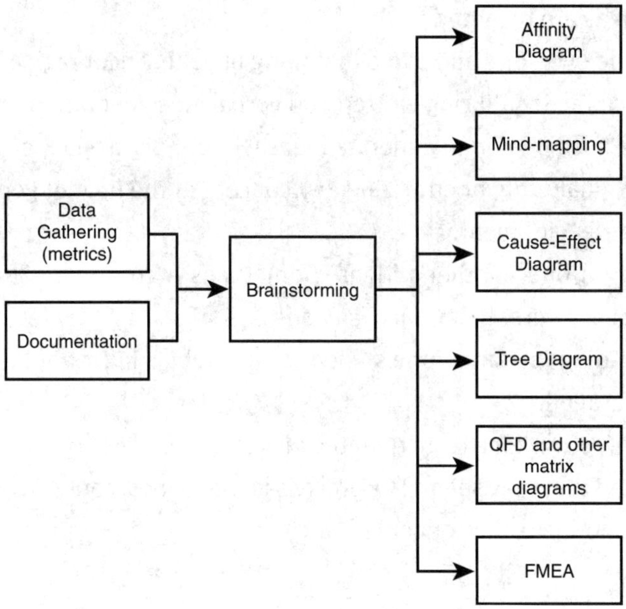

Figure B-6: Brainstorming Tool Linkage

C

Capability Analysis

See Also "Process Capability Analysis," **p. 486**.

Cause-and-Effect Diagram—7QC Tool

What Question(s) Does the Tool or Technique Answer?

*What are the **potential** causes of a problem or problematic outcome?*

Cause-and-Effect diagramming helps you to

- Organize ideas and understand the relationship between potential causes and a problem by formatting, arranging, organizing, and parsing potential causes into themes and sub-themes in preparation for a future cause identification effort.
- Stimulate thinking when developing the list of the potential sources of a problem.
- Guide concrete action and track the potential causes during an investigation effort to determine if the item significantly contributes to the problem.

Alternative Names and Variations

This tool is also known as:

- C and E diagram
- Fishbone or fishbone diagram
- Ishikawa diagram

Variations on the tool include

- Cause-and-Effect matrix or Cause-and-Effect Prioritization Matrix (**See Also** "Cause-and-Effect Prioritization Matrix," **p. 188**)
- Cause enumeration diagram
- Process Fishbone
- Time-delay Fishbone
- CEDAC (Cause-and-Effect Diagram and Cards)

- Desired-results Fishbone
- Reverse Fishbone diagram
- Cause and Prevention diagram
- Fault Tree analysis (**See Also** "Fault Tree Analysis," **p. 309**)

When Best to Use the Tool or Technique

Before any action is taken, this tool helps to organize and begins to analyze potential sources of problems.

The Fishbone is a good tool to use when multiple people (and/or disciplines) should be engaged in the problem-solving and there are many perspectives to capture.

Brief Description

The Cause-and-Effect diagram is one of the seven basic tools of quality. It goes by several names—a *Fishbone diagram* because a completed diagram resembles the skeleton of a fish and an *Ishikawa diagram* after its creator.

A professor at the University of Tokyo, Dr. Kaoru Ishikawa, developed the first Cause-and-Effect diagram in 1943. The original intent of the diagram was to sort out and depict the relationship among the several factors impacting Quality Control (QC), wherein the variables that cause dispersion, such as chemical composition, size of parts, or process workers, are called "factors" (causes). The quality characteristics describe the outcome, or effect, such as length, hardness, percentage of defects, and so on. As the tool grew in popularity, it was referred to as an Ishikawa diagram.

The Cause-and-Effect diagram is a member of the 7QC Tools (or *seven Quality Control tools*), attributed to Dr. Ishikawa. The 7QC Tools sometimes are called the *seven basic tools* because they were the first set of tools identified as the *core* quality improvement tools. Ishikawa's original 7QC Toolset includes 1) Cause-and-Effect diagram, 2) Check sheet (or Checklist), 3) Control charts, 4) Histogram, 5) Pareto chart, 6) Scatter diagram, and 7) Stratification. More recently, the 7QC Toolset is modified by substituting the Stratification technique with either a flowchart (or process map) or a run chart (or time series plot).

A Fishbone diagram is a focusing tool. It starts with identifying all the ideas for potential causes and then groups and categorizes the potential causes into themes. It takes a snapshot to hypothesize the collective potential causes—what the team thinks is currently happening.

It displays *potential* causes of a problem, with the potential causes depicted as "off-shoots" or bones of a fish stemming from the problematic outcome as the "head" of the fish. Increasing detail about the potential cause is displayed as the offshoot, or bone, branches further out from the main bone.

The relationships can be depicted as a matrix or Fishbone diagram, wherein the problem (or effect) is placed to the far right. The problem statement is placed in a box or diamond-shape to represent the head of a fish. This represents the "effect" that all the subsequent causes supposedly impact. The root cause themes and sub-themes are placed to the left of the problem.

There are two major types of Cause-and-Effect formats depending on the context in which the problem exists; one is a Dispersion, and the other is a Process format. There are some memory techniques that serve as checklists to trigger thinking about potential root cause categories—the 5Ms and P and the 4Ps. The former is typically applied to the Dispersion analysis, and the latter is applied to the Process approach.

Dispersion Analysis

Potentially, why does quality dispersion occur? This approach uses individual causes (or dispersion) grouped within a major cause category. The smaller categories on the skeleton of the fish drill down into sub-causes, identifying why this cause potentially happens.

Dispersion occurs typically due to differences in

- Raw materials' composition, size, and so on
- Tools, machinery, equipment, technology—with respect to their operating performance
- Work method, process or procedure—potentially because it could be incomplete, inaccurate, inflexible (accommodating change or a special occurrence), misunderstood
- Measurement—potentially from operator-to-operator error, operator-to-part, or part-to-part.

When using the Dispersion analysis approach, there is no right answer or *set* of categories names or themes. They should be modified to fit the situation or problem. However, a convention that often is used as a starting point is referred to by the acronym, *5Ms and P*.

5Ms and P Memory Triggers

A standard categorization technique for manufacturing industries is called **5Ms and P** [pronounced *"five Ms and a P"*] and can be used as a checklist to identify common cause themes:

- **M**achines (equipment, technology—hardware and/or software)
- **M**ethods (process, procedure, approach, policy, or practice)
- **M**aterials (raw materials, components, information, or data)

- **M**easurements (input and output metrics for quality, quantity, process performance metrics, calibration, and inspection)
- **M**other Nature (or a more encompassing term is *environment*), which includes external as well as workplace factors such as:
    - *Natural environment* [weather (and *acts of God*), temperature, humidity]
    - *Physical surroundings* (facilities, buildings, plant, workspace, office)
    - *Management surroundings* (organizational, social, political)
    - *Marketplace*
- **P**eople [those involved in the process (directly or indirectly)—customers, employees, managers, suppliers, partners, regulators and shareholders]

A B C D E F G H I J K L M N O P Q R S T U V W X Y Z

Sometimes this approach is called *The 6Ms*, where the term "People" (of 5Ms and P) is changed to **M**anpower.

Process Classification

The Process approach uses the major process steps, instead of major causal categories; however, this is rarely used. This diagram may be depicted in the Fishbone shape or as a Process map "spine" with potential causes linked to the appropriate process step.

When using the Process approach, again, there is no right answer or set of category names or themes. They should be modified to fit the situation or problem. However, the services industry has a convention that often is used as a starting point—*the 4Ps*.

4Ps Memory Triggers

Common categories that work well for the **service(s) industry** is called the **4Ps**, which represents

- **P**olicies (company and HR policies, including roles and responsibilities, performance metrics, reward and recognition, promotion, overtime, comp-time/vacation, and so on, and organizational—cultural, training versus apprenticeship, and so on)
- **P**rocedures (methodology, approach)
- **P**eople (employee profile, partners, sub-contractors, and so on)
- **P**lant/Technology (workspace, hardware, software, support tools)

Because the services industry is dependent on its human capital, the policies are critical to shaping its "personality" in the marketplace, and often it is the *make-or-break* for attracting the right talent. Hence, the Policy category often needs to be parsed further to get to the appropriate detail.

How to Use the Tool or Technique

When developing a Cause-Effect, Fishbone, or Ishikawa diagram:

Step 1. Agree on the topic or problem to be analyzed—for example, a quality characteristic that needs improvement. Use data specifically to describe the problem.

Step 2. Write the specific problematic outcome to the far right edge of the diagram and draw a box around the text to form the **"head" of the fish**.

Figure C-1 displays the general structure of the Cause-and-Effect diagram using the Fishbone analogy.

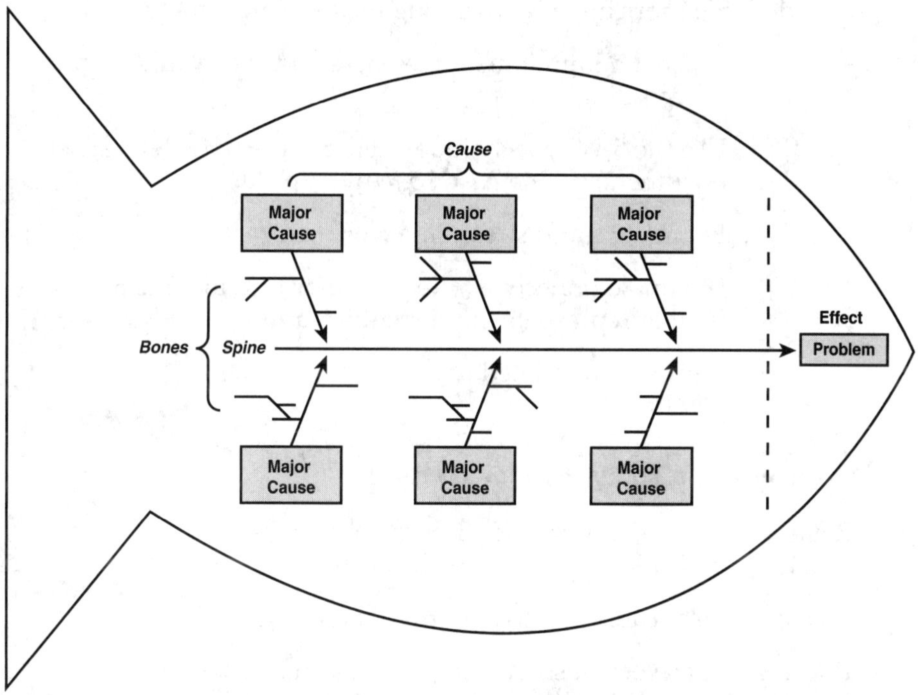

FIGURE C-1: Basic Cause-and-Effect Diagram (Fishbone) Using 5Ms and P for Dispersion Analysis

Step 3. Draw a line extending from the left edge of the fish head to the left edge of the diagram to represent **the spine**.

Step 4. Referencing data and experience, write the potential causes of the problem and group related topics. Have the participants bring to the meeting completed data collection sheets from their respective areas to inform the brainstorming activities. (**See Also** "Brainstorming Technique," **p. 168**)

Step 5. Name the major themes of each grouping (or high-level process steps).

Step 6. Draw a line extending from the spine for each theme and write the category name at the outermost end of the line (the end that is not attached to the spine) to represent a **"main" Fishbone**.

Step 7. Draw one branch or offshoot to the main Fishbone for each sub-theme (or sub-process step). Label the end of each branch with the sub-theme name to represent **smaller bones** and to show the linkage to the "higher level" category with which it is affiliated.

Step 8. Continue Step 7 to drill-down into as much detail as required, defining more bones and linking the relationship.

Step 9. Review the diagram for completeness and clarity and then modify accordingly.

  a. Use the 5-Why technique to ensure the potential cause is documented (**See Also** "5-Whys," **p. 305**).

  b. Fill in gaps and streamline wording.

  c. Eliminate unnecessary redundancy but document duplicity that represents the current situation.

  d. Eliminate unrelated items.

How to Analyze and Apply the Tool's Output

- Review the final diagram and discuss the themes thought to represent the most critical source of the problem.
- Look for repeat causes across multiple categories.
- Prioritize and select—use the diagram as a checklist to explore further into those critical themes. You could use a QFD tool to prioritize the potential causals. A less desirable selection approach could be a form of voting technique, such as weighted voting.
- Investigate or study the prioritized causes and/or critical themes to verify that they are in fact causals. Some tools and techniques to begin the analysis include the following:
  - Force-field analysis
  - Scatter Plots (**See Also** "Scatter Diagram—7QC Tool," **p. 640**)
  - Frequency Plots (**See Also** "Graphical Methods" and "Histogram —7QC Tool," **p. 323** and **330**, respectively)

- Tables of Results (for attribute data)
- ANOVA (Analysis of Variance) (**See Also** "Analysis of Variance (ANOVO)—7M Tool," **p. 142**)
- Regression Analysis (**See Also** "Regression Analysis," **pp. 571**)
- Time Trap Analysis
- Design of Experiment (DOE) (**See Also** "Design of Experiment (DOE)," **p. 250**)

Note

This tool documents POTENTIAL causes, not proven causes; so *improvement* action plans cannot be developed until the causes are verified. Seek the causes thoroughly using the Cause-and-Effect diagram. When the Cause-and-Effect is detected, check and record it on the diagram, by writing the date of occurrence and its measurement next to its corresponding "labeled bone." If the cause is not detected, then the "labeled bone" will lack any markings. This procedure highlights experienced data and focuses on the detected causes, which can lead to quicker error correction.

Examples

Dispersion Analysis example

An example of a Dispersion Analysis, as shown in Figure C-2, uses the "5Ms and P" technique to define the root causes as titles of the major bones off the spine of the Cause-and-Effect diagram. [Recall the 5Ms and P refer to **M**achine, **M**aterials, **M**ethod, **M**etrics, **M**other Nature, and **Peo**ple. See the "Hints and Tips" section later in this entry for more detail on 5Ms and P technique.]

Process Classification Analysis example

An example of a Process Classification Analysis, shown in Figure C-3, uses the high-level process steps as the titles of the major bones off the spine of the Cause-and-Effect diagram.

A B C D E F G H I J K L M N O P Q R S T U V W X Y Z

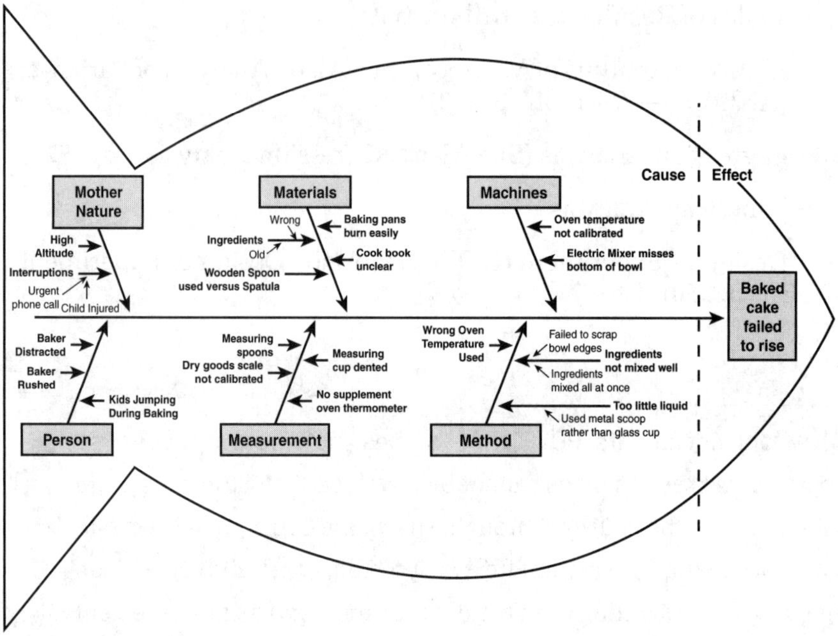

FIGURE C-2: Cause-and-Effect Diagram (Fishbone) Using 5Ms and P for Dispersion Analysis

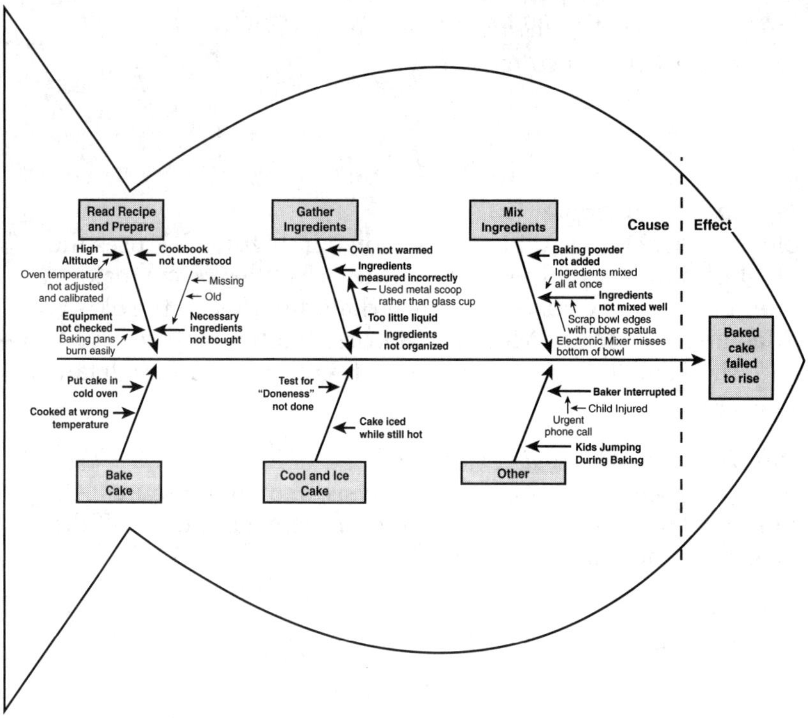

Figure C-3: Cause-and-Effect Diagram (Fishbone)—Using Process Steps for a Process Classification Analysis

Process Classification Analysis—a Second Example

Another example of a Process Classification Analysis, shown in Figure C-4, diagrams the high-level process steps as the actual spine of the Cause-and-Effect diagram. The bones off the spine align to the root causes directly linked to a particular process step.

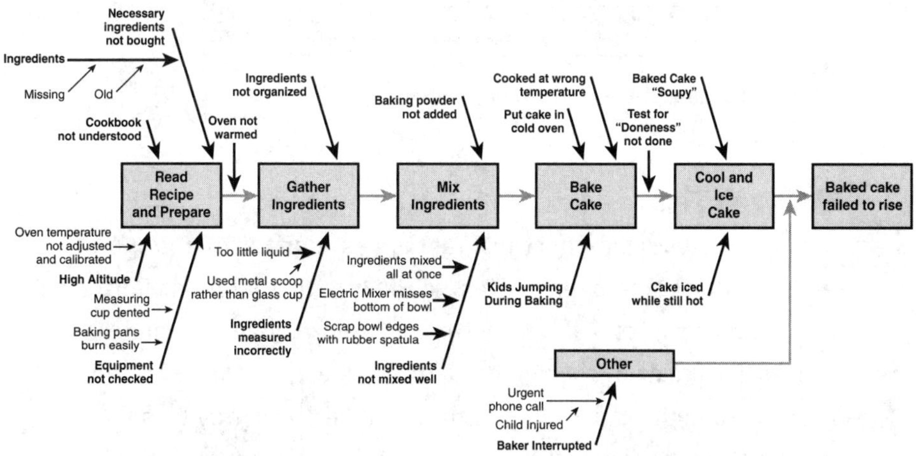

FIGURE C-4: Cause-and-Effect Diagram (Process Fishbone)—Using Process Steps for a Process Classification Analysis

Poorly Constructed C and E Diagram Example

A cause generally contains many complex elements. Therefore when building a Cause-and-Effect diagram, it is important to drill down into a cause to identify the potential root. The drill down can be achieved by using the 5-Whys technique. (**See** "5-Whys," **p. 305**) The following Fishbone diagram is an example of a poor one. Even though the "form" is correct, the diagram simply lists eight items. It lacks detail, which indicates potential lack of knowledge of the process. The Fishbone diagram should be detailed, not too generalized or shallow. The example of a poor Fishbone fails to describe why ingredients are missing and why the batter was not mixed well. The baker could have had the blender working overtime, but if the beaters were not touching all parts of the bowl and the mixing process lacked manually scraping of the sides of the bowl with a rubber spatula, the potential root causes behind why the batter was not mixed well are more obvious. Moreover, the poor Fishbone diagram contains some symptoms—"Baked Cake Soupy," rather than probing further to understand the driver as to why it's soupy and document the causes. A well-constructed Fishbone diagram documents all the possible causes in an organized way, depicting the relationship of the *major* causes themes to subsequent or more detailed causes. Figure C-5 displays a poorly constructed Cause-and-Effect diagram.

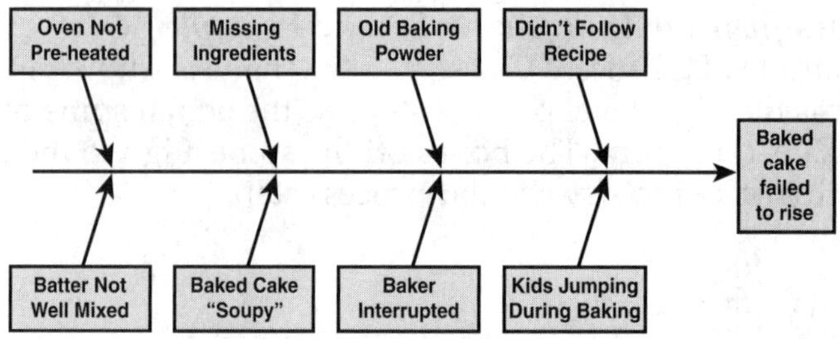

FIGURE C-5: Poor Cause-and-Effect Diagram (Fishbone)

Hints and Tips

Building a Cause-and-Effect diagram is best done with a group of people with diverse perspectives, to analyze the potential causes from multiple angles. Capture viewpoints from various roles and disciplines. Ensure everyone agrees on the problem statement first and focus on potential causes, not symptoms. Use the 5-Why technique to drill down and identify the potential source of the cause. (**See Also** "5-Whys," **p. 305**) Capture all ideas (information on "what," "where," "when," "how much..."). Consider all causes, not just those within the group's control.

Use data to specify the problem. Do not criticize or debate ideas during brainstorming or when representing them on the Cause-and-Effect diagram.

Histograms can reveal the frequency of defects caused by certain items. Simply **test for understanding** to clarify that an idea is communicated succinctly. Look for duplicates and aggregate them together if possible. Some causes may fit into more than one category (particularly people-related topics). Ideally a cause should be aligned with one major category, but if it is not clear where it best fits, represent it in multiple categories. With further analysis, eventually its best fit may narrow to one area. Take time in developing the diagram, ask as many people as possible, and avoid trying to complete it in one work session. Allow people to study a *work-in-process* diagram to trigger additional ideas. Prioritizing can be done later using a QFD. The appropriate number of categories depends on the problem and related potential causes. There is no prefect number. For a Process

Fishbone, the process depicted should be **high-level**. Therefore, ideally the number of steps is three to five and generally should not exceed 10; otherwise, it becomes too detailed and cumbersome. Category naming can be done two different ways: predetermined categories prior to the brainstorming portion or determined after brainstorming as a "fall-out" from the grouping portion. There are pros and cons to both approaches. The predetermined categories may expedite and better organize the work. If experience determines the categories selection, then the theme titles probably will not inhibit the generation of ideas. However, the addition of an "other" category also will help to prevent the stifling of ideas. If category selection occurs after the brainstorming, it takes time to organize and arrive at a common agreement on the theme. It involves testing for understanding and potentially regrouping and/or further consolidation of the ideas.

As a rule of thumb, the drill-down into sub-causes usually stops when a cause is controlled by a different level of management than the main causal bone. Common sense should rule. **Rarely are people the root cause of process problems.** Often people are compensating for something that is not working, confused by something that is misunderstood, or handling an exception to the rule. Probe further to understand the cause behind a person's given activity or behavior. Sometimes an "outside" person not involved in the development process can add a new perspective and identify gaps or patterns in the diagram.

Supporting or Linked Tools

Supporting tools that might provide input when developing a Cause-and-Effect diagram include

- Histogram (**See Also** "Histogram—7QC Tool," **p. 330**)
- Pareto chart (**See Also** "Pareto Charts—7QC Tool," **p. 445**)
- Process map to identify potential root causes (**See Also** "Process Map (or Flowchart—7QC Tool)," **p. 522**)
- VOC and VOB data (**See Also** "Voice of Customer Gathering Techniques," **p. 737**)

- 5Ms and P technique or its variants such as 6Ms and 4Ps (described in detail earlier in this section)
- 5-Why Technique (**See Also** "5-Whys," **p. 305**)
- Brainstorming technique (**See Also** "Brainstorming Technique," **p. 168**)

A B C D E F G H I J K L M N O P Q R S T U V W X Y Z

A completed Cause-and-Effect diagram provides input to tools such as

- Brainstorming potential solutions and concept generation methods
- Root Cause Analysis Techniques [such as Correlation and Regression, Hypothesis Testing, and Design of Experiment (DOE)] (**See Also** "Design of Experiment (DOE)," "Hypothesis Testing," and "Regression Analysis," **p. 250, 335**, and **571**, respectively)
- QFD (Quality Function Deployment) (**See Also** "Quality Function Deployment (QFD)," **p. 543**)
- FMEA (Failure Modes and Effect Analysis) (**See Also** Failure Modes and Effects Analysis (FMEA), **p. 287**)

Figure C-6 illustrates the link between the Cause-and-Effect diagram and its related tools and techniques.

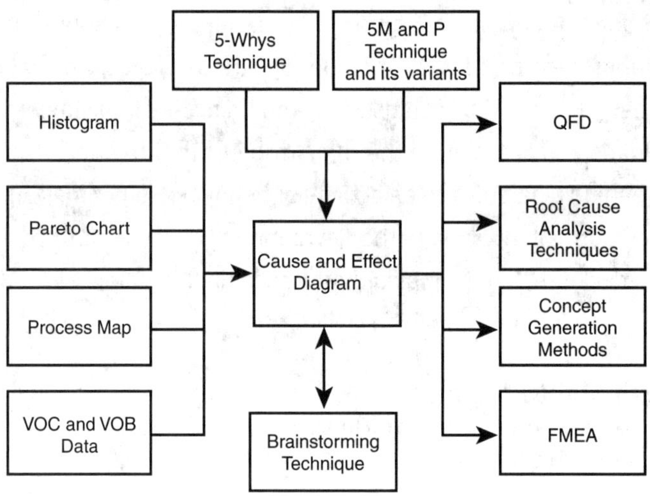

FIGURE C-6: Cause-and-Effect Tool Linkage

Variations

Cause-and-Effect Matrix

Links prioritize customer requirements with the steps of a process to understand which step(s) is more critical to meeting the customer needs. [*See the Cause-and-Effect Matrix entry in this section for more detail.*]

Cause Enumeration Diagram
This tool first starts with a brainstormed, exhaustive list of all the possible causes and then groups the items into affinities categories—thus enumerating a large number of likely causes. The very last step is drawing the Fishbone diagram or a matrix that considers the Cause-and-Effect relationship.

Benefit: The potential causes listed should be more comprehensive.

Weakness: It may be difficult to relate the large number of causes to the problematic outcome, making the diagram difficult to draw.

Cause and Prevention Diagram
This variation is used as part of risk mitigation planning. Figure C-7 illustrates the general structure of a Cause and Prevention diagram. *[See the "Cause and Prevention Diagram" entry in this section for more detail.]*

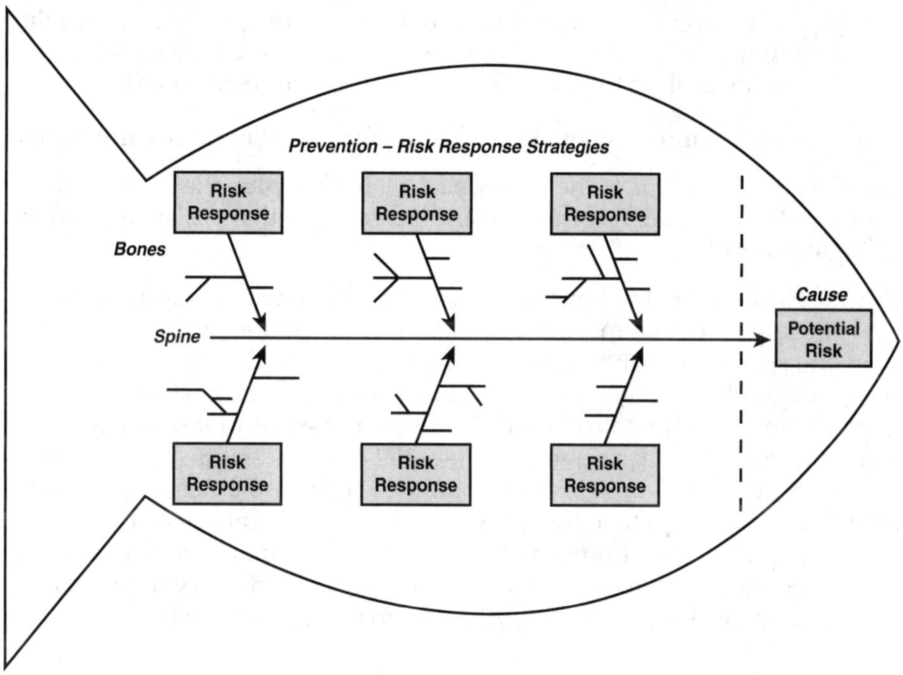

FIGURE C-7: Cause and Prevention Diagram (Fishbone variation)

CEDAC® (Cause-and-Effect Diagram and Cards)
Dr. Ryuji Fukuda (from Japan) developed this approach, which integrates problem-solving with the Fishbone diagram analysis, uses color cards, and prominently displays it to collect input from passers-by. The CEDAC approach includes the following steps:

Step 1. Problem is identified and an improvement target and metrics are selected and agreed to.

A B C D E F G H I J K L M N O P Q R S T U V W X Y Z

Step 2. Causes are brainstormed.

Step 3. Causes are sorted into categories.

Step 4. Using colored cards (or Post-It™ Notes); one color is assigned to a category.

Step 5. The causes identified within a category are written on a card with the assigned color—one idea per card. This step is repeated until all the ideas are documented on a card with category color assigned to it.

Step 6. The cards are placed on a Fishbone diagram.

Step 7. Improvement ideas for each card are written on a different color card and posted on the Fishbone diagram.

Step 8. The diagram is then placed in a high-traffic public area (similar to the time-delay Fishbone technique), wherein passers-by can write additional ideas on causes and/or improvements.

Step 9. Each improvement idea is studied for potential value contribution.

Benefit: Similar to that of the time-delay Fishbone, plus the color cards contribute an additional communication technique that assists with identifying patterns.

Weakness: Similar to the time-delay Fishbone, plus this technique adds a level of complexity that may or may not add value to the process of identifying and sorting potential causes. Moreover, the technique blends improvement ideas with the original intent of an Ishikawa diagram. Thus, the natural tendency of people to jump to solutions may be more difficult to overcome using this technique. The potential causes identified in a Cause-and-Effect diagram should be further studied (using other techniques such as Hypothesis testing) to identify the actual root causes and which have the greatest impact. If the improvement ideas are simple in nature, it will boost worker morale and require little investment (either budgetary or procedurally), and so by all means the suggestion can be implemented.

Warning

Caution should be taken that implementing simple band-aid improvements might merely treat the symptom temporarily and not address the underlying root cause, which can lull people into thinking they have "fixed" the problem because they have "done something." Keeping "busy" does not necessarily add value; avoid minutia.

Desired-results Fishbone

This technique uses the Ishikawa concept but uses a "desired result" as the focus, rather than a problem. Use the same procedure as when building a Cause-and-Effect diagram but write the desired outcome in the head of the Fishbone diagram and brainstorm ways to potentially achieve that goal.

Benefit: This technique helps to identify and organize a large number of improvement ideas.

Weakness: To avoid the natural tendency to "jump" to solutions before the root cause is understood, it is best to use this approach with new design or creation ideas, rather than solving a problem by rewording the problem in a "positive" way.

Fault Tree Analysis

This uses a top-down analytic method and Tree diagram structure with unique symbols to dissect a specific problem or problematic outcome into its related causes. (**See Also** "Fault Tree Analysis," **p. 309**)

Process Fishbone

This tool is used to analyze the potential causes of a problem within a process, typically a production or assembly, but can also be used with any services process. Start with a high-level process flow as the "main line" or "spine," with the problem identified in a box at the far right-end of the flow.

Draw the separate Fishbone (with its related potential causes) for each process step. Potential causes also can occur in the transition, or hand-off, between steps and should be drawn with a separate Fishbone.

Benefit: The tool follows the sequence of the process and thus is relatively easy to develop and easy for a "recipient" to understand.

Weakness: If similar potential causes recur throughout the process, they are repeatedly drawn. Thus, the compounding impact of a repeating cause or a combination of several related causes is difficult to illustrate.

Reverse Fishbone Diagram

This variation is used when identifying potential consequences to one or more solutions, improvement ideas, or possible actions. It also is called Backward Fishbone, Action and Effect diagram (AED), or Solution Impact diagram.

Similar to the desired-result Fishbone approach, the reverse Fishbone focuses on a proposed solution. Follow the basic Ishikawa diagram procedure but write the possible solution at the "head" of the fish. The bones identify possible effects the solution may have:

- You can initiate the brainstorming portion of the approach by answering, "What effect could this solution have?"

A B C D E F G H I J K L M N O P Q R S T U V W X Y Z

- Both positive and negative impacts of the solution should be explored.
- Categorizing techniques of the possible effects could include color-coding or marking the ideas to distinguish the positive from the negative.

If several options are being explored, develop one reverse Fishbone per possible solution.

Benefit: This approach leverages the knowledge of experienced people, particularly if a diverse cross-section of functional expertise participates.

Weakness: The results may fall victim to groupthink or paradigm paralysis, and potentially good solutions may be discounted too quickly, or a sub-optimal solution may slide through the scrutiny process too easily.

Time-delay Fishbone

This tool is used to collect input from people over time by posting a large (poster-size or bigger) Fishbone diagram in a high-traffic area (office break room or cafeteria) with pens available nearby. Passers-by are given a standing invitation to add to the diagram. Sometimes copies of the diagram posted in multiple locations are needed to capture a good cross-section of workers.

Benefit: More people can participate in building the diagram, and it extends beyond a meeting date and time, allowing people to reflect, research, and contemplate the causes over time while working. In addition, it can create some excitement (a "buzz") in the office as people contribute.

Weakness: Sometimes documenting the final Fishbone requires interpretation without testing to understanding with the author of the idea. Interpretation may be needed because symptoms (rather than causes) are posted, poor handwriting, unclear comments, or apparent redundancy. Moreover, finding an adequate central location for all to see can be difficult if the work environment includes several office travelers or "remote" workers.

Cause-and-Effect Prioritization Matrix

What Question(s) Does the Tool or Technique Answer?

Which key variables (process steps or process inputs) best meet customer requirements (or the key process output variables)?

A Cause-and-Effect Prioritization matrix helps you to

- Understand the impact or effect of key process elements cause to key outputs, as defined by customer priorities.

- Focus and prioritize activities according to what is important to customers, regardless if the resulting action improves a problem area or creates an innovation or delighter.

Alternative Names and Variations

This tool is also known as

- Cause-and-Effect matrix (or C and E matrix)
- Prioritization matrix

Variations on the tool include

- Cause-and-Effect diagram (**See Also** "Cause-and-Effect Diagram—7QC Tool," **p. 173**)
- Decision matrix, or Pugh matrix (**See Also** "Pugh Concept Evaluation," **p. 534**)
- QFD (Quality Function Deployment) or House of Quality (**See Also** "Quality Function Deployment (QFD)," **p. 543**)

When Best to Use the Tool or Technique

The Cause-and-Effect Prioritization matrix is an analysis tool to 1) evaluate the relative customer value of what is being provided today, 2) determine the biggest impact (or effect) on meeting customer requirements, and 3) determine where to potentially focus on opportunities for improvement.

This is best used when there are differing opinions about which process steps have the biggest impact on fulfilling requirements or when the number of input variables need to be narrowed.

Brief Description

The Cause-and-Effect Prioritization matrix identifies those potential root causes with the biggest impact on an effect, using customer requirements as its evaluation criteria. This analytical tool is similar to a Cause-and-Effect diagram in that it draws a relationship between the potential causes and an effect, but the C and E Matrix has an expanded purpose, which is to

1. Understand the relationship between output variables and customer requirements, based on what is important to the customer.

2. Understand the relationship between process steps (or input variables) and the key process output variables, based on what is important to the customer.

The C and E matrix examines the relationship of various process steps (or inputs) with prioritized customer requirements (or output variables). The purpose of this tool is to determine *which process steps (or inputs) have the strongest correlation to the requirements*, to determine areas of focus. This tool does not examine current performance.

This tool requires customers' input on not only their requirements, but also the relative priority ranking of each. Collecting this data is important to ensure the validity of the tool. If the data does not exist, the investment of time and money to collect this customer data is an opportunity to better understand customer needs and is a critical input that cannot be overlooked. The customer may be either external or internal.

If a known defect or problem is limited to a set of process steps (or input variables), the application of the C and E Prioritization matrix may be restricted only to those that are suspected to be causing variability (or nonconformance) to one or more of the key output variables. Hence, the C and E matrix prioritizes which of those process steps (or input variables) is likely to be causing the biggest impact.

How to Use the Tool or Technique

When developing a Cause-and-Effect Prioritization matrix, you must

Step 1. Identify key customer requirements and their relative importance (or ranking). Potential Voice of the Customer (VOC) information sources include

- Key outputs from the Process map
- VOC studies (focus groups, interviews, surveys, and so on)
- CTQ matrix or tree

Step 2. List the key customer requirements across the top of a grid as column headings (one topic per column).

Step 3. Place the customer's relative importance of each requirement in the row just beneath the column headings. These are collectively called the *Priority Score.* The scale usually ranges from 1 to 10 (with 10 being the most important or having the largest impact) and no duplicate ratings.

Step 4. Create the following extra columns in the grid:

A B C D E F G H I J K L M N O P Q R S T U V W X Y Z

- **Total Score**—as the last column in the grid, placed to the far right of the last key customer requirement column.
- **Process steps** (or inputs)—as the first column in the grid, placed to the far left of the first key customer requirement column.

Step 5. Identify all the process steps and key inputs from the Process map and list each one as an individual row heading, below the row containing the Priority Scores.

Step 6. Divide each open cell in half, with a diagonal line from the upper right corner to the lower left.

Step 7. Determine the *Correlation Rating*. Examine each step (or input) individually and evaluate how well the step (or input) influences or fulfills each customer requirement using the following rating scale:

- Blank or 0 = No correlation
- 1 = Remote correlation
- 3 = Moderate correlation
- 9 = Strong correlation

Step 8. Place the Correlation Rating score for each process step in the upper left of each cell.

Step 9. Within each cell, multiply the Correlation Rating score times the Priority Score and write that product in the lower right corner of the cell. *(Correlation Rating x Priority Score)* For example

- Step 1 and Customer Requirement 1 = (3 correlation x 10 priority) = 30
- Step 2 and Customer Requirement 2 = (9 correlation x 8 priority) = 72
- Step 3 and Customer Requirement 3 = (1 correlation x 5 priority) = 5; until...
- Step ***n*** : Customer Requirement 4 = (3 x 3) = 9

Step 10. Determine the *Total Score*. Add the multiplication products (in the lower right corner) across each row to determine the Total Score for each process step (or input). Record that number in the far right column. Continuing with the preceding example

- Step 1 Total Score = 30 + 0 + 5 + 0 = 35

- Step 2 Total Score = 0 + 72 + 0 + 9 = 81
- Step 3 Total Score = 90 + 0 + 5 + 27 = **112**; being the highest total
- Step *n* Total Score = 10 + 0 + 15 + 9 = 34

Step 11. Evaluate the Total Scores for each process step (or input) for reasonableness. The Total Scores should mirror the strength of the relative correlation that each process step (or input) has with the customer requirements. If something seems amiss, double-check to see if the following error conditions exist:

- Incorrect correlation rating score
- Missing process step or input variable
- Unrelated customer requirement (or output variables) with the process
- Unrecognizable linkage between the process and the customer requirements/deliverables

Table C-1 provides a Cause-and-Effect matrix template to display its generic structure using the example priority scores and calculations described in the preceding steps:

Table C-1: Generic Cause-and-Effect Matrix Template

Process Steps	Customer Requirement 1	Customer Requirement 2	Customer Requirement 3	Customer Requirement 4	Total Score
↓ (Priority Scores)	10	8	5	3	
Step 1	3 / 3 x 10 = 30	0 / 0	1 / 1 x 5 = 5	0 / 0	35 (30 + 5)
Step 2	0 / 0	9 / 9 x 8 = 72	0 / 0	3 / 3 x 3 = 9	81 (72 + 9)
Step 3	9 / 9 x 10 = 90	0 / 0	1 / 1 x 5 = 5	9 / 9 x 3 = 27	**112** (90+5+27)
• • •	SUMMATION →				
Step n	1 / 1 x 10 = 10	0 / 0	3 / 3 x 5 = 15	3 / 3 x 3 = 9	34 (10+15+9)

Correlation Scale: 0=None; 1=Remote; 3=Moderate; 9=Strong

How to Analyze and Apply the Tool's Output

- The highest ranked process steps (or inputs) indicate the strongest influence on meeting customer requirements.
- You might wish to indicate the top Total Scores (by circling, bolding, or changing the color of the number). Alternatively, display the findings in a Pareto chart visually to depict the relative contributions of the process steps (or inputs) to the customer requirements (or outputs).
- Determine the current performance to determine if there is room for improvement to identify the areas with the largest correlation and the poorest performance.
- Confirm the results and determine the magnitude of impact. Tools to support this next step include:
    - Stratified Dot Plots
    - Scatter Plots
    - Testing "quick fixes"
    - Multi-vari charts
    - Correlation analysis
    - Hypothesis testing (that is, Regression, Logistic Regression, ANOVA, or Chi-Square Test)
    - Design of Experiment
- The results can serve as input to developing a Control Plan or FMEA.

Examples

Cause-and-Effect Matrix Example

Table C-2 presents a sample Cause-and-Effect matrix that explores the relationship between the selling process steps for *computer consulting services* and customer requirements.

Table C-2: Cause-and-Effect Matrix Example for Computer Services Selling

Process Steps	Quality Product	Prompt Service	Professional & Knowledgeable People	Competitive Price	Total Score
	10	8	9	5	
Identify Prospect					
Understand Customer Needs	9 / 90		3 / 27	3 / 15	132
Present Value Proposition	9 / 90		3 / 27	9 / 45	162
Negotiate Deal				9 / 45	45
Deliver Offering	3 / 30	9 / 72	3 / 27	9 / 45	174
Support Offering	9 / 90	9 / 72	9 / 81	9 / 45	**288**

Poorly Constructed C and E Matrix Example

The following example examines the relationship between customer requirements for purchasing a high-end stereo versus the selling process steps at a retail store. Notice that many of the retail selling process steps have very low impact on the customer requirements. If the retail store owner wanted to better understand why the store was experiencing low sales, the owner might probe further into customer requirements that are more applicable to the selling process or examine other controllable processes at the store (such as brands inventoried versus those of discount stores). Moreover, when the requirements were gathered, the customer did not discriminate among all the needs, but simply gave them equal weight, which provides little insight. Table C-3 shows an ineffective use of a Cause-and-Effect matrix in that the answer is obvious and does not require a matrix to sort out the answer.

Table C-3: Ineffective Use of a Cause-and-Effect Matrix—Selling a High-end Stereo Example

Process Steps	Big Sound	Silver Exterior	Remote Control	Illuminated Dial	Lowest Purchase Price	Total Score
	10	10	10	10	10	
Identify Prospect						
Understand Customer Needs					1 / 10	10

Process Steps	Big Sound	Silver Exterior	Remote Control	Illuminated Dial	Lowest Purchase Price	Total Score
Present Value Proposition	1 / 10				3 / 30	**40**
Negotiate Deal					3 / 30	30
Install Stereo						
Service Stereo	1 / 10					10

Hints and Tips

The number of key customer requirements should be limited to about five. Select the appropriate customer requirements driving the set of process steps (or input variables) being evaluated. The relationship between the process steps (or inputs) and the Customer Requirements may produce a grid with about a half to two-thirds of the grid with empty cells (a blank influence rating) showing no correlation. If the grid contains a majority of cells filled in with Correlation Rating scores, then evaluate whether a relationship presumption is being forced. If the grid contains too few cells filled, examine whether the customer requirements are appropriate for the given process steps (or input variables).

The Correlation Rating simply determines the **magnitude of impact** or correlation between a process step (or input) and customer requirements (or outputs) and is indifferent to a positive or negative relationship. Hence the scale should always be positive number—**no** negative numbers. The Correlation Rating scale conventionally used is "Blank, 1, 3, 9" to create a spread that better distinguishes high correlation relationships from the remainder of the considerations. This tool evaluates the magnitude of impact between a process step (or input) and customer requirements (or outputs); it does NOT evaluate current performance.

Supporting or Linked Tools

Supporting tools that might provide input when developing a Cause-and-Effect Prioritization matrix include

- Key outputs from the process map (**See Also** "Process Map (or Flowchart)—7QC Tool," **p. 522**)
- VOC studies (focus groups, interviews, surveys, and so on) (**See Also** " Voice of Customer Gathering Techniques," **p. 737**)
- CTQ matrix or tree (**See Also** "Critical to Quality (CTQ)," **p. 242**)
- Fishbone diagram (**See Also** "Cause-and-Effect Diagram—7QC Tool," **p. 173**)

A B C D E F G H I J K L M N O P Q R S T U V W X Y Z

A completed Cause-and-Effect Prioritization matrix provides input to tools such as

- Stratified Dot Plots (**See Also** "Dotplot," **p. 280**)
- Scatter Plots (**See Also** "Scatter Diagram—7QC Tool," **p. 640**)
- Testing "quick fixes"
- Root Cause Analysis Techniques:
    - Multi-vari charts (**See Also** "Multi-vari Chart," **p. 439**)
    - Correlation and Regression analysis (**See Also** "Graphical Methods" and "Regression Analysis" **p. 323** and **571** respectively.)
- Hypothesis Testing (that is, Logistic Regression, ANOVA, or Chi-Square Test) (**See Also** Analysis of Variance (ANOVA)—7M Tool and "Hypothesis Testing" **p. 142** and **335** respectively.)
- Design of Experiment (**See Also** "Design of Experiment (DOE)," **p. 250**)
- FMEA (**See Also** "Failure Modes and Effects Analysis (FMEA)," **p. 287**)
- Control Plan (**See Also** "Matrix Diagrams—7M Tool," **p. 399**)

Figure C-8 illustrates the link between the Cause-and-Effect Prioritization matrix and its related tools and techniques.

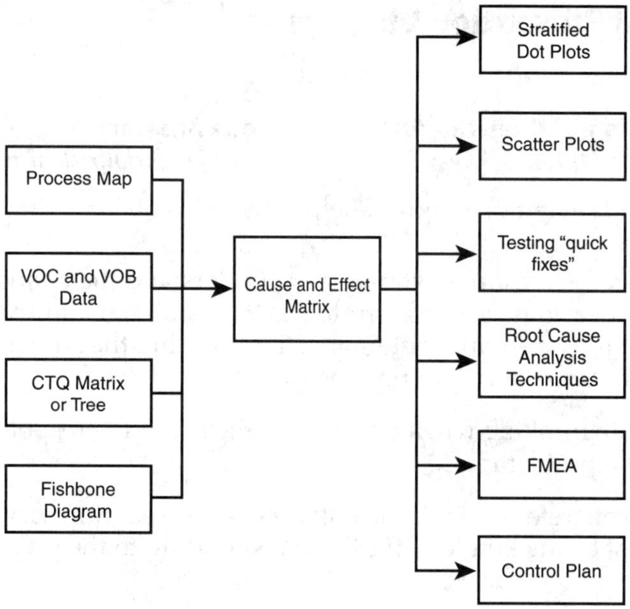

FIGURE C-8: Cause-and-Effect Tool Linkage

Variations

Cause-and-Effect Diagram

Depicts the relationship between a given problem and its potential causes. (**See Also** "Cause-and-Effect Diagram—7QC Tool," **p. 173** for more detail.)

Decision Matrix

Uses similar principles as the Cause-and-Effect matrix but has a slightly different purpose.

The Decision matrix (sometimes called the *Pugh matrix*) is a selection tool used to determine, from a list of options against a weighted set of criteria, which option to pick. (**See Also** "Matrix Diagrams—7M Tool," **p. 399** for details on different types of matrices.)

QFD (Quality Function Deployment)

QFD is a more in-depth tool than the Cause-and-Effect matrix but uses similar principles.

The QFD tool (sometimes called the House of Quality) translates customer requirements or needs (using CTQs—critical to quality) into actions and designs that build and deliver a quality offering (product and/or service). (**See Also** "Quality Function Deployment (QFD)," **p. 543**.)

Cause and Prevention Diagram

What Question(s) Does the Tool or Technique Answer?

What are the potential risk response strategies to a potential risk or failure?

A Cause and Prevention diagram helps you to

- Organize risk response strategies to understand the relationship between various response strategies to a potential failure (or risk) by formatting, arranging, organizing, and parsing the strategies into themes as part of risk mitigation planning.
- Stimulate thinking when developing the list of the responses to a potential problem.
- Guide concrete action. If the risk occurs, the diagram serves as a checklist or tracking tool for those responding to the given situation.

Alternative Names and Variations

Variations on the tool include:

- Cause-and-Effect diagram

When Best to Use the Tool or Technique

Before any action is taken, this tool helps to define and organize risk response strategies to a potential problem (or risk).

Brief Description

A preventative tool used for risk mitigation planning. It focuses people on the various response action plans, depending on the given situation, and is a variation on the Cause-and-Effect diagram (or Fishbone) tool. (**See Also** "Cause-and-Effect Diagram—7QC Tool," **p. 173**) Though the two tools have a similar structure, the Cause and Prevention diagram warrants its own entry to highlight its purpose as a *proactive risk tool*, versus the C and E matrix's post-mortem approach of examining potential root causes.

The Cause and Prevention diagram starts with the identification of a potential failure or problem and places it at the "head" of the Fishbone structure. Each major bone represents a major risk response category, stemming off the spine. The brainstorming technique is used to identify various risk response action plans. These response plans are mapped and organized with the corresponding risk response categories and depicted as the related "bones" stemming from the spine.

It can be depicted as a matrix or Fishbone diagram wherein the problem (or risk) is placed to the far right. The problem is placed in a box or diamond-shape to represent the head of a fish. The prevention strategy themes are placed to the left of the problem.

There are three major types of risk response classification:

1. **Acceptance**—Accept the consequences passively or actively; retain the risk.
2. **Avoidance**—Eliminate a specific threat, usually by eliminating the cause.
3. **Mitigation**—Reduce the expected monetary value of a risk event by reducing the probability of occurrence.
   a. Reduction—minimize its occurrence and effect
   b. Transfer—all or a portion of it to another party by using or implementing the following:
      i. Insurance for direct property damage
      ii. Indirect consequential loss (often performed by a contractor, for example, debris removal or equipment replacement)
      iii. Legal liability (design errors, public bodily injury, performance failure)
      iv. Personnel (employee bodily injury/Worker's Compensation)
      v. Employee replacement costs
      vi. Resulting business losses

How to Use the Tool or Technique

When developing a Cause and Prevention diagram, consider the following:

Step 1. Start with a prioritized list of potential risks or failures, which could come from an FMEA matrix. Agree on the risk or potential failure to be analyzed. (**See Also** "Failure Modes and Effects Analysis (FMEA)," **p. 287**)

Step 2. Write the potential risk to the far right edge of the diagram and draw a box around the text to form the "head" of the fish, as shown in Figure C-9.

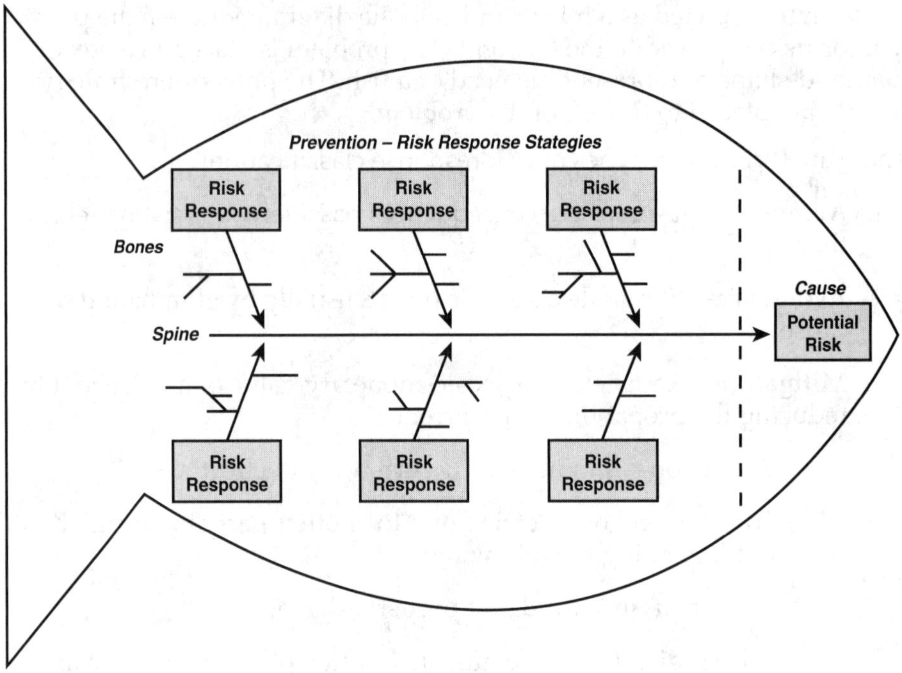

Figure C-9: Generic Cause and Prevention Diagram

Step 3. Draw a line extending from the left edge of the fish head to the left edge of the diagram to represent the spine.

Step 4. Agree on the major response categories: Acceptance, Avoidance, Reduction, Transfer, Other, and any additional major response approach theme.

Step 5. Reference policies, standard operating procedures and experience, brainstorm and write down the potential response action plans, and group them within the related major response categories.

Hints and Tips

Have the participants attending the meeting bring policies, procedures, instructions, and guidelines from their respective areas to inform the brainstorming activities. (**See Also** "Brainstorming Technique," **p. 168**)

Step 6. Draw a line extending from the spine for each theme and write a major response category name at the outermost end of line (the end that is not attached to the spine) to represent a "main" Fishbone.

Step 7. Draw one "branch" or offshoot to the main Fishbone and label the end of each branch with the specific response action plan idea to represent smaller bones and to show the linkage to the "higher level" category with which it is affiliated.

Step 8. Continue Step 7 to drill-down into as much detail as required defining more bones and linking the relationship.

Step 9. Review the diagram for completeness and clarity and then modify accordingly.

   a. Fill in gaps and streamline wording.

   b. Eliminate unnecessary redundancy.

   c. Eliminate unrelated ideas.

How to Analyze and Apply the Tool's Output

- Review the final diagram for completeness and clarity.
- Incorporate this Cause and Prevention diagram in the appropriate control documentation:
    - Risk Mitigation Plan
    - Transition Plan
    - Implementation Plan
    - Communication Plan
    - Control Plan
    - Standard Operating Procedure manual.

A B C D E F G H I J K L M N O P Q R S T U V W X Y Z

Hints and Tips

Prioritize the most likely to occur negative events and develop one Fishbone per potential major "negative" outcome. This is a good tool to develop an action plan for the high-priorities following an FMEA. (**See Also** "Failure Modes and Effects Analysis (FMEA)," **p. 287**)

It might help to categorize the different types of risk according to the cause or source of risks: Business Risks (chances for a profit/loss associated with any business endeavor) and Pure (Insurable) Risk (only a chance for loss).

Rank the risk according to the ability to manage effective responses: **Opportunity** (probability of occurrence certainty) and **Threat** (magnitude if occurred and the amount of information you have about the risk). Then look at possible cumulative effect of several risk events occurring in conjunction with each other and/or dependencies. Identify the Tools and Techniques/Processes to organize, document, and respond to risk if it occurs.

Key Questions to ask during the response plan generation step are as follows:

1. How could this risk be avoided?
2. Can this risk be reduced?
3. Can the risk be shared or transferred?
4. Should we face the risk, and if so, should scheduling and financial allowances be made?
5. Can we contain the risk?

Benefits: Can be a thorough, in-depth analysis of possible response strategies.

Weakness: Need multiple Fishbone diagrams for the multiple possibilities of risk occurrences. This requires time to develop the analysis, and the planning may never be needed. It may be difficult to prioritize and identify the possible negative outcomes.

Supporting or Linked Tools

Supporting tools that might provide input when developing a Cause and Prevention matrix include

- FMEA (Failure Modes and Effect Analysis) (**See Also** Failure Modes and Effects Analysis (FMEA), **p. 287**)

A completed Cause and Prevention matrix provides input to tools such as

- Checklist defining an action plan or standard response procedure (**See Also** "Checklists—7QC Tool," **p. 204**)
- Control Plan (**See Also** "Matrix Diagrams—7M Tool," **p. 399**)
- Risk Mitigation Plan (**See Also** "Risk Mitigation Plan," **p. 601**)
- Transition Plan (**See Also** "Matrix Diagrams—7M Tool," **p. 399**)
- Implementation Plan (**See Also** "Matrix Diagrams—7M Tool," **p. 399**)
- Communication Plan (**See Also** "Matrix Diagrams—7M Tool" **p. 399**)
- Standard operating procedure (SOP).

Figure C-10 shows the link between the Cause and Prevention diagram and its supporting tools and techniques.

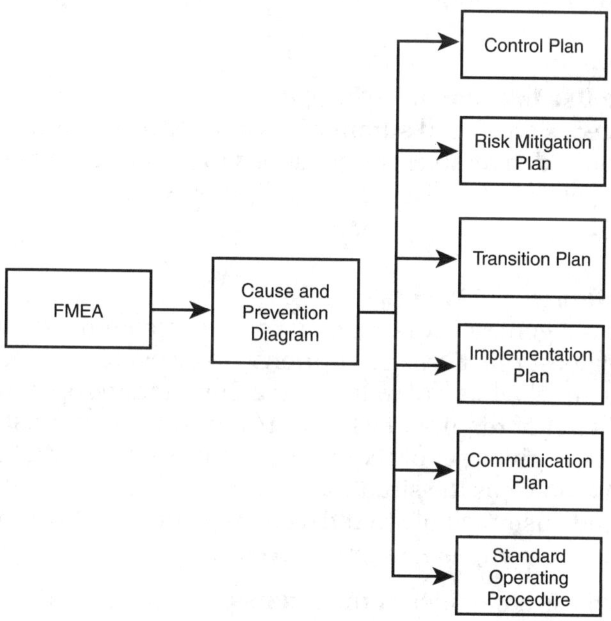

FIGURE C-10: Cause and Prevention Tool Linkage

Checklists—7QC Tool

What Question(s) Does the Tool or Technique Answer?

What activities or deliverables are required to meet requirements? Alternatively, what was observed?

Checklists help you to

- Outline (or list) the set of tasks or deliverables to be completed.
- Remind someone of the various components to fulfill requirements.
- List potential considerations to determine if appropriate or required for a given situation.
- Recap the standard operating procedure (SOP) activities or deliverables.
- Collect data about "how many" or "what type" of something occurred.

Alternative Names and Variations

This tool is also known as a *check sheet*.

When Best to Use the Tool or Technique

This tool **precedes** work and summarizes a set of pre-determined activities, action items, deliverables, or questions to be referenced before and/or during work.

Brief Description

The checklist is a generic tool that may provide recommendations or considerations of "best practices" or "options" to be evaluated for a given situation. Alternatively, the tool may outline the standard operating procedure (the set of required tasks) and/or outputs that must be completed to satisfy a given set of requirements. This generic tool can be as simple or as complex as the situation requires. It is a powerful tool in that it is flexible and customizable to reflect a range of complexity, depth, and breadth needed to complete a work procedure.

The checklist is a member of the original 7QC Tools, (or *seven Quality Control tools*), attributed to Dr. Kaoru Ishikawa. The 7QC tools sometimes are called the "seven basic tools" because they were the first set of tools identified as the *core* quality improvement tools. Ishikawa's original 7QC Toolset includes: 1) Cause-and-Effect diagram, 2) Check sheet (or checklist), 3) Control charts, 4) Histogram, 5) Pareto chart, 6) Scatter diagram, and 7) Stratification.

Structure the checklist to include the critical parameters of a task, such as the following:

- **Required procedure**—The sequence of activities or tasks to fulfill a certain standard or set of requirements. It can include who is to do what task or who is involved, as illustrated in Table C-4, and sometimes may include who the customers, decision-makers, and subject matter experts are.

Table C-4: Generic Procedure Checklist Example for Selling

Process Steps	Sales Rep	Sales Specialist	Field Engineer	Approver: District Manager
1. Identify Prospect	X			
2. Understand Customer Needs	X	X		
3. Develop Value Proposition, Price and negotiation boundaries	X	X		X
4. Present Value Proposition		X		X
5. Negotiate, if necessary	X	X		
6. Sign Contract				X
7. Deliver Offering			X	
8. Support Offering			X	

- **List of deliverables or outputs**—This can include critical features or functionality aspects. Can include *who* is accountable for producing the deliverable, as seen in Table C-5. Sometimes may include who are the (internal or external) customers and subject matter experts.

Table C-5: Generic Deliverables Checklist Example for Selling

Deliverables	Sales Rep	Sales Specialist	Field Engineer	Approver: District Manager
1. Prospect identified and entered into sales tracking database.	X			
2. Customer needs understood and documented	X			
3. Offering value proposition and price documented in presentation format.	X	consulted		approved
4. Negotiation boundaries defined.	informed	informed		X

continues

TABLE C-5: Continued

Deliverables	Sales Rep	Sales Specialist	Field Engineer	Approver: District Manager
5. Value Proposition presented.	X	support		
6. Negotiations conducted, if necessary, final agreement reached. Customers signature secured on contract		X		
7. Contract signed by manager, copy, then sent to customer.	X			approved
8. Offering delivered and installed			X	
9. Customer trained and offering supported.			X	

- **List of planning considerations (best practices or conditional what-ifs to evaluate prior to or during an activity)**—If the added complexity is useful, the potential conditions should include the corresponding next steps or alternatives, as seen in Table C-6.

TABLE C-6: Generic Planning Checklist Example for Selling

Considerations to select appropriate selling support literature:	
1. Identify industry segment	
2. Identify role of key contact	
3. Who has budgetary responsibility? What is the relationship with the key contact?	
4. What is the total department budget?	
5. Are the funds allocated for this initiative? If so, how much?	
6. What is the timeframe for the initiative?	
7. What are the key issues (pain) trying to be solved?	
8. What are the departmental near term and long term goals? How are they performing against those goals?	
9. Do other departments have similar issues?	
10. Are there any critical (internal or external) suppliers?	
11. Who are the (internal and / or external) customers? What are their requirements? How does the key contact know? How recent is the information?	
12 How predisposed is the key contact to our solutions (exploring options, not sure if will buy; interested but undecided which is the best approach; ready to buy)	

- **A collection tool**—To capture frequency of events or conditions needing to be tallied (attribute data), such as number of something (for example, defects, phone calls, and so on) or type of something (categories, names, colors), as illustrated in Table C-7.

TABLE C-7: Generic Tracking Checklist Example for Selling

Quarterly Selling Activity:	Q1	Q2	Q3	Q4
1. # incoming leads from trade shows and advertising				
2. # phone calls made				
3. # customer emails sent				
4. # exploratory appointments / meetings scheduled				
5. # Value proposition meetings conducted				
6. # deal negotiations in process				
7. # deals closed				
8. # customer contracts signed				
9. # offerings delivered and installed				
10. # of customer "relationship" building meetings conducted				

The power of the checklist comes about with active use. Its strongest attributes—simplicity, flexibility, and scalability—are also its weakness. It must be used and maintained to reflect the current thinking or *best practice* to reap its powerful benefits; otherwise, it could become obsolete or irrelevant. As with other soft tools, the checklist should be part of an evergreen process to maintain its applicability. The refresh process keeps people actively engaged using it because it will reflect the latest lessons learned and best practices.

Conjoint Analysis

What Question(s) Does the Tool or Technique Answer?

What product and/or services features and functionality do customers prefer and would be willing to buy?

A conjoint analysis helps you to

- Determine consumer preference and behavior by making trade-offs between choices.
- Understand price sensitivities to various product and/or services attributes and levels to determine the most appealing combinations.
- Characterize and verify customer requirements (both implicit and explicit).

When Best to Use the Tool or Technique

A conjoint analysis collects the Voice of the Customer (VOC) and derives preference profiles from a portfolio of options to predict buying behavior. The technique is best used during the design phase of a new product and/or services offering or when an existing offering needs to be modified or refreshed to help drive sales and market share. It can also be used during the define phase of a project to determine what is important to the customer.

Brief Description

The conjoint analysis technique was developed in the early 1970s as a market research tool to assess the potential profitability, sale revenue, and market share for a new or modified offering. It employs a structured methodology to determine consumer preference and buying behavior by studying the joint effects of multiple attributes. The term "conjoint" means two or more items joined together or combined. The research subjects are asked to select their preference given a set of alternatives. The trade-off decisions indicate price sensitivity and determine the most appealing combination of attributes for an offering.

The conjoint analysis reveals the underlying preferences that guide purchasing decisions. It determines which features to offer that will appeal to maximum number of potential purchasers. It also identifies segments that will prefer a particular set of features and measure how desirable that combination may be to target.

The conjoint model decomposes a potential product and/or services offering into a few major attribute categories, wherein each attribute is further decomposed into different levels. The levels can be non-metric and metric levels. The potential offering scenarios are created by assembling different attribute combinations (or *bundles*) using a fractional factorial design (from the DOE technique) to create a subset of the possible options. The research respondents evaluate each "test" alternative combination to determine their preferences. This selection asks the respondents to trade-off various attributes for others to decipher **relative** degrees of importance and/or price sensitivities. The conjoint analysis defines a set of *part-worths* (also referred to as *utilities*) for each attribute to understand the relative importance of each feature and the value of its different levels when making a purchasing decision.

There are three different conjoint analysis methods: The *adaptive conjoint*, which is probably the most popular approach, employs a rough ordering of importance and then uses pair-wise tradeoffs to compute the part-worth. The *hybrid model* weighs each attribute from zero to ten, and then 100 points are allocated across all attributes and respective levels. The *bridging design* is used for a large number of attributes, and it evaluates bundles on a subset of attributes common across several respondents.

A B C D E F G H I J K L M N O P Q R S T U V W X Y Z

The conjoint analysis examines the relative value of attributes considered jointly versus in isolation. This technique's power is that it more closely aligns to real-world trade-off scenarios, using a statistical approach to economically evaluate a fraction of the possible combinations, ultimately to predict consumer choice among multi-attribute offerings.

How to Use the Tool or Technique

The conjoint analysis procedure involves three steps, which build on the DOE fundamentals. As with a traditional DOE, the important and often the most time-consuming activity in the process is the planning step. The general conjoint analysis procedure is as follows:

Step 1. **Design conjoint study.**

a. **Select attributes relevant to product (or service) category.**

This refers to appearance, size, or price of an offering, for example. A good source of input is from the various data collection methods such as focus groups of targeted customers. Also consider utilizing secondary sources of data, such as desktop research, Web blogs, and periodicals.

Consider whether the target audience is segmented. If so, unique market segments may require additional or unique attributes.

b. **Select levels for each attribute.**

A level represents the values or options for each attribute, such as the different appearance options, different size options, or the different price options. The more options or levels introduced into the study, the greater the complexity for the respondents to select an option, and the greater the complexity of the test design and analysis.

Try to include the same number of levels for each attribute to simplify evaluation for respondents and to avoid misleading results on the importance of attributes. Consider picking similar ranges found in existing products or services to compare with their "delighters" and "dissatisfiers" (from a Kano Analysis). Figure C-11 illustrates potential attributes and respective levels for a new computer concept. Notice that although the design has a relatively large number of attributes, it features an equal number of levels per attribute, thereby simplifying the analysis.

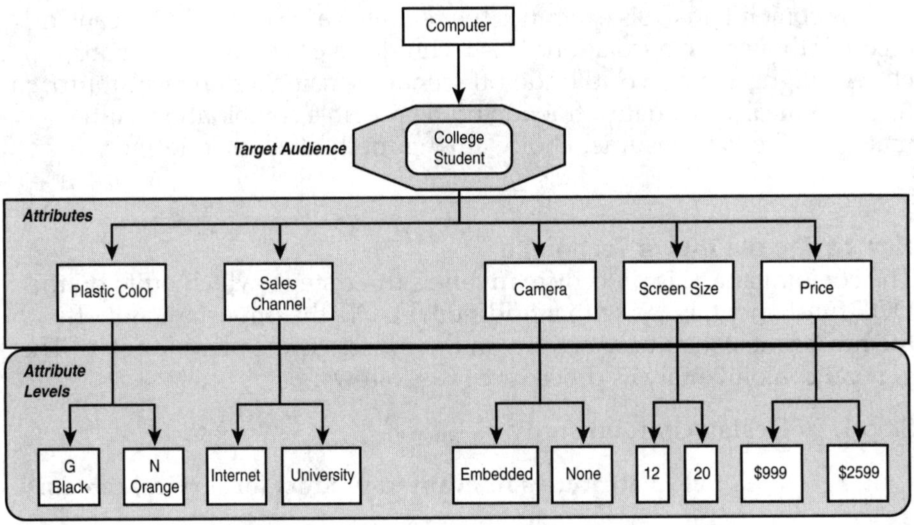

FIGURE C-11: Computer Example for Conjoint Attributes and Levels

c. **Develop relevant option bundles (or combinations) to be evaluated.**

Determine appropriate *bundles of attribute* profiles to define a potential offering to test. A bundle contains one option (or level) of each attribute selected for analysis. The combination of attribute options within a bundle defines the product or services offering.

The complete set of combinations defines the possible design options to be tested, using a fractional factorial (orthogonal array) design to reduce the number of evaluation options instead of full factorial design with all possible combinations. (**See Also** "Design of Experiment (DOE)," **p. 250**)

If every attribute at every level were to be tested in the computer example shown in Figure C-11 (a full factorial design), the test would involve 64 different possible combinations (2x2x2x2x2x2). If the design were run at half-fractional factorial DOE, the number of unique runs reduces to 32. If the Taguchi Orthogonal Array were used, only 16 runs would be required.

The experimental design is based on the amount of acceptable confounding balanced by the available time, resources, and funding. In DOE terminology, the "resolution" must reflect the business conditions.

A B C D E F G H I J K L M N O P Q R S T U V W X Y Z

To double-check the number of runs required for a design, using MINITAB, follow the procedure outlined as follows:

- **Full Factorial Design** (for 6 attributes, 2 levels each)—The drop-down menu to highlight: ***Stat>DOE>Factorial>Create Factorial Design...*** to open the main screen. Select ***2-level factorial (default generators)*** and enter 6 into the Number of Factors dialog box. Click OK. The resulting session window display 64 runs.

- **Half-Factorial Design** (for 6 attributes, 2 levels each)—The drop-down menu to highlight: ***Stat>DOE>Factorial>Create Factorial Design...*** to open the main screen. Select ***2-level factorial (default generators)*** and enter 6 into the Number of Factors dialog box. Click the Designs... button and select the ***1/2 fraction 32 VI 2**(6-1)*** design option and click OK to close the Design window. In the main screen, click OK again. Click OK. The resulting session window displays 32 runs. (Note this is a Resolution IV design.)

- **Taguchi Orthogonal Array Design** (for 6 attributes, 2 levels each)—The drop-down menu to highlight: ***Stat>DOE>Taguchi>Create Taguchi Design...*** to open the main screen. Select ***2-Level Design,*** and enter 6 into the Number of Factors dialog box. Click the Designs... button and select the ***L16 2** 6*** design option and click OK to close the Design window. In the main screen, click OK again. The resulting session window displays 16 runs.

The MINITAB Worksheet can be used as the design structure and the data collection sheet for the respondents' "preference score." The design structure defines how to combine the various attributes and respective levels to ensure randomness. For those fractional factorial experiments, the design structure also identifies which runs to select while maintaining balance.

In addition, these procedures are illustrative and lack "replicates." **Replicates should be added to the design** to make the experiment more robust and improve the experiment's ability to predict.

If the design involved three levels for each attribute, the complexity would be compounded, resulting in 729 runs for a full factorial design (3x3x3x3x3x3); however, an Orthogonal Array design would result in only 27 runs.

Step 2. **Obtain data from a sample of respondents.**

a. **Choose the format to present the bundles.**

Select the form in which the respondents will receive the bundled options. Typically, the bundles can be represented as a verbal presentation, written description, pictorial presentation, or samples (that is, proto-types).

Often a "card" is used to display the configuration of bundled attributes and appropriate levels. Each run requires a unique "card," whereby the set of cards represent the full portfolio of multiple configurations options. The respondent, then, sorts the cards in rank-order from highest preference to lowest.

b. **Design a data collection procedure.**

The respondents assign part-worth for each level of each attribute.

Rank-order approach: The product bundles could be rank-ordered (for example, one to sixteen), with one representing the highest or most preferred rank and sixteen the lowest.

Rating approach: If a ratings scale is used, respondents evaluate each product on scale of 0 to 100 points with larger sums indicating preference. The advantage of this approach is that the *Least Squares Regression* can be applied with dummy variables to compute part-worth functions. [A constant sum also can be allocated (100 points, for example) across 16 bundles.]

c. **Select computation method for obtaining part-worth function.**

The data is entered in the data collection worksheet of a statistical software application for analysis (MINITAB).

Obtain the part-worth functions by computing the preference score for an attribute and computing the average scores of the bundles that contain that attribute. For example, if four bundles contain one attribute, add up the respondent's preference scores for those four bundles and divide by four to calculate its average. This computed number represents the part-worth.

The part-worth model is one of the simpler models to define the various attributes. There are also linear models (using vectors) and quadratic models (using ideal-points).

Step 3. **Evaluate product design options.**

a. **Segment customers based on part-worth functions.**

Determine how to arrange and aggregate responses. Cluster the *part-worth* data to see if any patterns emerge. Cluster according to one of the following options:

- the individual responses
- the pooled responses from all the participants into a one *part-worth* function
- similar preferences

Let the *part-worth* data reveal the number of segments (or latent class), if any. Heterogeneous grouping assumes all customers belong to same segment but differ as specified in *part-worth* distribution.

b. **Design market simulations** to assess likely success of a proposed new product concept under various simulated market conditions.

c. **Select choice rule** to transform part-worth into choices customers are likely to make to determine the best product and/or services offering. The choice rules include

- **Maximum Utility**—Compute the market share by counting the number of customers for whom that offering provides maximum utility and divide by the number of customers in the study.
- **Share of Utility**—Select the higher the utility of a product corresponds with the higher likelihood that a customer will choose it.
- **Logit Choice**—Determine the proportion of times that product has maximum utility; the brand with the maximum utility varies randomly.
- **Alpha**—Weight the maximum utility and the share of preference wherein the chosen weight ensures that the simulated market share reflects the actual market share of existing products.

Examples

Laptop Computer Functions Example

This scenario involves a computer manufacturing's marketing department trying to increase market share and brand loyalty by wooing a new target audience— high school students preparing to go to college. The target market wants mobility, "hip" design and functionality, and plenty of storage capacity for its large music files—all in all, the best value package. The marketing department wants to leverage current technology but create a new image that convinces the consumers that the product meets their needs.

To keep the example simple, presume the high school student population is homogeneous and that a sample size of three is sufficient. The study's design involves five attributes with two levels and two replicates. (Given the two replicates, each respondent will complete the study twice; however, the card deck must be placed in a different random order between runs.)

Design Elements—The computer attributes of interest are

- **Plastic color (2 levels)**—Graphite Black; Neon Orange
- **Sales Channel (2 levels)**—Internet; University
- **Camera (2 levels)**—Embedded; None
- **Screen Size (2 levels)**—12 inches; 20 inches
- **Price**—$999; $2599

Figure C-11 illustrates the attributes and their respective numbers of levels, which is unequal across the options.

The total possible number of combinations for these five attributes at two levels is 32 (2x2x2x2x2), and with two replicates, the runs increase to 64. Given economic and time pressures, the study will use a half-fractional factorial, Resolution V design, requiring only 32 runs.

Using MINITAB to build the study's worksheet, access the main screen by using the following drop-down menus: ***Stat > DOE >Factorial >Create Factorial Design....*** Enter the study's attributes and level information in the main screen and its three "options" screens, as illustrated in Figure C-12. (**See also** "Design of Experiment (DOE)" **p. 250** for more details.)

The respondents will rate the multiple configuration options using "cards." Each card will contain a unique configuration option. The MINITAB Worksheet displays the bundle alternatives. The final Session Window and Worksheet for this example are shown in Figure C-13. Note that the Worksheet displays the attributes levels as coded integers (1, 2, 3, 4).

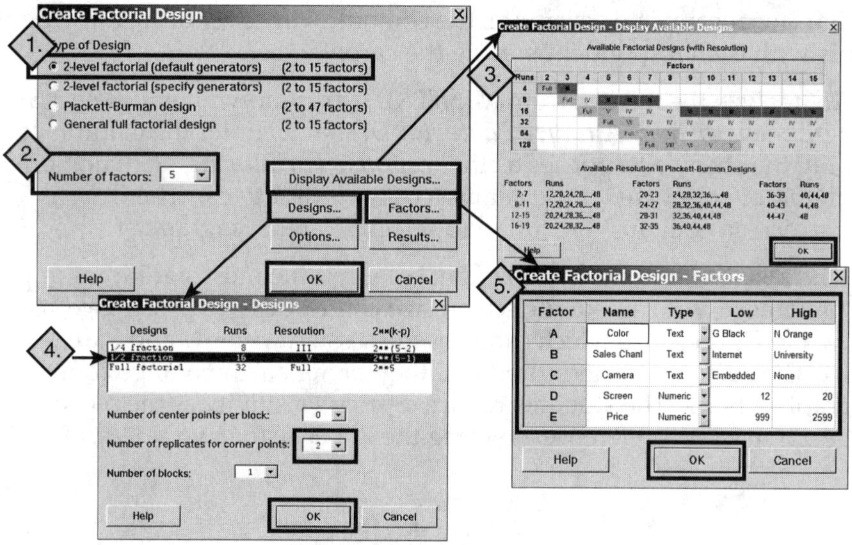

FIGURE C-12: MINITAB Set-up for Computer Example

Label the Worksheet column C10 as the "Response" column. Create the 32 response cards, following the design in the MINITAB Worksheet. (One card would feature Neon Orange, Internet sales, Embedded camera, 20 inch screen for $2599 to mirror the first line in the MINITAB Worksheet.) Run the study with multiple respondents and record the results on the MINITAB Worksheet.

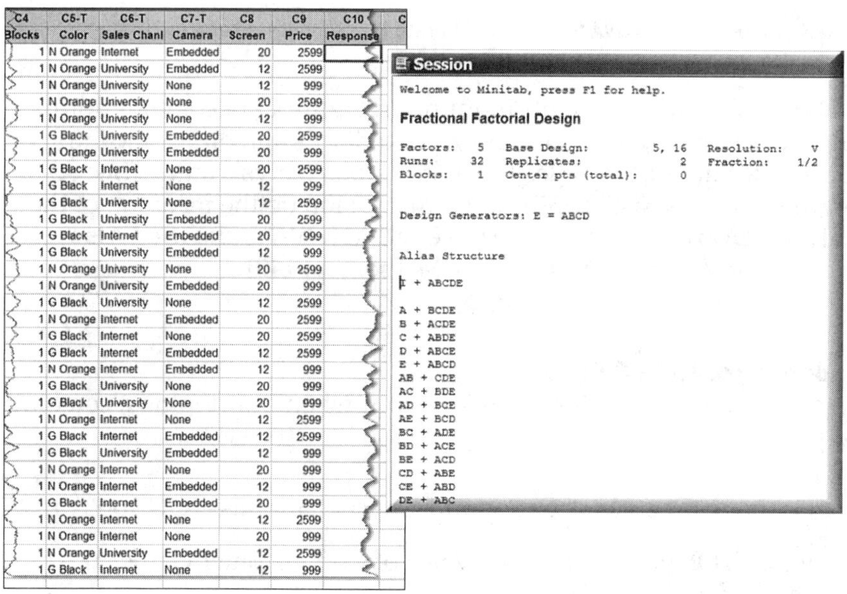

FIGURE C-13: MINITAB Worksheet and Session Window for Computer Example

For purposes of this example, the conjoint will cover only four respondents with each respondent sorting the cards twice.

To analyze the results, from MINITAB's drop-down menu, select ***Stat > DOE > Factorial > Analyze Factorial Design....*** Enter the column containing the study's responses as the ***response variable*** in the dialog box. Click OK or select any of the analysis options of interest. The Graphs option screen contains the option to select a Pareto diagram.

The results of this study show that color has the most significant impact on preference, with all but the screen size contributing to preference, as shown in Figure C-14, which captures the session window and Pareto chart. The respondents indicated that they prefer a neon orange computer to graphite black. The students reported price sensitivity as the second largest impact on preference, favoring the lower of the two prices, at $999.

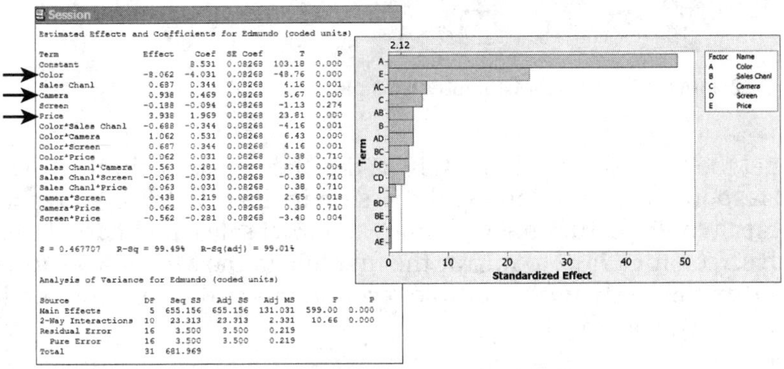

FIGURE C-14: MINITAB ANOVA Session and Pareto for Computer Example

The Session Window Output found in Figure C-14 contains additional terms not defined within this entry. For a discussion on the meaning of R-Sq and R-Sq(adj), reference the Regression Analysis entry. (**See Also** "Regression Analysis," **p. 571**) For a discussion on the meaning of the Analysis of Variance terms, reference the ANOVA entry. (**See Also** "Analysis of Variance (ANOVA)—7M Tool," **p. 142**)

Supporting or Linked Tools

Supporting tools that might provide input when developing a conjoint analysis include

- Fishbone (**See Also** "Cause-and-Effect Diagram—7QC Tool," **p. 173**)
- VOC/VOB data (**See Also** "Voice of the Customer Gathering Techniques," **p. 737**)
- CTQs (**See Also** "Critical to Quality (CTQ)," **p. 242**)

- QFD (**See Also** "Quality Function Deployment (QFD)," **p. 543**)
- Solution selection matrix (**See Also** "Solution Selection Matrix," **p. 672**)

A completed conjoint analysis provides input to tools such as

- QFD (**See Also** "Quality Function Deployment (QFD)," **p. 543**)
- FMEA (**See Also** "Failure Modes and Effects (FMEA)," **p. 287**)
- Pugh matrix (**See Also** "Pugh Concept Evaluation," **p. 534**)

Figure C-15 illustrates the link between the conjoint analysis and its related tools and techniques.

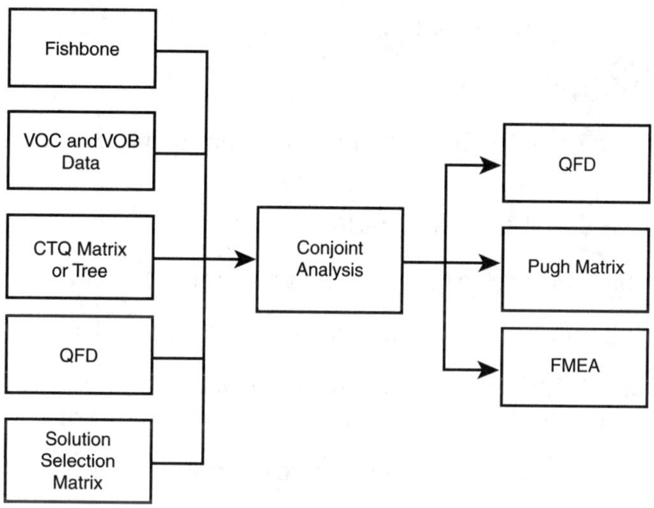

FIGURE C-15: Conjoint Analysis Tool Linkage

Variations

DOE (**See Also** "Design of Experiment (DOE)," **p. 250**)

Control Charts—7QC Tool

What Question(s) Does the Tool or Technique Answer?

How is the process behaving; is it in statistical control?

Control charts help you to

- Determine whether a process is stable (in statistical control) over time.

- Measure, monitor, and control processes.
- Determine what is a special cause event (unexpected variation) that needs to be identified and eliminated.
- Decipher common cause variation (predictable and expected) or the inherent variation in a process versus unexpected variation.
- Predict changes and indicate the need to improve process performance or determine if an *improvement* actually reduces process variation. (Key triggers include quality, cost, and capacity.)

Alternative Names and Variations

This tool is also known as

- Statistical process control (SPC) charts
- Process behavior charts

Variations on the tool that depend primarily on the type of data being plotted include

- Charts for variable data include
    - X-bar and R chart (or averages and range)
    - X-bar and S chart (or averages and standard deviation)
    - I and MR chart (or "*individuals and moving range*", or I-MR, X-chart, X-R chart, IX-MR chart, XmR, moving range)
    - Moving average-moving range chart (or MA-MR)
    - Target charts (or difference charts, deviation charts, or nominal charts)
    - CUSUM charts (or "*cumulative sum*")
    - EWMA (or "*exponentially weighted moving average*")
    - Multivariate chart (or Hotelling T^2)
- Charts for attribute data include
    - p-chart (or "proportion" or "percentage" chart)
    - np-chart (or number within a proportion (or affected units))
    - c-chart (or count chart)

    - u-chart (or unit chart/counts-per-unit)
    - P Prime SPC chart
- Charts for either type of data include
    - Short run charts (or stabilized charts/Z-charts)
    - Group charts (or multiple characteristic charts)

When Best to Use the Tool or Technique

Control charts should be used on a regular basis as part of the monitoring and management of a process to control variation. They ensure process stability over time—**a prerequisite for process capability analyses**. Control charts also aid in distinguishing between special cause variation and common cause variation as a guide for management.

Brief Description

A control chart often is referred to as a *super-charged* run chart or *time series plot* used to monitor a process over time. It is a frequency distribution plotted over time in the sequence that the data occurred or was produced and adds three reference lines for interpreting patterns in the data. These reference lines comprise a centerline (the average or mean) and control limits—an upper and lower control limit. The control limits are defined as three-sigma on either side of the collected data's mean. Why three standard deviations? It balances the likelihood of a false signal and maintains sensitivity to detect real signals. A *Statistical Process Control (SPC)* is the application of statistical methods to identify and control special cause variation in a process.

Control charts are a graphical tool used to monitor changes that occur within a process by distinguishing variation that is inherent in the process (*common cause*) from variation that indicates a change in the process (*special* or *assignable cause*). This change may be a single point or a series of points in time—each is a signal that something is different from what was previously observed and measured. Unusual variation is signaled by any point outside or specific patterns within the control limits.

Control limits are established at a distance of three standard deviations on either side of the centerline, or mean, of data plotted on a control chart. **Do not confuse control limits with specification limits.** Control limits help to identify special cause variation and confirm stability over time—the Voice of the Process (VOP) metric. Specification limits describe conformance to customer expectations—a Voice of the Customer (VOC) metric, which are utilized in process capability charts.

A B C D E F G H I J K L M N O P Q R S T U V W X Y Z

Recall that a "bad" part is determined by 1) dispersion in the process, where it is wider than the specified spread, or 2) off-target, where the process variation could be "skinnier" than the control limits but have drifted (that is, are off-center). (**See Also** "Process Capability Analysis," **p. 486**) Therefore, it is important to note that whether or not a process is meeting its control limits depends on the desired outcome itself, whether

- larger is better,
- smaller is better, or
- nominal (target) is best.

As just stated, the type of control chart primarily depends upon the type of data being plotted. *Variable data* is quantitative results or data where measurements are used for analysis. Sometimes variable data is called *continuous* data because it is measured on a continuous scale such as temperature, distance, cycle time, profit, or mass and pressure, rather than in discrete units or yes/no options. Ranks are also a type of variable data, such as the customer satisfaction scores on a scale of one to ten. *Variable data control charts come in pairs,* with the top chart plotting the average or centering of the process data, and the bottom plot displaying the range of the data distribution. Continuous data is used to create an I-MR chart (or X-MR), X-bar and R chart, and X-bar and S chart.

Attribute data, on the other hand, is qualitative, rather than quantitative in nature. It can be counted for recording and analysis and is measured strictly by either conforming or not. Therefore, a specification is imbedded in the response as to whether or not a criteria was met. Attribute data (also called "discrete" data) is a characteristic that may take on only one value (0 or 1, for example) and be counted in discrete units such as items or and yes/no options. Discrete examples include binary responses, such as "yes/no" or "pass/fail," and counted data, such as number of defects. Attribute data is used to create p-charts (percent chart), np-charts (number of affected units chart), c-charts (count chart), u-charts (counts-per-unit chart). Figure C-16 summarizes the breakdown of data types into the specific control chart produced.

In comparison, variable data is more informative than attribute data, as attribute data is more limited in describing a process. Plus, the sample size required for attribute data must be larger than an equivalent variable data measurement to be statistically significant. In the context of data type, control charts measure variation between samples for attribute data and variation both between and within sample groups over time for variable data.

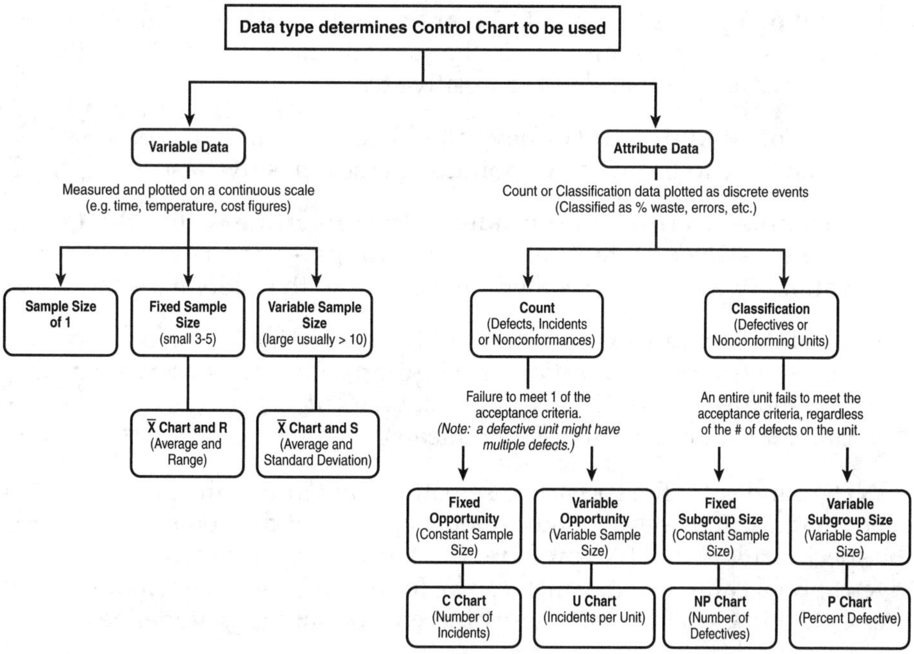

FIGURE C-16: Summary Tree of Different Control Charts

Guidelines or tests exist to identify unexpected patterns in data, wherein it is unlikely that these patterns occurred by chance alone. These special cause guidelines presume the observations are independent. Independence means the value of the given data point is not influenced by the value of another data point. If data are not independent, the data values will not be random. This means the rules for determining special cause variations cannot be applied (because they are based on rules of statistical probability). Different statistical authorities vary slightly on the exact guideline details; however, in general there are eight tests for special cause events. MINITAB's guidelines cover the following scenarios:

1. **1 point beyond control limit (3σ)**—Detects a shift in the mean, an increase in the standard deviation (σ), or a single aberration in the process. Check your R-chart to rule out increases in variation.

2. **9 points in a row on one side of the mean**—Detects a shift in the process mean.

3. **6 points in a row steadily increasing or decreasing**—Detects a trend or drift in the process mean. Small trends will be signaled by this test before the first test.

4. **14 points in a row alternating up and down**—Detects systemic effects, such as two alternately used machines, vendors, or operators.

5. **2 out of 3 points in a row @ 2 σ or beyond**—Detects a shift in the process average or increase in the standard deviation (σ). Any two out of three points provide a positive test.

6. **4 out of 5 points @ 1 σ or beyond**—Detects a shift in the process mean. Any four out of five points provide a positive test.

7. **15 points in a row on both sides of the centerline within the 1 σ "zone"**—Detects stratification of subgroups—appears when observations in a subgroup come from sources with different means.

8. **8 points in a row on both sides of the centerline all beyond 1σ zone**—Detects stratification of subgroups when the observations in one subgroup come from a single source but subgroups come from different sources with different means.

MINITAB indicates the special cause pattern by plotting the unexpected event in red and places a footnote alongside the red data point to reference the specific rule. MINITAB provides the specific rule information about a given special cause event, which can be found on its session window. Figures C-17 and C-18 illustrate these different special cause guidelines.

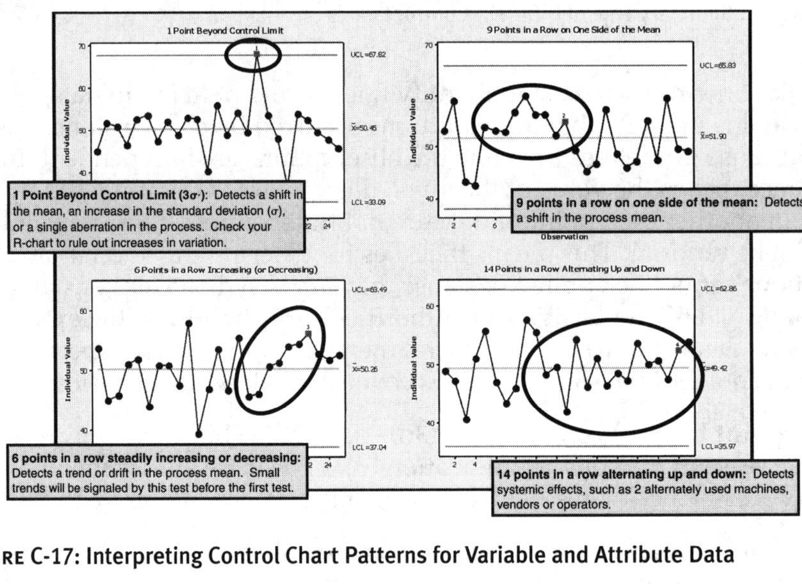

FIGURE C-17: Interpreting Control Chart Patterns for Variable and Attribute Data

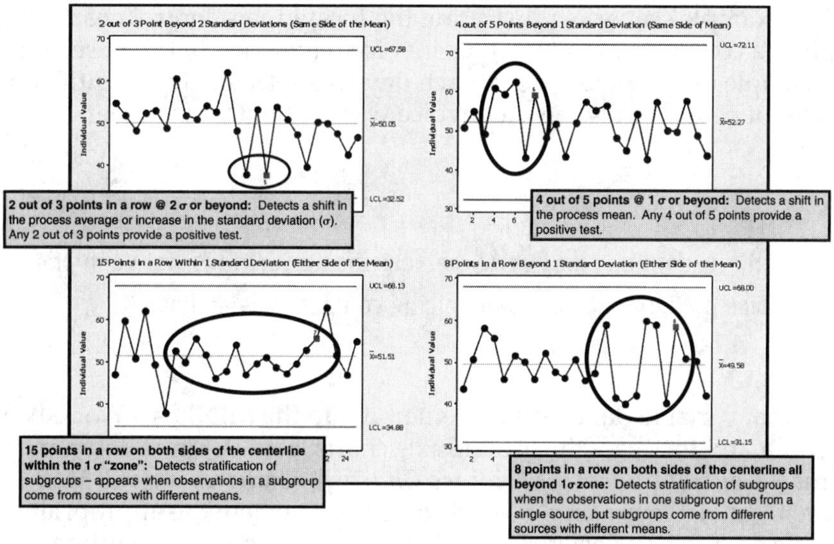

FIGURE C-18: Interpreting Control Chart Patterns for Only Variable Data

How to Use the Tool or Technique

When creating control charts, they can be drawn by hand on graph paper using special control limit factors unique to what is being plotted—or created using statistical software. For illustration purposes, MINITAB will be used in this section to plot each of the different types of control charts.

Individuals Control Chart

The Individuals control chart is one of the most commonly used tools, and variable data is used to construct it. It can go by several names other than Individuals chart—I-MR, Individuals-Moving Range, X-MR, or chart of individuals. I-MR charts are used in the following scenarios:

- Low volume manufacturing process (such as commercial airplanes)
- Batch-type processes (such as producing a vaccine)
- Infrequently reported transactional processes (such as monthly financials)

Scenarios wherein rational sub-grouping exists, I-MR charts do **not** apply.

I-MR charts assume that the observations are independent from one another. If the data are not normal, they need to be transformed using a constant such as lambda. MINITAB applies the Box-Cox as the default transformation function to the non-normal by simply selecting the utility. It is a good chart to use to gain a quick understanding of the process to help support improvement and control efforts.

The example scenario, illustrating the Individuals control chart, involves a company monitoring its invoice payments. In this case, a random sample of one paid invoice each day was selected and identified the number of days from when the invoice was sent to the customer.

> **Note**
>
> In this example, **smaller is better**; the fewer days between the invoice paid, the better the invoicing company's cash flow.

Given that variable data has been entered into the MINITAB Worksheet, the command to access the Individuals charts from its drop-down menu is ***Stat > Control Charts > Variables Charts for Individuals > I-MR....*** Figure C-19 displays sample MINITAB screens where the appropriate variable data set is selected (Area 1) and x-axis labels are identified (Area 2) to produce the final Individuals control chart in Figure C-20.

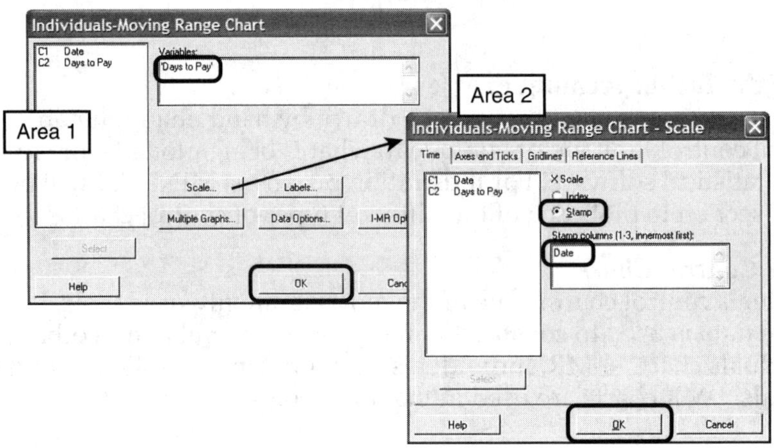

FIGURE C-19: Example MINITAB I-MR Main Screen

Notice that the top graph plots the *individual* data points, in this case the days it took for an invoice to be paid over time. The mean (denoted as x-bar) is 31.8 days, and the UCL (upper control limit) is 64.15 days, and the LCL (lower control limit) is -0.55 days. There is one data point out of control, as indicated by the point above the UCL with "1" footnote beside it. The special cause guideline that a MINITAB reference is ***"1 Point Beyond Control Limit (3σ):** Detects a shift in the mean, an increase in the standard deviation (σ), or a single aberration in the process. Check your R-chart to rule out increases in variation."* Figure C-21 provides this example's MINITAB session window.

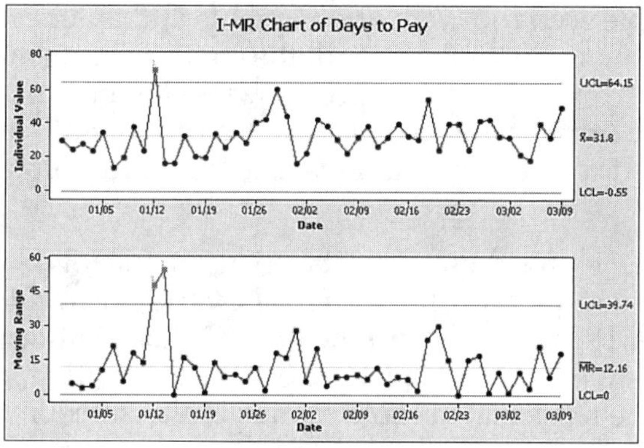

FIGURE C-20: Example MINITAB I-MR Control Chart

It is best to determine whether a process is in or out of control by checking the *moving range* chart because it is the average moving range (MR-bar) that defines the width of the control limits for both graphs. The bottom graph displays the "*moving range*" (MR), which is the difference between two consecutive individual data points, hence the MR graph has one less data point than the Individuals chart. The average moving range (denoted as MR-bar) is 12.16 days, and the UCL is at 39.74 days, and the LCL is at 0 days. There are two data points out of control (calculated by the absolute difference between the one out of control individual data point and those points collected on either side of it). They are indicated by the two points above the UCL with "1" footnote beside them—the same rule referenced in the top chart. Figure C-21 provides this example's MINITAB session window.

Session

I-MR Chart of Days to Pay

Test Results for I Chart of Days to Pay

```
TEST 1. One point more than 3.00 standard deviations from center line.
Test Failed at points:  10
```

Test Results for MR Chart of Days to Pay

```
TEST 1. One point more than 3.00 standard deviations from center line.
Test Failed at points:  10, 11
```

FIGURE C-21: Example MINITAB I-MR Session Window

Recalculated Control Limits

Control limits may be recalculated if and only if the process was intentionally modified; otherwise, the control limits should hold constant. If that is the case, the data collection sheet used to monitor the process should indicate when the special cause event occurred—the data of the changed process.

If this example scenario were unacceptable and an improvement to enhance cash flow were implemented, the control chart should indicate a "special cause" event because the process was changed. In this example, let's say that the process improvement was a prominent "reminder" notice of the 45-day policy on the subsequent invoices after the March timeframe.

Follow the same procedure just detailed, using the following MINITAB drop-down menu sequence: ***Stat > Control Charts > Variables Charts for Individuals > I-MR…***. Next, select the I-MR Options button and click the Stages tab, as shown in Area 3 of Figure C-22. The resulting Individuals control chart with the recalculated control limits is illustrated in Figure C-23.

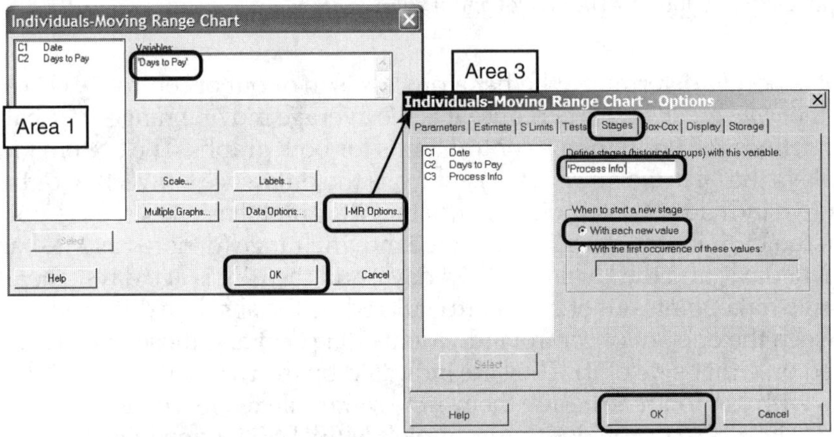

FIGURE C-22: Example MINITAB I-MR Main Screen and Stages Tab

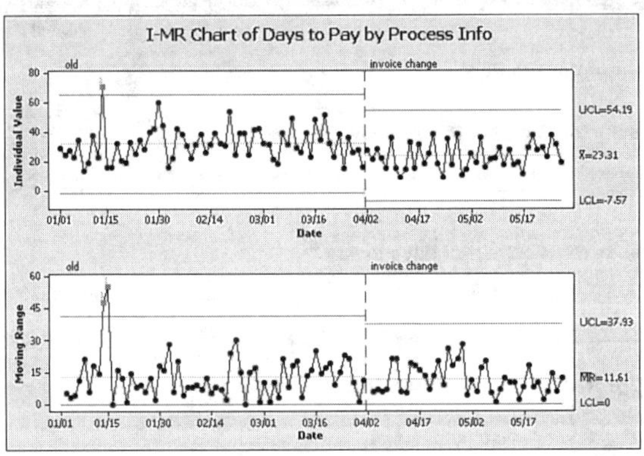

FIGURE C-23: Example MINITAB I-MR Control Chart with Recalculated Control Limits

Did the process improve, knowing that smaller is better? Notice that both the top and bottom graphs have a vertical dashed line to indicate where the process change occurred. Both of their means and the control limits became smaller. In the top graph, the mean fell to 23.31 days (from 31.8), the UCL decreased to 54.19 days (from 64.15), and the LCL at -7.57 days (from -0.55). In the new period to the right of the dashed line, no points are identified as out of control. Similarly, in the bottom graph, the average moving range (MR-bar) fell to 11.61 days (from 12.16), the UCL at 37.93 days (from 39.74), and the LCL constant at 0 days—again with no points out of control. In conclusion, the improvements successfully reduced cash flow on average by 8.49 days.

Normal Versus Non-normal Data—Normal Probability Plots

Individuals control charts assume that the data are distributed normally; otherwise, it must be transformed. There are several ways to check if the data distribution is normal:

1. Visually, by examining a frequency plot such as a histogram, Dotplot, or Boxplot. Figure C-24 illustrates a normally and non-normally distributed histogram. A definitive conclusion using only visual graphs is difficult. (**See Also** "Graphical Methods," **p. 323**)

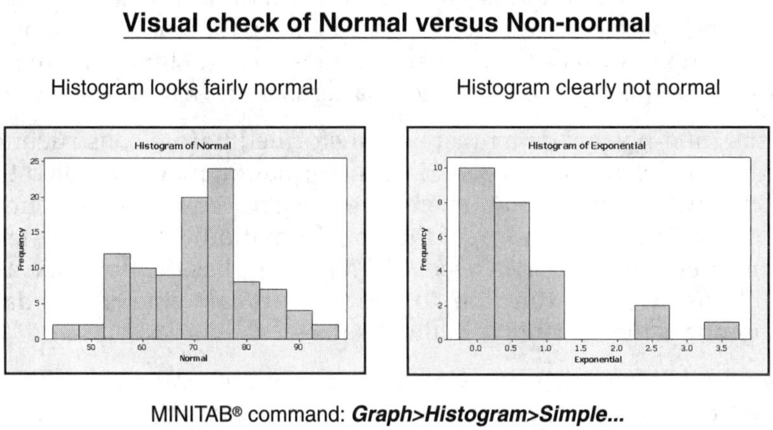

FIGURE C-24: Histograms of Normal and Non-normal Data

2. Statistically, by running a normal probability plot. If data are normal then a normal probability plot will show a linear relationship.

   Using MINITAB to create a normal probability plot, simply select the following commands from its drop-down menu: ***Graph > Probability Plot > Single…*** Select the data to be analyzed and place it in the dialog box. Click OK. The resulting graph may look similar to either half of Figure C-25—showing normal data on the left and non-normal data on the right of the figure.

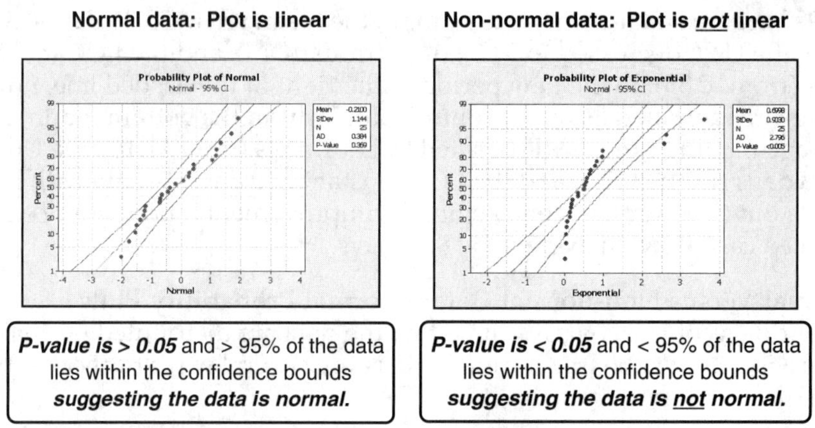

FIGURE C-25: Histograms of Normal and Non-normal Data

Transformation of Non-normal Data (Plus Normal Probability Plots)
The control limits on an Individuals charts are based on a normal distribution. If the data are not normal, then the control limits are not appropriate for assessing special cause variation. A transformation of the data usually will correct non-normality (symmetry) issues. Control limits can then be used to assess special cause variation patterns in the transformed data.

Hence, non-normal data must be transformed **before** constructing an Individuals control chart. A generic transformation method, called the *Box-Cox* (power) transformation, changes the shape of the exponential data to look normal. The form of the transformation is $Y_T = Y^\Lambda$, wherein the constant of lambda (Λ) is used. MINITAB applies the Box-Cox, as the default transformation function, to find the optimal value of lambda (Λ) that minimizes the variability in the data.

The MINITAB applies Box-Cox method from several areas within the application software. It can be accessed directly through the main menu drop-down: ***Stat > Control Charts > Box-Cox Transformation…*** Another way to access the Box-Cox method is within the control chart area of the software with the following the drop-down procedure: ***Stat > Control Charts > Variables Charts for Individuals > I-MR….*** Next, select the I-MR Options… button and click the Box-Cox tab, (the tab just to the right of the Stages tab shown in Area 3 of Figure C-22). Two sample Individuals control charts, pre- and post-Box-Cox transformation, can be found in Figure C-26. Notice how the non-normal, non-transformed chart on the right shows the data "squished" toward the top of the graph—the data is skewed left. After the transformation, the data looks more randomly and evenly distributed (without a pattern) about the mean.

Non-normal data *before*
Box-Cox transformation

Non-normal data *after*
Box-Cox transformation

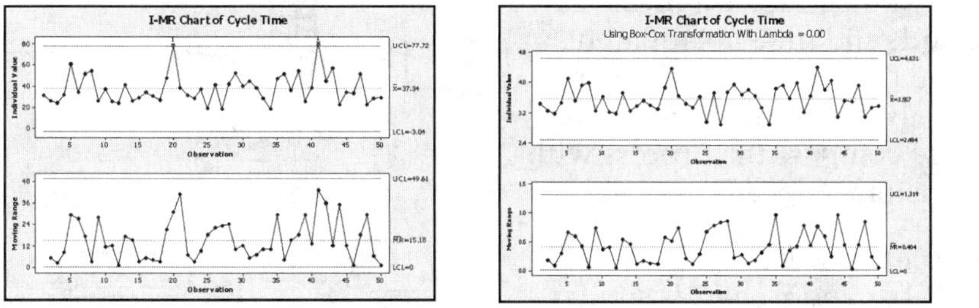

FIGURE C-26: Comparison of Non-normal Data not Transformed and Transformed Using Box-Cox

X-bar and R Chart

Variable data create the X-bar and R charts to comprised of a paired set of *averages and range* control charts. Typically these charts are used for high-volume processes, wherein sampling occurs on a regular time-based interval. The averages data is derived from a subgroup sampling scheme, such as a sample of two to five items taken at regular intervals, and the average of that subgroup represents a control chart data point. The criteria for sampling frequency focuses on ensuring the sample consists of homogeneous items (low variability) within the sample and allows for greater variability from sample to sample.

> **Note**
>
> The *Central Limit Theorem* purports that "averages" vary less than individual data points and tend to be normally distributed as the sample size increases. A distribution of the averages of sampling subgroups exhibits a tighter spread around the mean (standard deviation) than a distribution plot of the individual data points. Hence, testing for normally distributed data is unnecessary for this type of control chart.

The X-bar and R chart works for both normal and non-normally distributed data, wherein the process contains little to no changes and produces the data frequently. Manufacturing is a common process for X-bar and R charts since a sample of a few items can represent several hundred pieces.

Given that variable data has been entered into the MINITAB Worksheet, the drop-down menu sequence of commands is: ***Stat* > *Control Charts* > *Variables Charts for Subgroups* > *Xbar-R*...** Figure C-27 displays a sample MINITAB screen where the appropriate variable data set is selected (Area 1) to produce the final X-bar and R control chart in Figure C-28.

The conclusions from studying Figure C-28 might entail the fact that the process appears stable and random—no apparent special causes. Hence, a capability analysis could be conducted next to compare the process with the customer specifications.

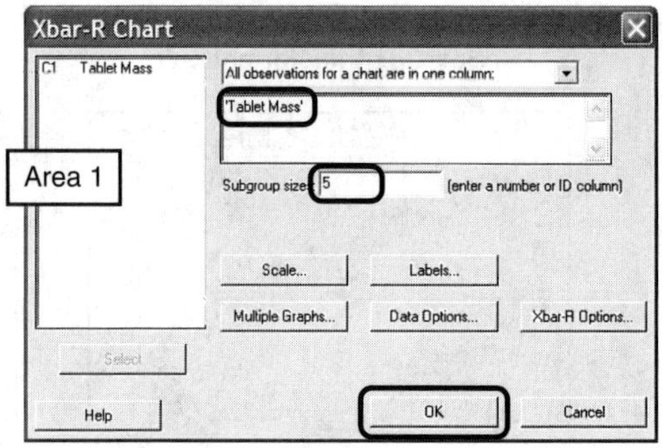

FIGURE C-27: Example MINITAB X-bar and R Main Screen

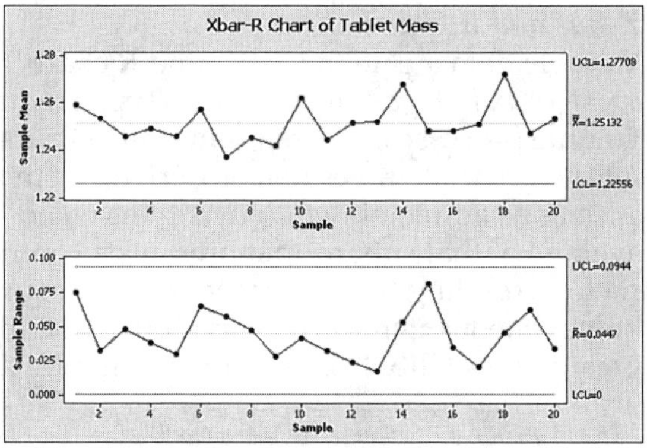

FIGURE C-28: Example MINITAB X-bar and R Control Chart

X-bar and S Chart

The X-bar and S chart is similar to the X-bar and R charts except that the sample data's standard deviation is plotted, instead of its range, to assess the sample to sample variability. These charts frequently are used when subgroup size is larger than the X-bar and R scenarios (that is, the sample size (n) is greater than five (> 5)). When the sample size is ten or more, the range is no longer an efficient statistic for measuring variability; hence, the less sensitive standard deviation calculation is used.

The X-bar and S chart work for both normally and non-normally distributed variable data. It is commonly used to quickly detect very small process changes. Healthcare scenarios often call for utilization of X-bar and S charts.

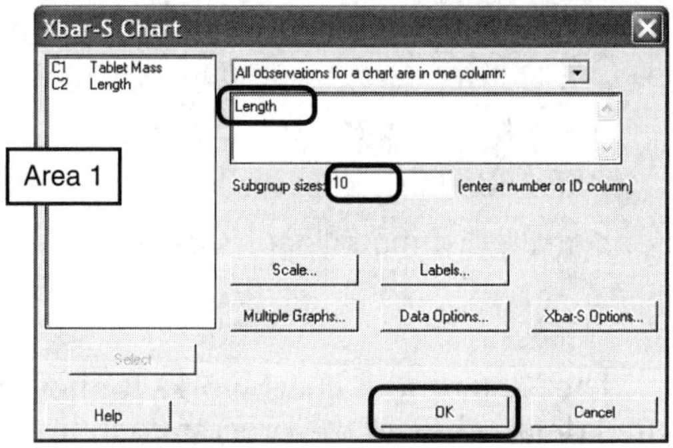

FIGURE C-29: Example MINITAB X-bar and S Chart Main Screen

Given that variable data has been entered into the MINITAB Worksheet, the drop-down menu sequence of commands is: ***Stat > Control Charts > Variables Charts for Subgroups > Xbar-S...*** Figure C-29 displaysa sample MINITAB screen where the appropriate variable data set is selected (Area 1) to produce the final X-bar and S control chart in Figure C-30.

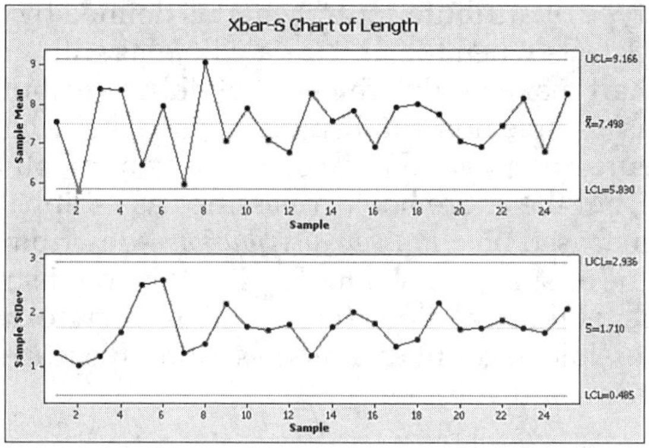

FIGURE C-30: Example MINITAB X-bar and S Control Chart

Figure C-30 displays a general, fairly random pattern. The sample sets two and eight in the top charts may require some investigation to verify the lack of any special cause event.

Attribute Control Charts

Attribute data contains less information and less granularity than variable data. Attribute data defines if a product is working or not working, but it fails to describe the magnitude of the failure. Attribute data may describe that the battery life failed, but it lacks the discriminating information as to how far from target. Moreover, attribute data represents "after-the-fact" information, causing the process players to be reactive. If an item fails to meet specifications, it needs to be scrapped or reworked. When possible, move upstream in the process to try to identify leading indicators (variable data) that enable proactive work to prevent failed products **before** they occur.

Only four of the eight guidelines to determine special cause variation are applicable for attribute control charts. Figure C-17 displays the appropriate four tests for attribute data. Attribute control charts only feature one graph and do not require normal data. Hence, MINITAB's attribute control chart screens lack the final four special cause tests and the Box-Cox transformation option.

There are two types of control charts developed from attribute data. Data depicting the number or percentage of completely "bad" (or good) units—known as *defective* or *non-conforming units*—make up the first category to construct either the p or np charts. A defective is a unit that fails to satisfy requirements due to one or more defects. Hence, the entire product is classified as "good" or "bad," and "rejected" or "un-shippable" describes a bad unit. These type of charts are based on a *binomial distribution*—a two-state scenario with constant probability of either state occurring.

The second type of attribute control chart is defined by collected data about flaws or defects when the unit (as a whole) is still "acceptable." The c-chart and u-chart make up this category of defect attribute control charts. A *defect* is a single non-conforming quality characteristic. Hence, the data may represent a folded page of a book coming off the printing production line, but the entire book is classified as "shippable." These type of charts are based on a *Poisson distribution*—describing the area of opportunity (time or space, for example). The Poisson distribution represents fairly large total populations (measured as discrete data) that are difficult or impossible to count and models random occurrences over time, distance, area, and volume.

Figure C-31 summarizes the scenario that requires which attribute chart.

	Fixed Sample Size	Varying Sample Size	Supported Distribution
Defects	c	u	Poisson
Defective	np	p	Binomial

FIGURE C-31: Summary of Attribute Control Charts

p-Chart

The p-chart uses binary attribute data to monitor defective units. It is the most sensitive attribute control chart. The data may contain pass/fail of a test or errors detected or not detected. The p-chart plots the *proportion* (p) of data for either aspect (passed or failed) of criteria; hence the chart is based on the binomial distribution. A p-chart can accommodate unequal group sizes of non-conforming or defective items. Often the data is collected as a percentage or fraction of the total lot and may contain multiple types of non-conformity. The data collection sheet needs to identify the different types of non-conformity to better determine the cause(s) of variation.

Given that variable data has been entered into the MINITAB Worksheet, the drop-down menu sequence of commands is: ***Stat > Control Charts > Attribute Charts > P...*** Figure C-32 displays sample MINITAB screen where the appropriate attribute data set is selected (Area 1) to produce the final p-chart in Figure C-33.

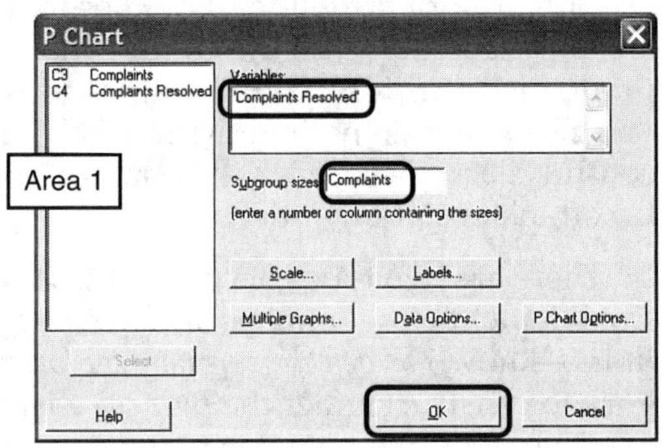

FIGURE C-32: Example MINITAB p-Chart Main Screen

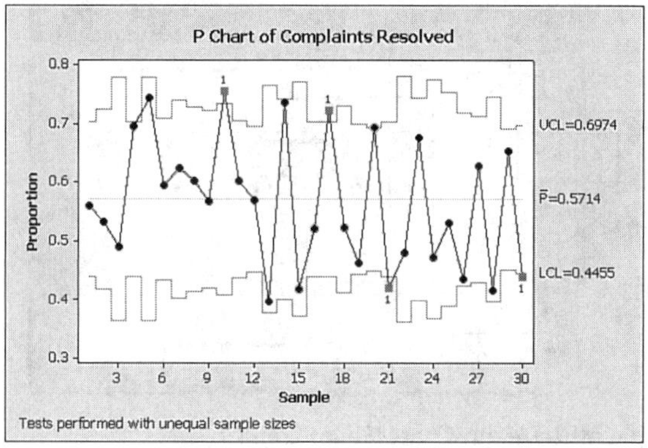

FIGURE C-33: Example MINITAB p-Chart

Figure C-33 illustrates the classic "city skyline" of the unequal sample size proportional display of a p-chart. Notice the vertical axis is labeled "proportion" because the chart uses percentage data to accommodate the unequal sample sizes. The changing limits is also a function of the changing sample size. Statistically there is more confidence with more data, thus tighter control limits. For smaller sample sizes, the confidence decreases; therefore, the control limits loosen (or get larger). Figure C-33 contains four flagged data points that require further investigation.

P Prime

The P Prime SPC chart is a variant that basically combines the Individual-Moving Range (I-MR) chart and a p-chart. This chart is used when the p-chart produces wrong control limits (that is, compresses) because it is not adequately showing the *within-sample variation*, typically resulting from a very large sample size with very few defects. This chart corrects the p-chart by using the moving range to adjust the control limits with an estimate of the within-sample variation. Dr. David Laney (of Stamford University) created this tool and has written much on the topic.

np-Chart

The np-chart uses binary attribute data to monitor defective units. The np-chart uses the same data structure as a p-chart (hence the binomial distribution applies); however, it plots the number of event occurrences rather than proportions. The equation can be written as

 # occurrences = (subgroup size) x (proportion) or np = n*p

Different from the p-chart, the np-chart's subgroup size must remain constant.

Given that variable data has been entered into the MINITAB Worksheet, the drop-down menu sequence of commands is: ***Stat > Control Charts > Attribute Charts > NP...*** Figure C-34 displays sample MINITAB screen where the appropriate attribute data set is selected (Area 1) to produce the final np control chart in Figure C-35.

A B C D E F G H I J K L M N O P Q R S T U V W X Y Z

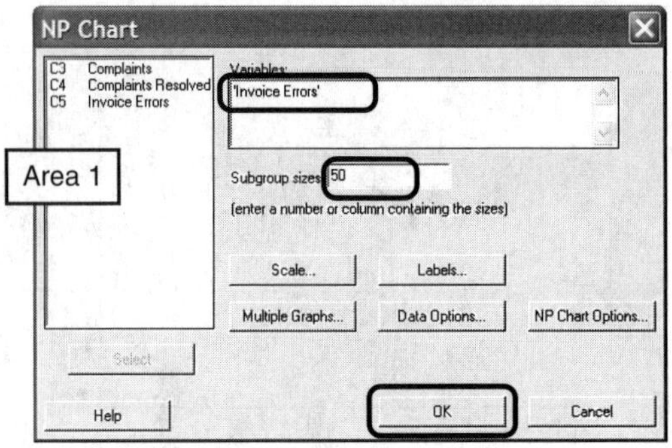

FIGURE C-34: Example MINITAB np-Chart Main Screen

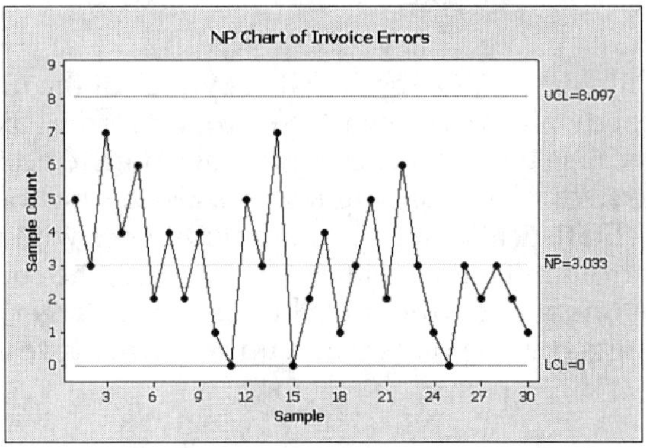

FIGURE C-35: Example MINITAB np-Chart

Figure C-35 illustrates the sample np-chart. Notice that the vertical scale on the graph is *count* versus the p-chart's proportion. The attribute data charted in Figure C-35 is Invoice Errors; therefore, smaller is better, and the graph indicates an improvement over time. A question arises as to whether this improvement is by random chance alone. It should be investigated, and if changes have been made, they should be standardized in order to maintain this downward trend.

c-Chart

The c-chart monitors the count (c) of defects in a process when an individual item may have multiple defects. The Poisson distribution is the basis for the construction of the chart, assuming an equal number of opportunities for defects must be reasonable; hence, the subgroup size must remain constant.

Given that variable data has been entered into the MINITAB Worksheet, the drop-down menu sequence of commands is: ***Stat > Control Charts > Attribute Charts > C…*** Figure C-36 displays a sample MINITAB screen where the appropriate attribute data set is selected (Area 1) to produce the final c control chart in Figure C-37.

Figure C-37 illustrates the sample c-chart. Notice that the vertical scale on the graph is counted similar to the np-chart. There is one point out of control in this example, where bad solder joints is plotted. In this case, smaller is better, so the upward trend also may cause concern worth investigating. In addition, the upward trend includes seven points above the mean count, which is approaching the rule of "9 data points above (or below) the centerline" and may indicate a shift change.

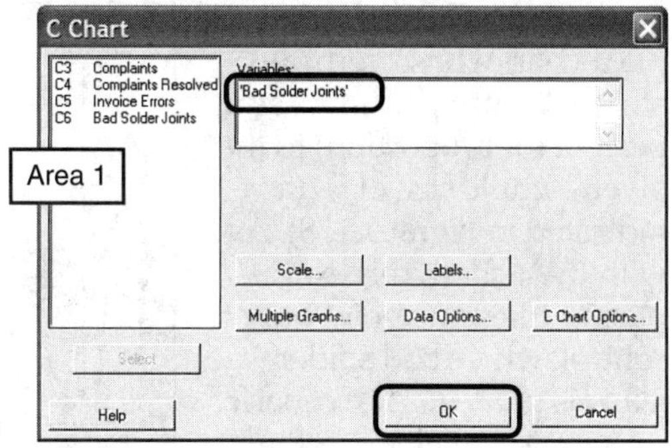

Figure C-36 Example MINITAB c-Chart Main Screen

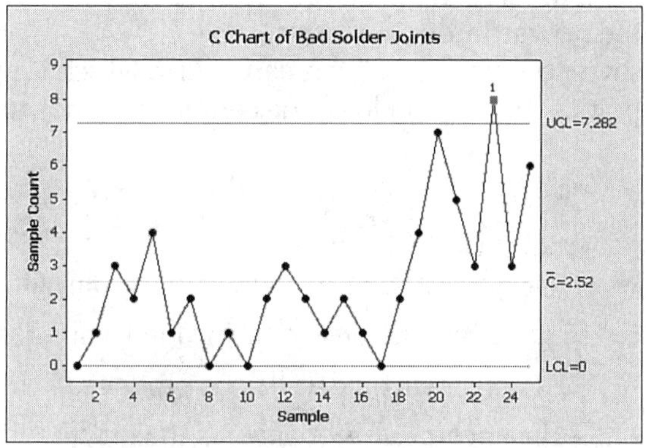

Figure C-37 Example MINITAB c-Chart

u Chart

The u-chart uses the same data structure as a c-chart except that the number of defects per unit (u) is plotted instead of the counts of defects. Again, the Poisson distribution is the basis for the construction of this chart, so it assumes an equal number of opportunities for defects must be reasonable. The u-chart, considering it monitors the number of defects per unit, can accommodate unequal subgroup sizes.

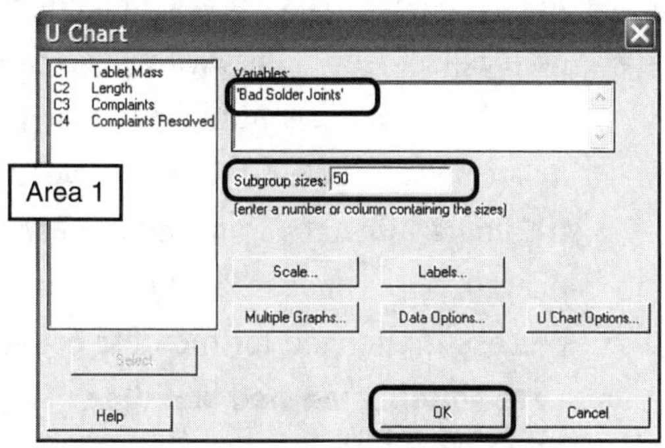

Figure C-38: Example MINITAB u-Chart Main Screen

Given that variable data has been entered into the MINITAB Worksheet, the drop-down menu sequence of commands is: ***Stat > Control Charts > Attribute Charts > U...*** Figure C-38 displays sample MINITAB screen where the appropriate attribute data set is selected (Area 1) to produce the final u control chart in Figure C-39.

Figure C-39 shows the sample u-chart whose vertical scale is "count per unit," as a percent (or proportion) to the given sample size of 50 (as indicated in Figure C-38). As with the c-chart, this example indicates one point out of control, where bad solder joints is plotted. The scenario calls for a "smaller is better," so the upward trend also may cause concern worth investigating. In addition, the upward trend includes seven points above the mean count, which is approaching the rule of "9 data points above (or below) the centerline" and may indicate a shift change.

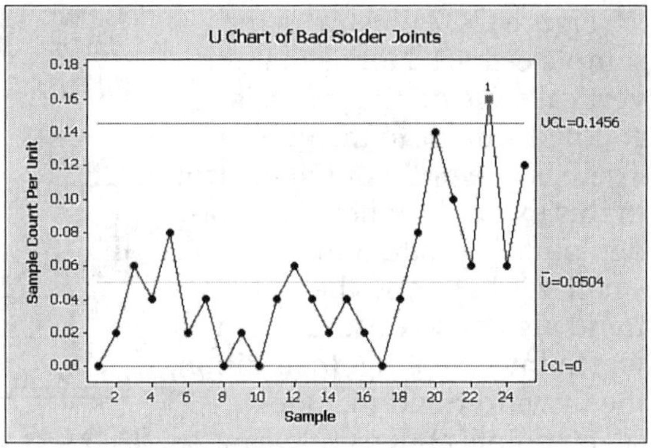

FIGURE C-39: Example MINITAB u-Chart

Hints and Tips

The process control limits are the only limits that belong on a control chart; **not** customer specification limits. Control charts should be constructed real time as data is produced in the process, not from inspection records. *Control* means the process is consistent, not necessarily that the results are meeting requirements. *Statistical control* means there are no special causes affecting a process—as indicated by randomly dispersed points within the control limits around the average line. Points within control limits that indicate a trend, shift, or instability are special causes requiring investigation. Points outside the control limits should be removed from control limit calculations (but still plotted) once cause is identified. Change control limits **only** when the process is changed for the data collected after the process change.

- **Sampling method and plan**
  - To establish inherent variation and allow process to run without sampling.
  - **Determine sample size**
    - **Attribute at least 50.** (c and u-charts, samples to average 5+ defects)
    - Variable at least three to five

- Ensure random samples
- Consistent conditions (same machine, operator, lot, and so on) when sampling
- Collect 20 to 25 different sample groups before calculating control limits.
- Patterns determine sampling frequency. If process is in "control," frequency can be reduced.

Supporting or Linked Tools

Supporting tools that might provide input when developing a control chart include

- Data collection sheets (**See Also** "Data Collection Matrix," **p. 248**)
- Graphical tools—particularly frequency plots such as a histogram, Dotplot, or Boxplot (**See Also** "Graphical Methods," **p. 323**)
- Normal probability plot (Figure C-25)

A completed control chart provides input to tools such as

- Fishbone diagram (**See Also** "Cause-and-Effect Diagram—7QC Tool," **p. 173**)
- Root cause analysis techniques (**See Also** "Hypothesis Testing" and "Regression Analysis" **p. 335** and **571**, respectively.)
- FMEA (**See Also** "Failure Modes and Effects Analysis (FMEA)," **p. 287**)
- Brainstorming technique (**See Also** "Brainstorming Technique," **p. 168**)
- Concept generation methods
- Control plans (**See Also** "Matrix Diagrams—7M Tool," **p. 399**)
- Risk mitigation plan (**See Also** "Risk Mitigation Plan," **p. 601**)
- Transition plan (**See Also** "Matrix Diagrams—7M Tool," **p. 399**)
- Implementation plan (**See Also** "Matrix Diagrams—7M Tool," **p. 399**)
- Communication plan (**See Also** "Matrix Diagrams—7M Tool," **p. 399**)
- Standard operating procedure (SOP)

Figure C-40 illustrates the link between a control chart and its related tools and techniques.

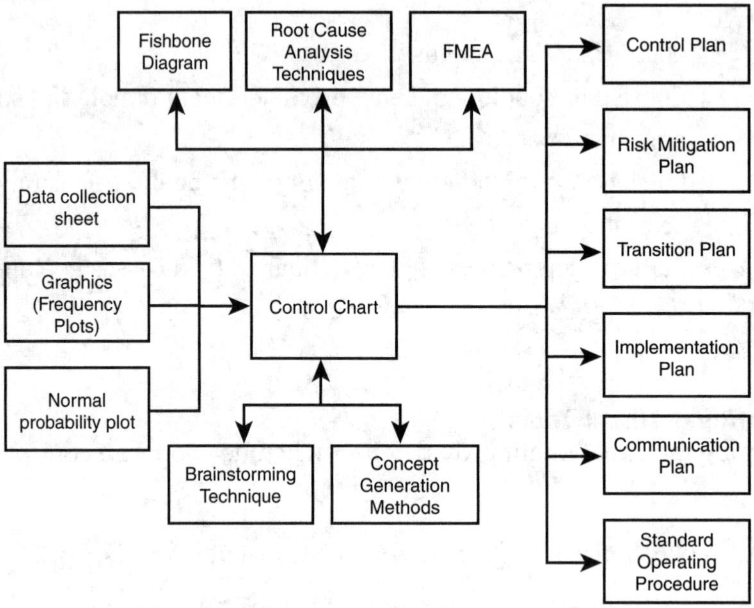

FIGURE C-40: Control Chart Tool Linkage

Cost/Benefit Analysis

What Question(s) Does the Tool or Technique Answer?

What is the payback time period for an investment?

A cost/benefit analysis helps you to

- Understand if and when net cash benefits are equal to or begin to out-weigh the net costs or outflows.
- Evaluate if the project is worthwhile and whether or not to proceed with an investment.

Alternative Names and Variations

This tool is also known as

- Benefit cost analysis

Variations on the tool include

- Internal Rate of Return (IRR)
- Return on Investment (ROI)

- Return on Assets (ROA)
- Net Present Value (NPV)

When Best to Use the Tool or Technique

Conducting a cost/benefit analysis is an important management tool to determine whether to proceed with a project and/or an improvement investment. This technique should be utilized at the onset of a project, during its planning stage, and re-evaluated throughout the project lifecycle (including the project close) to monitor the project's financial performance and healthiness and validate assumptions.

Brief Description

The cost/benefit analysis serves two key purposes: 1) determining financial healthiness of a decision and 2) helping to monitor risk of success and sustainability. The technique compares the cost or expenses associated with a process or project and contrasts it with its respective benefit to the organization. The cost/benefit analysis provides input to go/no go decisions and risk mitigation planning. If the benefits outweigh the costs, it generally leads to a "go forward" decision.

Conducting a cost/benefit analysis requires input from those familiar with the process, often supported by a finance expert if not an active member of the project team. The data may include hard and/or soft costs, which should be categorized based on accounting principles employed by the organization. *Hard costs* and savings easily are associated with money and are evident on the bottom-line of a profit and loss (P&L) statement. *Soft costs* or savings imply financial benefit and can be quantifiable in monetary terms but have a more nebulous impact on the bottom-line P&L. Soft benefits include time freed-up to do other activities and improved satisfaction and morale. However, depending on the root cause of a problem, some improvements such as improved organization or tidying-up may be a soft or hard benefit, given that those improvements are part of the Lean 5S strategy and may have direct positive safety or waste reduction impacts that are traceable to the bottom-line. (**See Also**, "5S," in Part I "Lean and Lean Six Sigma," p. **29**) Thus, categorizing the hard and soft money should be done in conjunction with the organization's financial expert(s).

How to Use the Tool or Technique

Conducting a cost/benefit analysis generally consists of the following steps:

Step 1. Gather and categorize the financial and non-financial impact of the project or potential investment as to the potential costs and savings.

a. Document assumptions about the project/investment. They may include current problem, its metrics, the period of time involved, the root causes, any boundary conditions, the proposed improvements/investments, and expected impact (both positive and negative, financial and non-financial).

b. Gather required cost factors: materials, labor, other resources and the cost of capital (interest rate or financing charge) if an investment is under consideration and identify when they will be incurred, as a one-time event or ongoing.

c. Identify the benefits as monetary, time, or other and identify when they will be realized, as a one-time event or ongoing.

Step 2. Using the cost and savings factors, develop a calendar that schedules the net project gain (or loss) and calculate appropriate financial ratio(s).

Step 3. Decide whether to proceed with or kill the project/investment or make appropriate modifications to the assumptions and recalculate the financial ratios.

The various financial ratios used in a cost/benefit analysis include the following calculations. Consult with the organization's financial expert to assist in the selection of the appropriate ratio.

Return on Assets (ROA)
ROA = Net Income divided by Total Assets, where net income equals the expected revenue or earnings.

Return on Investment (ROI)
ROI = Net Income divided by Total Investment, where net income equals the expected revenue or earnings.

Net Present Value (NPV)
NPV = Sum of the ratios of cash flow in a time period divided by one plus the cost of capital for that same time period. The NPV formula is

$$NPV = \sum_{t=0}^{n} \left(\frac{CF_t}{(1 + r)^t} \right)$$

A B C D E F G H I J K L M N O P Q R S T U V W X Y Z

Where "n" is the number of time periods, "t" is the specific time period, "r" is the cost of capital (interest rate) for the given time period, and "CF" represents the cash flow in that time period.

Internal Rate of Return (IRR)

The IRR is used to evaluate a portfolio of projects. The IRR of one project (or opportunity) is compared with other projects. The go/no go decision for a project is based on the relative IRR ratios among a portfolio of opportune projects; wherein the larger IRR is preferred.

The IRR calculation defines the cost of capital ("r") at the point when the NPV equals zero. The exact calculation is an iterative process to determine the exact IRR. The IRR formula is

$$\text{IRR} = \text{NPV} @ 0 = \sum_{t=0}^{n} \left(\frac{CF_t}{(1+r)^t} \right)$$

Where "n" is the number of time periods, "t" is the specific time period, "r" is the cost of capital (interest rate) for the given time period, and "CF" represents the cash flow in that time period.

Payback Period

This is a simple, thus commonly used cost/benefit calculation. It defines the fulcrum point at which the cash flow becomes positive, when the net cash benefits (inflows or revenues) outweigh the net costs (outflows).

The calculation for the breakeven point usually ignores the time value of money and simply determines the cumulative net cash flow over time:

Payback = Investment divided by Cash Inflow

Where the investment is both any initial one-time payment and any incremental ongoing outlays of cash. The cash inflow is any resulting savings or revenues on a one-time and ongoing basis.

Supporting or Linked Tools

Supporting tools that might provide input when developing a cost/benefit analysis include

- Data collection sheet (**See Also** "Data Collection Matrix," **p. 248**)

A completed cost/benefit analysis provides input to tools such as

- Project charter (**See Also** "SMART Problem and Goal Statements for a Project Charter" for more on project charter, **p. 665**)
- Phase-gate reviews
- FMEA and risk mitigation planning (**See Also** "Failure Modes and Effects Analysis (FMEA)," **p. 287**)
- Control plan (**See Also** "Matrix Diagrams—7M Tool," **p. 399**)

Figure C-41 illustrates the link between a cost/benefit analysis and its related tools and techniques.

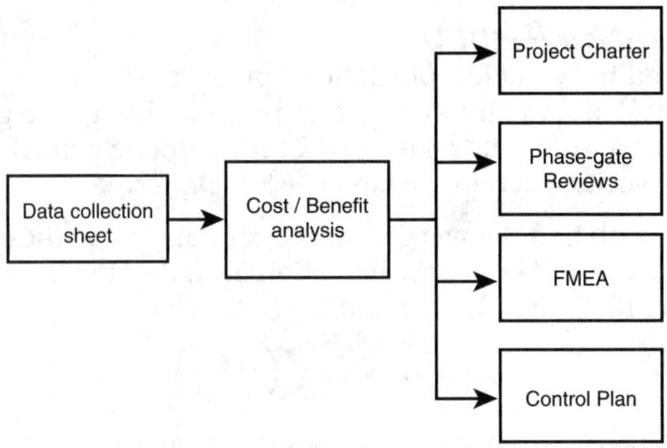

FIGURE C-41: Cost/Benefit Analysis Tool Linkage

Critical Path Method (CPM)

See Also "Activity Network Diagram (AND)—7M Tool," **p. 127**

Critical-to-Quality (CTQ)

What Question(s) Does the Tool or Technique Answer?

How does my work relate to the customer requirements, and how do I know when I have fulfilled them?

CTQ helps you to

- Understand customer general requirements in more specific terms.
- Translate customer requirements into specific, actionable, measurable language for those who work in the process (process worker, for example).

When Best to Use the Tool or Technique

The Critical-to-Quality (CTQ) elements of a customer requirement should be defined as early in the project as possible, immediately following the Voice of the Customer (VOC) gathering activities.

Once customer input has been gathered and categorized, the requirements should be translated into specific, measurable terminology, referred to as a CTQ. CTQ can be used to evaluate how well the current process works or how well the current deliverables meet customer requirements. CTQ also can be used to select improvement options, such as process improvements or new features/functionality.

Brief Description

CTQ is a simple, yet powerful tool that *translates* customer needs into a *meaningful, measurable,* and *actionable* metric for the person or persons doing the work needed to deliver the requirement.

Customers typically describe their requirements using vague words or generalities. In addition, customers' descriptions may or may not mirror one another's value system or criteria given a strong likelihood of different perspectives. In the example of prepared food, the requirement of food to "taste good" may not be perceived the same from person to person. If the company making prepared food wants to meet customer requirements, it needs to understand the characteristic of "tastes good" and its qualities. To do so, the company needs to drill down and dissect its meaning and interpret the quality of "tastes good" until it arrives at meaningful and actionable terms for the process of cooking, packaging, and delivering the food to the marketplace. To eliminate ambiguity, CTQs translate customer needs into internally meaningful, specific, and measurable terms .

Understanding and interpreting a customer's requirements is important because they define or help identify the important elements to satisfy their own needs—that which is critical to providing or delivering the quality they expect. CTQs serve as a bridge between the internal process and its deliverables and customer satisfaction. Therefore, an accurate translation is critical to understanding the customer's perspective. A fully developed CTQ has four elements: 1) output characteristics, 2) output metric (or "Y"), 3) target value, and 4) specification/tolerance limit. The inclusion of these four elements eliminates ambiguity around the interpretation of the customer needs to help the business achieve customer satisfaction.

In a Six Sigma project, essentially there may be two types of CTQs. The first and most prevalent type addresses the process and its outputs that the project aims to improve. The second type of CTQ describes what must be done to meet the requirements of the project itself. This is a one-time set of CTQs needed to accomplish the specific project deliverables and may or may not exist. Ultimately, a CTQ translates a VOC requirement into a metric that is actionable by either a process player or project team member.

How to Use the Tool or Technique

Developing a Critical-to-Quality metric is fairly simple. The key is to arrive at a measure that is meaningful and actionable. Follow the subsequent procedural guidelines:

Step 1. Identify the customers and their requirements.

Gather the customer needs as verbatim input (using their language). If a large number of requirements exist, often it helps to organize the VOC input into an Affinity diagram. (**See Also** "Affinity Diagram—7M Tool," p. **136**)

Step 2. Translate the customer need into a CTQ.

- Addressing one customer need or category at a time, begin to parse it down into more specific meaningful descriptions, translating what is critical to the customer ultimately into an actionable metric for either a process player or project team member.
- Continue to break down and translate the VOC need until a **meaningful** and **actionable** measure exists for each of the process players involved in delivering that requirement. Stop when the last level of detail is measurable.
- Repeat this step until a CTQ is aligned with each VOC requirement or category.
    - A given CTQ can be aligned with multiple VOC requirements.
    - A given VOC need can elicit multiple CTQs to address multiple dimensions or multiple process player roles.

Step 3. Organize and document the drill-down work into a matrix or a Tree diagram, linking the CTQ to the original VOC requirement.

Step 4. Validate the CTQs with the customer to ensure proper understanding and translation of the original requirement.

Examples

CTQ Tree

Figure C-42 illustrates a CTQ tree for a restaurant patrons' food requirements. Notice that one customer requirement, "Good Food," generated two CTQs: "order taken correctly" and "cooked correctly." Each of those CTQs link to a different process player. The order taking CTQ links to the wait staff, whereas the cooking properly CTQ links to the chef (or kitchen staff).

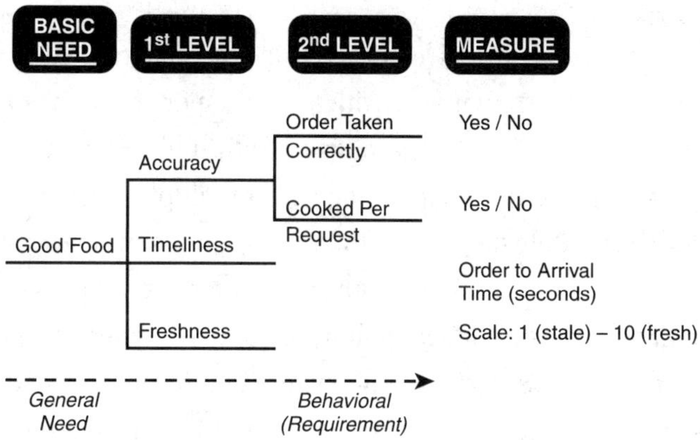

FIGURE C-42: Sample CTQ Tree

CTQ Matrix

Figure C-43 illustrates another example of a restaurant patron's food requirements displayed as a CTQ matrix.

VOC	Issue Translation	Key Item	CTQ	Measure
I like my food hot.	Cooking	Temperature	Temperature when fully cooked	Fahrenheit
	Delivery	Quick Service	Transfer Time from kitchen to table	seconds

FIGURE C-43: Sample CTQ Matrix

Hints and Tips

Start with customer needs. Use nouns without adjectives; adjectives can introduce ambiguity, allowing for multiple interpretations or perspectives (that is, how hot is hot; what is good; how blue is blue). Using a CTQ tree approach, the branches to the right show increasing detail of that requirement, **not** a new requirement. The number of levels (in a matrix) or branches (in a tree) needed to define a CTQ depends on how specific the requirement is to the required work. For example, if the requirement is an internal customer requirement from an area manager, more than likely, the resulting CTQ would require

continues

A B C D E F G H I J K L M N O P Q R S T U V W X Y Z

A B C D E F G H I J K L M N O P Q R S T U V W X Y Z

less granular parsing to arrive at something meaningful, measurable, and actionable, than compared with a requirement from the external customer. A given CTQ can be aligned with multiple VOC requirements. A given VOC requirement may elicit multiple CTQs to achieve and address multiple dimensions or multiple process player roles, as illustrated in Figure C-42. CTQs centered around the process or product and/or services offering ultimately should be aligned to a process player's role and reflected in commensurate performance documents such as personnel job appraisal and standard operating procedure (SOP).

Process-specific CTQs: A CTQ developed for a process or product becomes part of standard work wherein the generic project requirements translate into something meaningful, measurable, and actionable for the process player and can become part of his/her job performance metrics. Process CTQs are relevant until customer requirements or the process is changed. An improvement project may aim to define, clarify, or refine the process CTQs as part of its improvement deliverables and incorporate them as part of the transition and control plan.

Project-specific CTQs: A CTQ can be developed for a project (for example, an improvement project), wherein it is achieved by completion of the project, thereby satisfying a customer requirement. Hence, the generic project requirements translate into something meaningful, measurable, and actionable for the project team member. Upon project completion, that CTQ is no longer relevant.

While on a project, if both process and project-specific CTQs exist, keep them separated to avoid confusion.

Supporting or Linked Tools

Supporting tools that might provide input when developing a CTQ include

- VOC and VOB data (**See Also** "Voice of the Customer Gathering Techniques," **p. 737**)
- Current Process map (**See Also** "Process Map (or Flowchart)—7QC Tool," **p. 522**)

A completed CTQ matrix or tree provides input to tools such as

- QFD (**See Also** "Quality Function Deployment (QFD)," **p. 543**)
- Root cause analysis techniques
- Concept generation methods
- FMEA (**See Also** "Failure Modes and Effects Analysis (FMEA)," **p. 287**)
- Standard operating procedure (SOP)

Figure C-44 illustrates the link between a CTQ and its related tools and techniques.

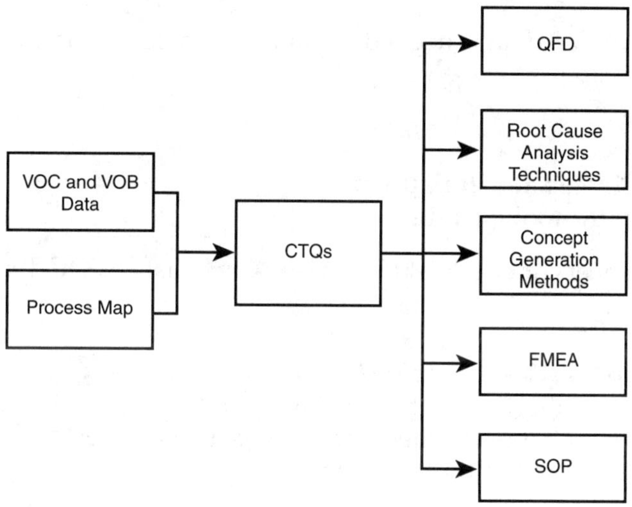

FIGURE C-44: CTQ Tool Linkage

D

Data Collection Matrix

What Question(s) Does the Tool or Technique Answer?

What information is available about the current process, product, or services offering?

A data collection matrix helps you to

- Plan and organize current data sources and collection plan.
- Identify what data is needed, what is available, where to get it, and who is responsible to get it.

Alternative Names and Variations

Variations on the tool include

- Check sheets or checklist (**See Also** "Checklists—7QC Tool," **p. 204**)

When Best to Use the Tool or Technique

Develop a data collection matrix during both the project planning phase to decipher what data is needed and what is available, as well as the data collection phase to monitor the activity.

Brief Description

A data collection matrix is often overlooked because it is a simple tool, and its value goes unnoticed. As is true with all matrix tools, the data collection tool power unveils itself when used in concert with other tools. Use the matrix to plan and organize the data collection process for the project. Near completion of the project, the project team may need to transition the responsibilities to the process players for ongoing operations. If so, then the data collection matrix describes the critical variable to monitor. Hence, the document serves as part of the project's control plan deliverable.

A data collection matrix should be **customized** to the unique needs of the organization and the type of data needed to understand its critical parameters. However, a fundamental structure provides a good starting point from which to modify. The tool provides a roadmap as to what data is required and organizes that which is collected to ensure nothing is missing or there's no duplicity. At a minimum, the matrix should identify the following items:

- The metric of interest (the CTQ)
- The official source of the data to ensure integrity and avoid duplications or voids in the data sources
- The supplier of the data, who measures it, and how
- Specifications or targets and/or operational definition

The data collection matrix is the "master plan" for a process' data collection (and/or sampling plan). It identifies the various sources, including supporting tools that contain the actual data of interest. Another related but separate tool is the actual data collection sheet. This document is used to gather and record the data collected and often is referred to as a "checklist." Similar to the data collection matrix, the checklist should be tailored to the work scenario and becomes the basic elements of a "frequency" checklist. (**See Also** "Checklists—7QC Tool," p. **204**)

Examples

Data Collection Matrix

Figure D-1 shows a sample data collection matrix template.

Data Collection Matrix

Process Name:

Process Owner: ______	Prepared by: ______	Page: ___ of ___
Contact Info: ______ Int/Ext ___	Approved by: ______	Document No: ______
Location: ______	Approved by: ______	Revision Date: ______
Area: ______	Approved by: ______	Supersedes: ______

Process Step	CTQ		Metric	Data Type	Operational Definition	Specification/ Requirement		Measurement Method (how measured and collected)	Sample Size	Frequency	Who Measures (supplier)	Data Source and Location (System, location, owner)	Reporting Method and Customer	SOP Reference
	KPIV	KPOV				USL	LSL							

FIGURE D-1: Sample Data Collection Matrix

Supporting or Linked Tools

Supporting tools that might provide input when developing a data collection matrix include

- Standard operating procedure (SOP)
- Process maps (**See Also** "Process Map (or Flowchart)—7QC Tool," **p. 522**)
- CTQs (**See Also** "Critical to Quality (CTQ)," **p. 242**)

A completed data collection matrix provides input to tools such as

- Data collection checklists (**See Also** "Checklists—7QC Tool," **p. 204**)
- Graphical tools (**See Also** "Graphical Methods," **p. 323**)
- Process maps (**See Also** "Process Maps," **p. 522**)
- Control plan (**See Also** "Matrix Diagrams—7M Tool," **p. 399**)

Figure D-2 illustrates the link between a data collection matrix and its related tools and techniques.

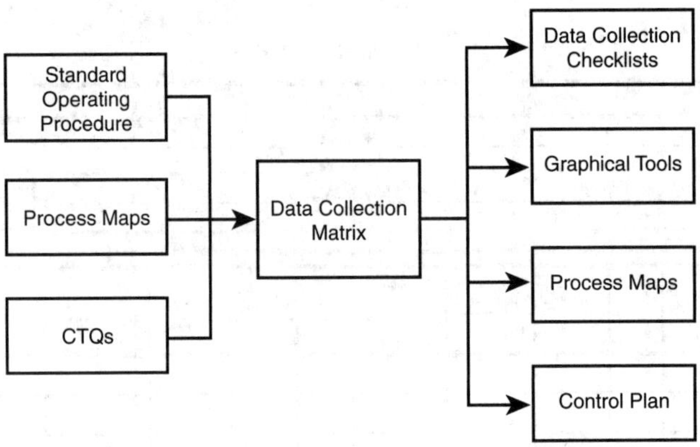

FIGURE D-2: Data Collection Matrix Tool Linkage

Design of Experiment (DOE)

What Question(s) Does the Tool or Technique Answer?

Which variables or combination of variables proves the best in producing the desired results, based on experimental data?

DOE helps you to:

- Identify which factors are most important in producing the desired outcome.
- Determine the optimal "setting" for the critical factors.

Alternative Names and Variations

This tool is also known as:

- D of E
- Designed experiments

Variations on the tool include

- Conjoint Analysis (**See Also** "Conjoint Analysis," **p. 207**)

When Best to Use the Tool or Technique

The scientific investigative nature of a DOE provides actual information about a causal relationship in a more efficient and effective manner than *one factor at a time*, also known as OFAT [pronounced "*O-fat*"]. A DOE is useful when selecting the key factors affecting the desired outcome when trying to choose between alternatives and when trying to model a system to identify its optimal settings.

Within a Six Sigma project, the DOE technique helps to answer several questions and thus has several applications. The DOE approach helps to identify the most important variables needed to produce a desired outcome. Typically, a DOE can be used during the design and optimize phases of a DFSS project, and the analyze and improve phases of a DMAIC project, depending on the key business question being asked.

Brief Description

Design of Experiments (DOE) is a structured and efficient approach to investigate Cause-and-Effect relationships between input variables (referred to as "X"s) and outputs (referred to as "Y"s) within a process. George Box claimed that "to determine what happens with a process when you interfere with it, you have to interfere with it, not passively observe it." In other words, to understand how best to produce the desired results, "adjust" the critical variables in a system to understand the consequences or impact of the action.

Through careful planning, a well-constructed DOE tests several factors (Xs) simultaneously. The resulting analysis explains the impact on the desired outcome of not only each individual factor (called a *main effect*), but also the **interaction** of the multiple factors (the combined or confounded effect). A test variable that proves not to be highly significant does not mean that it is insignificant, so examining the interaction between variables sometimes reveals important insights.

DOE as an investigative technique is more efficient and effective as compared with the "trial-and-error" and the one factor at a time (OFAT) approaches. The DOE technique's unique features include the following:

- It is a **balanced** experiment allowing multiple factors to be studied simultaneously.
- **Interactions** are comprehended (detected and measured).
- Outputs are controlled, minimizing or eliminating the effect of **noise** factors through **blocking** techniques and quantifying experimental **error**.
- Fewer experiments are required, using sound scientific and mathematical methodology.

The DOE has two primary advantages:

- **Real-time data** from the DOE technique actually demonstrates a **Cause-and-Effect** relationship—adjust a critical variable (X) to observe the outcome (Y). In the regression model, the technique relies on **predicted** data (that is, educated guesses) from either historical or happenstance scenarios where the relationship is theorized.
- The **structured and active experimentation** technique from a DOE provides more information about the causal relationship versus a model-building technique using less-structured, happenstance data such as OFAT or trial-and-error.

Alternative Experimental Techniques

In a controlled lab environment of a middle and high school science laboratory classroom, with a known (or predetermined) outcome, the teacher wants to demonstrate the response. This is the scenario most people relate to when thinking about experimentation. However, recall the conditions. The discovery process is controlled for the students; the critical variables and the outcome are known (by the teacher and the textbook author); and the time allotted for the procedure is reduced to fit the class period. The common techniques used individually or in combination by classroom teachers include: *One Factor At A Time* (OFAT), *Stick With A Winner*, *Implement Many Solutions* (process), or test at extremes (design). However, in most real-world situations in business, the critical factors may not be known (or isolated), and the Cause-and-Effect relationship might be only a hypothesis (not proven). To fully understand the power of the DOE technique, let us contrast it with these three different alternatives.

One Factor at a Time
The OFAT technique depends on pure luck to find the optimal combination and settings of multiple factors. It is a classic experimental approach (often taught in secondary school and college science labs), wherein only one factor is the focus, and all else should be held constant. However, in a complicated process, it is extremely difficult, if not impossible, to hold constant. This technique's shortcomings also include that it requires a significantly large number of experiments to study the effects of all the input factors. Plus, the technique may never reveal the optimum combination of the multiple variables. Moreover, the OFAT cannot distinguish any interaction between factors, uncovering if the behavior and optimal setting of one factor may depend on another. Thus often OFAT conclusions are wrong, misleading, or statistically inconclusive. It produces little usable data, after relatively large commitments of time and effort, and the observed effects often are unexplainable.

For illustration purposes, the OFAT experimental scenario involves identifying if two components aid in accelerating the dissolving time of an instant drink mix—temperature and type of liquid. The experimental design examines the temperature at two levels (room temperature and boiling), and two types of liquid (tap water and carbonated water). Figure D-3 displays this scenario for the instant drink mix design and its two adjustable factors; wherein the "-" symbol indicates the *base configuration* (room temperature and tap water), and the "+" symbol indicates the *adjusted factor* (boiling temperature and carbonated water).

The experiment design adjusts one factor at a time, holding all other factors at the base level. The experiment starts by adjusting the liquid type only and holding the temperature steady at room temperature. Thus, the first test is run with faucet water as the base configuration, and the second test is run with carbonated water. The results showed that the instant drink mix dissolved in 45 seconds in the tap water (Run 1) and 41 seconds in the carbonated water (Run 2). Because the carbonated water dissolution time is faster, the third run maintains the carbonated water and adjusts the second variable using the boiling temperature. The results of Run 3 for carbonated water at a boiling temperature is 47 seconds, slower than Run 2 for carbonated water at room temperature. It is believed that Run 2 produced the fastest dissolving time.

Given time and resource pressures, the experiment stops at this point and concludes that the combination of carbonated room temperature water (Run 2) is the optimal variables setting for the fastest dissolving time. The final combination of tap water at boiling temperature is never tested. If Run 4 were run, the dissolution time would have been 37 seconds, as shown in Figure D-3. The wrong conclusion is drawn. The conclusion depended on which test combination ran first and at what point the experiment ceased. If a scenario involved more than two factors that needed testing, the likelihood of coming to the incorrect conclusion using this technique increases.

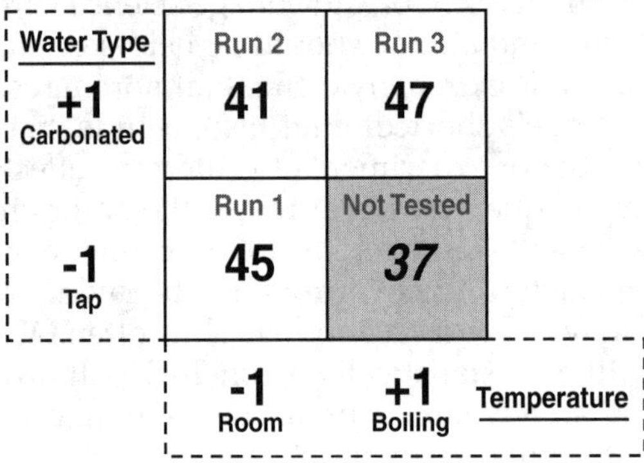

FIGURE D-3: Results of the OFAT Instant Drink Mix Example

Stick with a Winner

For illustration purposes, the *Stick with a Winner* experimental scenario involves building a paper helicopter to determine which design descends the slowest when dropped from a six-foot high stairwell landing. The design contains multiple components that can be adjusted to obtain the slowest speed. For this scenario, six factors will be tested: the type of paper it is made of, the weight of an added paper clip, the wing length, the wing width, the body length, and the body width.

To start the Stick with a Winner technique in this experiment, the paper helicopter's base construction is noted, and the time it takes to descend is recorded. Next, each component is changed one at a time, and descent time is recorded. **If the change produces the desired result, the factor is left at that level; otherwise revert back to the previous setting for the remainder of the experiment.** In this example, if the resulting time is slower, then hold (or stick) with the adjustable factor at that level. Figure D-4 displays this scenario for the base helicopter design and its six adjustable factors wherein the "+" symbol indicates the one factor adjustment, and the "-" symbol indicates the base construction.

Design	Paper Type	Paper Clip	Wing Length	Wing Width	Body Length	Body Width
Base	-	-	-	-	-	-
Paper Change	+	-	-	-	-	-
Add Paper Clip	-	+	-	-	-	-
Wing Length Change	-	-	+	-	-	-
Wing Width Change	-	-	-	+	-	-
Body Length Change	-	-	-	-	+	-
Body Width Change	-	-	-	-	-	+

FIGURE D-4: Paper Helicopter Stick with a Winner Design

After running the OFAT experiment, the times were recorded, as shown in Figure D-5. The slowest descending time was 2.7 seconds for the combination of the new paper type and longer wing length. However, notice not all of the different combinations were tested. Based on the previous instant drink OFAT experiment, what is the likelihood that this 2.7 second combination is the optimal one? More than likely, a better combination exists, but we cannot be certain.

Design	Paper Type	Paper Clip	Wing Length	Wing Width	Body Length	Body Width	Flight Time
Base	-	-	-	-	-	-	2.2
Paper Change	+	-	-	-	-	-	2.4
Add Paper Clip	+	+	-	-	-	-	1.9
Wing Length Change	+	-	+	-	-	-	2.7
Wing Width Change	+	-	+	+	-	-	2.5
Body Length Change	+	-	+	-	+	-	2.4
Body Width Change	+	-	+	-	-	+	2.5

FIGURE D-5: Results of the Paper Helicopter Stick with a Winner Example

Implement Many Solutions or Test at Extremes

Often project improvement teams generate multiple solutions to enhance the process and implement these simultaneously for quick results. Given time and resource pressures, these circumstances frequently occur and describe the typical conditions for an Implement Many Solutions (process) or test at extremes (design) situation. Continuing with the paper helicopter scenario from the previous example, the approach would examine the extreme conditions only to select the combination that produces the slowest descent time. Per the design approach, not all the potential combinations would be tested. Regardless of the resulting times, the likelihood of discovering the best configuration is slim with this all-

or-nothing approach. Figure D-6 illustrates the test at extremes approach for the paper helicopter example.

Design	Paper Type	Paper Clip	Wing Length	Wing Width	Body Length	Body Width
Standard	-	-	-	-	-	-
Change All Parameters	+	+	+	+	+	+

FIGURE D-6: Paper Helicopter Test at the Extremes Design

DOE Terminology

To understand the DOE technique, one first must understand some of its unique concepts and terminology. The DOE terms listed as follows represent the fundamentals, but know that there are whole college-level, semester-long, courses in the DOE techniques, and this entry is just skimming the surface to explain how to conduct some simple DOE tests. The key DOE terminology includes

- **Alias**—A synonym for confounded.
- **Blocking**—Method of negating the effect of uncontrollable factors (such as noise) or factors not of interest. It reduces unwanted variability in the response due to a *nuisance variable* (a factor impacting the response but not of interest). For example, depending on the business questions some blocking variables may include: shifts, day, location, batch, machine, operator, tenure, or age.
- **Design Structure**
    - **Balanced design**—When the test design involves an equal number of runs for each combination of each setting (or level) for each factor.
    - **Orthogonal design (or array)**—When the test design reflects a balanced matrix that avoids any confounding of the main effects or any interactions. A full factorial is described as balanced and orthogonal because the design represents an equal number of runs for each level and combinations of all the factors.
- **Confounded**—When the effects of two or more factors are indistinguishable from one another (also called alias).

- **Factor** (x)—There are two types of factors (**See Also** "Y=f(x)," **p.** 758):
    - **Control Factor**—An input of a process having an effect on a response and whose value can be easily selected.
    - **Noise Factor**—An input to a process having an effect on a response but is not easily managed.
- **Factorial Design Structure**
    - **Full Factorial**—A design wherein all the possible combinations of all levels of all factors are tested; none are omitted. A full fractional design for a two-level (indicated by either + or -), three-factor (A, B, and C) is shown here:

Run	A	B	C
1	-	-	-
2	-	-	+
3	-	+	-
4	+	-	-
5	-	+	+
6	+	+	-
7	+	-	+
8	+	+	+

This design can be denoted as 2^3 indicating two-levels for three-factors (2x2x2, for example), represented by a total of eight possible combinations to represent a full factorial. (The notation n^k denotes n-levels for "k" factors.)

    - **Fractional Factorial**—A design describing fewer combinations than a full design are tested. For example, if a half-fractional design for a two-level (indicated by either + or -), three-factor (A, B, and C, for example) would have only four runs (*half* of the full factorial design $2^3 = 8$) of the possible eight total combinations. (The notation n^k denotes n-levels for "k" factors.)

For example, a half-factorial 2^3 design might select the odd runs of the full factorial design example:

A B C D E F G H I J K L M N O P Q R S T U V W X Y Z

Run	A	B	C
1	-	-	-
2	-	-	+
3	-	+	-
4	+	-	-
5	-	+	+
6	+	+	-
7	+	-	+
8	+	+	+

- **Interaction**—The effect of a single factor (X) on the output (Y) is dependent on another factor. There are two types of interaction: "synergistic" (mild to moderate and useful in its effect) and "anti-synergistic" (strong and disruptive in its effect).

  The interactions represented in a full fractional design for a two-level (indicated by either + or -), three-factor (A, B, and C, for example) is shown here with two-way interactions AB, AC, and BC, and a three-way interaction ABC:

> **Note**
>
> Notice the "+" and "-" signs of the main effects (A, B, or C) are multiplied together to denote the appropriate interactions (AB, AC, BC, ABC). Recall the rules of multiplying signs: (-) x (-) = (+); (-) x (+) = (-); and a (+) x (+)=(+).

Run	A	B	C	AB	AC	BC	ABC
1	-	-	-	+	+	+	-
2	-	-	+	+	-	-	+
3	-	+	-	-	+	-	+
4	+	-	-	-	-	+	+
5	-	+	+	-	-	+	-
6	+	+	-	+	-	-	-

Run	A	B	C	AB	AC	BC	ABC
7	+	-	+	-	+	-	-
8	+	+	+	+	+	+	+

- **Levels**—Settings or value for a given factor. A two-level setting often is indicated by a "+" to represent the high setting and "-" for the low setting.
- **Main Effect**—The effect of a single factor (X) that is independent of any other factors.
- **Orthogonal Array (OA)**—A design that is balanced wherein each factor level appears the same number of times. Originally defined by Taguchi. The simplest is an L4 OA design (four trial runs) for two levels:

	Column 1	Column 2	Column 3
Trial 1:	+	+	+
Trial 2:	+	-	-
Trial 3:	-	+	-
Trial 4:	-	-	+

Notice that each factor (+ and -) appears twice within each column of data.

- **Randomization**—Method of assigning the order in which the treatment combinations are measured. Minimizes the effect (bias) of a "lurking' (unaccounted for) variable.
- **Replication**—Repeating all treatment combinations more than once (not repeating a measurement of a single treatment combination). Provides a better estimate of random error.
- **Response** (Y)—A measurable output of a process.
- **Resolution Level**—The amount of confounding found in a design, either main effects with interactions or interactions with interactions, based on the number of factors and levels in a design, wherein the higher the resolution, the less confounding exists. Selecting the appropriate resolution level depends upon the tolerable level of uncertainty about the experimentation balanced against both time and economic constraints on the number of experiments (runs) that can be conducted.

Table D-1 summarizes some of the various resolution levels wherein "Full" represents a full factorial design with no confounding. Resolution II (or R_{II}) contains some of the *main effects confounding with one another* (or 1+1=II), which is undesirable. Resolution III (or R_{III}) is a

A B C D E F G H I J K L M N O P Q R S T U V W X Y Z

fractional factorial design with no main effects confounding each other, but the *main effects are confounded with two-factor interactions* (or 1 main effect + 2 factor interactions = 1+2 = III). Resolution IV (or R_{IV}) is a fractional factorial design with no main effects and two-factor interactions are confounded each other, but *two-factor interactions may be confounded with each other* (or 2+2=IV). A Resolution V has no confounding with main effects and two-factor interactions, but *two-factor interactions may be confounded with three-factor* and higher interactions or *confounding of a main effect and a four-factor interaction* (or 2+3=V=1+4). This matrix can also be found within MINITAB or in the appendix of many DOE textbooks.

The confounding found in the following example of a half-factorial design of a two-level, three-factor, 4-run (out of a possible eight combinations) structure contains three sets of confounding:

	Number of Factors						
Runs	**2**	**3**	**4**	**5**	**6**	**7**	**8**
4	Full	III					
8		Full	IV	III	III	III	
16			Full	V	IV	IV	IV
32				Full	VI	IV	IV
64					Full	VII	V
128						Full	Full

Run	A	B	C	AB	AC	BC	ABC
2	-	-	+	+	-	-	+
3	-	+	-	-	+	-	+
4	+	-	-	-	-	+	+
8	+	+	+	+	+	+	+

The confounding scenarios (wherein the runs are exactly the same) represented in the above example are

- Column "A" is confounded with the "BC" column.
- Column "B" is confounded with the "AC" column.
- Column "C" is confounded with the "AB" column.

Each of the main effects (A, B, and C, for example) is confounded with a 2-way interaction factor (perhaps BC, AC, and AB), hence, this is a Resolution III design, which is a low (and often undesirable) resolution.

- **Treatment Combination**—Combination of control factors and levels, sometimes simply called *runs* or *combinations*.
- **Treatment Order**
  - **Run Order**—When the test is conducted in a random order to eliminate bias.
  - **Standard Order**—When the test is conducted such that its treatment combinations are in order.

A DOE design could represent a single type of experiment or a suite of different types of experimentations wherein the carefully designed combination of tests builds a deeper understanding of the process. Not surprisingly, there are various classes of designed experiments, each having strengths for use in certain applications. The main categories are

- **Screening Design**—To sort through several variables using a two-level design (minimum and maximum) to identify the critical ones. They allow for the estimation of main effects only and are useful when faced with many potential factors to assess. Screening DOEs typically use a Resolution III design and nearly always are followed by a higher resolution design to select the optimal level of selected significant variables.
- **Full-Factorial Design**—To test all the possible combinations and levels of the factors. It allows for the estimation of all effects so there is no confounding of effects. Due to size of experiment, it is useful with only a small number of factors (typically less than 6).
- **Fractional Factorial Design**—To test a subset of a full-factorial design, to save on time, resources, and money, and is one of the most common types of DOE. Often follows after a screening design to consider potential interactions. It allows for the estimation of main effects and **some** interactions. It typically uses a Resolution IV and V design but can be folded over to increase resolution and simplify confounding structure.
- **Optimizing DOE**—To determine optimum levels of the variables, using more complex designs such as Response Surface methodology or **Ev**olutionary **Op**eration (EVOP).
- **Response Surface Design**—To discover the shape of the response using geometric concepts to depict the relationships being discovered. Often used when a response needs to be optimized. It is able to estimate nonlinear (usually quadratic) effects. Because of size of experiment, only a small number of factors can be considered (generally less than eight). Often preceded by a screening design or fractional factorial design in order to focus on a few main factors of interest.
- **Robustness (Taguchi) Design**—To find a region of low variability in the response and then moving the response to minimum, target, or maximum requirements, ultimately to optimize signal-to-noise ratios.

A B C D E F G H I J K L M N O P Q R S T U V W X Y Z

- **Confirming DOE**—Method that tests the factors at the *optimum* level to verify that the prediction matches reality.
- **Mixture Design**—To test variables expressed as proportions of a blend, which when summed would equal 100%, as the whole blend. Used when there is a dependency structure within the factors. This design type frequently is found in chemical applications or other recipe-type applications.
- **Conjoint Analysis**—To investigate the attribute data associated with customer preference and purchasing behavior. (**See Also** "Conjoint Analysis," **p. 207**)

How to Use the Tool or Technique

The methodology, in its most basic sense, involves four key steps:

Step 1. Plan the experiment

Step 2. Conduct the experiment

Step 3. Analyze the data

Step 4. Draw conclusions

Step 1 Plan the Experiment

A majority of the effort in conducting a DOE should be budgeted for the planning phase.

The planning tasks include

A. Identifying the process for experimentation and the experiment's objective(s) or hypotheses.

B. Defining the response variables (Ys).

C. Gathering the knowledge about the process (via Process maps, Cause-and-Effect diagrams, and so on).

D. Developing a list of process variables (Xs) and classifying them as independent, dependent, uncontrollable, or noise factors.

E. Assigning the appropriate level to each variable. To simplify the experiment, a two-level (high/low) might a good initial setting.

F. Assessing the measurement system. If the measurement process is new, run several trials to become comfortable with the measurement system and validate its accuracy prior to running the experiment. (**See Also** "Conjoint Analysis," **p. 207**)

G. Selecting the appropriate design.

H. Conducting a "sanity" check on the treatment combinations to ensure a reasonable experimental region.

I. Determining the appropriate sample size. (**See Also** "Sampling," **p. 618** for more information in determining the correct sample size.)

The DOE Planning deliverables include

- Specified hypothesis of interest
- Experimental protocol documented
- All treatment combinations specified
- Run order of treatment combinations defined

Generic 2^3 Design Example Using MINITAB

Set up a generic MINITAB DOE design by following this procedure:

1. Open the MINITAB application software; from its toolbar, select the following sequence: ***Stat > DOE > Factorial > Create Factorial Design....*** If the scenario of a 2-level, 3-factor design were continued (a 2^3 design), the MINITAB dialog boxes are illustrated in Figure D-7.
2. Select the ***2-level factorial (default generators) (2 to 15 factors)*** because this example involves a 2-level factorial design, indicated by Area 1 in Figure D-7.
3. In the Number of Factors area, select **3** from the drop-down list for the three factors in the example (that is, A, B, and C), indicated by Area 2 in Figure D-7.
4. Select the Designs... button on the main dialog box, and a second dialog box opens called Create Factorial Design—Designs, indicated by Area 3 in Figure D-7.

   In this case, because there are only three factors and two levels, the only options are "Resolution III" or "Full."

> **Note**
>
> As needed, select the number of center points, replicates (typically three or more are selected, but to keep the example simple, one is selected), and the number of blocks (one is selected to represent different operators who will be involved in the experiment).

a. Select ***Full factorial 8 Full 2**3*** because the example will test all possible combinations of the three factors at two levels, versus a fractional-factorial design. Leave the other options at their default levels.

b. Select **OK** in the Create Factorial Design—Designs dialog box.

c. Select **OK** in the main Create Factorial Design dialog box.

A MINITAB Worksheet opens with the specified experimental design as its structure. This window is to be used as the data collection template in the experiment.

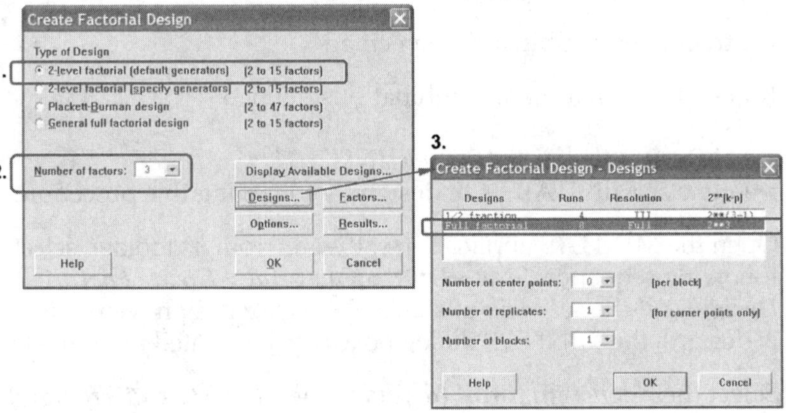

FIGURE D-7: *Creating the 2^3 Design in MINITAB*

After OK is selected in both dialog boxes, a MINITAB Worksheet will be created. Figure D-8 displays a sample one for this scenario.

Worksheet 1 ***

↓	C1	C2	C3	C4	C5	C6	C7
	StdOrder	RunOrder	CenterPt	Blocks	A	B	C
1	6	1	1	1	1	-1	1
2	1	2	1	1	-1	-1	-1
3	2	3	1	1	1	-1	-1
4	8	4	1	1	1	1	1
5	3	5	1	1	-1	1	-1
6	4	6	1	1	1	1	-1
7	7	7	1	1	-1	1	1
8	5	8	1	1	-1	-1	1

FIGURE D-8: *Sample MINITAB Worksheet Contents in (Random) Run Order*

Note

The data in your "StdOrder" (C1) column more than likely is different from that in Figure D-8 because the order is generated randomly to eliminate testing bias.

Recall, that this MINITAB Worksheet serves as the DOE "data collection form" referred to in the subsequent DOE Step 2a instructions.

Step 2 Conduct the Experiment

To conduct the DOE, the tasks include

A. Utilize a data collection form that mirrors the design's resolution and run order, including replications. (See Figure D-06.)

B. Actively collect the data.

C. Note any unusual outcomes.

The DOE conducting deliverables include

- Completed and documented experiment results.

Catapult Example Using MINITAB

For purposes of example, a catapult experiment will serve as the DOE for the remainder of this section. Most Six Sigma practitioners experience a catapult at some point in their careers because it is such a good teaching tool. Figure D-9 provides a simple illustration of a catapult device.

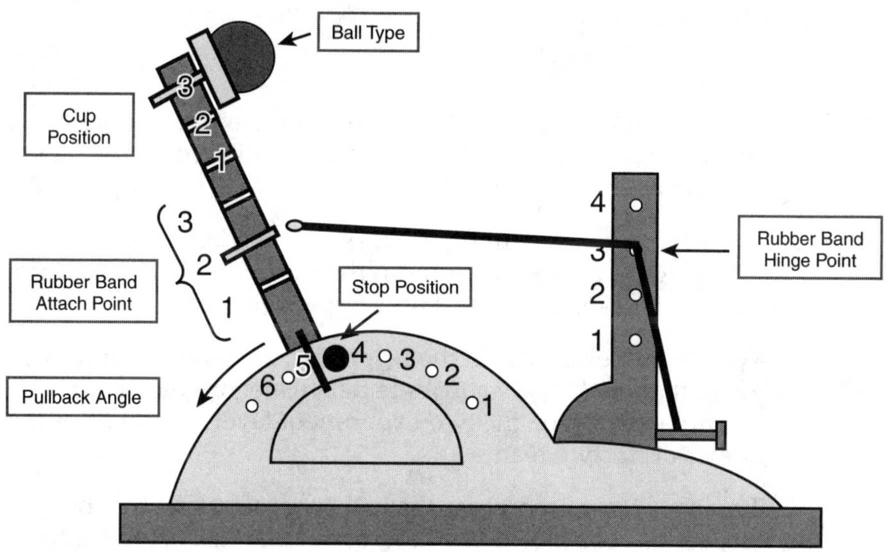

FIGURE D-9: Illustration of a Simple Catapult

Figure D-9 illustrates the multiple variables that can be adjusted to determine the optimal distance that the ball can be projected. For purposes of this illustration, the three factors will be assigned as follows:

- A = "Team" conducting the experiment, of which there are two in this example, with its two levels indicated as team "A" or "B."
- B = "Projectile" type, specifying what type of ball is used in the experiment. In this case, the two levels (or types) of balls are "plastic" or "golf."
- C = "PBA" (Pull Back Angle) on the catapult arm, with its two levels identified as "40" and "50" degrees as indicated on the catapult device itself.

This "factor" information can be entered into the MINITAB by selecting the Factors button on the Create Factorial Design dialog box and entering the preceding data, as shown in "Area 4" of Figure D-10.

Note

If the MINITAB Worksheet window is in view, a shortcut to return to that dialog box from the MINITAB Worksheet is by simultaneously hitting the Ctrl+E keys on your computer keypad. The most recent active dialog box should appear.

1. Select Ctrl+E simultaneously on the computer keypad. Notice the prior Steps 1, 2, and 3 data still are selected. If not, please re-enter the example information stated earlier.
2. Select the ***Factors*** button on the MINITAB dialog box (shown as "Area 4" in Figure D-10), and enter in the experiment design by providing the specific input variable and respective levels:
    a. Type `Team`, select ***Text*** in the respective drop-down window, given that the factor is attribute data (category name, for example), and type the two experiment levels **A** and **B** for the names of the two teams.
    b. Type **Projectile** to represent the ball-type, select ***Text*** in the respective drop-down window, given that the factor is attribute data (category name), type the two experiment levels **Plastic** and **Golf** for the two types of balls to be used.
    c. Type **PBA** to represent the "pullback angle" in the experiment, select ***Numeric*** in the respective drop-down window, given that the factor is continuous data (inches, for example), and type the two experiment levels **40** and **50** for the pullback angle distance on the catapult.

d. Select **OK** in the Create Factorial Design—Factors dialog box.

e. Select **OK** in the main Create Factorial Design dialog box.

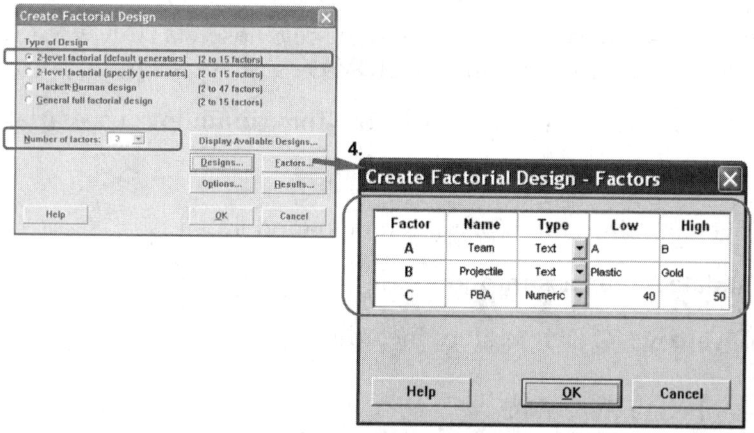

FIGURE D-10: Defining the Catapult Factors of a 2^3 Design in MINITAB

For purposes of illustration, two teams conducted the catapult experiment, following the RunOrder instructions from the MINITAB Worksheet for the prescribed factors at the prescribed levels. Recall that the RunOrder is randomized by MINITAB to eliminate bias, such that the teams A and B alternate randomly, and their respective experimental set-up is randomly alternated.

The teams recorded their results of the catapult experiment in the MINITAB Worksheet (serving as the data collection form). Figure D-11 shows a MINITAB Worksheet containing the recorded data in column C8, "Distance," where the distance of the two different projectiles were recorded in inches for the eight runs.

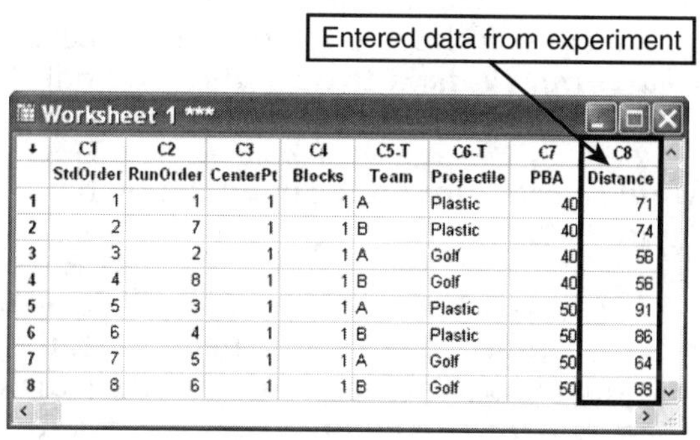

↓	C1	C2	C3	C4	C5-T	C6-T	C7	C8
	StdOrder	RunOrder	CenterPt	Blocks	Team	Projectile	PBA	Distance
1	1	1	1	1	A	Plastic	40	71
2	2	7	1	1	B	Plastic	40	74
3	3	2	1	1	A	Golf	40	58
4	4	8	1	1	B	Golf	40	56
5	5	3	1	1	A	Plastic	50	91
6	6	4	1	1	B	Plastic	50	86
7	7	5	1	1	A	Golf	50	64
8	8	6	1	1	B	Golf	50	68

FIGURE D-11: MINITAB Worksheet Showing Sample Catapult 2^3 Results

A B C D E F G H I J K L M N O P Q R S T U V W X Y Z

Step 3 Analyze the Experiment

To analyze the DOE results, the tasks include

A. Plot the data.

B. Separate sources of variability in the response via ANOVA. (**See Also** "Analysis of Variance (ANOVA)—7M Tool," **p. 142**)

C. Distinguish special cause variation from common cause variation.

D. Create plots of factor effects.

E. Develop a simplified (parsimonious) model.

F. Validate model assumptions.

The DOE analyzing deliverables include

- Graphed of the results
- Modeled assumptions

Catapult Example Using MINITAB (continued)

For purposes of the catapult example using MINITAB, we first generate the graphical plots that show the impact of each factor (or input variable—the two teams doing the experiment, pullback angle, and type of ball) on the response variable (distance the ball traveled) to see which one had the greatest impact. We will consider both of the input variables and the interaction of the three. Using MINITAB, the sequence from the toolbar would be ***Stat > DOE > Factorial > Factorial Plots...***

First generate the plots that show the impact of each input factor on the response by selecting the following items in the MINITAB DOE screens:

1. Using the completed MINITAB Worksheet with the recorded experimental data; select the following toolbar sequence: ***Stat > DOE > Factorial > Factorial Plots*** ... to open the Factorial Plots main screen.

2. Select the ***Main Effects Plot*** on the main dialog box, indicated by Area 5 in Figure D-12.

   a. Notice that as you select the checkbox, its respective ***Setup***... button becomes highlighted. Select the ***Setup***... button, and a Factorial Plots—Main Effects dialog box opens, as shown by Area 6 in Figure D-12.

b. Select the response variable in the far left of the screen and double-click it to insert it into the ***Response*** dialog area.

c. Find the screen's ***Available*** list of main effects variables located in the center of the screen and select all the factors by clicking the > button to select one at a time or by clicking the **double-stacked** > button to select all at once.

d. Click **OK**.

3. Select the ***Interaction Plot*** on the main dialog box, also indicated by Area 5 in Figure D-12.

a. Notice that as you select the checkbox its respective ***Setup...*** button becomes highlighted. Select the ***Setup***... button, and a Factorial Plots—Interactions dialog box opens, as shown by Area 6 in Figure D-12.

b. Same as the preceding step, select the response variable in the far left of the screen and double-click it to insert it into the ***Response*** dialog area.

c. Find the screen's ***Available*** list of main effects variables located in the center of the screen and select all the factors by clicking the > button to select one at a time or by clicking the **double-stacked** > button to select all at once.

d. Click **OK**.

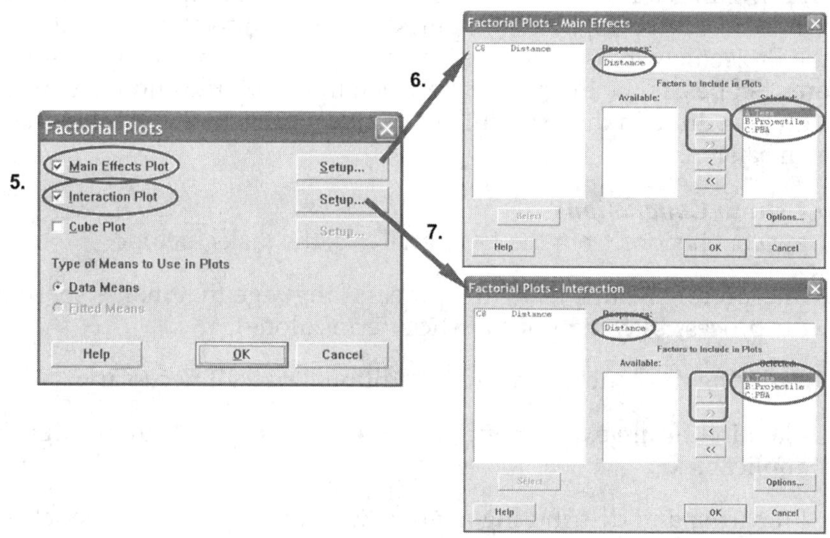

Figure D-12: Analyzing the Catapult Factors of a 2^3 Design in MINITAB

The resulting graphical analysis from MINITAB produces two graphs, one for the main effects and one for the interactions among them, as shown in Figure D-13.

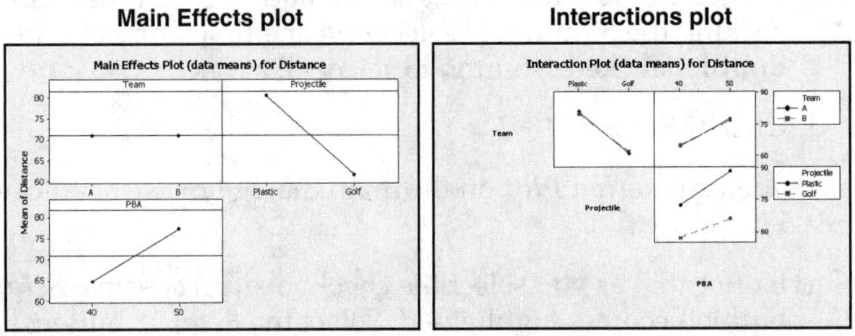

FIGURE D-13: Graphical Analysis of the Catapult Factors of a 2^3 Design in MINITAB

What do these graphs indicate? The greater the slope in the line (that is, the less horizontal and more vertical the line is oriented), the greater the impact.

In the Main Effects plot, the line between team A and B is horizontal, indicating no difference between the two teams. The Projectile (ball-type) and PBA (pullback angle) graphs both indicate a difference in effect between their two respective levels.

In the Interactions plot, both the Team versus Projectile-type (in the upper-left corner) and the Team versus PBA (in the upper-right corner) feature two lines almost on top of one another indicating little ability to discriminate. However, in the graphical segment in the lower-right corner for the Projectile versus PBA, it features two separate lines with slightly different slopes. Is the interaction between the Projectile and PBA large enough to include in the experimental model? This question is answered in the next step.

Step 4 Draw Conclusions

To draw conclusions from the DOE analysis, the tasks include

A. Interpreting results using the process language that makes sense to the process players (not statistical terms alone).

B. Verifying the conclusions with additional runs, if necessary.

C. Planning future experiments, based on new the new knowledge, if applicable.

D. Preparing and distributing a report with the findings and conclusions, including what new knowledge was learned about the process and any lessons learned from using the DOE technique.

The DOE Conducting deliverables include

- Documented report with key findings, conclusions, and lessons learned.

Catapult Example Using MINITAB (continued)
MINITAB will answer the question of statistical significance for the factors that were graphed as part of Step 3.

To analyze the data and plots, follow the procedure in the MINITAB DOE screens:

1. Select the following MINITAB toolbar sequence: ***Stat > DOE > Factorial > Factorial Design*** ...to open the main Analyze Factorial Design screen.

2. Select the response variable from the list of options on the left of the screen by double-clicking the correct factor or clicking the Select button. Notice it will populate in the **Response** dialog box. In this example, "Distance" is selected.

3. Click the Graphs button to open the ***Analyze Factorial Design—Graphs*** screen, indicated by Area 7 in Figure D-14.

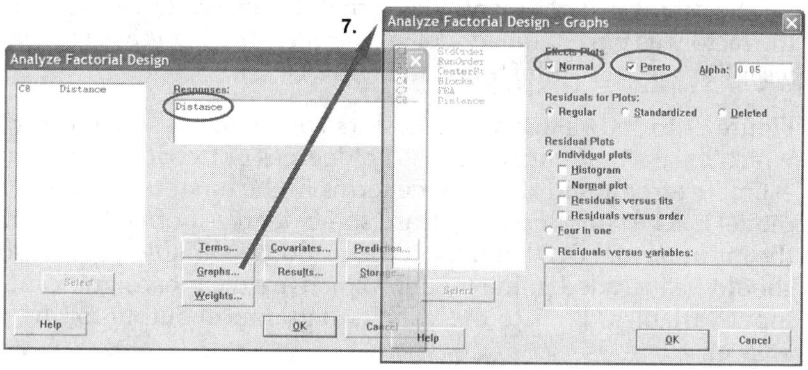

FIGURE D-14: Analyzing MINITAB Graphs of the Catapult Example (2^3 Design)

   a. Select two graphical tools of the Effects Plots checkboxes, ***Normal*** and ***Pareto***, to use to analyze the data. Keep all other defaults selected, as shown by Area 7 in Figure D-14.

   b. Click **OK**.

4. MINITAB produces two outputs: the Session Window Output to understand the amount of error produced in the model, shown in Figure D-15, and the Normal plots and Pareto graph, shown in Figure D-16.

A B C D E F G H I J K L M N O P Q R S T U V W X Y Z

```
Factorial Fit: Distance versus Team, Projectile, PBA

Estimated Effects and Coefficients for Distance (coded units)

Term                    Effect     Coef
Constant                         71.000
Team                    -0.000   -0.000
Projectile             -19.000   -9.500
PBA                     12.500    6.250
Team*Projectile          1.000    0.500
Team*PBA                -0.500   -0.250
Projectile&PBA          -3.500   -1.750
Team*Projectile*PBA      3.500    1.750

S = *

Analysis of Variance for Distance (coded units)

Source              DF   Seq SS   Adj SS   Adj MS  F  P
Main Effects         3  1034.00  1034.00  344.833  *  *
2-Way Interactions   3    27.00    27.00    9.000  *  *
3-Way Interactions   1    24.50    24.50   24.500  *  *
Residual Error       0        *        *        *
Total                7  1086.00
```

Current model for all terms (factors) selected.

There are no degrees of freedom remaining to estimate error after fitting all model terms.

FIGURE D-15: MINITAB Session Window of the Catapult Example (2^3 Design)

The ANOVA is used to analyze the model. (**See Also** "Analysis of Variance (ANOVA)—7M Tool," **p. 142.**)

The MINITAB Session Window Output provides the current model for the catapult experiment. Recall that this example selected all of the factors as part of the model, as per *Step 3) "Analyze the Experiment", sub-step 3,"* (illustrated in Figure D-12).

Figure D-15 shows that model results with zero "Residual Error", which means that the model has no degrees of freedom remaining to estimate error after fitting all the terms in the model. Hence, the model has be simplified to release some degrees of freedom. Only the input factors (X) that impact the response variable (Y) the most should be included in the model. Which are the most significant input variables? Look to the MINITAB graphical output for the answer.

Next, examine the MINITAB graphical output, shown in Figure D-16.

As shown in Figure D-16, both the Normal plot and Pareto chart depict that the Projectile type (Golf or Plastic ball, indicated by "B") Is significant in impacting the distance of the projectile when tossed from the catapult. Hence, it should be included in the model. The Pareto chart also indicates that the Pullback Angle (Factor "C") is very close to having a statistically significant influence on the distance output; therefore, it could be included in the model as well.

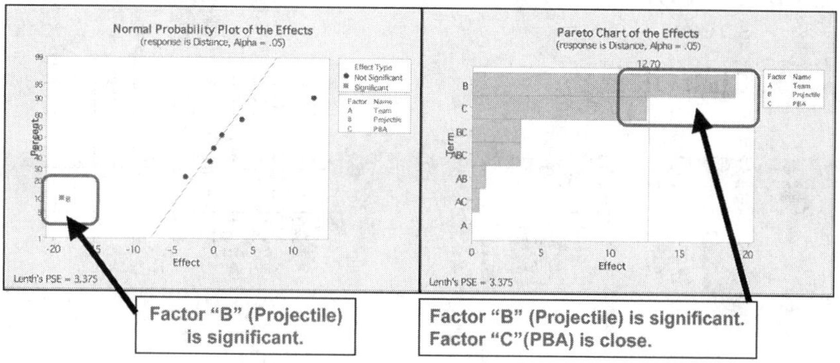

FIGURE D-16: MINITAB Normal and Pareto Graphs for the Catapult Example (2^3 Design)

The interaction of any main effect showing significance must also be included in the model. In this case, that would be the 2-way interaction of the Projectile/Pullback Angle. The 3-way interaction of Projectile/Pullback Angle/Team appears equally significant when comparing the lengths of the Pareto chart bars, so it should be included in the model.

As a result, the model to analyze next should involve the following factors: Projectile (B), Pullback Angle (C), and Projectile/Pullback Angle interaction (BC) and the Projectile/Pullback Angle/Team interaction (ABC).

Next re-run the simplified model with only the four factors: B, C, BC, and ABC, as illustrated in Figure D-17.

5. Select the following MINITAB toolbar sequence: ***Stat > DOE > Factorial > Factorial Design*** ... to open the main Analyze Factorial Design main screen.
6. Select the Terms button to open the Analyze Factorial Design—Terms screen.
    a. Select the four terms (B, C, BC, and ABC) and place in the Selected Terms dialog box.
    b. Click OK to return to the main Analyze Factorial Design main screen.
7. Select the Graphs button to open the Analyze Factorial Design—Terms screen.
    a. Select the Four in One Residual Plots graphs button to view all the graphical tool analyses in one screen.

b. Click **OK** to return to the main Analyze Factorial Design main screen.

c. Click **OK** again.

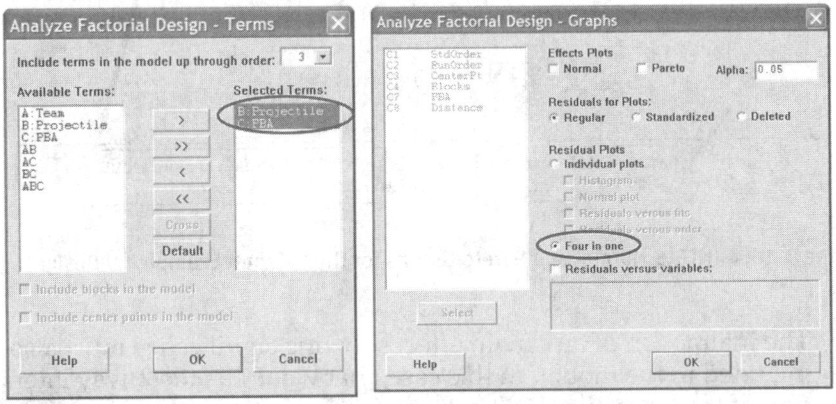

Figure D-17: Re-run MINITAB Analysis on Simplified Model for the Catapult Example (2^3 Design)

The resulting MINITAB Session Window Output is shown in Figure D-18, and its graphical analysis is found in Figure D-19.

Session Window Output

```
Factorial Fit: Distance versus Projectile, PBA

Estimated Effects and Coefficients for Distance (coded units)

Term          Effect     Coef  SE Coef       T      P
Constant               71.000    1.135   62.57  0.000
Projectile   -19.000   -9.500    1.135   -8.37  0.000
PBA           12.500    6.250    1.135    5.51  0.003

S = 3.20936   R-Sq = 95.26%   R-Sq(adj) = 93.36%

Analysis of Variance for Distance (coded units)

Source          DF   Seq SS   Adj SS   Adj MS      F      P
Main Effects     2  1034.00  1034.50  517.250  50.22  0.000
Residual Error   5    51.50    51.50   10.300
  Lack of Fit    1    24.50    24.50   24.500   3.63  0.129
  Pure Error     4    27.00    27.00    6.750
Total            7  1086.00

Estimated Coefficients for Distance using data in uncoded units

Term           Coef
Constant    14.7500
Projectile -9.50000
PBA         1.25000
```

The effect of going from Plastic to Golf reduces Distance by 19 inches

Changing Pullback Angle from 40 to 50 degrees increases Distance by 12.5 inches.

5 degrees of freedom to estimate error after fitting all model terms

Figure D-18: MINITAB Session Window of the Re-run Catapult Example (2^3 Design)

This example's simplified, 4-factor model can be written as

Y = 71 - 9.5*Projectile + 6.25*PBA

This simplified model has sufficient degrees of freedom (that is, five) to estimate error.

The p-value (or probability of being significant at the 5% alpha default level) is a test for significance. Examining the p-value in the far right column of the Session Window Output indicates that for this simplified model, the Projectile is at 0.000, and the PBA is at 0.003, both of which are the critical value of < 0.05, hence they are significant. (**See Also** "Hypothesis Testing," p. **335** for additional information on p-values.)

Notice that the Session Window Output also provides the coefficient of determination ("R-Sq," or R-squared or R^2), indicating how well the overall "goodness of fit" for the model. The R-square is defined as the ratio of explained variation to unexplained variation in the model. In the case of this example, the simplified model has an R-square at 95.26%, which is good.

The **coefficient of determination** is bounded between zero and one, or 0% and 100%. It measures the percent of variability in the response explained by the model with the predictor. The "rule of thumb" to determine the R-square's acceptability, or strength of the relationship, includes

- Mechanistic and Physical models:
    - Excellent = 90 to 100%
    - Good = 75 to 89%
    - Weak = 60 to 74%
- Behavioral and Human models:
    - Excellent = 70 to 100%
    - Good = 50 to 69%
    - Weak = 30 to 49%

Hence, the projectile and the pullback angle are useful in predicting the distance. By referencing the "Effects" column in the Session Window Output, the effect on distance is modeled. The golf ball travels 19 inches farther than the plastic golf ball. Pulling the catapult's arm 10 degrees (from 40 to 50) increases the distance by 12.5 inches, or pulling it back by simply one degree increases the distance by 1.25 inches.

The simplified model's four-in-one graphical output indicates that the residual analysis passes model assumption tests, as shown Figure D-19. The data falls close to the normal probability plot. However, there is a slight smile-shape to the Residuals Versus the Fitted Values chart, suggesting a quadratic term. The graph should look random, without any pattern at all. If the BC interaction is placed back into the model, and the MINITAB analysis is re-run, the slight smile pattern becomes more random. **See Also** "Regression Analysis," **p. 571**.

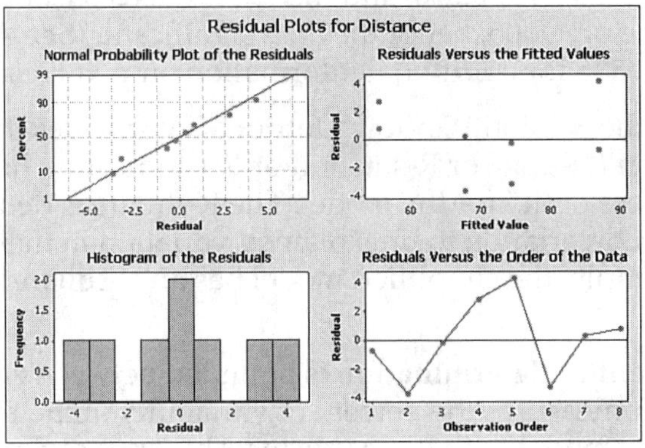

FIGURE D-19: MINITAB Residual Plots for the Re-run Catapult Example (2^3 Design)

Typically a two-level model (as in the case of this catapult example) is insufficient to model a non-linear event. The best approach is to start with a two-level model first to eliminate insignificant factors. Then conduct a three-level experiment on only those remaining critical variables. Notice that the number of test runs in a three-level full factorial model increases. For a three-factor, two-level (2^3) design, the number of runs is eight (2x2x2). For a three-factor, three-level (3^3) design, the number of full factorial runs is 27 (3x3x3).

Hints and Tips

Planning a DOE is the most critical step in the process, and an appropriate amount of time should be taken to understand the variables of interest.

Planning a DOE

Brainstorm a list of key independent and dependent variables with people knowledgeable of the process. Use Process Knowledge to select the factor settings. Try to select extreme settings (high and low) to begin to distinguish boundary conditions. Understand the process conditions and standard operating procedures. Determine any constraints—cost, time, materials, resources (including people). Consider conditions, and if these elements represent a non-linear effect, a large number of runs may be needed to define the curvature of the model, which often requires a larger budget.

Rule of Thumb Sequence of the Different Types of DOEs

Type of DOE:	Screening	Fractional Factorial	Full Factorial	Response Surface
Usual # of Factors	> 4	3–15	1–5	< 8
To Identify	Most important factors (the vital few)	Some interactions	Relationships among all factors	Optimal factor settings
To Estimate	Crude direction for improvement (linear effects)	All main effects and some interactions	All main effects and all interactions	Curvature in response, empirical models
Typical Resolution	III	IV+	Full	N/A

Due to the rapid increase in the number of experimental runs, full factorials are rarely constructed for more than four factors. What is gained by having additional degrees of freedom? The ability to estimate higher-order interactions. It is a rare process where three-factor interactions are of interest. A highly fractionated design is considered a screening design.

Rule of Thumb for Full Factorial DOEs

Number of Factors (k)	Number of Runs (2^k)
1	2
2	4
3	8
4	16

continues

Number of Factors (k)	Number of Runs (2k)
5	32
6	64
7	128
8	246
9	512
10	1024

Fractional Factorial DOEs Are Used When Time and Resources Prohibit a Full Factorial Design

If there are four factors of interest but you only have budgeted resources sufficient for eight runs, consider either a half-fraction of a 2^4 full factorial or better yet start with a 2^3 full factorial design. The 2^3 full factorial approach is an 8-run experiment and avoids missing any important information versus the half-fractional design.

Confounding Rule of Thumb

If a main effect accounts for most of the variability, the effects of large interactions (for example, a 3-way or larger) may be ignored when estimating the model.

Conduct Experiment

Conduct the experiment in random order to minimize bias.

Randomized Replication

Randomized replication of runs provides more data to better determine the inferential model.

The Prediction Model

The prediction model for coded versus uncoded units will be different, but the output range is the same. *Coded units* enable an easy comparison of the factors and their levels (e.g. "+" for the high setting, and "-" for the low setting).

Verify Conclusions

Verify conclusions by replicating experiments if necessary.

A B C D E F G H I J K L M N O P Q R S T U V W X Y Z

Supporting or Linked Tools

Supporting tools that might provide input when developing a DOE include

- Process maps (**See Also** "Process Map (or Flowchart)—7QC Tool," **p. 522**.)
- Brainstorming (**See Also** "Brainstorming Technique," **p. 168**)
- Affinity diagrams (**See Also** "Affinity Diagram—7M Tool," **p. 136**)
- Cause-and-Effect diagrams (**See Also** "Cause-and-Effect Diagram—7QC Tool," **p. 173**)
- QFD or House of Quality (**See Also** "Quality Function Deployment (QFD)," **p. 543**)
- Noise diagrams

A completed DOE provides input to tools such as

- Control plans (**See Also** "Matrix Diagrams—7M Tool," **p. 399**)
- FMEAs (Risk mitigation plans) (**See Also** "Failure Modes and Effects Analysis (FMEA)," **p. 287**)
- Transition plans (**See Also** "Matrix Diagrams—7M Tool," **p. 399**)
- Implementation plans (**See Also** "Matrix Diagrams—7M Tool," **p. 399**)
- Communication plans (**See Also** "Matrix Diagrams—7M Tool," **p. 399**)
- Standard operating procedure (SOP) documentation

Figure D-20 illustrates the link between a DOE and its related tools and techniques.

See Also "Conjoint Analysis," **p. 207**.

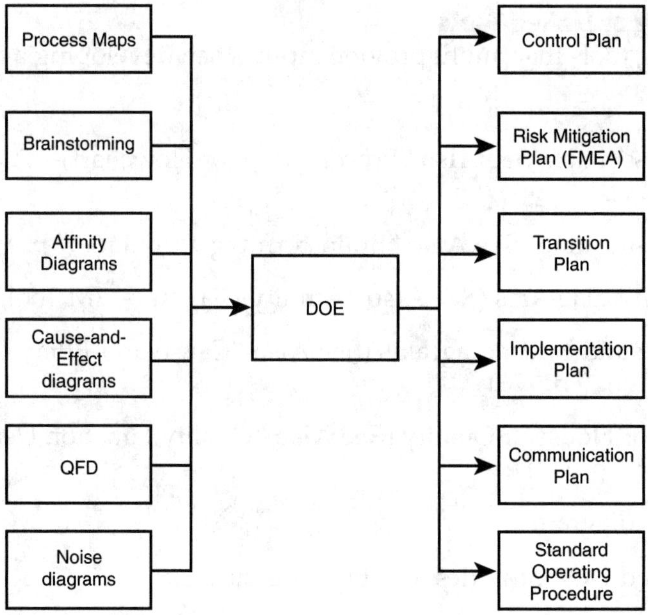

FIGURE D-20: DOE Tool Linkage

A B C D E F G H I J K L M N O P Q R S T U V W X Y Z

Dotplot

What Question(s) Does the Tool or Technique Answer?

How is the data of a process distributed; what does the data set look like?

A Dotplot helps you to

- Graphically display a data set to determine whether a process is distributed approximately normally.
- Determine the shape of the data—skewed or not, location of central tendency, and spread.

Alternative Names and Variations

This tool is also known as

- Dot chart

Variations on the tool include

- Histogram (**See Also** "Histogram—7QC Tool," **p. 330**)
- Boxplot (**See Also** "Boxplot—Graphical Tool," **p. 165**)

When Best to Use the Tool or Technique

The old adage holds true, "*A picture is worth a thousand words.*" A Dotplot depicts the shape of the data. Prior to applying any analytical tools and techniques to a data set, its shape must be understood to select the appropriate analysis.

Brief Description

A Dotplot is a type of frequency plot that visually displays how often something occurs. The data is collected in consecutive order of occurrence in a process and is plotted on a graph with the y-axis labeled *frequency* or *count* and the x-axis marked in the unit of measure for the variable of interest.

Each individual data point is plotted on a Dotplot graph, so it requires less data than its kin, the histogram. This graphical method is appropriate for stratified data and multiple measurements (or variables) for one or more types of data. Preferably, a Dotplot best handles 15 or less data points, otherwise the graph contains more dots than can be distinguished. If the data set contains between 20 and 50 data points, consider using a Boxplot. More than 50 data points requires a histogram.

Dotplots utilize either categorical or numerical (continuous, quantitative, interval, or ratio) data. For numerical data, this graphical method identifies the location of central tendency (mean, median, and mode), the *spread* (standard deviation, variance, range), and the *shape* (bell-shaped, skewed, binomial) of the distribution. For categorical-type data, stratification can be illustrated via unique shapes or shading for a dot.

How to Use the Tool or Technique

Use the following procedure to develop a Dotplot using MINITAB. Given that numeric data has been entered into the MINITAB Worksheet, select the following commands from its drop-down menu: ***Graph > Dotplot > One Y > Simple....***

Figure D-21 displays sample MINITAB screens where the appropriate graph type is selected (Area 1), and the appropriate variable data of interest (that is, Yield) are selected (Area 2) to produce the final sample Dotplot found in Figure D-22.

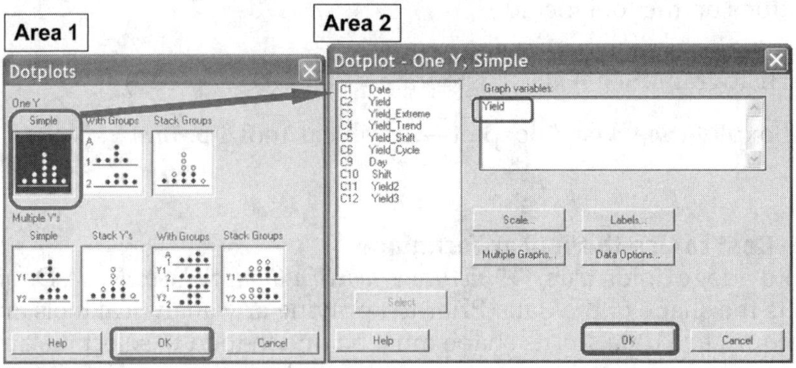

FIGURE D-21: Example MINITAB Dotplot Main Screen

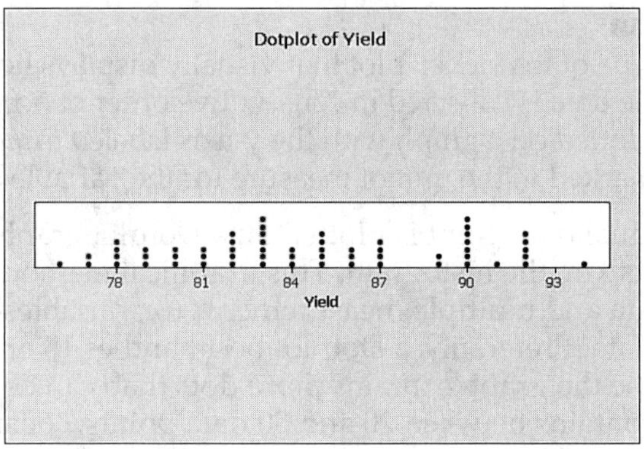

FIGURE D-22: Example MINITAB Dotplot

Figure D-22 illustrates how each data point is displayed in a Dotplot. If the data contains stratified information, the figure lacks the detail to distinguish one category from another.

Plotting Stratified Data of One Variable

To examine the stratification within a data set, select ***With Groups*** in MINITAB's Dotplot main screen to delineate that stratified data exits.

For example, Figure D-24 illustrates the scenario wherein the data set also contains shift and location information about the produced yield. Select the following MINITAB commands from its drop-down menu: ***Graph > Dotplot > One Y > With Groups***..., as shown in Figure D-23, identify the data to be plotted (shown in Area 2), and identify the data to be stratified (shown in Area 3) to produce the final sample Dotplot found in Figure D-24.

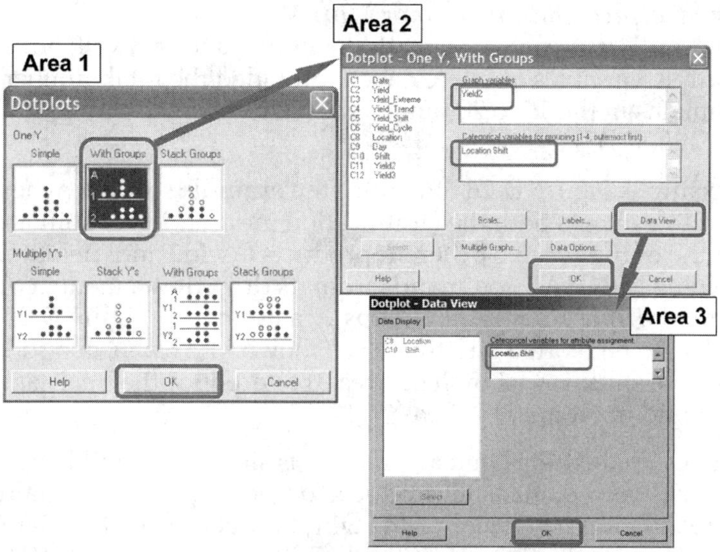

FIGURE D-23: Example MINITAB Dotplot With Groups Main Screen

Figure D-24 illustrates how there appears to be similar variability across Shifts 1, 2, and 3 and between Plants A and B. However, Plant A looks to be producing about a 10% higher yield than Plant B.

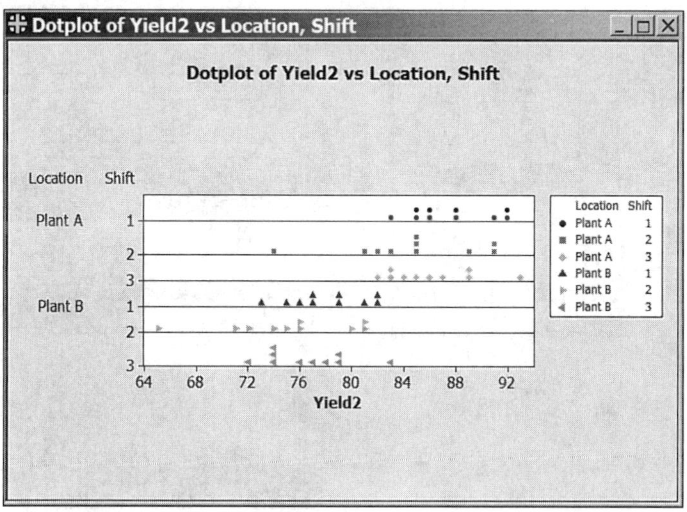

FIGURE D-24: Example MINITAB Dotplot with Groups

Plotting Stratified Data of Two or More Variables

Dotplots handle not only stratification within a data set well, but also display multiple variables of interest well, provided the total number of data points remains in the 15 to 20 range. MINITAB refers to multiple variables as *Multiple Y's*.

For example, Figure D-26 illustrates the example scenario wherein the data set also contains both the stratification information of shift and location and the output of two different products (Yield2 and Yield3). Select the following MINITAB commands from its drop-down menu: ***Graph > Dotplot > Multiple Y's > With Groups***..., as shown in Figure D-25, identify the data to be plotted and the scale (shown in Area 2), and identify the data to be stratified (shown in Area 3) to produce the final sample Dotplot found in Figure D-26.

Figure D-26 illustrates that again there is similar variability across shifts and between plants and products (or yields). There are many mean differences: Plant A has lower yields with Product 3 (Yield3); Plant B has differences among the three shifts with Shift 1 being the best and Shift 2 being the worst. (**See Also** "Stratification—7QC Tool," **p. 697**, for more detail on stratifying data.)

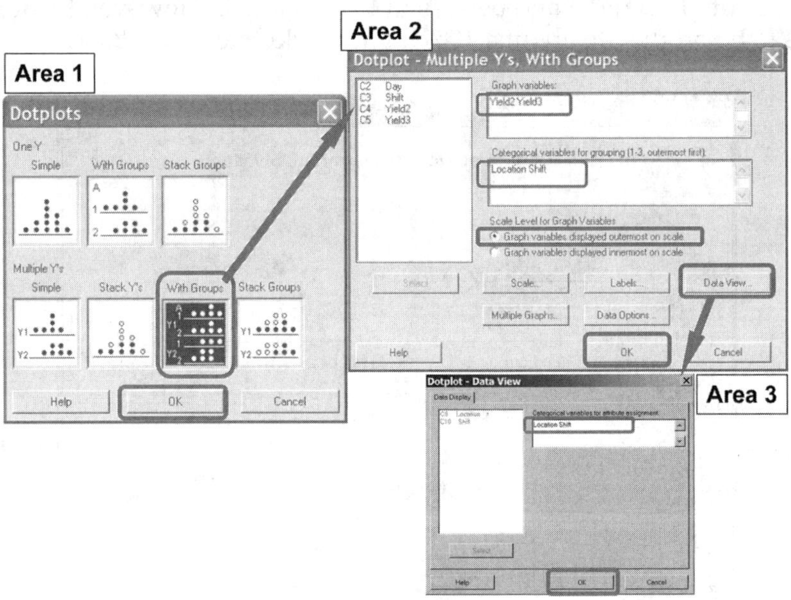

FIGURE D-25: Example MINITAB Dotplot Multiple Y's and with Groups Main Screen

A B C D E F G H I J K L M N O P Q R S T U V W X Y Z

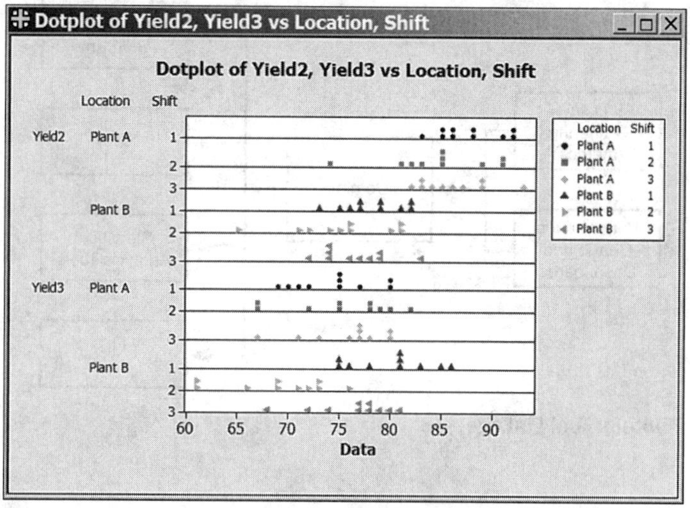

FIGURE D-26: Example MINITAB Dotplot Multiple Y's and with Groups

Supporting or Linked Tools

Supporting tools that might provide input when creating a Dotplot include

- Data gathering plan to collect the appropriate metrics (**See Also** "Data Collection Matrix," **p. 248**)
- Performance charts and dashboards

A Dotplot can provide input to tools such as

- Cause-and-Effect diagram. (**See Also** "Cause-and-Effect Diagram—7QC Tool," **p. 173**)
- QFD (**See Also** "Quality Function Deployment (QFD)," **p. 543**)
- Statistical analysis tools and other root cause analysis techniques (**See Also** "Hypothesis Testing" and "Regression Analysis," **p. 335** and **571**, respectively.)
- FMEA with follow-on action planning (**See Also** "Failure Modes and Effects Analysis (FMEA)," **p. 287**)

Figure D-27 illustrates the link between a Dotplot and its related tools and techniques.

A B C D E F G H I J K L M N O P Q R S T U V W X Y Z

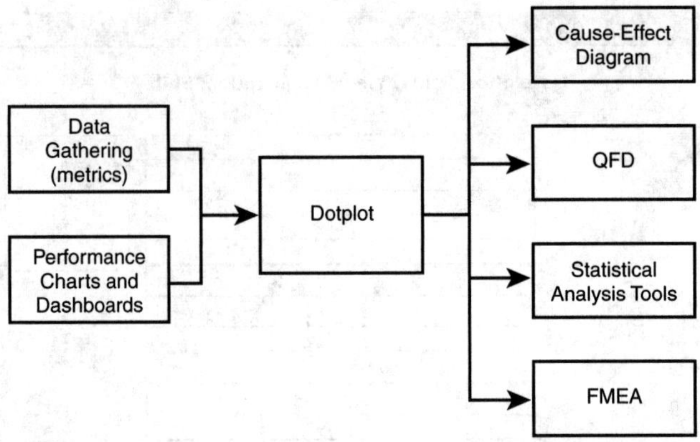

FIGURE D-27: Dotplot Tool Linkage

Failure Modes and Effects Analysis (FMEA)

What Question(s) Does the Tool or Technique Answer?

What can go wrong? What can be done to prevent or minimize it? What are the ***potential*** *failures that could occur, and what is the best response (action) plan to minimize its impact if it does happen?*

An FMEA helps you to

- **Identify** what could go wrong.
- **Analyze** the magnitude of potential risk associated with a possible problem or failure event.
- **Prioritize** the potential problems based on a three-fold approach integrating the ability to detect their occurrence, their likelihood of occurrence, and potential *severity of impact* if they do happen.
- **Document** an action plan if the failure event occurs, particularly those posing the biggest risk.

Alternative Names and Variations

This tool is also known as

- FMEA

Variations on the tool include

- Design FMEA
- Healthcare FMEA
- Product FMEA
- Process FMEA
- Systems FMEA

When Best to Use the Tool or Technique

FMEA is a proactive tool. To remain competitive and prevent possible failures, this tool is used in risk planning and management, often prior to implementing something new—a change or improvement. The FMEA serves as an integral tool used in Six Sigma projects to evaluate and reduce the risk in a new process. It is also used in ongoing operations to evaluate risks and develop contingencies. The FMEA should be part of an evergreen process and a critical component of a control management plan.

Brief Description

The **Failure Mode and Effects Analysis** (FMEA) is a popular and well-documented tool that uses a systematic method to define a problem-prevention strategy when implementing a new and/or managing a process, project, system, or offering of product and/or services. *Failure mode* means the various ways something could go wrong. *Effect Analysis* describes the potential impact of a failure if it occurs and the potential counter-measures to put in place to minimize or prevent the impact. It features different evaluation criteria scales customized for different situations. If constructed properly, using data as input, this analysis can represent a time-consuming process, thus the contingency planning portion of the process focuses on the highest priority risks.

The FMEA tool is designed to identify potential failure modes early to minimize their effect (or their impact) if they happen. It aids in developing a robust design and respective control plans. FMEA identifies potential critical and significant characteristics in a design. The tool establishes a priority for actions and documents the rationale used.

The approach used to develop an FMEA is frequently referred to as a *bottoms-up* systems analysis, trying to identify possible failure modes before they occur. This is a complementary analysis to the Fault Tree Analysis (FTA), which employs a tops-down method.

The United States military developed the tool in the 1940s, and NASA used it in the space program in the 1960s. Initially, the tool had two primary purposes: 1) to analyze possible failures of product components (hardware, software) and 2) to analyze the functionality (what something does) relative to its designed performance and output.

Now multiple industries have embraced it such as manufacturing, high-tech, and healthcare. The FMEA popularity has expanded its application to existing or newly developed product (hardware, software, or system), service, process, or project.

The purpose of an FMEA is to prevent problems with ongoing operations and implementing improvements or newly designed items or events. A well-constructed FMEA produces the following deliverables:

- List of potential failure modes
- List of potential critical and potential significant characteristics
- List of effects
- List of causes
- Documentation of current controls
- Requirements for new control plans

- Documentation and prioritization of improvement activities
- Documentation of the history of improvements

Figure F-1 provides a blank FMEA template that contains the deliverables listed here, labeled with the primary steps involved in building its content. The next section, "How to use the Tool or Technique," covers this entire procedure.

FMEA Template

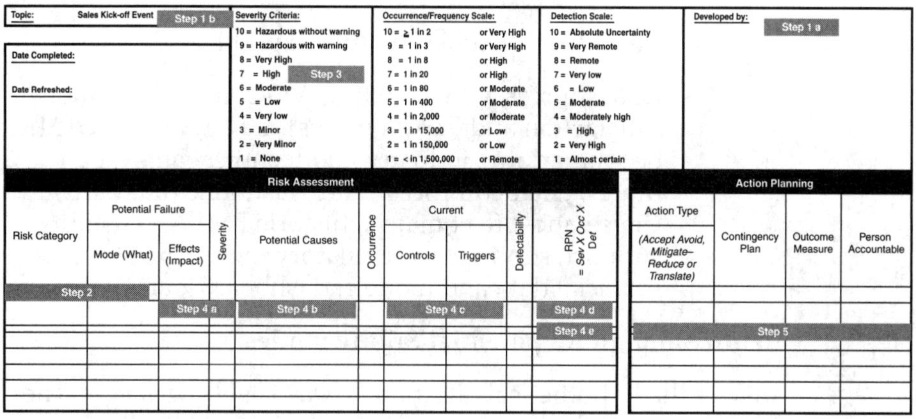

Figure F-1: Blank FMEA Template

How to Use the Tool or Technique

After the initial planning and organization activities, there are basically four major deliverables to develop an FMEA document:

1. Failure modes identified
2. Risk categorizations defined for Severity, Occurrence, and Detection
3. Effects and risk impact determined
4. Contingency action plan developed

The general 5-step procedure is as follows:

Step 1. **Plan and Organize for an FMEA.**

a. **Assemble diverse team** to develop the FMEA, ideally comprised of multiple disciplines, backgrounds, experience, roles, rank, and years of experience. Record their names and function.

b. Gain agreement on **target topic**, and associated outputs, metrics, and/or critical parameters of the target topic.

c. **Gather "background" data** from a variety of sources including

- Historical documents on current topic when examining an existing product, process, or service
- Historical documents on similar topics if analyzing a new design, process, service, or project
- Insights from subject matter experts
- Data sources, which include VOC data (Customer satisfaction and VOC surveys), CTQs, QFD, SWOT, benchmarking, industry and technical analyst critiques, detailed Process map, Fishbone diagrams, Cause-and-Effect matrix, historic FMEAs, customer support, service and regulatory logs, control charts, financial data and models, dashboards, and so on

Step 2. **Identify and group potential failure modes.**

Answer the hypothetical question, *"What could go wrong?"* The team references the background information and its diverse perspectives and employs any or all of the following approaches: 1) Brainstorming, 2) Affinity diagramming, and 3) Tree diagramming.

a. **Record** the failure modes and corresponding category (or group) on an FMEA template. Figure F-2 shows the Risk Assessment portion of an FMEA template.

Risk Assessment									
Risk Category	Potential Failure		Severity	Potential Causes	Occurrence	Current		Detectability	RPN = Sev X Occ X Det
	Mode (What)	Effects (Impact)				Controls	Triggers		

FIGURE F-2: Sample Risk Assessment Portion of an FMEA Template

Step 3. **Define the three evaluation criteria** to be used and record them on the FMEA as a reference key or footnote. *(Customize the evaluation criteria definition and scales to what is the most appropriate for the situation and the industry.)*

a. **Severity Rating (SEV)**—1 to 10, for example, where 10 indicates *most severe.*

b. **Occurrence or Frequency (OCC)**—1 to 10, for example, where 10 represents the *highest likelihood of occurrence*. The occurrence criteria can indicate:

   i. Number of times the failure may occur

   ii. Time period (measured in appropriate increments: minutes, days, months, and so on)

   iii. Percent probability or terms describing probability

c. **Detection Rate (DET)**—1 to 10, where 10 represents the *lowest detectability.*

Figure F-3 provides an example scale for Evaluation Criteria portion of an FMEA:

Severity Criteria:	Occurrence/Frequency Scale:		Detection Scale:
10 = Hazardous without warning	10 = ≥ 1 in 2	or Very High	10 = Absolute Uncertainty
9 = Hazardous with warning	9 = 1 in 3	or Very High	9 = Very Remote
8 = Very High	8 = 1 in 8	or High	8 = Remote
7 = High	7 = 1 in 20	or High	7 = Very low
6 = Moderate	6 = 1 in 80	or Moderate	6 = Low
5 = Low	5 = 1 in 400	or Moderate	5 = Moderate
4 = Very low	4 = 1 in 2,000	or Moderate	4 = Moderately high
3 = Minor	3 = 1 in 15,000	or Low	3 = High
2 = Very Minor	2 = 1 in 150,000	or Low	2 = Very High
1 = None	1 = < in 1,500,000	or Remote	1 = Almost certain

FIGURE F-3: Sample FMEA Rating Criteria

If unique evaluation criteria are used, gain agreement on the scale and criteria and then document it on the FMEA template.

Step 4. **Determine effects and risk impact.**

a. Evaluate the potential impact (or effect/consequences) of each failure mode, using financial models and sensitivity studies and document and rate the Severity according to the agreed-to rating. Record both items on the FMEA template.

b. Identify the potential causes for each failure mode and rate the likelihood of occurrence (or frequency). Record both items on the FMEA template.

c. Identify the controls and/or indication triggers (if any) and rate the ability to detect each failure mode. Record both items on the FMEA template.

d. Calculate the Risk Priority Number (RPN) for each failure mode effect and record it on the FMEA template.

   i. *RPN Calculation* is the product of multiplying the three risk scores together: **Severity x Occurrence x Detection Rate = RPN**

   ii. The RPN number is a guide and requires judgment and scrutiny when determining the relative importance of any given failure mode.

e. Identify the high-priority failure modes by selecting the largest RPNs or those with the highest possible Severity or Occurrence rating, particularly if it impacts the customer.

   i. Sort the FMEA by RPN, highest to lowest.

Step 5. Develop the contingency action plan for each of the failure modes with the highest priority or risk as a contingency and record it on the FMEA.

> **Note**
>
> Those items with medium to low priorities should remain listed on the FMEA but do not require an Action Plan be developed for them unless the have particularly high probability of occurrence or evoke "special interest."

A B C D E F G H I J K L M N O P Q R S T U V W X Y Z

a. Provide immediate attention to any failure identified with the highest priority.

b. Decide which action plan is appropriate for each failure mode cause. There are three major Action Types or Risk Response classifications:

    i. **Acceptance**—Accept the consequences passively or actively; retain the risk.

    ii. **Avoidance**—Eliminate a specific threat, usually by eliminating the cause.

    iii. **Mitigation**—Reduce the expected monetary value of a risk event by reducing the probability of occurrence.

    - **Reduction**—Minimize its occurrence and effect.
    - **Transfer**—All or a portion of it to another party via
        - Insurance for direct property damage
        - Indirect consequential loss (often performed by a contractor—debris removal or equipment replacement, for example)
        - Legal liability (design errors, public bodily injury, performance failure)
        - Personnel (employee bodily injury/Worker's Compensation)
        - Employee replacement costs
        - Resulting business losses

c. Develop the specific contingency plan or action that addresses the root cause.

d. Identify the outcome measurements to ensure the recommended contingency was achieved per plan; used to analyze or test success per the plan.

e. Assign **one and only one** person to be accountable for carrying out the recommended contingency plan if needed.

f. Figure F-4 provides a sample Action Planning section of an FMEA template.

Action Planning			
Action Type *(Accept, Avoid, Mitigate --* Reduce or Transfer)	Contingency Plan	Outcome Measure	Person Accountable

FIGURE F-4: Sample Action Planning Portion of an FMEA Template

Step 6. Refresh and re-evaluate the FMEA periodically (updating it with new threats) or re-score it as failures occur and action is taken.

How to Analyze and Apply the Tool's Output

Evaluate internal and external risk position and document contingency plans if the risk were to happen. And then take action; prevent possible risk or respond to the occurrence of one. This is when the value of the FMEA analysis comes to light. (See **Hints and Tips** that follow for additional detail.)

- The Occurrence number (OCC) is the primary item to reduce.
- After an action has been implemented, update the Risk Priority Number (RPN).
- Typically, the severity number stays the same.
- Detection actions alone are taken only as a last resort.

Periodically refresh and update the document as part of an evergreen process.

- Communicate and distribute the FMEA to appropriate parties.
- Incorporate the FMEA into the control plan.

Examples

FMEA Example—Acme Inc. Sales Kick-off Event

Acme Inc. FMEA Template

Topic: Sales Kick-off Event	Severity Criteria:	Occurrence/Frequency Scale:		Detection Scale:	Developed by:
Date Completed: 10/6/2006	10 = Hazardous without warning	10 = ≥1 in 2	or Very High	10 = Absolute Uncertainty	George S.
	9 = Hazardous with warning	9 = 1 in 3	or Very High	9 = Very Remote	Sam T.
	8 = Very High	8 = 1 in 8	or High	8 = Remote	Lois M.
	7 = High	7 = 1 in 20	or High	7 = Very low	Bill M.
Date Refreshed:	6 = Moderate	6 = 1 in 80	or Moderate	6 = Low	Laura H.
	5 = Low	5 = 1 in 400	or Moderate	5 = Moderate	Julie H.
	4 = Very low	4 = 1 in 2,000	or Moderate	4 = Moderately high	
	3 = Minor	3 = 1 in 15,000	or Low	3 = High	
	2 = Very Minor	2 = 1 in 150,000	or Low	2 = Very High	
	1 = None	1 = < in 1,500,000	or Remote	1 = Almost certain	

Fisk Assessment											Action Planning			
Risk Category	Potential Failure		Severity	Potential Causes	Occurrence	Current		Detectability	RPN = Sev X Occ X Det		Action Type (Accept Avoid, Mitigate–Reduce or Translate)	Contingency Plan	Outcome Measure	Person Accountable
	Mode (What)	Effects (Impact)				Controls	Triggers							
Materials	Kick-off Agenda not available	Speakers miss event	8	Lack of management agreement	4	VP Sales inspection	4Q Outlook Meeting	2	64					
		Participants miss event	8	Lack of management agreement	2	VP Sales inspection	4Q Outlook Meeting	2	32					
	Kick-off Notebooks printed incorrectly	Participant satisfaction	6	Presentation materials not finalized in time to meet printer's lead-time.	6	1-week lead time policy	Microsoft Outlook task reminder	2	72					
Method	Mgmt. disagree on date	Participants unavailable	8	Conflict with own team meeting	5	VP Sales inspection	4Q Outlook Meeting	2	80					
	Budget not agreed to	Cost overruns	8	Unclear whose budget will cover event (HQ versus Region)	3	VP Sales inspection	4Q Outlook Meeting	2	144		Avoid	Sr Field Mgmt. to establish a policy	Policy communicated	VP of Finance
	RSVPs not received	Cost overruns	4	Invitations sent during year-end vacations.	3	4-week lead time policy	Microsoft Outlook task reminder	2	72					
	Participants not surveyed at end	Next event reports same mistakes	5	Nobody assigned the task.	8	None		6	240		Avoid	Assign a person to the task	100% participants surveyed	VP of HR
Mother Nature	Bad Weather	Unable to travel to/ from meeting	8	Winter time frame	8	None		9	576		Reduce	Accommodate early travel time when needed	Voice mail; when needed	VP of HR
	Meeting Facility too small	Lack of participant's attention	7	More people attend than invited (2nd level management)	3	VP Sales to communicate policy	4Q Outlook Meeting and Invitation Memo restate policy	6	126		Avoid	Turn away "uninvited" guests	Person turned away at meeting door	VP of HR
		Participants dissatisfied	8	All participants did not RSVP	3	None		7	168		Avoid	Call those who did not RSVP	100% invites accounted for	Natasha G., of HR

FIGURE F-5: Example: Acme Inc. FMEA for Sales Kick-off Event

Hints and Tips

Fully develop the list of failure modes first before jumping to fixes or causes. Otherwise a significant potential risk may be overlooked because many potential failures build on one another.

Include subject-matter experts and those with no knowledge of the topic under review to help develop a list of potential failures. Those with little to no knowledge provide a fresh perspective on the topic, and at times their seemingly naïve questions give insight into a root cause.

Consider assigning team roles to those assembled to develop the FMEA and avoid having one person assigned multiple roles. The recorder of ideas should not participate in the idea creation to avoid issues such as power of the pen and lack of time to think before contributing. (**See Also** "Brainstorming Technique," **p. 168**)

There can be multiple failures to any one item or process step being examined, and they should be listed and analyzed separately. There can be multiple effects for any one failure and they should be listed and analyzed separately. Definitions of and scale for Severity, Occurrence, and Detection should be modified to fit the appropriate situation or industry. The highest score(s) indicate the need for significant contingency planning. Typical default scale is 1 to 10, where 10 represents the *most severe*, *highest likelihood of occurrence*, and *lowest detectability*.

FMEA templates should reflect the purpose of the risk analysis (product, process, healthcare) and the needs of the organization. Some FMEA templates also add an Action Results section to the far right of the template that often includes items such as:

- Action Taken and Date (what)
- Management Approval (who) and Date
- Recalculation Columns for Post-Action: Severity, Occurrence, Detection, and Risk Priority Number (RPN)
- Implement whichever approach best reinforces the importance of using the FMEA as an evergreen document.

Some industries associated with products or processes that involve safety issues or compliance with government regulations may need special controls to monitor potential risks. In that case, some add a "Critical Characteristic" column to the FMEA template to capture the specific measurements or indicators required.

When adding the Critical Characteristics column, a new risk calculation is introduced, called *Criticality*, which is determined by multiplying the Severity times the Occurrence. Both of these columns are often placed just to the right of the RPN column.

Similar to the RPN number, the Criticality is a guide and requires judgment and scrutiny when determining the relative importance of any given failure mode.

A well-constructed FMEA is time-intensive to build and requires thorough research, exploration, and idea generation; hence, many think this tool is best suited for non-redundant systems or projects.

Note

Risk Priority Number (RPN)

There is no RPN threshold value needed to classifying a potential failure mode as high risk.

There is no set RPN value that triggers a mandatory "Recommended Action" needed. Likewise, there is no minimum RPN value below which the team is automatically excused from action planning.

Caution

Don't focus on the Risk Priority Number (RPN) alone because other important items may be overlooked, such as those with a high Severity or of particular importance to the customer or a regulatory agency.

Taking Action

1. To reduce RPN, focus on reducing probability of occurrence, not on increasing the ability to detect an error. Lowering the likelihood of an error occurring is a measurable, proactive change to a process.
2. Avoid reducing RPN with only detection actions because often the implementation that improves the ability to detect an error involves increased inspection (reactionary) methods. This seldom adds any value to the process.
    a. 100% inspection is only (at best) 80% effective.
    b. Reducing RPN with detection does not eliminate the failure mode, nor does it reduce the probability of causes.
    c. Detection of 10 is not bad if the Occurrence is ranked as 1.
3. Techniques for generating risk ideas include
    a. Interview Experts (Delphi technique—panel of experts, convergent solutions, for example)
    b. Group Techniques
        i. Brainstorming—different starting point options
            - Start with prior project review of lessons learned.
            - Start with literature review (divided among team, bring summary and inferences to team).

- Same industry
- Similar industry
- Trend/Thought leader
- Start with category types of Risk Identification: Internal/External.

ii. Project team review of project deliverables

- Team define High/Medium/Low Risk
- Review WBS for risk items
- Review project schedule and estimate for risk items
- Review with subject matter experts/stakeholders
- Compare to similar projects

iii. Combination (experts and internal brainstorming)

- Independent events
- Invite Experts to brainstorming

iv. Tools and Techniques

- Checklists (by sources of risk)
- Flowcharting to show how various elements of a system relate to Cause-and-Effect risks
- Interviewing various stakeholders

4. **Risk Identification**: Define the source, potential of occurrence, risk symptoms, and inputs to other processes. Developing a checklist of potential failure modes that is refreshed and improved over time with experience can only help provide a *trigger* for idea generation. Not all items may be applicable in each situation, but the checklist can serve as a thought-starter. The following list is intended simply for that purpose.

a. **Internal**

i. Equipment/Hardware/Technical:

- State of the Art
- Physical Properties
- Material Properties

- Material Availability
- Processes (design, operations, maintenance, and so on)
- Integration
- Testing
- Security
- Safety
- Personnel
- Business
- Others…

ii. Software/Technical

- (See level i.)

iii. People (labor supply/issues)

- Along the value chain

iv. Work Environment—Internal/Non-Technical

- Political
- Organizational
- Process
- Safety
- Financial (including cash flow)

b. **External**

i. Marketplace

- Competition
- Suppliers—along the supply chain (people, processes, technology, tangibles, and intangibles)
- Customers
- Non-customers
- Political
- Economic

A B C D E F G H I J K L M N O P Q R S T U V W X Y Z

ii. Legal

- Licenses/patent rights
- Lawsuits
- Contractual agreements/breeches

iii. Weather/Mother Nature/Natural Disasters

vi. Government and Regulatory

A B C D E F G H I J K L M N O P Q R S T U V W X Y Z

Supporting or Linked Tools

Supporting tools that might provide input when developing an FMEA include

- Graphical tools (**See Also** "Graphical Methods," **p. 323**)
- Process map (**See Also** "Process Map (or Flowchart)—7QC Tool," **p. 522**)
- VOC and VOB data (**See Also** "Voice of Customer Gathering Techniques," **p. 737**)
- Control chart (**See Also** "Control Charts—7QC Tool," **p. 217**)
- Brainstorming (**See Also** "Brainstorming Technique," **p. 168**)
- 5-Whys technique (**See Also** "5-Whys," **p. 305**)
- 5M-and-P technique and its variants (**See Also** "Cause-and-Effect Diagram—7QC Tool," **p. 173**)

A completed FMEA provides input to tools such as

- Control chart (**See Also** "Control Charts—7QC Tool," **p. 217**)
- Brainstorming (**See Also** "Brainstorming Technique," **p. 168**)
- 5-Whys technique (**See Also** "5-Whys," **p. 305**)
- 5M-and-P technique and its variants (**See Also** "Cause-and-Effect Diagram," **p. 173**)
- Control plan (**See Also** "Matrix Diagrams," **p. 399**)
- Transition and/or Implementation plan (**See Also** "Matrix Diagrams," **p. 399**)

Figure F-6 illustrates the link between an FMEA and its related tools and techniques.

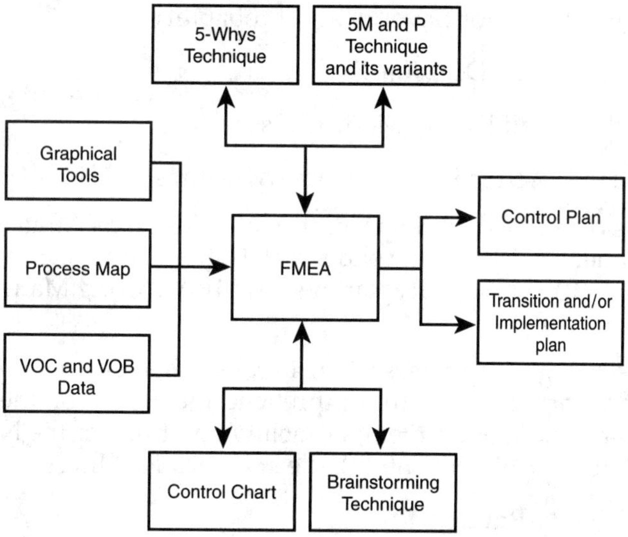

Figure F-6: FMEA Tool Linkage

Variations

Design FMEA

FMEA examines potential risks associated with the design and development of a new product, service, or process prior to launch or implementation. Its risk categories include

- Environmental changes such as competitive and technology threats
- Transition and implementation issues (ownership, training, problem and support escalation, documentation, and control plan)
- Marketplace acceptance
- Stability and quality issues of the new offering (including safety hazards)

Healthcare FMEA (HFMEA)

HFMEA is used by hospitals, healthcare centers, and clinics involved in patient safety issues and consists of a unique five-step process:

Step 1. Define the Topic

Step 2. Assemble the Team

Step 3. Graphically Describe the Process

Step 4. Conduct the Analysis

a. List Failure Modes

b. Determine Severity and Probability

c. Use the Decision Tree

d. List all Failure Mode Causes

Step 5. Identify Actions and Outcome Measures

This tool was developed by the Department of Veterans Affairs National Center for Patient Safety (www.patientsafety.gov), and its template is supported by HFMEA Severity Ratings, HFMEA Scoring Matrix, and HFMEA Decision Tree.

HFMEA scoring criteria uses a 1 to 4 scale. When evaluating each failure mode, the scoring is used to comprehend the impact on the patient outcome, visitor, staff, equipment, or facility and fire. Per the NCPS (National Center for Patient Safety), the standard HFMEA scale is

1. Event Severity Rating:
    - Catastrophic (4)
    - Major (3)
    - Moderate (2)
    - Minor (1)
2. Probability Rating:
    - Frequent: (4) Likely to occur immediately or within a short period (*several times in 1 year*).
    - Occasional: (3) Probably will occur *(1–2 times per year)*
    - Uncommon: (2) Possible to occur *(sometime in 2 to 5 years)*
    - Remote: (1) Unlikely to occur *(sometime in 5 to 30 years)*

HFMEA Combined Scoring
Figure F-7 provides the Healthcare FMEA risk rating criteria, showing a "combined" approach.

HFMEA Blank Template
Figure F-8 illustrates a blank HFMEA template. [Source: www.patientsafety.gov]

Probability	Severity			
	Catastrophic (4)	Major (3)	Moderate (2)	Minor (1)
Frequent (4)	16	12	8	4
Occasional (3)	12	9	6	3
Uncommon (2)	8	6	4	2
Remote (1)	4	3	2	1

Scores of 8 to 6 require significant contingency planning

Source: www.patientsafety.gov

FIGURE F-7: HFMEA Rating Criteria

HFMEA™ Subprocess Step Title and Number													
HFMEA™ Step 4 - Hazard Analysis									HFMEA™ Step 5 - Identify Actions and Outcomes				
		Scoring			Decision Tree Analysis								
Failure Mode: First Evaluate failure mode before determining potential causes	Potential Causes	Severity	Probability	Haz Score	Single Point Weakness?	Existing Control Measure?	Detectability	Proceed	Action Type (Control, Accept, Eliminate)	Actions or Rational for Stopping	Outcome Measure	Person Responsible	Management Concurrence
	→												

Source: www.patientsafety.gov

FIGURE F-8: HFMEA Blank Template

Process FMEA

A Process FMEA examines potential risks or failures associated with an **existing, operating process** (typically a transactional or highly-repetitive operational process). It often follows the process flow through all steps in sequential order or analyzed in reverse order.

Risk categories include

5Ms-and-P: machines (equipment and technology), materials, methods (process steps), measurements, Mother Nature (environment, surroundings), people (or manpower). (**See Also** "Cause-and-Effect Diagram—7QC Tool," **p. 173**)

This tool can be applied to a project, following the project plan and analyzing its key milestones and deliverables.

Risk categories could project plan milestones, deliverables, and the nine knowledge areas of project management:

- Scope
- Time
- Cost/Budget
- Quality
- Risk Management
- Human Resources
- Procurement (including partners and suppliers)
- Communications
- Integration

A B C D E F G H I J K L M N O P Q R S T U V W X Y Z

A Process FMEA sometimes is referred to as a *Service FMEA* (wherein the process is delivering a service). Whichever "label" is given to the FMEA, the approach still involves similar risk categories (5Ms and P and 4Ps) and should suit the particular industry requirements. (**See Also** "Cause-and-Effect Diagram—7QC Tool," **p. 173**)

Product FMEA

A Product FMEA examines potential risks or failures associated with an **existing product or services offering** (including consultative services, maintenance services, and so on). Risk categories include

- Product features and functionality, robustness, manufacturing costs, and warranty and related maintenance, support, safety costs
- 5Ms-and-P: machines (equipment and technology), materials, methods, measurements, Mother Nature (environment, surroundings), people (or manpower)
- 4Ps: policies, procedures, people, plants (environment, surroundings)

(**See Also** "Cause-and-Effect Diagram—7QC Tool," **p. 173**)

A Product FMEA sometimes is referred to as a *Service FMEA* (where the offering is the service provided). Whichever "label" is given to the FMEA, the approach still involves similar risk categories (5Ms and P and 4Ps), and should suit the particular industry requirements. (**See Also** "Cause-and-Effect Diagram—7QC Tool," **p. 173**)

Systems FMEA

Examines potential risks associated with the design and development of systems and subsystems prior to launch or implementation.

A B C D E F G H I J K L M N O P Q R S T U V W X Y Z

5-Whys

What Question(s) Does the Tool or Technique Answer?

What is the root cause of this problem or problematic outcome?

The 5-Whys help you to

- Quickly focus on discovering the root cause
- Understand the relationship between a problem and its different causes

Alternative Names and Variations

This tool is also known as

- Why-why
- 5W

Variations on the tool include

- 5W2H or 5Ws and 2Hs
- Cause-and-Effect diagram; Ishikawa diagram; Fishbone diagram (**See Also** "Cause-and-Effect Diagram—7QC Tool," **p. 173**)
- Fault tree analysis (**See Also** "Fault Tree Analysis (FTA)," **p. 309**)

When Best to Use the Tool or Technique

Before any action is taken, this tool helps identify the source(s) of problem, rather than the symptom.

It is often used in conjunction with the development of a Fishbone diagram when trying to analyze and prioritize multiple problems and their root causes, rather than superficial symptoms.

Particularly useful when the problems involve human behavior or interactions.

Brief Description

The 5-Whys technique asks *Why* five times to drill down and identify the potential root cause. Applying this method prevents the analysis from stopping with *symptoms*, which leads to superficial solutions.

This is a simple technique—easy to use and often gets to the heart of the issue without statistical analysis.

How to Use the Tool or Technique

When using the 5-Whys technique

1. Select a potential key causal (often from a Cause-and-Effect diagram or Pareto). (**See Also** "Cause-and-Effect Diagram—7QC Tool," **p. 173**)
2. Ensure everyone on the team has a common understanding of what that cause means.
3. Ask *Why* this outcome occurs (or could occur)?
4. Select one of the reasons for the *Why* in number 3 and ask again: *Why does that occur?* or *Why does this situation cause the problem?*
5. Continue asking *why* until the potential root cause is reached and enough information is gathered, such that it is actionable.

How to Analyze and Apply the Tool's Output

- Review the information gathered and develop an appropriate action plan to remedy the root cause.
- Document the lessons learned so that the error is not repeated in the future.

Examples

A pricing analyst fails to notify management of an error in last month's report...

- *Why was the error not reported?*
    - Because I didn't know it was my responsibility.
- *Why did you not know it was your responsibility?*
    - Because we implemented a new process last month, and I presumed that finance was responsible.
- *Why did you think finance was responsible?*
    - Because finance was responsible when we used the former procedure.

- *Why did you not understand the procedure change?*
    - The change was not documented in the new procedure, and it was not discussed in the transition training workshop.

> **Hints and Tips**
>
> The number "5" is **not** a magical number. Stop whenever a potential root caused is reached, and the team can act on the information gathered. And as information is gathered, record any potential solution ideas, but do not stop the inquiry process until the root cause is reached and understood. Documentation of findings is important; the responses can be recorded as a list, Tree diagram, Fishbone diagram, or in table format. People are seldom the root cause of a problem—often their behavior is caused by something (poorly documented or communicated procedure, response to a special cause situation, and so on). Continue to ask *why* even if a person or role is identified.

Supporting or Linked Tools

Supporting tools that might provide input when using the 5-Whys include

- Histogram (**See Also** "Histogram—7QC Tool," **p. 330**)
- Pareto chart (**See Also** "Pareto Chart—7QC Tool," **p. 445**)
- Process map (**See Also** "Process Map (or Flowchart)—7QC Tool," **p. 522**)
- VOC and VOB data (**See Also** "Voice of Customer Gathering Techniques," **p. 737**)
- Brainstorming (**See Also** "Brainstorming Technique," **p. 168**)
- Cause-and-Effect diagram (**See Also** "Cause-and-Effect Diagram—7QC Tool," **p. 173**)
- 5MandP technique and its variants (**See Also** "Cause-and-Effect Diagram—7QC Tool," **p. 173**)

A completed 5-Whys provides input to tools such as

- Brainstorming (**See Also** "Brainstorming Technique," **p. 168**)
- Cause-and-Effect diagram (**See Also** "Cause-and-Effect Diagram—7QC Tool," **p. 173**)

A B C D E F G H I J K L M N O P Q R S T U V W X Y Z

- 5MandP technique and its variants (**See Also** "Cause-and-Effect Diagram—7QC Tool," **p. 173**)
- QFD (**See Also** "Quality Function Deployment (QFD)," **p. 543**)
- Other root cause analysis techniques (**See Also** "Hypothesis Testing," and "Regression Analysis" **p. 335** and **571**, respectively.)
- Concept generation methods (**See Also** "Failure Modes and Effects Analysis (FMEA)" "Hints and Tips" section, **p. 287**)
- FMEA (**See Also** "Failure Modes and Effects Analysis (FMEA)," **p. 287**)

A B C D E F G H I J K L M N O P Q R S T U V W X Y Z

Figure F-9 illustrates the link between the 5-Whys and its related tools and techniques.

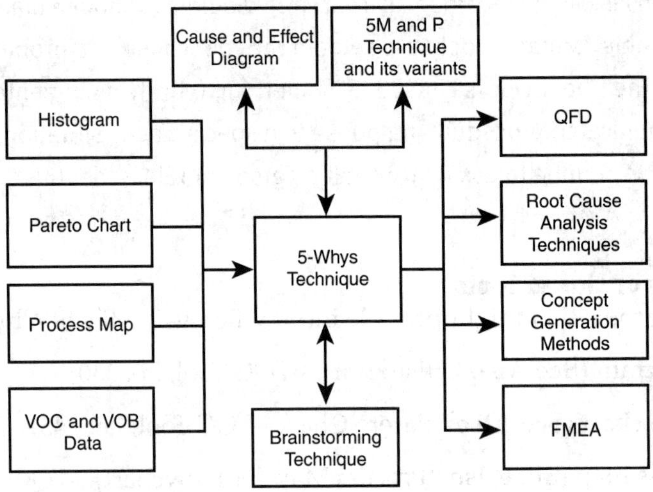

FIGURE F-9: 5-Whys Tool Linkage

Variations

5W2H

This uses the basic 5-Whys technique, but the following questions are asked **instead** of the five whys:

1W. Who?

2W. What?

3W. When?

4W. Where?

5W. Why?

1H. How?

2H. How much?

The 5W2H technique is often used when analyzing an improvement solution or when planning a new product, service, process or project initiation, or design and implementation.

The *how* and *how much* questions could be referring to cost or time or *how many* resources.

Cause-and-Effect Diagram; Ishikawa Diagram; Fishbone Diagram
The tool helps to organize and begins to analyze potential sources of a problem. (**See Also** "Cause-and-Effect Diagram—7QC Tool," **p. 173**)

Fault Tree Analysis
This tool uses a top-down analytic method and tree diagram structure with unique symbols to dissect a specific problem or problematic outcome into its related causes. (**See Also** "Fault Tree Analysis (FTA)," p. 309)

Fault Tree Analysis (FTA)

What Question(s) Does the Tool or Technique Answer?

What are the potential root causes of a single problem or problematic outcome, and how can they be prevented or mitigated? (Typically applied to a design of a system, product, or process that involves human interaction.)

Fault Tree analysis helps you to

- Quickly focus on discovering the root cause to prevent a single failure
- Understand the relationship between a problem and its different causes, to develop proactive risk mitigation strategies

Alternative Names and Variations

This tool is also known as

- FTA

Variations on the tool include

- Failure Modes and Effect Analysis (FMEA) (**See Also** "Failure Modes and Effect Analysis (FMEA)," **p. 287**)
- Cause-and-Effect diagram; Ishikawa diagram; Fishbone diagram (**See Also** "Cause-and-Effect Diagram—7QC Tool," **p. 173**)
- 5-Whys (**See Also** "5-Whys," **p. 305**)

When Best to Use the Tool or Technique

Before any action is taken, this tool helps identify the source(s) of problem, rather than the symptom, and categorizes them to assist in developing preventative strategies. This technique is best done early in the design stage, examining the robustness of a design (typically of a system or program). FTA is a narrow analysis, used to study one failure at a time.

Often it used in conjunction with the development of an FMEA when trying to analyze and prioritize multiple problems and their root causes, rather than superficial symptoms. It is particularly useful when the problems involve human behavior or interaction.

Brief Description

The Fault Tree Analysis (FTA) uses a *top-down* analytic method and Tree diagram structure to dissect and define the relationship of related causes of a specific problem or problematic outcome, using a unique set of classification symbols. The approach starts with a failure (either potential or existing) and probes backward toward the fundamental events or root causes. This FTA approach is frequently referred to as a *top-down* systems analysis.

In contrast, the FMEA analyzes an entire process or system to understand the causes of many potential failure modes and employs a bottoms-up method, starting with potential failure modes and then determining the possible consequences of a given failure, if it occurs. An FTA can be used within an FMEA process.

Using a vertical "tree" structure, constructed from top-to-bottom, the FTA graphically drills down into what could go wrong or what has gone wrong. Its unique symbols set it apart from other approaches that analyze root causes of failures or potential failures. The FTA unique symbols divide into two categories: 1) Event Symbols and 2) Gate Symbols. The symbols found next are the most commonly used; however, there are additional symbols for more complex situations. The Event symbols represent a dissection of elements describing what has or could occur and are hierarchically displayed from the key problem being analyzed. The Gate Symbols represent logical operators that link higher level and lower level elements in a hierarchical depiction of events. *[Symbols produced using Microsoft Office Visio software.]*

The Fault Tree analysis symbols include

Event Symbols

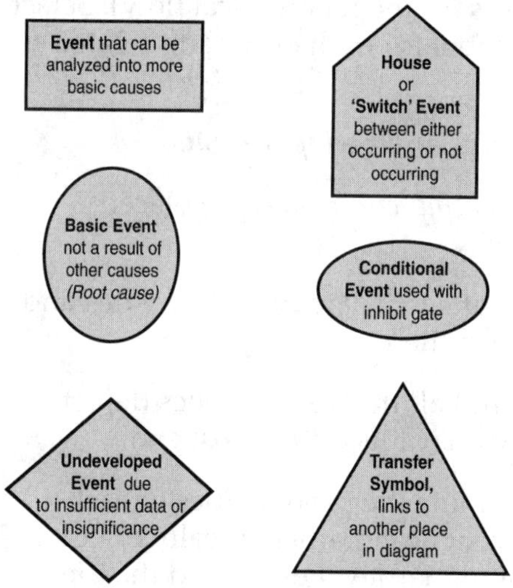

Gate Symbols

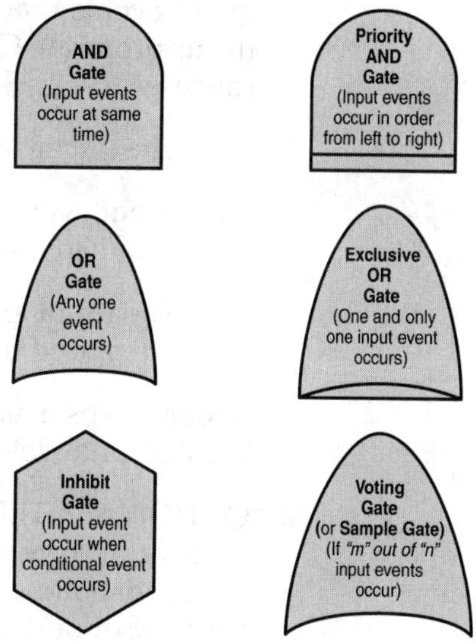

How to Use the Tool or Technique

When using a Fault Tree analysis

Step 1. Identify the key failure event to be studied, often called a *primary* or *top* event, including the boundaries that limit the analysis.

Step 2. Draw a rectangle at the top center and label it with this key failure event.

Step 3. Examine the system, design, or process and identify related events and elements to the key failure event being analyzed, using the appropriate background information and documentation, particularly a Process map or flow diagram if studying a process.

Step 4. Starting with the primary failure event, construct a hierarchical relationship of related events and elements.

  a. For each event, determine if it is a basic failure, or if can it be analyzed further for its own causes.

    i. If it is a basic failure, draw a circle around it.

    ii. If it is not a basic failure, draw a rectangle around it.

    iii. If appropriate, use the other symbols available to better define this element, such as an undeveloped event.

b. For each event, determine how it is related to the subsequent events that it causes (in the hierarchical flow). Select the appropriate Gate Symbol for each event and related causes.

   i. The lower-level events are the *input events.*

   ii. Input events cause *output events* and are placed above in the hierarchy.

   iii. The input and output events are linked by gates and are placed in between them.

c. Repeat Steps a. and b. until all the tree branches depict basic or undeveloped events at its outer-most end.

d. OPTIONAL STEP: Determine the probabilities of each basic or undeveloped event and mathematically calculate the probabilities of each higher-level event and the top event. (Supporting software can assist in this calculation.)

Step 5. Risk Analysis: Analyze the Fault Tree diagram to understand its multiple relationships and define strategies to prevent potential events that lead to failures.

a. Use the Gate Symbols to assist in determining the type of relationship and efficiently determine appropriate preventative strategy. (Reference the following Hints and Tips section for suggested strategies for each Gate Symbol.)

b. Focus first on the most likely to occur events.

Step 6. Document these strategies in a contingency action plan, describing what to do if a failure occurs and who is accountable for taking action.

How to Analyze and Apply the Tool's Output

Take action to prevent possible risk or respond to the occurrence of one. This is when the value of the FTA comes to light. (**See Also** "Failure Modes and Effects Analysis (FMEA)," "Hints and Tips" section, **p. 287**, for additional detail on risk analysis and management.)

Also periodically refresh and update the document as part of an evergreen process.

- Communicate and distribute the FTA's contingency action plan to appropriate parties.

- Incorporate the FTA and its contingency action plan into the control plan.

Document any lessons learned after action is taken and update the FTA and its contingency action plan to prevent similar events in the future.

Examples

Sample FTA

Figure F-10 illustrates an FTA about a sales prospect who refuses to see a sales representative.

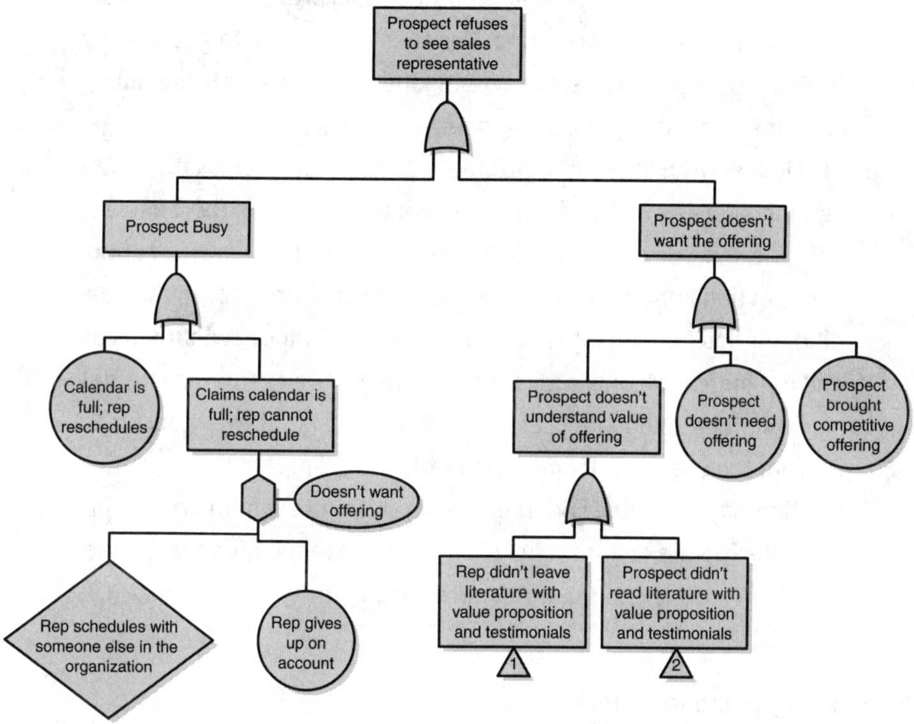

Figure F-10: Sample Fault Tree Analysis

> ## Hints and Tips
>
> **Potential Prevention Strategies for Each Gate Symbol**
>
> **AND Gate:** Eliminate at least one of the input events (given that they all must occur at the same time) to prevent the outcome from occurring.
>
> *continues*

OR Gate: Eliminate as many of the input events as possible (because any one of the input events could occur). Best to prioritize contingency actions on the most likely event to occur first. The more input events that are prevented, the more the overall outcome risk is minimized.

Inhibit Gate: Eliminate either the input event or the related conditional event to prevent the outcome from occurring.

Priority AND Gate: Eliminate any of the input events to disrupt the sequence of input events from occurring, thus preventing the subsequent outcome from occurring. Another strategy is to change the sequence of input events, considering the occurrence of the resulting outcome depends on the set sequence occurring.

Exclusive OR Gate: Similar to the recommended "OR Gate" strategy, eliminate as many of the input events as possible (because any one of the input events could occur). It's best to prioritize contingency actions on the most likely event to occur first. Given the condition of this gate dictates "one and only one" input event to occur, and not more than one to cause an output failure event, the strategy also might be to ensure that MORE than ONE input event occurs to prevent the overall outcome event from occurring.

Voting Gate: Starting with the most likely input event to occur, eliminate the input event until only "m-1" events remain (one less event than "m"), to prevent the outcome from occurring.

Supporting or Linked Tools

Supporting tools that might provide input when developing a Fault Tree Analysis include

- Histogram (**See Also** "Histogram—7QC Tool," **p. 330**)
- Pareto chart (**See Also** "Pareto Chart—7QC Tool," **p. 445**)
- Process map (**See Also** "Process Map," **p. 522**)
- VOC and VOB data (**See Also** "Voice of Customer Gathering Techniques," **p. 737**)
- 5-Whys (**See Also** "5-Whys," **p. 305**)

- 5MandP technique (**See Also** "Cause-and-Effect Diagram—7QC Tool," **p. 173**)
- Brainstorming (**See Also** "Brainstorming Technique," **p. 168**)

A completed Fault Tree analysis provides input to tools such as

- Brainstorming (**See Also** "Brainstorming Technique," **p. 168**)
- QFD (**See Also** "Quality Function Deployment (QFD)," **p. 543**)
- Root cause analysis techniques (**See Also** "Hypothesis Testing" and "Regression Analysis," **p. 335** and **571**, respectively.)
- Concept generation methods (**See Also** Failure Modes and Effects Analysis (FMEA) in "Hints and Tips" section, **p. 287**)
- FMEA (**See Also** "Failure Modes and Effect Analysis (FMEA)," **p. 287**)

Figure F-11 illustrates the link between a Fault Tree analysis and its related tools and techniques.

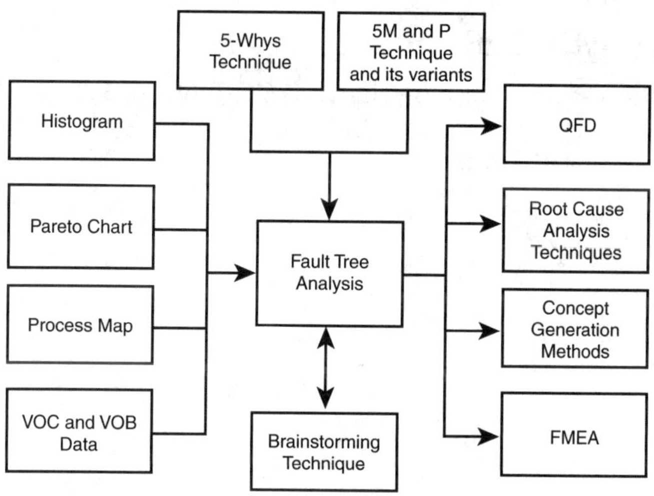

FIGURE F-11: Fault Tree Analysis Tool Linkage

Variations

Failure Modes and Effect Analysis

(**See Also** Failure Modes and Effects Analysis (FMEA), **p. 287**)

- Uses a bottoms-up analytic method to identify potential failures of a system, product, or process and defines their related causes (effects) and action plan to minimize or eliminate risk.

A B C D E F G H I J K L M N O P Q R S T U V W X Y Z

- FMEA often is applied to an entire system, product, or process, and the FTA can be used within it.

Cause-and-Effect Diagram; Ishikawa Diagram; Fishbone Diagram
(**See Also** "Cause-and-Effect Diagram—7QC Tool," **p. 173**)

- The tool helps to organize and begins to analyze potential sources of a problem.

5-Whys
(**See Also** "5-Whys," **p. 305**)

- This technique asks the question *why* to drill down and identify the potential root cause and prevent being satisfied with "symptoms," which leads to superficial solutions.

Fishbone Diagram—7QC Tool

See Also "Cause-and-Effect Diagram," **p. 173**.

Flowchart—7QC Tool

See Also "Process Map (or Flowchart)—7QC Tool," **p. 522**.

G

Gantt Chart

What Question(s) Does the Tool or Technique Answer?

How long will the project take? What are the key milestones, and when should you expect them?

A Gantt chart helps you to

- Organize the project's activities.
- Display and communicate the project's planned activities and key milestones, as well as its status on the actual progress to-date.

A B C D E F G H I J K L M N O P Q R S T U V W X Y Z

Alternative Names and Variations

This tool is also known as

- Milestone chart
- Project bar chart or (misnomer as "project plan")
- Activity chart

When Best to Use the Tool or Technique

The Gantt chart tool is an important project management tool and is used throughout a project. Initially, a Gantt chart is developed during the planning phase of a project and is used to monitor progress throughout the duration of a project.

A Gantt chart summarizes and communicates the planned project activities and the progress to complete them, thus is included in review meetings and status reports. It is a good summary tool to document the project schedule and file with the project "close-out" documentation as well.

Brief Description

The Gantt chart, named after Henry Gantt, is an effective project schedule summary communication tool. It displays project activity as a single horizontal bar chart, typically with time as the horizontal-axis and the planned activities as the vertical-axis. The left-most edge of the bar defines the *start date,* and right-most edge defines the *end date* of an activity. The portion of the bar containing unique shading represents the progress to-date, status, or percent-complete of a given activity. Often a diamond-shape or similar symbol indicates a key milestone. The person held accountable for an individual activity may be documented adjacent to the respective bar on the chart as well.

A Gantt chart is simply constructed and easy to understand. The tool is easy to use, and it quickly conveys the important project management information to an executive level. The critical path can be indicated on the chart by bolding its associated activity-bars.

Tools often confused with the Gantt chart are the Activity Network Diagram (AND) and PERT charts. If further analysis of bottlenecks and interdependencies is required, the AND is a more appropriate tool. A traditional Gantt chart lacks information about any interdependency between and among tasks; however, current software packages expand the original design with the important information. The PERT and CPM charts better reflect any activities' linked relationships. The PERT, AND (or CPM), and Monte Carlo simulations are well-suited tools for predicting variations in expected completion time of a project. (**See Also** "Activity Network Diagram (AND)—7M Tool," "PERT (Program Evaluation and Review Technique) Chart" and "Monte Carlo Simulation" **p. 127, 453,** and **431**, respectively, for more details.)

A B C D E F G H I J K L M N O P Q R S T U V W X Y Z

Example

There are several software packages that can be used to build a Gantt chart, such as Microsoft Project, Primavera, or Open Workbench, and one of the newest entries to the market is Minitab Quality Champion® used to plan and manage a project. Figure G-1 displays a simple Gantt chart using Minitab Quality Champion® application software.

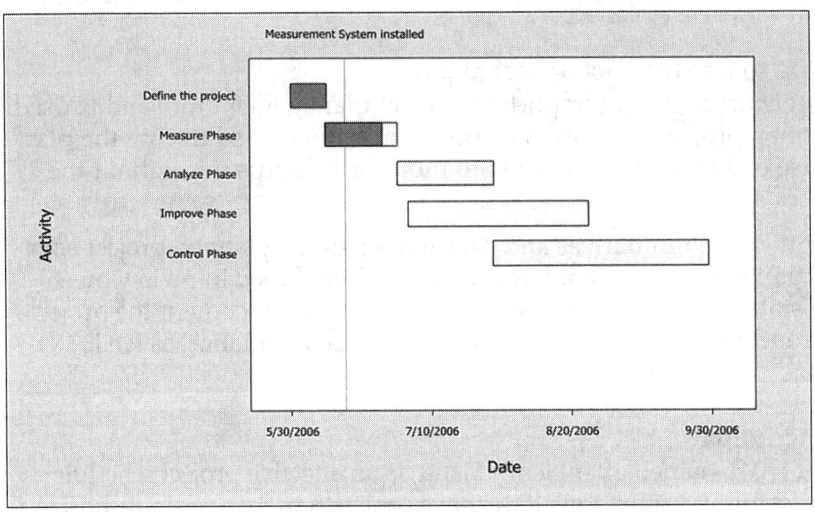

FIGURE G-1: Sample Gantt Chart (from Minitab Quality Champion®)

Notice in the Figure G-1 example that each bar represents a cluster of activities (or phases). The open, or un-shaded, bars represent planned phases, and the shaded portion of the top two bars depicts that portion of the project already completed. Hence, this sample project shown in Figure G-1 is close to completing the Measure phase of the five-phase project.

Supporting or Linked Tools

Supporting tools that might provide input when developing a Gantt chart include

- VOC (**See Also** "Voice of Customer Gathering Techniques," **p. 737**)
- Activity List
- Brainstorming (**See Also** "Brainstorming Technique," **p. 168**)
- Activity Network diagram (**See Also** "Activity Network Diagram (AND)—7M Tool," **p. 127**)
- Monte Carlo Simulation (**See Also** "Monte Carlo Simulation," **p. 431**)

A completed Gantt chart provides input to tools such as

- Activity Network diagram (**See Also** "Activity Network Diagram (AND)—7M Tool," **p. 127**)
- Monte Carlo Simulation (**See Also** "Monte Carlo Simulation," **p. 431**)
- Project plan and status report
- Project close-out documentation

Figure G-2 illustrates the link between a Gantt chart and its related tools and techniques.

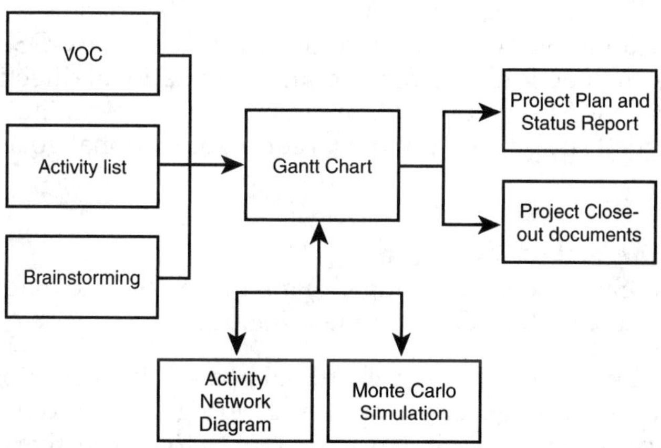

FIGURE G-2: Gantt Chart Tool Linkage

GOSPA (Goals, Objectives, Strategies, Plans, and Actions)

What Question(s) Does the Tool or Technique Answer?

How do the activities of a function, program, or project align with the overall organizational strategy?

GOSPA helps you to

- Develop a planning methodology used to align actions with organizational direction.
- Communicate direction and synergy with organizational goals.
- Audit specific activities to ensure linkage with strategic direction.

When Best to Use the Tool or Technique

Use the GOSPA technique at the beginning of the planning phase of a program or project and monitor its progress against the document throughout execution of a given plan.

Brief Description

GOSPA is a top-down planning technique that starts with the highest-level organizational goals and defines how individual groups support specific objectives. GOSPA is a simple acronym that stands for **G**oals, **O**bjectives, **S**trategies, **P**lan, and **A**ctions. This planning methodology summarizes an organization's direction and defines how supporting programs and/or projects relate to it.

In the context of Six Sigma for Marketing (SSFM), business, marketing, sales, and customer value chain professionals rely on the GOSPA technique to link their activities to the overall organizational direction to meet a specific goal. GOSPA helps to ensure that the goals of an individual program (or project) support that of the larger organizational goals.

How to Use the Tool or Technique

Develop a GOSPA document using either a matrix or Tree diagram. The guidelines for using the GOSPA method include

1. **Goals**—Document the organizational goals using directional, qualitative, broad, and general language. Every goal should have at least one objective; typically it may have two to four objectives.
    - For example, "Become a global leader in..."

2. **Objectives**—Identify specific goals that are measurable and provide an aggressive but achievable target.
   - For example, "$7million growth in sales by 2010…"
3. **Strategies**—Define the specific approach to achieve the objectives. **The strategy links the objective to a plan.** Consider the various alternative means to achieve the objective and select the best approach. Anticipate potential issues or changes that might emerge and identify an appropriate contingency plan.
   - For example, 1) Develop new products for current market, 2) Introduce product A into new market segment, and 3) Maintain annual expenditures to prior year's level.
4. **Plan**—Devise a supporting plan to implement a specific strategy that includes specific tangible elements, involving near-term actions and a measurable timeframe.
   - For example, 1) Initiate a product development team, 2a) Modify product A and select a new color for the cover and its packaging, 2b) Develop marketing collaterals and an advertising campaign that highlight value proposition to the new market segment, 2c) Train the sales people, 3a) Issue policy memo restating year's level as the current year expenditure targets, and 3b) Communicate a new presidential recognition program to celebrate those departments that successfully implement process improvements.
5. **Actions**—Expand the plan with details about *who* is accountable to do *what* of the specific plan, by *when*, *why*, and for *how much* (budget).

Figure G-3 displays the example GOSPA scenario as a matrix.

Figure G-4 uses a Tree diagram to illustrate the multi-dimensional hierarchical relationship that could result from an organization with three goals.

Hints and Tips

An organization should limit its goals to four or five. Ideally a goal represents two to four different objectives. Goals and strategies are **qualitative**. Objectives, plans and actions are **quantitative**. To help visualize the hierarchical supporting linkage, map the relationships using a Tree diagram or matrix.

Goal	Objective	Strategy	Plan (what)	Actions: Who	When	Why	How Much
Become a global leader in commercial batteries.	$7 million growth in sales by 2010	Develop new products for current market	Initate a product development team	R. Griggs	Completed by August 2008	New technology available	$1,500,000
		Introduce product A into new market segments	Modify product A and select a new color for the cover and its packaging	E. Kullberg	Completed by June 2007	Unique segment requirements and differentiation	$250,000
			Develop marketing collaterals and advertising campaign the highlight value proposition to the new market segment	F. DeYoung	Completed by June 2007	Unique segment requirements	$250,000
			Train the sales people	B. Skea	Completed by July 2007	Unique segment requirements	$300,000
		Maintain annual expenditures to prior year's level	Issue policy memo restating year's level as the current year expenditure targets	R. Keller	Completed by January 2007	Maintain year spend	N/A
			Communicate a new presidential recognition program to celebrate those departments that successfully implement process improvements	N. Minocha	Completed by January 2007	Implement Six Sigma methodology	$100,000

Figure G-3: Example GOSPA Matrix

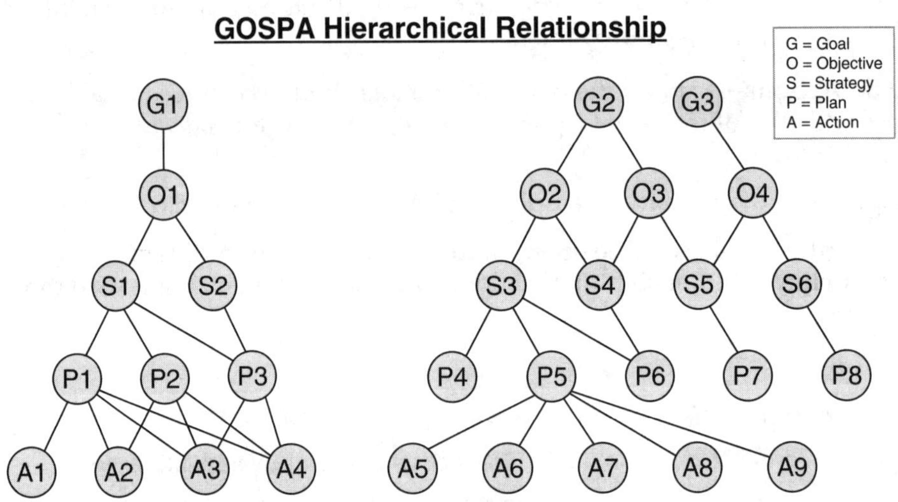

Figure G-4: GOSPA Multi-tiered Relationship

A B C D E F G H I J K L M N O P Q R S T U V W X Y Z

Supporting or Linked Tools

Supporting tools that might provide input when using the GOSPA technique include:

- VOB (**See Also** "Voice of Customer Gathering Techniques," **p. 737**)
- Brainstorming (**See Also** "Brainstorming Technique," **p. 168**)
- SMART (**See Also** "SMART Problem and Goal Statements for a Project Charter," **p. 665**)

A completed GOSPA matrix provides input to tools such as:

- SMART (**See Also** "SMART Problem and Goal Statements for a Project Charter," **p. 665**)
- Project Charter (**See Also** "SMART Problem and Goal Statements for a Project Charter," **p. 665**)
- FMEA (**See Also** "Failure Modes and Effects Analysis (FMEA)," **p. 287**)

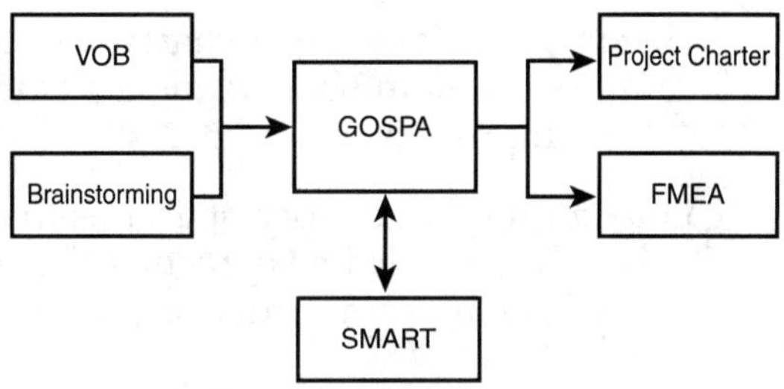

Figure G-5: GOSPA Tool Linkage

Figure G-5 illustrates the link between a GOSPA matrix and its related tools and techniques.

Graphical Methods

What Question(s) Does the Tool or Technique Answer?

What does the data look like?

Graphical methods help you to

- Describe what the data looks like (good communication tool).
- Identify trends and stability over time—the patterns or tendencies.
- Display the data distribution over its range (the spread).
- Isolate unusual observations (often called "special cause" variation).
- Detect relationships among variables.
- Support business decisions, by showing the data in a graph or chart.

Types of Graphical Methods

Figures G-6 and G-7 show the various types of graphs, interspersed with a list of graphical methods:

- **Bar chart**—A category or type of graph, wherein the height of a bar represents frequency of occurrence in summary or interval groupings. Bar charts can be individual, multiple (or clustered bars), or stacked bars representing groups of data.
- **Boxplot** (or Box-and-Whisker Plot)—Shows summary frequency in quartiles. (**See Also** "Boxplots—Graphical Tool," **p. 165**)
- **Dotplot** (or Dot graph) —Shows frequency in detail. (**See Also** "Dotplot," **p. 280**)
- **Histogram** (a type of bar chart)—Used with quantitative, continuous data to show summary frequency of interval or ratio data. (**See Also** "Histogram—7QC Tool," **p. 330**)
- **Line graph**—A category of graphs that shows the "connected" frequency as a single line depicting the area under the curve of a single variable, along some order (time or distance for example).

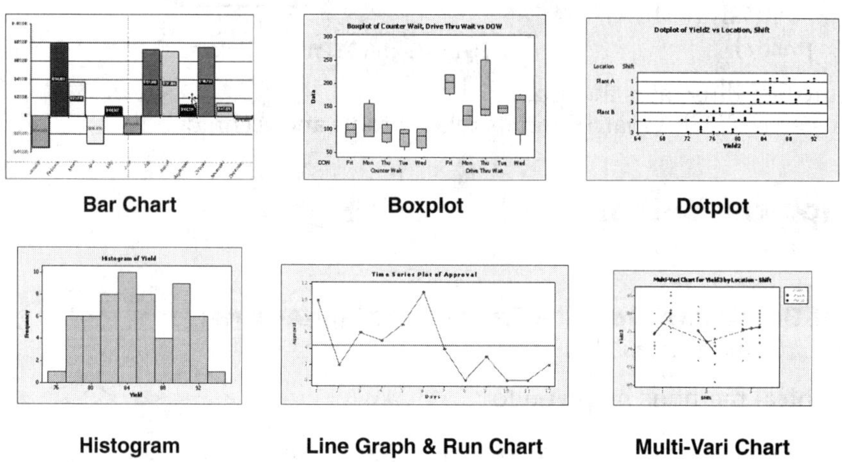

FIGURE G-6: Examples of Graphical Methods—Part 1 of 2

- **Multi-vari Plot**—Similar to a scatter plot; displays the relationship between variables, but this diagrams one variable's relationship (on the y-axis) with many other variables (depicted on the x-axis). (**See Also** "Multi-vari Chart," **p. 439**)
- **Pareto chart**—A specialized bar chart that shows frequency and accumulation of nominal data. (**See Also** "Pareto Charts—7QC Tool," **p. 445**)

A B C D E F G H I J K L M N O P Q R S T U V W X Y Z

- **Pie chart**—Used to show proportions often in business reports and presentations (but seldom in Six Sigma).
- **Run chart**—A type of line graph that shows a variable over time (time sequenced). It also is sometimes called Trend chart, Time Series plot, or Control chart. (**See Also** "Run Chart—7QC Tool," **p. 611**)
- **Scatter Plot**—Shows the relationship between the distributions of two different numeric variables—whether or not they are correlated. (**See Also** "Scatter diagram—7QC Tool," **p. 640**)
- **Stem-and-Leaf**—Displays the distribution of a sample as a set of numbers grouped by tens. Also called a stem-plot.

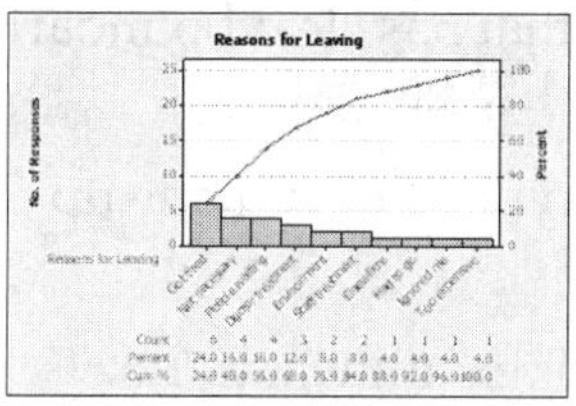

Pareto Chart

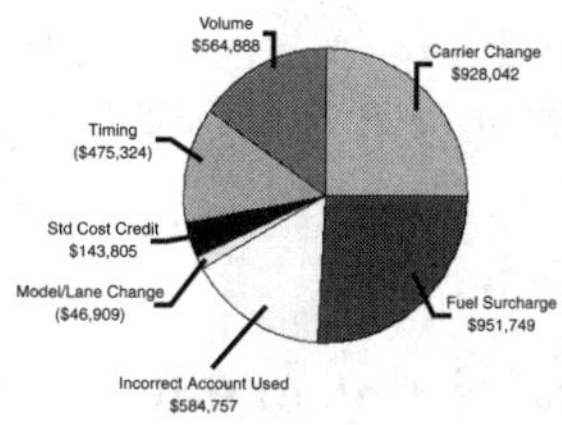

Pie Chart

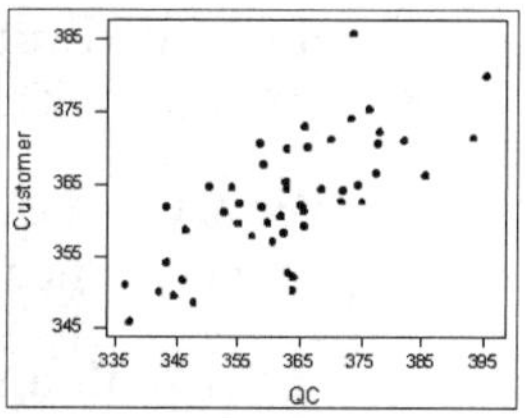

Scatter Plot

0	6																						
1	0	1	3	4	5	5	6	6	6	6	6	7	7	7	7	7	8	8	9				
2	0	0	0	0	1	1	2	6															
3	2	5	7	8																			
4	4	4	5	6	6	7	7	7	7	7	7	7	8	8	8	8	8	8	8	9	9	9	
5	0	0	0	1	2																		
6	3	4	5																				
7	1	5																					
8	0	2																					
9	5																						

Stem & Leaf Chart

FIGURE G-7: Examples of Graphical Methods—Part 2 of 2

When Best to Use the Tool or Technique

The old adage holds true, *a picture is worth a thousand words*. The graphical method is the picture that visually displays the distribution of a set of data or compares multiple sets of data.

Brief Description

The terms graph and chart often are used interchangeably. The rule of thumb is, "The simpler the better." Graphs are an effective way to present data, and they assist in communicating key causes and potential recommendations.

The type of data collected defines the type of graph used to depict the data. Data can be categorized into two primary classifications—either discrete or continuous data. Continuous data provides the most amount of information because it has limitless number of possible values between two given contiguous values. Continuous data is said to have no boundaries between values. It can be divided by two and still make sense. Often times, collecting and measuring continuous data requires more effort (time and money) to study compared with attribute or discrete data. Examples of both kinds of data include

- **Continuous or variable data**—A quantitative type of data that can take on a range of numeric values and is comprised of "non-counting" intervals such as time, height, weight, length, and temperature.
- **Discrete or attribute data**—A type of data that has clear boundaries between values. There are several types of discrete data:
    - **Binary**—A qualitative or categorical type of data made up of two classifications
        - **With a value rank or order—**For example, yes/no, pass/fail, good/bad, on/off, agree/disagree, and good/better/best.
        - **Without a value rank**—For example, male/female and heads/tails.
- **Counts**—A type of "quantitative" discrete data containing rank order or ordinal numbering (treated as a numeric scale of intervals). Number of defects (integers), days in the week, age in only years, number of wins or losses, income, and proportions are examples.
- **Nominal or categorical data**—Attribute data that is descriptive, not numeric, with **more than two categories**, for example, names, phone numbers, colors, type of car, capital cities, and states.

  This type of attribute data also can be equated to discrete values—for example, to distinguish among people, such as sales representatives, but maintain some anonymity, they could be coded and referred to a discrete number whereby Sally = 1, John = 2, Neeta = 3, and Vladimir = 4.

Several tools are listed in Table G-1, which summarizes the rule of thumb guidelines to select which type of graphical method to use when.

TABLE G-1: Rule of Thumb Summary of Graphical Methods

Graphical Tool	Purpose	Data Type	Average Sample Size
Boxplot	Depict a sample's frequency distribution as a summary box of quartile data. Used to identify the location of median, the spread (range), and the shape (skewed) of the distribution.	Continuous Quantitative	16-49
Dotplot	Depict a sample's frequency distribution, showing the detail of each data point. Used to identify the location of mean, median and mode, the spread (standard deviation, variance, range), and the shape (bell-shaped, skewed, binomial) of the distribution.	Continuous Quantitative	≤15
Histogram	Draw a sample's frequency distribution as a bar chart. Used to identify the location of mean, median and mode, the spread (standard deviation, variance, range), and the shape (bell-shaped, skewed, binomial) of the distribution.	Continuous Quantitative	50
Multi-vari plot	Draw the relationship between one variable with many other variables, to examine the differences in their means and variability.	Continuous Discrete	15 per variable

continues

A B C D E F G H I J K L M N O P Q R S T U V W X Y Z

TABLE G-1: Continued

Graphical Tool	Purpose	Data Type	Average Sample Size
Pareto chart	Draw a special type of bar chart frequency distribution showing the number of occurrences of events depicted and sorted as bars and a cumulative line showing the percent contribution of each event. Used to identify the largest occurrence or impact (80%) accounted for by the smallest number of events (20%)—the 80/20 rule or the Pareto principle, attributed to an Italian economist, Vilfredo Pareto.	Count (Attribute)	30
Run chart	Illustrate a variable over time, to detect trends, shifts, cycles, or extreme values in the data. Variable of interest plotted on y-axis and time on the x-axis.	Continuous Quantitative	25
Scatter Plot	Display the relationship between two variables	Continuous Discrete	15

The graphical tools listed in Table G-1 also have their own article found in this section of the book, describing the "how to" details.

Supporting or Linked Tools

Supporting tools that might provide input when using graphical methods include

- Data gathering plan to collect the appropriate metrics
- Performance charts and dashboards to study historical trends

A completed set of graphical methods can provide input to tools such as

- Cause-and-Effect diagram (**See Also** "Cause-and-Effect Diagram—7QC Tool," **p. 173**)

- QFD (**See Also** "Quality Function Deployment (QFD)," **p. 543**)
- Statistical analysis tools and other root cause analysis techniques (**See Also** "Statistical Tools," **p. 684**)
- FMEA with follow-on action planning (**See Also** "Failure Modes and Effects Analysis (FMEA)," **p. 287**)

Figure G-8 illustrates the link between Graphical Methods and its related tools and techniques.

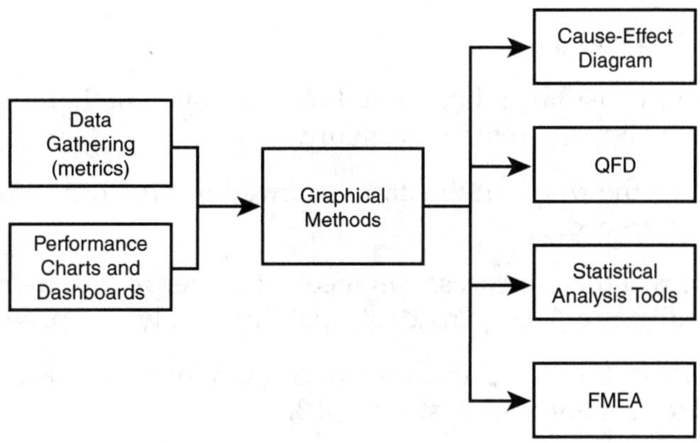

FIGURE G-8: Graphical Methods Tool Linkage

A B C D E F G H I J K L M N O P Q R S T U V W X Y Z

H

Histogram—7QC Tool

What Question(s) Does the Tool or Technique Answer?

How is the data of a process distributed; what does the data set look like? Are the data normally distributed?

A histogram helps you to

- Graphically display a data set to help indicate whether a process is distributed approximately normally.
- Determine the shape of the data—skewed or not, location of central tendency, spread.
- Analyze whether a process can meet customer requirements (that is, the spread). **See Also** "Process Capability Analysis," **p. 486**.
- Understand if the output from two or more processes is different. (**See Also** "Hypothesis Testing," p. **335**)

Alternative Names and Variations

This tool is also known as

- Frequency plot
- Area chart
- Bar chart or bar graph (for category data)

Variations on the tool include

- Dotplot
- Boxplot

When Best to Use the Tool or Technique

The old adage holds true, "a picture is worth a thousand words." A histogram depicts the shape of the data. Prior to applying any analytical tools and techniques to a data set, its shape must be understood to select the appropriate analysis.

Brief Description

A histogram is a type of *frequency plot* to visually display how often something occurs. The data is collected in consecutive order of occurrence in a process and plotted on a graph with the y-axis labeled "frequency" or "count" and the x-axis marked in the unit of measure for the variable of interest.

A histogram plots preferably 50 to 100 numerical data points (continuous, quantitative, interval, or ratio). This graphical method identifies the location of central tendency (mean, median, and mode), the *spread* (standard deviation, variance, range), and the *shape* (bell-shaped, skewed, binomial) of the distribution.

Two histogram guidelines distinguish it from other frequency plots. First, the data must be grouped into equal intervals, demarcated by the width of the bars. The data is sorted in order and equally divided to define the bar-width intervals. The length of a bar is proportional to the number of data points within each interval. Second, the graph displays the bars adjacent to one another, without any gap between them, representing a continuous scale on the x-axis. The center of the interval equals the midpoint of the bar. Graphing and statistical software packages such as MINITAB and Microsoft Excel employ these two guidelines in the background, making it easy for the user to produce a histogram.

Some histograms feature a "smoothing" line, which connects the top of a bar's midpoint to its adjacent bars. This smoothing line makes it easier to determine the area under the curve; hence, it is often called an *Area chart*.

The histogram is a member of the 7QC Tools (or *seven "Quality Control" tools*) attributed to Dr. Kaoru Ishikawa. The 7QC Tools sometimes are called the "seven basic tools" because they were the first set of tools identified as the *core* quality improvement tools. Ishikawa's original 7QC Toolset includes: 1) Cause-and-Effect diagram, 2) Check sheet (or checklist), 3) Control charts, 4) Histogram, 5) Pareto chart, 6) Scatter diagram, and 7) Stratification. More recently, the 7QC Toolset has been modified by substituting the Stratification technique with either a flowchart (or Process map) or a run chart (time series plot).

A *bar graph* or *chart* is a special kind of histogram to plot the frequency of categorical data. The bars for this type of graph should be drawn with a "gap" in between them because the x-axis scale is **not** continuous. Additional stratification can be illustrated by unique shading for each stratification. Stratified bars can be stacked or clustered side-by-side within each category.

The *Pareto chart* is a special kind of bar chart. The Pareto displays the categories ranked from highest to lowest frequency, and a line connects each of the bars to indicate the cumulative total, which adds to 100%. (**See Also** "Pareto Chart—7QC Tool," **p. 445**)

How to Use the Tool or Technique

Use the following procedure to develop a histogram using MINITAB. Given that numeric data has been entered into the MINITAB Worksheet, select the following commands from its drop-down menu: ***Graph > Histogram > Simple....***

Figure H-1 displays sample MINITAB screens where the appropriate graph type is selected ("Area 1") and the appropriate variable data of interest (for example, Yield) are selected ("Area 2") to produce the final sample histogram found in Figure H-2.

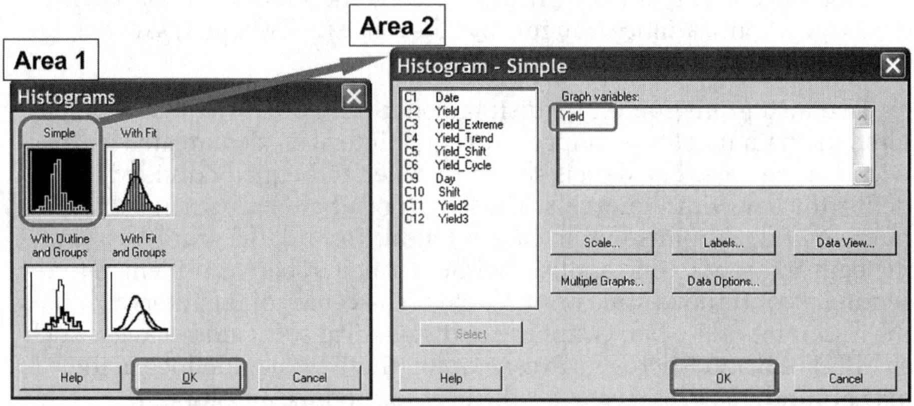

FIGURE H-1: Example MINITAB Histogram Main Screen

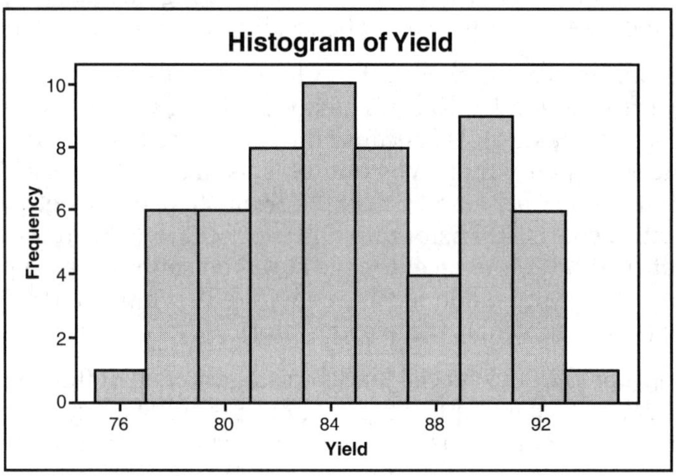

FIGURE H-2: Example MINITAB Histogram

Based on the Figure H-2 graph, one could expect the process to produce a yield in the mid-80s, with it ranging from the upper-70s to lower-90s. If a yield outside of this range occurred, we might investigate whether a special cause was present. The histogram appears to be fairly normally distributed, and we would suspect that as more data were gathered, the graph would become more bell-shaped. **(See Also** "Control Charts—7QC Tool," in the "Normal versus Non-normal Data" section for a discussion on statistical tests of normal distributions, **p. 217**)

How to Analyze and Apply the Tool's Output

A histogram that approximates a single, uni-modal, bell-shaped distribution is said to be a "normal", as the data are normally distributed. By chance alone, the data points are equally likely to occur on either side of the mean, such that the distribution is symmetrical about the mean. Bell-shaped symmetry contains a mean, median, and mode that equal one another. A normal histogram represents a stable, predictable process.

A histogram shape with a particularly large "spike" in one of the bars typically represents a special cause variation and should be investigated. Special cause events should be identified and eliminated.

A binary outcome produces a bimodal distribution, such as "yes/no" or "fail/pass." Bimodal distribution also could represent a segmented data set that might best be stratified, such as a two-shift operation. A bimodal histogram also may represent a stable process.

A skewed distribution represents a particularly long tail on either side of the mode—an asymmetrical curve. A negatively (or left) skewed curve features a long tail toward the "negative" side of the peak—off-center. The opposite is true for a positively skewed curve, which has a tail trailing off to the right of the peak. Distributions with a few extreme data points produce a skewed-shaped graph. An example often sited is the average home price in the United States with a few homes priced over $1 million.

If a process variation exceeds the upper and lower specification limits (that is, USL and LSL), then the process requires improvement modifications to reduce the variation. Inspection or audits may artificially cause the process to "squeeze" within the specification limits, thereby producing a truncated histogram with its tails being "artificially" cut off. Figure H-3 features various histogram shapes.

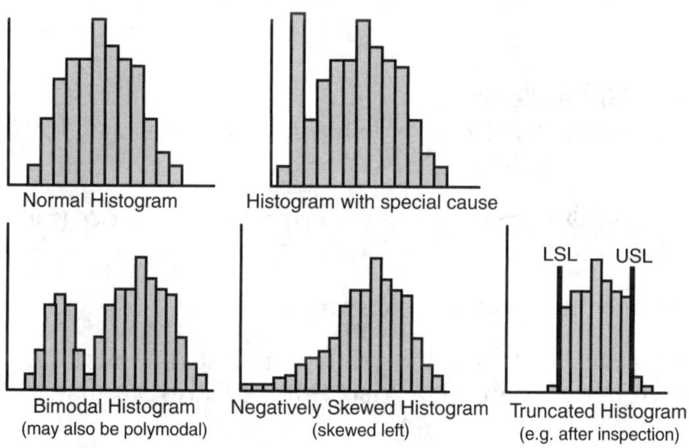

FIGURE H-3: Examples of Histogram Shapes

Hints and Tips

Histograms plot **numeric** data, and their bars are contiguous or adjacent to one another because their x-axis scale is continuous.

Bar charts plot **categorical** data and are depicted with "gaps" in between the bars because the x-axis contains the categories—non-continuous data. A Pareto is a special type of bar chart.

The histogram's visual approach allows estimation of

- Central tendency—mean, median, mode
- Spread—range, standard deviation, variance
- Shape—bell-shaped or not (skewed or bi-modal); unusual variation or truncation
- Kurtosis—a measure of the amount that the middle of the curve is flattened or peaked.

A bell-shaped (so called "normal") histogram characterizes a stable process in that the process is predictable. If the histogram appears to be distributed normally, it should be double-checked statistically for more information. If using MINITAB, its *Normal Probability Plot* or *Descriptive Statistics* applications provide the necessary statistics to verify statistical normality. If the data are normal, a larger number of statistical analytical tools become options as a "next step." (**See Also** "Control Charts—7QC Tool," in the "Normal versus Non-normal Data" section for a discussion on statistical tests of normal distributions, **p. 217**)

Supporting or Linked Tools

Supporting tools that might provide input when creating a histogram include

- Data gathering plan to collect the appropriate metrics (**See Also** "Data Collection Matrix," **p. 248**)
- Performance charts and dashboards
- Monte Carlo simulation, the probability outcome distributions displayed as a histogram (**See Also** "Monte Carlo Simulation," **p. 431**)

A histogram can provide input to tools such as

- Hypothesis testing, examining the frequency of sample of data (**See Also** "Hypothesis Testing," **p. 335**)

Figure H-4 illustrates the link between a histogram and its related tools and techniques.

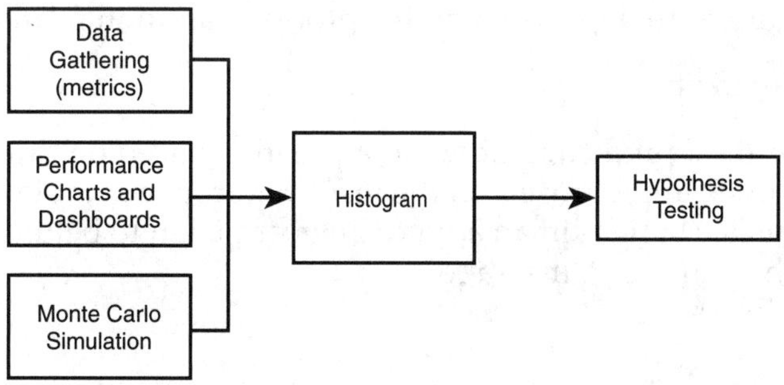

FIGURE H-4: Histogram Tool Linkage

House of Quality (HOQ)

See Also Quality Function Deployment (QFD), **p. 543**

Hypothesis Testing

What Question(s) Does the Tool or Technique Answer?

Is this population (or sample) of data different from another by chance alone or because of an outside influence?

Hypothesis testing helps you to

- Determine if one population (or sample) of items is statistically different from another.
- Understand if there is a statistical difference between two sets of data, either due to common cause (or chance alone) or because something has influenced the process.
    - Use as a Root Cause Analysis technique.
- Infer whether two sets of data are similar or different; determining if there is a relationship between them or not. Identify if one set of data is related to or is different from another set of data.

- Make decisions...
    - Decide if the "improvement" made to the product or process produced a statistically different result and is it worth the investment to continue to implement it.
    - Determine if the different defect types are independent of when in the production process they occur or if the defect categories are dependent on the process variation from day-to-day.
- Make inferences...
    - Make a prediction about a population based on sample data. For example, based on the feedback from a sample audience, make changes in an advertising campaign to better segment the marketing message.

Alternative Names and Variations

This technique uses several different statistical tests. Some of the more common ones include

- Chi-Square Test
- Student t-Test, or "t-Test", or Paired t-Test
- Z-test
- F-Test or ANOVA (Analysis of Variance)

(Table H-1 provides a more complete list of the different tests.)

Variations on the tool include

- A type of "inferential" statistics, wherein collected data is organized, analyzed, and interpreted so as to make predictions about a population or relationship.

When Best to Use the Tool or Technique

Hypothesis testing generally happens at two different times. First, after data has been collected, a hypothesis emerges as to the root causes (the critical variables) behind why the process or product performance is what it is. More questions arise to understand what the **k**ey **p**rocess **i**nput and **o**utput **v**ariables (KPIV and KPOV) are. Second is after a process or product modification has been made and the changed state is hypothesized to be improved from the prior state.

Brief Description

Hypothesis testing primarily answers the question of cause-and-effect relationship—a Root Cause Analysis technique. Sir Ronald Fisher, an early 1900s scientist and mathematician, developed several statistical principles to conclude causal relationship with a degree of confidence (or significance). His work established the *alpha-level* (α-level) standard (a default of 0.05) to represent the point beyond which the hypothesis gets accepted if supported by sufficient evidence. The larger the sample, the more confident one can conclude about the hypothesis. However, the practicality of time and economics often restricts the sample size. Hypothesis testing takes this balance into account by determining the amount of acceptable error.

Hypothesis tests use information contained in sample data to make conclusions about process or population characteristics. A formed theory, or hypothesis, may question whether two sets of data are related to one another; thus, it is part of the same population. If the theory tries to show they are different, generally it is because of chance alone or because one set of data had something influence it, causing a change. For example, a theory about whether soybeans grown in Asia are statistically different from soybeans grown in the United States. Do they have different genetic profile? Do the soil, climate conditions, and growth enhancers from the different regions influence how human digestion breaks down and uses its chemical components? Hypothesis testing can help answer these questions. Another example is a theory comparing two different processes with different timing and procedure for harvesting and cooking soybeans to see which process yields the most statistically nutritious by-products. Which process is better? Were the improvements worth the investment for a national rollout? Does the new process produce a better yield? Again, Hypothesis testing can help to answer these questions.

Hypothesis testing follows similar logic to that of the U.S. judicial system when drawing conclusions about evidence. A defendant is presumed innocent until proven guilty. The prosecutor carries the burden of proof to produce sufficient evidence (data) to reject the presumption of innocence. The presented evidence must be "beyond a reasonable doubt" to be considered valid. If sufficient evidence exists, a "guilty" verdict results. However, if the defendant is **not** declared "innocent," the final verdict merely states that there was "insufficient evidence" to convict. In terms of Hypothesis testing, the presumption of innocence is called the *Null hypothesis*.

Hypothesis Testing Terminology

Hypothesis testing utilizes a variety of special terms to describe its principles, including the following:

- **Hypothesis**—A theory requiring proof to declare if true or false.

- **Hypothesis test**—One of the statistical tests to declare if a theory is true or false. The appropriate test to use is based on both the type of data available and kind of decision required. Each test has a unique formula to evaluate the test frequency distribution relative to the calculated test statistic (or p-value).

  If the area under the curve beyond the test statistic is smaller than the significance level (alpha), then the test statistic probably did not come from the known distribution. Figure H-5 shows the Alternative Distribution's test statistic (its mean or μ_a) to the right of the upper confidence limit indicating that p-value is less than the alpha. Examples of common statistical tests include z, t-test, F and Chi-square. (**See Also** "Analysis of Variance (ANOVA)—7M Tool," for more details, **p. 142**)

A B C D E F G H I J K L M N O P Q R S T U V W X Y Z

> **Note**
>
> Alpha equals the shaded area to the right of the upper confidence limit.

- **Test statistic**—A calculated number that summarizes sample data and used to analyze an hypothesis. Compare the test statistic to the appropriate hypothesis test's critical value (or "t-critical") to decide to reject or not reject the Null hypothesis. Examples of a *statistic* include average (or mean), variance, and proportion.

- **Null hypothesis (H_0)**—A statement of what typically is desired to be disproved through sample data. In general, the statement about two sets of data (or a sample data set versus a population) are said to exhibit "no difference due to chance alone," denoted by the symbol "H_0" and illustrated in Figure H-5.

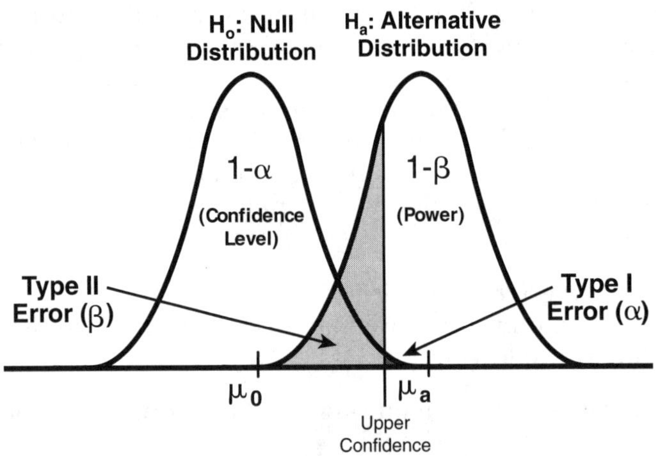

FIGURE H-5: Generic Hypothesis Testing Schematic

- **Alternative hypothesis(H_a)**—A statement of what is desired to be concluded through sample data. The alternative statement is dependent on the Null hypothesis. Hence, a "true" alternative hypothesis statement comes about only if the Null hypothesis is proven false. It implies that the two sets of data (or a sample data set versus a population) exhibit a difference due to a "real effect," not chance, denoted by the symbol "$\mathbf{H_a}$."
    - If the Null is true, the correct wording is that there is "insufficient evidence to reject the Null hypothesis" or we are "unable to reject" it.
    - If the Null is false, the correct wording is, "there is sufficient evidence to reject the Null hypothesis."

    Figure H-5 illustrates the two distributions contrasting a "generic" Null and alternative hypothesis.

- **Confidence interval**—The probability range within which the population data (parameters) exists. This region can be described two ways, either a single limit (an upper or lower) or a two-sided limit (both an upper and lower). Outside these limits is defined as the *rejection region* or *critical region*, wherein the "test variable" is said to **not** be part of the population. Sometimes these limits are referred to as the *t-critical* as they demarcate the rejection boundaries. Figure H-5 illustrates a one-tail upper limit.

    A commonly used confidence interval is 95%, wherein 95 times out of 100 expect to be right, versus five times out of 100 expect to be wrong. This means that if sample data (or test statistic) were drawn from a population distribution, there is a 95% confidence that the test statistic will be contained within this range.

- **One versus two-tailed tests**—Hypothesis statements are worded as pairs—the Null and alternative hypothesis—to represent the probability of all the possible outcomes. In general, the Null hypothesis contains an "equal to" statement, and the alternative statement is the theory to be proven.

    One-tail test—Seeks to answer if the test statistic is less than or greater than the known distribution. An alternative hypothesis stated as a "greater than" test is a right-tail test. Conversely, a "less than" alternative hypothesis statement is a left-tail test.

    - **Right-tail test—$\mathbf{H_0}$:** H_a *mean* $\leq H_0$ *mean*; $\mathbf{H_a}$: H_a *mean* $> H_0$ *mean*.

        Example of a written word statement: $\mathbf{H_0}$: The average speed of Product B is **less than or equal to** that of Product A; $\mathbf{H_a}$: The average speed of Product B **is greater than** that of Product A.

- **Left-tail test—$\mathbf{H_0}$—H_a *mean* $\geq H_0$ *mean*; $\mathbf{H_a}$: H_a *mean* $< H_0$ *mean*.**

  Example of a written word statement: $\mathbf{H_0}$—The average speed of Product B is **greater than or equal to** than that of Product A; $\mathbf{H_a}$: The average speed of Product B **is less than** that of Product A.

> **Note**
>
> This left-tail test scenario dictates that **bigger is better.**

Two-tailed test—Merely asks if the test statistic is different from the known distribution; it is indifferent to whether it is larger or smaller. Hence, the alternative hypothesis statement is worded as a "not equal to" and as a two-tailed test because either tail would satisfy the hypothesis.

- **Two-tailed test—$\mathbf{H_0}$—H_a *mean* $= H_0$ *mean*; $\mathbf{H_a}$: H_a *mean* $\uparrow H_0$ *mean*.**

  Example of a written word statement: $\mathbf{H_0}$: The average speed of Product B is **not different** from that of Product A; $\mathbf{H_a}$: The average speed of Product B **is (detectably) different** from that of Product A.

> **Note**
>
> This two-tailed test scenario dictates that **nominal is best.**

Figure H-6 illustrates one versus two-tailed tests.

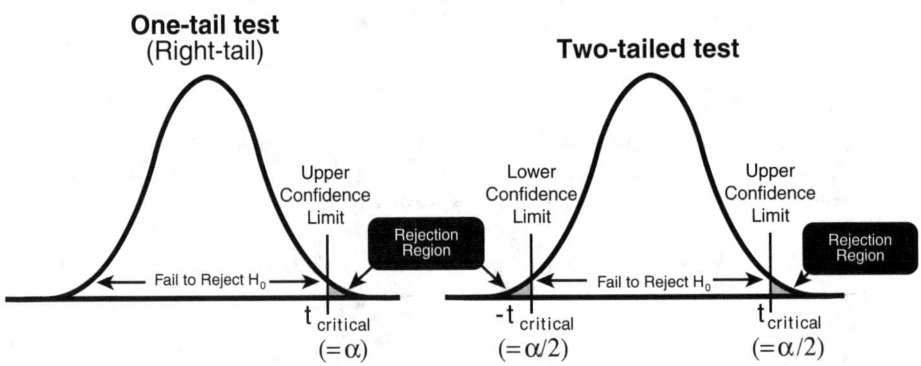

FIGURE H-6: Generic One versus Two-tailed Hypothesis Test

- **Confidence level**—Defines making a correct test decision (correctly failing to reject the Null hypothesis) because the Null is true. The "double-negative" wording is awkward English, but this language adheres to the probabilistic-nature of hypothesis testing, similar to the U.S. Judicial system. It is determined by the formula ***one minus alpha*** (or 1- α). Often the default is set at 95% as an industry standard. Figure H-5 illustrates a confidence level of a generic hypothesis test schematic.

- **Significance level**—The pre-defined level of certainty required to "test" such that the observed effect (a test statistic or "test factor") **actually caused** the response or is different from the known distribution, and not by chance alone, denoted by the term alpha (α).

  This significance level (alpha) determines the probability that a test statistic is part of the test known distribution—the distribution representing Null hypothesis. This probability relates to the *p-value.*

  Often the default alpha = 0.05 (or 5%), but sometimes 0.01 (1%) or 0.10 (10%) are used.

  - If the hypothesis test requires a one-tail test (wherein only one confidence level is used), then the default alpha is 5%.

  - If the hypothesis test is a two-tailed test (wherein both the upper and lower confidence intervals are used), the default alpha is evenly split between the two tails $\pm$ 2.5% (0.05/2 = 0.025).

  Confidence level, significance level, and confidence interval are related. With the significance level (alpha) set at 5% and the confidence level is 95% (one minus alpha), then the data that falls within the 95% confidence interval are said to have occurred by chance alone (no special cause effecting the outcome).

Figure H-5 illustrates an alpha in a "generic" hypothesis test schematic.

- **P-value**—The p-value serves the decision criteria to interpret a hypothesis test. A "goodness of fit" statistic represents the **probability** that a test statistic is part of the known distribution (the Null hypothesis). It is calculated by the area of the distribution beyond the test statistic. The smaller the p-value, the more confidence exists that the effect is real and not by chance alone.

  If the **p-value $\leq$ 0.05, then reject H_0**. Hence, the test statistic is located in the tail of the distribution beyond the confidence limit (or t-critical), in the rejection region. This provides sufficient evidence to suggest that the test statistic is not part of the known distribution and the H_0 is false. The likelihood of being wrong by rejecting the Null hypothesis is less than 0.05.

  - The *Type I Error* is no bigger than 0.05. (Most equate the p-value with the Type I Error.)

A B C D E F G H I J K L M N O P Q R S T U V W X Y Z

If the **p-value > 0.05, then do <u>not</u> reject H_0**. Hence, the test statistic is located inside the confidence interval of the known distribution, and **not** in the rejection region. There is insufficient evidence to suggest that H_0 is false, and the test statistic appears to be part of the known distribution.

- Perhaps with more evidence (that is, data), rerunning the test may indicate that the H_0 is false. However, the additional data (or increased sample size) is needed. (Type II Error is too large or Power is too low.)

A B C D E F G H I J K L M N O P Q R S T U V W X Y Z

- **Power**—Defines making a correct test decision—correctly rejecting the Null hypothesis because the Null is false. Power represents the *ability to detect a difference when one exists*. It is determined by the formula ***one minus beta*** (or 1- β). Often the default is set at 80%. Figure H-5 illustrates how Power is associated with the Alternative Hypothesis distribution of a generic hypothesis test schematic.
- **Errors in Hypothesis testing**—Hypothesis testing involves both samples and probabilities; therefore, at times incorrect conclusions are drawn. Two types of hypothesis errors exist, and both represent a different type of risk. The appropriate risk levels are selected prior to testing the hypothesis with sample data—namely the significance level (alpha) and beta.
  - **Type I Error** arises when incorrectly rejecting the Null hypothesis, when it is actually true. Thus based on the test statistic, the conclusion rejects the Null hypothesis, but in truth, it should be accepted. Type I Error equates to the *alpha* (α) or *significance level*, whereby the generally accepted default is 5%.

    This error type is also known as *producer's risk* because the consequence of this incorrect decision of thinking a part or product is bad causes it to be scrapped or reworked, when in fact the item actually is acceptable.

    Alternatively, the confidence level for making the correct decision by failing to reject the Null hypothesis, when in fact the Null is true, is calculated by ***one minus alpha***. This can be found in the upper left corner of both the quadrants found in Figures H-7 and H-8.

    Hence, total probability (p) of correctly deciding that a Nullhypothesis is true (that is, 100% or 1) *equals* the probability of correctly failing to reject the Null (*1 minus alpha*) *plus* the probability of incorrectly rejecting the Null (*alpha*)—or 1 = (1-α) + (α) = confidence level *plus* the Type I Error.

- **Type II Error** occurs when the test decision incorrectly accepts the Null hypothesis. Based on the test statistic, the final decision fails to reject the Null when it is actually false. Type II Error also is called *"beta"*(β), and the default is typically set at 20%.

  This error type is also known as *consumer's risk* because the manufacturer ships what it thinks is a good product, but in truth, the customer receives a bad (or defective) product.

  Alternatively, the Power of making the correct decision to reject the Null hypothesis, when in fact the Null is false, is calculated by ***one minus beta***. This can be found in the lower right corner of both the quadrants found in Figures H-7 and H-8.

  Hence, *total probability (p)* of correctly deciding that a Null hypothesis is false (that is, 100% or 1) equals the probability of correctly rejecting the Null (*1 minus beta*) plus the probability of incorrectly accepting the Null (*beta*)—or $1 = (1 - \beta) + (\beta) =$ Power *plus* the Type II Error.

- **Interdependency and Risk Level**—In general, the intention is to minimize both alpha and beta. However, given a constant sample size, alpha and beta are inversely related, such that minimizing alpha increases beta. As Figure H-5 illustrates, if the upper confidence level setting moves left to shrink the Type II Error (or *beta*), the Type I Error (or *alpha*) increases.

Note

Increasing the sample size for a given alpha-level decreases beta. The sample size is impacted by the two error types (alpha and beta), the standard deviation and whether there is a detectable difference when one exists (Power, for example). (**See Also** Sampling, **p. 618**)

The scenario (or situation) of the theory helps to prioritize between the two errors. Setting an acceptable level of risk is determined by the seriousness of the error. If the product affects the customers' health, the default of 20% probably is unacceptable, and this Type II Error would be set much lower.

Figure H-7 illustrates the various possible decision outcomes versus the truth, including the two error types and the consequences, using the U.S. Judicial system, and Figure H-8 uses a product scenario.

A B C D E F G H I J K L M N O P Q R S T U V W X Y Z

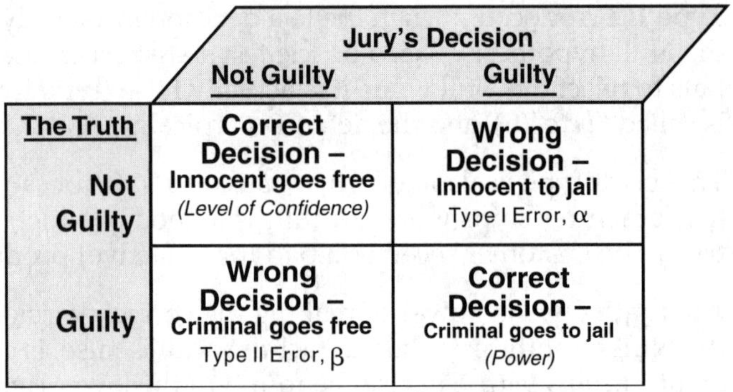

FIGURE H-7: US Judicial System Example of Decision Outcomes

Figure H-8 represents a similar decision outcomes quadrant as Figure H-7 but includes the distribution curves similar to those found in Figure H-5 to show how the test statistic (identified by an "X" on the x-axis) relates to the curves and the different decision options. If the calculated test statistic (X) falls to the right of the "t-critical" (or upper confidence limit) in the tail, then the test results produce a p-value that falls into the *rejection region*, thereby rejecting the Null hypothesis.

Using a generic two-product example for a one-tail test, the scenario theorizes that (the new) Product B works better (faster) than Product A. Knowing that the Alternative Hypothesis should represent the theory, the pair of hypothesis statements for Figure H-8 can each be written one of two ways as a one-tail test:

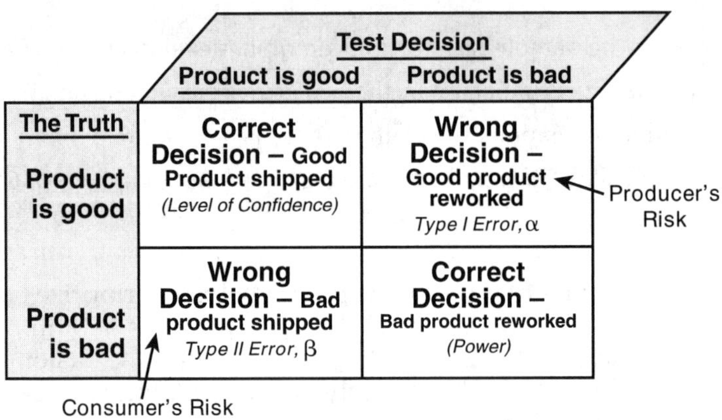

FIGURE H-8: Generic Product Example of Decision Outcomes

Option 1:

- Null Hypothesis (**Ho**): **Product A ≥ Product B**
- Alternative Hypothesis (**Ha**): **Product A < Product B**

Option 2:

- Null Hypothesis (Ho): Product A—Product B = 0
- Alternative Hypothesis (Ha): Product B—Product A > 0

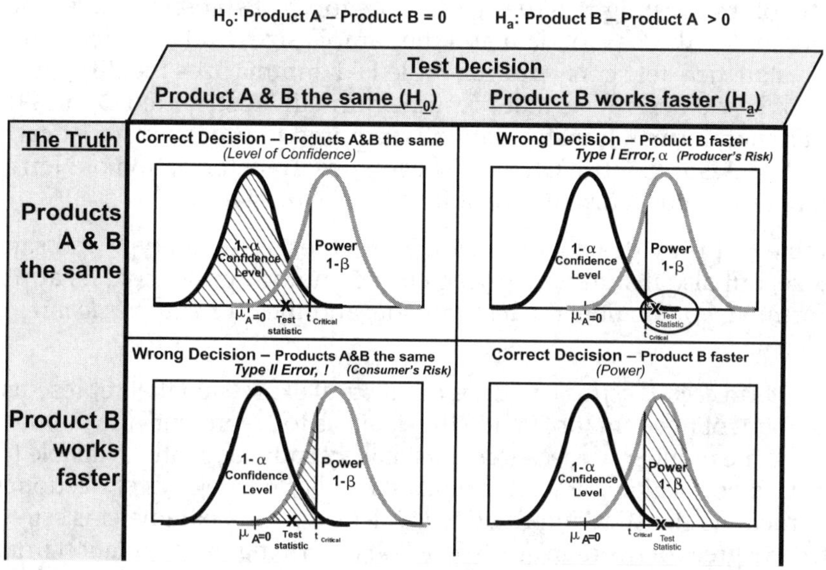

FIGURE H-9: Generic Two-Product Example of Decision Outcomes with Distribution Curves

In this two-product scenario, *bigger is better*, indicating that Product B works faster than A. Presume that the hypothesis test set the significance level at 95% for this one-tail test, with the risk levels set at their respective defaults—alpha at 5% and beta at 20% . Hence, if the hypothesis test results show that the p-value < 0.05, then reject H_0 and take action to ship the new product (Product B) because it works faster than Product A.

In Figure H-9, the upper left quadrant illustrates when the hypothesis test concludes with 95% confidence that the Null is true; and the p-value is greater than or equal to 0.05. In the lower right quadrant, the hypothesis test provides sufficient evidence to reject the Null, the p-value is less than 0.05, and the test conclusion is correct. *Power* is set at 80% (that is, one minus beta of 20%) to detect a distinguishable difference between the two products.

In both the upper right and lower left quadrants of Figure H-9, the risk of making a wrong decision is illustrated. The upper right quadrant shows the 5% probability (that is, 1 minus the 0.95 significance level) of making a Type I Error of incorrectly rejecting the Null hypothesis, when the test results-produced p-value is less than 0.05; however, in truth the products worked the same. The lower left quadrant shows the 20% probability of making a Type II Error of incorrectly accepting the Null

A B C D E F G H I J K L M N O P Q R S T U V W X Y Z

hypothesis when the test results produced p-value is greater than 0.05; however, in truth Product B worked faster than A.

Different Hypothesis Tests

Hypothesis testing is actually the "category" of statistical tools, and depending on the type of data and the sample size available, determines which statistical test gets applied. Table H-1 summarizes the different types of Hypothesis Tests that presume normally distributed data. If the data are not normal, transformation is required to use these statistical tests. (**See Also** "Control Charts," section on "Transformation of Non-normal Data (plus Normal Probability Plots," **p. 444**)

Table H-1 provides not only the different tests by data type and sample size, but also it summarizes each test common application, its applicable formulas to calculate the test statistic, and the *degrees of freedom* if applicable.

The term *degrees of freedom* (df) is a statistical expression that represents the amount of freedom (or "float") the data has to "represent" the population. It is the number of measurements that are independently available to estimate a population parameter. As data is collected, degrees of freedom (df) are earned. As statistical information is calculated (or a parameter is estimated), degrees of freedom are "spent" on describing the population from the sample data. The mean is a calculated statistic, and it uses up one degree of freedom, resulting in (n-1) degrees of freedom, where "n" represents the sample size. If a population average is represented by four numbers, there is freedom for the first three numbers to be whatever they want to be, but the fourth number must be "dedicated" (as a calculation) to achieve the same population average. Hence, that fourth number does **not** have the freedom to be whatever it wants to be, and so a four-number sample (n=4) has three degrees of freedom to describe the average (for example, df = n-1; 4-1 = 3).

Manual calculations require the appropriate hypothesis test formula to compute and compare the specific sample data set's test statistic with the test's critical value. Manual calculations also require referencing of the statistical test's distribution table to derive the appropriate critical value for the specific scenario. These distribution tables are provided as a reference toward the back of the book. (See the "Statistical Distribution Tables," in Appendix A, **p. 939**.) However, statistical software packages, such as MINITAB, have automated the calculations and negate the need for the formulas and distribution tables. Table H-1 summarizes the commonly used Hypothesis tests, including the type of data required for the test, the recommended sample size, its appropriate application, its test statistic formula, and degrees of freedom formula.

TABLE H-1: Reference Summary of Common Hypothesis Tests

Hypothesis Test	Data Type and Sample Size	Application	Test Statistic Formula	Degrees of Freedom
t-Test (or Student t)	Continuous or Variable. Sample size is less than 30 (n<30); but ideally at least 20 (20<n<30).	• Comparing one mean to a standard—historical average or target; sometimes called ***One-sample t-Test***.	$t = \frac{\overline{X} - \mu}{\frac{s}{\sqrt{n}}}$	n – 1
		• Comparing two independent means; sometimes called ***Two-sample t-Test***.	$t = \frac{\overline{X}_1 - \overline{X}_2}{Sp\sqrt{1/n_1 + 1/n_2}}$ where $Sp = \sqrt{\frac{(n_1-1)S_1^2 + (n_2-1)S_2^2}{(n_1+n_2-2)}}$	$n_1 + n_2$ -2
			$t = \frac{\overline{X}_1 - \overline{X}_2}{\sqrt{S_1^2/n_1 + S_2^2/n_2}}$	*Determined by the Welch-Satterthwaite approximation*
		• Comparing the difference between two dependent sample means with paired or matched data; sometimes called ***Paired t-Test***.	$t = \frac{\overline{d}}{S_d/\sqrt{n}}$ $S_d = \sqrt{\sum \frac{(d_1 - \overline{d})^2}{n-1}}$	n - 1

continues

TABLE H-1: Continued

Hypothesis Test	Data Type and Sample Size	Application	Test Statistic Formula	Degrees of Freedom
Z-test	Continuous or Variable. Sample size is 30 or more (n≥30).	• Comparing one mean to a standard	$z = \frac{\overline{X} - \mu}{\frac{\sigma}{\sqrt{n}}}$ or sample std. deviation	N.A.
		• Comparing two means.	$z = \frac{\overline{X}_1 - \overline{X}_2}{\sqrt{\sigma_1^2/n_1 + \sigma_2^2/n_2}}$ or sample std. deviation	
		• Comparing two proportions.	$z = \frac{p - \pi_0}{\sqrt{\pi_0(1 - \pi_0)/n}}$ For sample data: $z = \frac{p_1 - p_2}{\sqrt{\overline{p}(1 - \overline{p})((1/n_1) + (1/n_2))}}$ $\overline{p} = \frac{n_1 p_1 - n_2 p_2}{n_1 + n_2}$	
		• Comparing the difference between two means with paired or matched data.	$z = \frac{\overline{d}}{S_d/\sqrt{n}}$	

Hypothesis Test	Data Type and Sample Size	Application	Test Statistic Formula	Degrees of Freedom
Analysis of Variance (ANOVA) or F-test	Continuous or Variable (One-Way ANOVA: Continuous Y and Discrete (count) Xs). Sample size is 20 or more per group (n≥20 per group).	• Comparing two or more means; called ***ANOVA***. • Comparing two or more variances sometimes to quantify the variation between and within factors (e.g. Repeatability and Reproducibility). Called ***Test for Equal Variances*** or ***F-test***.	$F = \frac{(S_1)^2}{(S_2)^2}$	$n_1 - 1$ $n_2 - 1$
Chi-Square Test ($\chi 2$)	Attribute (category/ nominal and count/ ordinal). Sample size is five or more per cell of a contingency table or cross-tab matrix of data (n≥5 per cell).	• Comparing two or more proportions. • Tests sample variance against known variance (the population) to determine if a data set fits a known distribution; called a ***Goodness of Fit*** test. • Compares observed and *expected frequencies of test* outcomes to determine if probabilities of one variable are associated with a second; called ***Test for Independence***.	$\chi^2 = \frac{(n-1)(s^2)}{\sigma^2}$ $\chi^2 = \sum \frac{(O-E)^2}{E}$ where O = observed; E = expected frequencies. (*Test for Independence calls for a right-tailed test.*)	n – 1 (r –1)(c-1); *where r = row; c = column*

> **Note**
>
> Some of the symbol definitions include: μ = population mean; $\overline{X}$= "x-bar" = sample mean; σ = population's standard deviation; s = sample's standard deviation; n = sample size; d = difference between two sample means; "d-bar" = mean of d; Πo = hypothesized population proportion; p = proportion in sample; and p-bar = average of two sample proportions.

A *Contingency table* is a matrix used to organize categorical data in preparation for analysis such as a Chi-Square Hypothesis test. The columns and rows represent two different ways to analyze the same set of data and are used to calculate degrees of freedom (df = [(# rows–1) x (# columns –1)], for example).

Guidelines for Hypothesis Statements

Statistical hypothesis testing applies to many different situations, and accordingly the correct wording of the statements varies as well. As a rule of thumb, the *Null hypothesis statement generally contains an "equal to"* (=) in it (including "greater than or equal to" (≥) or "less than or equal to" (≤)). The following table summarizes the appropriate hypothesis test wording for each of the different types of hypothesis tests. Table H-2 provides general guidelines for the wording of both the Null and Alternative hypothesis for each test.

Table H-2: Hypothesis Statement Guidelines

Type of Test	Null Hypothesis (H_o)	Alternative Hypothesis (H_a)
• Normal Probability Plot	Data are normal. (that is, mean = mode = median)	Data are not normal.
• Anderson-Darling Test for Normality (Descriptive Statistics)	Data are normal. (that is, mean = mode = median)	Data are not normal.
• Student t-Test	The mean of the sample distribution is the same as the reference mean.	The means are not equal.
• One-way ANOVA Table	There is no difference between the means.	There is at least one mean different from the others.

Type of Test	Null Hypothesis (H_0)	Alternative Hypothesis (H_a)
• Two-way ANOVA Table	There is no difference between the means.	There is at least one mean different from the others.
• Balanced ANOVA Table	The factor has no effect on the response.	The factor does have an effect on the response.
• Mood's Median Test (non-parametric)	Medians are equal.	At least one median is different from the others.
• Kruskal-Wallis Test (non-parametric)	Medians are equal.	At least one median is different from the others.
• Chi-Square	There is no relationship between the variables.	There is a relationship between the variables.
• Correlation	There is no correlation between the variables (correlation = 0).	There is correlation between the variables.
• Regression Analysis (Starting at the Top of the Table)	Coefficient is equal to zero.	Coefficient is not equal to zero.
• Regression Analysis (Starting at the Bottom of the Table)	All coefficients are equal to zero, except b0. ...OR... The slope of the line is equal to zero.	At least one of the coefficients is not equal to zero. ...OR... The slope of the line does not equal zero.
• Lack of Fit Test	The proposed model fits the data.	The proposed model does not fit the data.
• Homogeneity of Variance (Note: Use F-Test or Bartlett's Test for normal data; Levene's Test for non-normal data.)	The variances are equal.	At least one variance is different.

See Also "Control Charts—7QC Tool," for a discussion on *Normal Probability*, **p. 217**; "Analysis of Variance (ANOVA)—7M Tool," for a discussion on the different ANOVA tests, **p. 142**; "Scatter Diagram—7QC Tool," for additional discussion on correlation, **p. 640**; and "Regression Analysis," **p. 571.**

A
B
C
D
E
F
G
H
I
J
K
L
M
N
O
P
Q
R
S
T
U
V
W
X
Y
Z

How to Use the Tool or Technique

A summary Hypothesis testing procedure involves the following activities:

Step 1. **Define the hypothesis and its conditions** to be studied.

a. Identify the problem or issue to be studied.

b. Define the objective: Is bigger better; smaller better; or nominal (a target) best? Is this a one-tail or two-tailed test?

> **Hints and Tips**
>
> Determine type of data available to study and state the Null hypothesis. A *Null hypothesis* is a statement of **no difference** between the before and after conditions (for example, defendant being not guilty in court).

c. State the alternative hypothesis—typically the theory to be tested.

d. Determine the practical difference that the hypothesis test should detect—the Confidence Level (1 – *alpha*) (or the critical value; t-critical) to define the rejection region and the Power (1–*beta*).

e. Establish the risks for the test—*alpha* and *beta*. Recall alpha default is 5% and beta default is 20%.

Step 2. **Prepare to run the study.**

a. Based on type of data available, select type of Hypothesis test.

b. Determine the sample size needed to obtain the desired risk.

c. Collect the samples.

Step 3. **Conduct Hypothesis test and draw conclusions.**

a. Calculate the test statistic to determine a p-value. (Usually this is achieved by using computer software. If calculated manually, the appropriate statistical table for each test must be referenced.)

b. Compare the p-value to the decision criteria (t-critical) and determine if the Null hypothesis is to be rejected or not.

Hints and Tips

If the test statistic falls in the rejection region (that is, outside the confidence interval), then reject the Null hypothesis.

Draw the distribution curve with the upper and/or lower confidence limits (t-critical) delineated and indicate where the test statistic falls to visualize if it is located in the rejection region or not, as shown in Figure H-6.

Examples

The following section demonstrates the most common Hypothesis tests using both manual calculations and the MINITAB software package, except ANOVA. The ANOVA (or F-test) hypothesis test can be fairly involved and warranted a separate entry. (**See Also** "Analysis of Variance (ANOVA)—7M Tool," **p. 142**)

One Sample t-Test Example

Scenario: We recently purchased a new hybrid automobile. The manufacturer promoted fuel efficiency at 45 MPG city and 60 MPG highway. We have recorded the miles traveled between each fill-up and the number of gallons consumed and calculated the fuel efficiency for the first 25 trips to the pump. Based on our careful driving techniques (such as anticipating stops and gently accelerating at starts), we hope our mileage is better than the manufacturer's claim. We estimate that our driving is evenly split (50/50) between city and highway; thus our theory is that our mileage is better than 52.5 MPG [(45 + 60)/2].

Hypothesis Statements and Conditions

- $\mathbf{H_0}$: The car gets the mileage advertised by the manufacturer (μ = 52.5 MPG).
- $\mathbf{H_a}$: The *car gets more than* the mileage advertised by the manufacturer (μ > 52.5 MPG).
- Theory to question is a one-tail (right) test; wherein "bigger is better."
- Set confidence level at 95%; *alpha* at default 5%; *beta* at default 20%.

- Sample size (n) is 25. The following data was collected: 52.1; 49.6; 56.2; 54.7; 47.4; 54.8; 50.6; 57.1; 55.9; 47.5; 48.9; 49.5; 56.4; 47.1; 52.2; 54.0; 54.0; 55.5; 53.0; 53.0; 52.0; 55.0; 56.0; 54.3; 55.3.

MINITAB Commands

- After the data is entered into a MINITAB Worksheet, select the following sequence from its drop-down menu: ***Stat > Basic Statistics > 1-Sample t....***
- Select the sample data from its worksheet; set the test mean in the 1-Sample t main screen as shown in "Area 1" of Figure H-10.
- Set risk levels in the "Options" screen for the Confidence level, leave the Confidence at the default of 0.95, and select the "greater than" test for the Alternative hypothesis statement, as shown in "Area 2" of Figure H-10. Select OK on both the Options and main screens to run calculations.

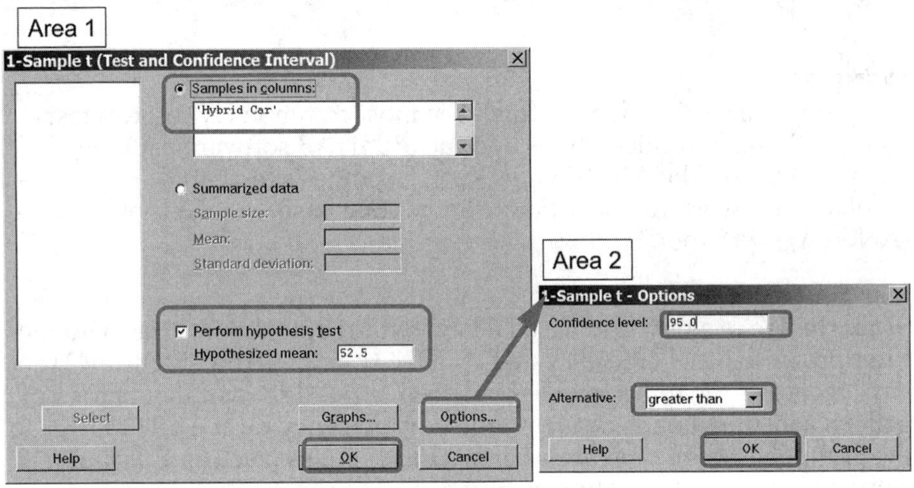

Figure H-10: MINITAB One Sample t-Test Example

Conclusions Using MINITAB Sample Output

- The output displayed in the MINITAB session window provides the p-value. Figure H-11 annotates MINITAB's output to highlight the conclusions.
- The p-value of 0.269 is greater than the set alpha of 0.05; hence, we cannot reject the Null hypothesis. We conclude that there is **not** enough evidence to say that the new hybrid gets better mileage than the manufacturer's claim.

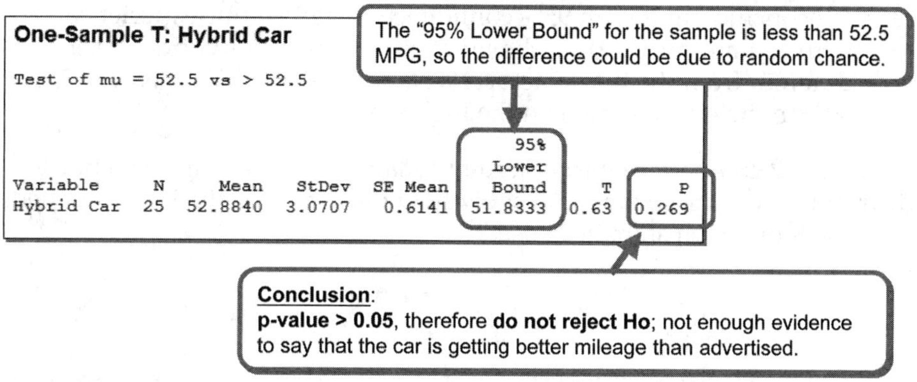

Figure H-11: MINITAB One Sample t-Test Session Window Example

- The sample's 95% Lower Bound of 51.8333 is less than population's 52.5 MPG mean, so the sample mean of 52.884 is likely different from 52.5 because of random chance alone.

Manual Calculations

- To determine the t-test critical value, look up *alpha* in a *t-Distribution statistics table.*

> **Note**
>
> The example is a one-tail test, with alpha at 0.05, and the degrees of freedom (n–1) is 24. The critical value found in the t-Distribution Table is 1.711. (**See Also** "t Distribution Critical Values," p. **939** in Appendix A, "Statistical Distribution Tables" of the Appendix.)

- Using the Student t formula and the following statistics from sample data to calculate the test statistic: sample mean (x-bar) = 52.884; population mean (52.5); standard deviation = 3.0701; n = 25.

$$t = \frac{\overline{X} - \mu}{\frac{s}{\sqrt{n}}}$$

- The resultant test statistic equals 0.625, which rounds to 0.63 and matches MINITAB's output.

Conclusions Using Manual Calculations

- The test statistic of 0.63 does not fall into the rejection region; it is less than the critical value of 1.711. Fail to reject the Null hypothesis.

- We conclude that with 95% confidence, based on our experience with 25 tanks of gas, that the mileage for our new hybrid car is no different from the manufacturer's claim of 52.5 MPG when driving 50% in the city and 50% on the highway.

Figure H-12 compares a *p-value* and *test statistic* drawn on a frequency t-distribution for this example to show how the same conclusion is reached regardless of which statistic is used.

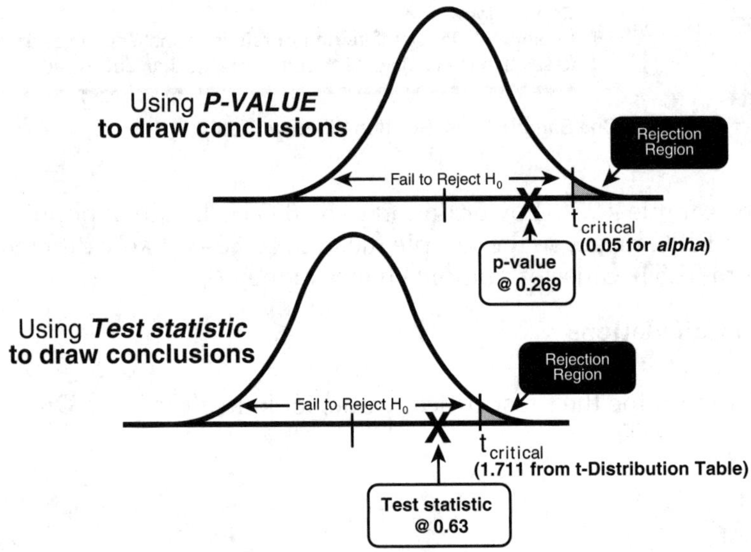

FIGURE H-12: Compare P-value and Test Statistics to Draw Conclusions

Paired t-Test Example

Scenario: An advertising firm wants to compare two different collaterals to see if its new version (called Material 2) more effectively communicates the message to its target audience. It decides to survey five customers to compare effectiveness of the new versus the old version. The data collected are the number of negative comments about the two pieces. Material 1 refers to the old content.

Consider sources of variability. In addition to the two different training materials, this scenario has other sources of variability, which are not of interest. The different learning styles and backgrounds represent a potential source of variability among the students.

Effective planning isolates these variability sources of interest (testing the training materials on two different groups) for further study. The concept of blocking in a Hypothesis test removes unwanted sources of variation (called noise) from the study. Blocking improves the signal-to-noise ratio, thereby improving the ability to detect a difference if it exists. Conducting a Paired t-Test accounts for blocking,. The test compares two sample groups, wherein the sample size of each group is less than 30.

> **Note**
>
> As the sample size approaches 30, the t-distribution approaches a normal distribution.

Other examples of paired data include

- Comparing scores on tests both pre-and post-training sessions
- Comparing the operators' efficiency ratings both before and after training
- Comparing machine throughput both before and after a software upgrade
- Comparing the durability of different (product) materials under varying conditions

Hypothesis Statements and Conditions

- **H_0:** Material 2 is the same as Material 1; no distinguishable difference ($\mu_2 = \mu_1$).
- **H_a:** Material 2 is different from Material 1 ($\mu_2 \neq \mu_1$).
- Theory to question is a two-tailed test; wherein "nominal is best."
- Set confidence level at 95%; *alpha* at default 5%; *beta* at default 20%.
- Sample size (n) is 25. The following data was collected:

Subject	Material 1	Material 2
Skip	6	7
Kathy	12	7
Eduardo	15	10
Neeta	8	4
Kevin	7	4

MINITAB Commands

- After the data is entered into a MINITAB Worksheet, select the following sequence from its drop-down menu: ***Stat > Basic Statistics > Paired t….***

- Select the sample data from its worksheet in the Paired t main screen as shown in "Area 1" of Figure H-13.
- Set risk levels in the Options screen; leave the Confidence level at the default of 0.95, leave the Test mean at the default of 0.0, and select ***not equal*** for the Alternative hypothesis statement, as shown in "Area 2" of Figure H-13.
- Select "OK" on both the "Options" and main screens to run calculations.

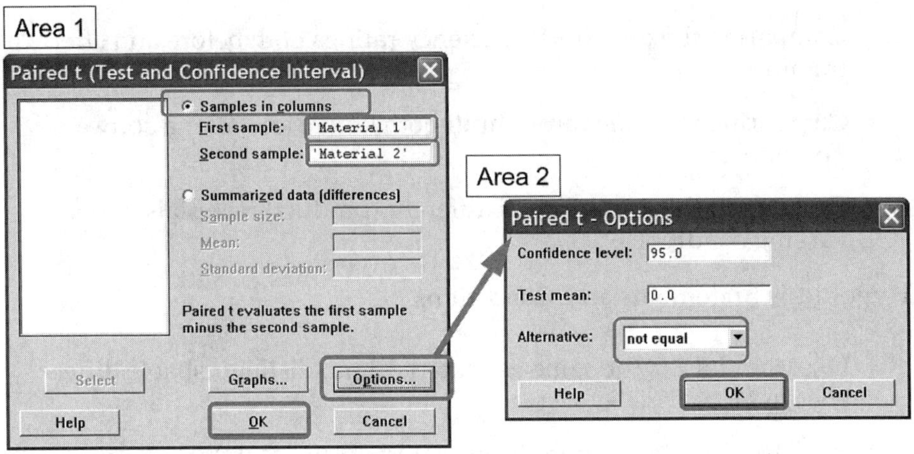

FIGURE H-13: MINITAB Paired t-Test Example

Conclusions Using MINITAB Sample Output

- The output displayed in the MINITAB session window provides the p-value. Figure H-14 annotates MINITAB's output to highlight the conclusions.

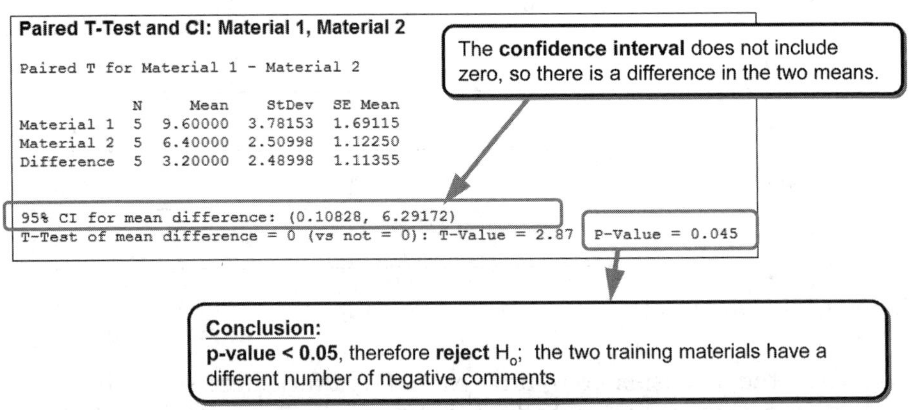

FIGURE H-14: MINITAB Paired t-Test Session Window Example

- The p-value of 0.045 is less than the set alpha of 0.05; hence, we reject the Null hypothesis. We conclude that there is sufficient evidence to say that the new advertising collateral is different from the original.

Manual Calculations

- To determine the t-test critical value, look up alpha in a t-Distribution statistics table. (**See Also** "t Distribution Critical Values," p. **939** in Appendix A, "Statistical Distribution Tables.")

> **Note**
>
> This example is a two-tailed test with alpha at 0.05 split between the two tails (0.05/2 = 0.025), and the degrees of freedom (n-1) is 4. The critical value found in the t-Distribution Table is 2.776. (**See Also** "t Distribution Critical Values," p. **939** in Appendix A, "Statistical Distribution Tables.")

- Calculate the difference between the pair of data per subject and calculate the average difference for the set of data.

Subject	Material 1	Material 2	Difference (d)	$S_d = (d_i - \text{d-bar})^2/\text{n-1}$
Skip	6	7	-1	17.64/4 = 4.41
Kathy	12	7	5	3.24/4= 0.81
Eduardo	15	10	5	3.24/4= 0.81
Neeta	8	4	4	0.64/4 = 0.16
Kevin	7	4	3	0.04/4 = 0.01
Sum:			16	6.2
Calculation:			(16/5=**3.2 *d-bar***)	(*Sq. root of* 6.2 = **2.489979 S_d**)

- Using the Paired t-test formula and the following statistics from sample data to calculate the test statistic: sample difference mean (d-bar) = 3.2; standard deviation of difference = 2.48998; n = 5.

$$t = \frac{\bar{d}}{S_d/\sqrt{n}}$$

$$S_d = \sqrt{\sum \frac{(d_i - \bar{d})^2}{n-1}}$$

A B C D E F G H I J K L M N O P Q R S T U V W X Y Z

- The resulting test statistic equals 2.87368, which rounds to 2.87, and matches MINITAB's output.

Conclusions Using Manual Calculations

- The test statistic of 2.87368 falls into the rejection region; it is greater than the critical value of 2.776. Therefore, reject the Null hypothesis.
- We conclude that with 95% confidence, based on our experience with five customers, that the new advertising collateral is different from the old version.

Figure H-15 compares a *p-value* and *test statistic* drawn on a frequency t-distribution for this example to show how the same conclusion is reached regardless of which statistic is used.

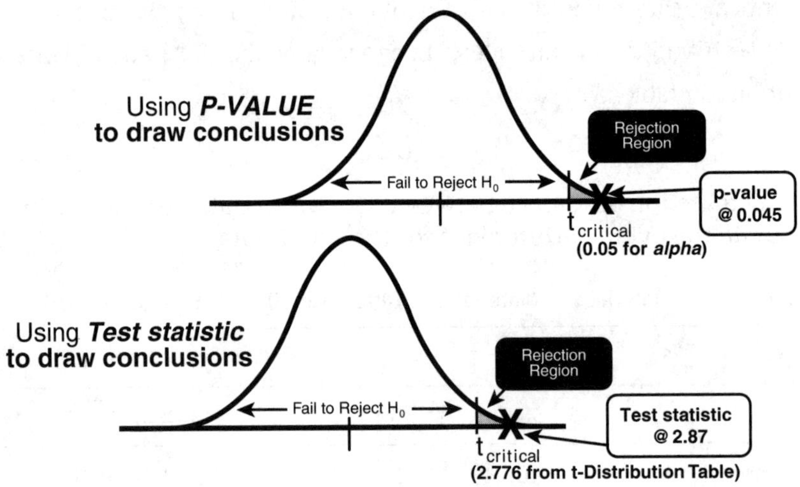

Figure H-15: Compare P-value and Test Statistics to Draw Conclusions for Paired t-Test Example

Chi-Square (χ^2) Test Example (pronounced "ky" square)
Scenario: A marketing department conducts a survey of premier customers to understand their preference for a new service offering it just announced. The survey was distributed by email and paper. A total of 200 customers replied. Some of the respondents chose to respond by email, and some replied by paper using the "snail-mail" approach. Marketing questioned whether the customers' response differed depending on

which of these two methods (paper or email) they used (*Test for Independence*).

Attribute data: If response data is attribute (counts of occurrences or defects) and we wish to make **comparisons** of these proportions across **multiple groups** or strata, then we use a **Chi-Square test** for independence. Examples of typical Chi-Square scenarios include

- Product (or services) preference compared with gender type, age, or income category.
- Defect types compared with categories of time (such as Week 1, 2, and so on) or compared with suppliers.
- Cycle time compared with machine type, shift, type of process, or job type.

Hypothesis Statements and Conditions

- $\mathbf{H_0}$: The number of positive responses ("yes") (or the number of non-responses "no") is independent of paper and email communication approaches ($p_2 = p_1$).

> **Note**
>
> In the preceding formula, "p" represents "proportions."

- $\mathbf{H_a}$: The number of positive responses (or the number of non-responses) differs by paper and email; they are *not* independent ($p_2 \neq p_1$).
- Theory to question is a *test for independence*, hence use a right-tail Chi-Square test.
- Set confidence level at 95%; alpha at default 5%; beta at default 20%.
- Sample size (n) is 200. The survey results are

Response	Paper	Email	Total
"Yes"	45	62	107
"No"	55	38	93
Total	100	100	200

> **Note**
>
> The collected data is arranged in a table, often called a *contingency table*, *cross-tab matrix*, or *row and column analysis*.

MINITAB Commands

- After the data is entered into a MINITAB Worksheet, select the following sequence from its drop-down menu: ***Stat > Tables > Chi-Square Test (Table in Worksheet)....***
- Select the sample data from its worksheet in the Chi-Square Test main screen and select OK to run the calculations, as shown in Figure H-16.

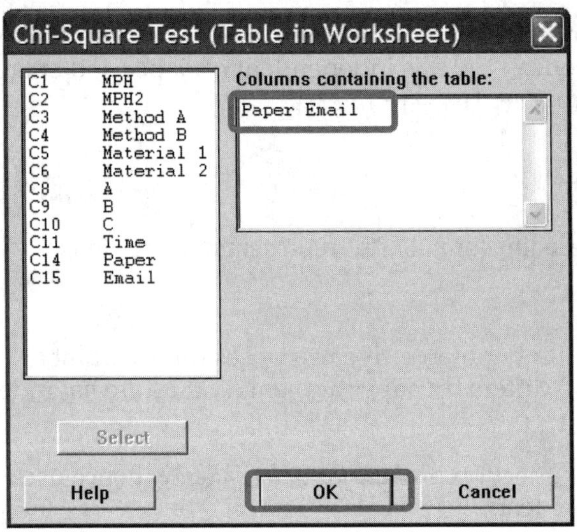

FIGURE H-16: MINITAB Chi-Square Test Example

Conclusions Using MINITAB Sample Output

- The output displayed in the MINITAB session window provides the p-value. Figure H-17 annotates MINITAB's output to highlight the conclusions.

A B C D E F G H I J K L M N O P Q R S T U V W X Y Z

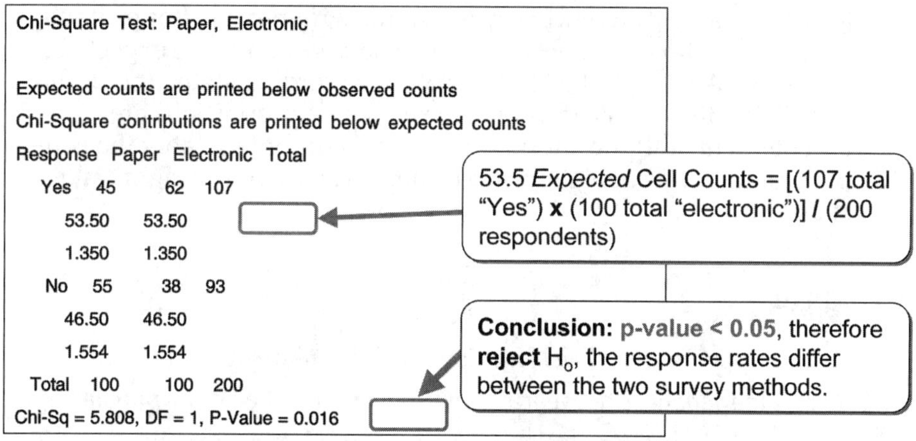

FIGURE H-17: MINITAB Chi-Square Test Session Window Example

- The p-value of 0.016 is less than the set alpha of 0.05; hence, we reject the Null hypothesis.
- We conclude that with 95% confidence, there is sufficient evidence to say that customer responses in the email communication were statistically different from those who selected the paper "snail" mail approach. The response and communication method selected are not independent; there is a relationship. The customers who responded by email were more likely to respond "yes" to preferring the new service than the customers who chose to reply by paper.

Manual Calculations

- The shape of the Chi-Square distribution depends on the degrees of freedom. As the sample size gets larger and the degrees of freedom are 15 or higher, the shape approaches normal distribution. Otherwise, the Chi-Square distribution is a right (or positive) skewed curve with asymmetrical tails, as illustrated in Figure H-18.
- The Chi-Square *test for independence* generally is a right-tail test (α), to

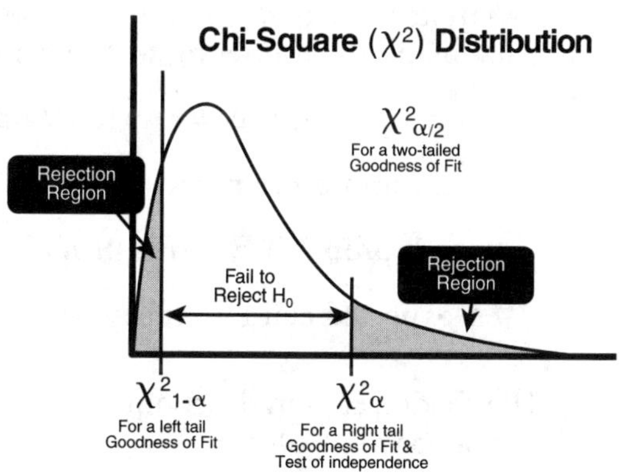

FIGURE H-18: Illustration of a Chi-Square Distribution

look for a large difference in the formula's numerator (for example, the squared difference between the observed minus the expected frequencies). Although the Null hypothesis wording states "no difference," that the observed versus the expected proportions are independent, and the alternative states "a difference" that the comparison indicates a relationship or dependency, only a right-tail test is used.

> **Note**
>
> The appropriate *alpha*-level for a *Goodness of Fit* Chi-Square test, comparing the variances between a sample and that of the known population, depends on the Null and alternative hypothesis statements.

- To determine the Chi-Square critical value, look up *alpha* with the corresponding degrees of freedom in a Critical Values of the Chi-Square-Distribution statistics table. The degrees of freedom (df) formula is "number of rows minus one times the number of columns minus one," or (r-1)(c-1).
    - For this example, use a right-tail test, with alpha at 0.05.
    - The df calculation is (2 rows–1)(2 columns–1) = (1 x 1) = 1, which matches MINITAB's Session window output.
    - The critical value found in the Chi-Square Distribution Table for *alpha* of 0.05 and df of 1 is 3.84.*]* (**See Also "Chi-Square (χ^2) Distribution Critical Values," p. 939** in Appendix A, "Statistical Distribution Tables.")
- Calculate the expected value (or frequency) for each cell of the set of data, using the following formulas:
    - **row probability = row total/grand total**
    - **column probability = column total/grand total**
    - **cell probability = row probability x column probability**
    - **expected cell frequency = cell probability x grand total**

  The shaded cells in the contingency table represent those containing the actual observed data.
- Double-check the math by ensuring that the summed rows equal the grand total and the summed columns equal the grand total.

Response	Paper Actual	Paper Expected (probability)	Email Actual	Email Expected (probability)	Total	Total Expected (probability)
"Yes" responses	**45**	(0.535 × 0.5 = 0.2675) × 200 = **53.5**	**62**	(0.535 × 0.5 = **0.2675**) × 200 = **53.5**	**107**	(107/200= **0.535**) × 200 = **107**
χ^2 for Yes		$(45\text{-}53.5)^2/53.5$ = **1.35**		$(62\text{-}53.5)^2/53.5$ = **1.35**		**2.7**
"No" responses	55	(0.465 × 0.5 = **0.2325**) × 200 = **46.5**	**38**	(0.465 × 0.5 = 0.2325) × 200 = **46.5**	**93**	(93/200 = **0.465**) × 200 = **93.0**
χ^2 for No		$(55\text{-}46.5)^2/46.5$ = 1.55376		$(38\text{-}46.5)^2/46.5$ = 1.55376		**3.108**
Total responses	**100**	(100/200 = 0.50) x 200 = **100**	**100**	(100/200 = 0.50) x 200 = **100**	**200**	(200/200 = 1.00) x 200 = **200**
Total χ^2		**2.904**		**2.904**		**5.808**

A B C D E F G H I J K L M N O P Q R S T U V W X Y Z

- Using the Chi-Square (χ^2) Test for Independence formula, calculate the test statistic.

$$\chi^2 = \sum \frac{(O - E)^2}{E}$$

where O = Observed (or actual) frequency, and E = Expected frequency

- The resultant Chi-Square (χ^2) test statistic equals 5.808, which matches MINITAB's session window output.

Conclusions Using Manual Calculations

- The Chi-Square test statistic of 5.808 is greater than the critical value of 3.84; therefore, it falls into the rejection region. Reject the Null hypothesis.
- We conclude that with 95% confidence, based on our experience with 200 respondents, that the number of positive responses (or the number of non-responses) differs by paper and email communication method. The response and communication method are **not independent**.

Hints and Tips

All hypothesis tests are constructed in a similar manner. The Null hypothesis states that there is no difference between the parameters of interest. The Alternative hypothesis states that there is a difference between the parameters of interest.

Rule of thumb: The Null hypothesis (H_o) usually is worded as "equal to" (including less than and equal to or more than and equal to). Hence, the alternative hypothesis (H_a) generally contains "not equal to."

Rule of thumb: The underlining hypothesis, what is hoped to be true, generally worded as the alternative hypothesis (H_a). So generally, start by stating the alternative (H_a) first and then state the Null hypothesis (H_o) afterward.

The Difference between Directional versus Non-Directional Tests

Non-directional or two-tailed test—If you want to determine if **two things are different,** then the Alternative hypothesis is considered *non-directional.* The alternative hypothesis will use non-equivalence language (mathematically a "not equal" sign is used ($\neq$ or <>).

Directional or one-tail test—If you are interested in stating that **something is smaller than or larger** than a quantity, then the Alternative hypothesis is called *directional*. The alternative hypothesis will consist of a comparison that is either less than or greater than a quantity of interest (mathematically this equates to using a < or > symbol).

Hints and Tips

Decision Criteria—If this ***p-value*** is smaller than the pre-defined level of significance (typically $\alpha = 0.05$), then the Null is rejected.

The p-value is the actual Type I Error based on the sample data.

A signal-to-noise ratio is calculated with a probability (p-value) assigned to it, designating the likelihood that this size of ratio could occur by random chance.

Supporting or Linked Tools

Supporting tools that might provide input when developing an hypothesis test include

- Data Gathering techniques (**See Also** "Data Collection Matrix," **p. 248**)
- Performance charts and dashboards
- Graphical tools (**See Also** "Graphical Methods," **p. 323**)
- Cause-and-Effect diagram (**See Also** "Cause-and-Effect Diagram—7QC Tool," **p. 173**)
- Process map (**See Also** "Process Map (or Flowchart)—7QC Tool," **p. 522**)
- Hypothesis
- QFD (**See Also** "Quality Function Deployment (QFD)," **p. 543**)
- Solution selection matrix (**See Also** "Solution Selection Matrix," **p. 672**)
- Brainstorming (**See Also** "Brainstorming Technique," **p. 168**)

A completed hypothesis test provides input to tools such as

- QFD (**See Also** "Quality Function Deployment (QFD)," **p. 543**)

- Solution selection matrix (**See Also** "Solution Selection Matrix," **p. 672**)
- Brainstorming (**See Also** "Brainstorming Technique," **p. 168**)
- Control Plan (**See Also** "Matrix Diagrams—7M Tool," **p. 399**, for a brief discussion on control plan.)
- FMEA (**See Also** "Failure Modes and Effects Analysis (FMEA)," **p. 287**)

Figure H-19 illustrates the link between an hypothesis test and its related tools and techniques.

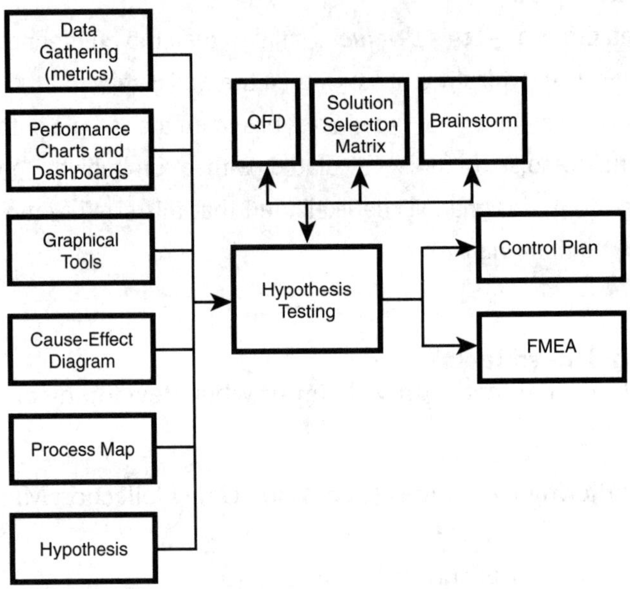

Figure H-19: Hypothesis Test Tool Linkage

I

Interrelationship Diagram—7M Tool

What Question(s) Does the Tool or Technique Answer?

How do the various cause-and-effect ideas relate to one another in a complex situation?

An Interrelationship diagram helps you to

- Understand the relationship between topics of a complex situation
- Decipher the relationship of intertwined topics to identify the area of greatest impact for improvement

Alternative Names and Variations

This tool is also known as

- Relationship diagram or digraph
- Network diagram

Variations on the tool include

- Matrix relations diagram

When Best to Use the Tool or Technique

The tool is best used on more complex issues wherein the exact causal relationship or inner-dependencies are difficult to discern.

Brief Description

The type of diagram maps the various links, using directional arrows to indicate source of a cause (at the base of the arrow) and impact (at the point of the arrow). Developing an Interrelationship diagram involves a team, preferably a cross-functional one, using creative problem-solving to construct the diagram. This tool builds on work done in other tools such as Affinity diagrams, Tree diagrams and Cause-and-Effect diagrams. Upon completion, the tool maps the connections of the various cause-and-effect relationships to highlight the hub of greatest activity. The team then analyzes the network of relationships to identify the key causes.

This tool is a member of the 7M Tools, attributed in part to Dr. Shewhart, as *seven "management" tools*, or sometimes referred to as the *7MP*, for *seven management and planning tools*. These 7M tools make up the set of traditional quality tools used to analyze quantitative data. The 7M toolset includes: 1) Activity network diagrams or Arrow diagrams, 2) Affinity diagrams, 3) Interrelationship digraphs or Relations diagrams, 4) Matrix diagrams, 5) Prioritization matrices, often replacing the more complex Matrix data analysis, 6) Process decision program charts (PDPC), and 7) Tree diagrams. *The Quality Toolbox*, by Nancy Tague, presents the 7M tools ranked from those used for abstract analysis to detailed planning: Affinity diagram, Relations diagram, Tree diagram, Matrix diagram, Matrix data analysis (commonly replaced by a more simple Prioritization matrix), Arrow diagram, and Process Decision Program Chart (PDPC). (**See Also** "References," in Appendix C, **p. 981**)

How to Use the Tool or Technique

The Interrelationship diagram is built much like an Affinity diagram. The best development technique involves materials such as sticky notes or index cards to capture one idea per sticky note or card. The procedure involves the following steps:

Step 1. **Prepare to build the diagram.**

Identify the complex topic of interest and gather supporting or input documents including those with text or verbal data. Examples of different inputs include: written reports, Affinity diagrams, Tree diagrams, and Fishbone (or Cause-and-Effect) diagrams.

Step 2. **Construct topic cards.**

Document one idea or topic on either a single piece of paper (or index card) or sticky note. Accomplish this by referencing supporting documents and brainstorming additional ideas. If a Fishbone or Tree diagram is referenced, select the most granular or detailed level of information. Place only one idea per piece of paper (index card or sticky note).

Note

As a short cut, if the data already exists in a text document, simply cut the document into strips, such that one strip of paper contains only one written idea.

Step 3. **Group related ideas.**

Working with one documented idea at a time, place the idea card or sticky on the working surface (a flip chart or table). Try to place the topic near a related idea. If no related themes exist, place it on the working surface with enough empty space around it to allow room for related cards to be added and for arrows to be drawn. Repeat this step until all cards (or stickies) are placed.

> **Hint**
>
> S huffling the cards (or stickies) and randomly addressing each idea sometimes triggers creative thinking about related topics.

Step 4. **Draw the relationship arrows.**

Review and discuss how an idea is related to another. Draw the relationship arrow by starting with the causal topic as the base of the arrow and connecting it to the effected topic (cause→ effect). Repeat this step until each card has been reviewed and all the relationship arrows drawn.

a. There are no restrictions on number of arrows coming into or out of any one card; however, there should be no bi-directional (or two-ended) arrows.

b. Take the time to reflect and review the diagram. Often the revisions are as insightful as the original draft.

c. **Option**: Once the diagram is finalized, record the number of incoming and outgoing arrows on a lower corner of each card (or sticky) to assist with the prioritization step (for example, "1/5" for one incoming and 5 outgoing arrows).

d. A large number of arrows stemming from a topic indicates a root cause or driver. Several arrows pointing to an item indicates a key outcome.

Step 5. **Prioritize and assign accountabilities.**

Identify those topics with the most outgoing arrows and with the most incoming arrows. Draw conclusions about those item(s) requiring the most attention. Select the top priority items (usually the top 2 or 3), encircle the cards (or stickies), assign each top priority to one person accountable for resolution, and write the person's name on the card.

> **Hint**
>
> Highest priority may involve severity of impact, rather than highest number of inbound or outbound arrows. Thus, double-check the prioritization list and gain agreement on those requiring immediate action.

Figure I-1 illustrates a completed Interrelationship diagram to look at the Cause-and-Effect relationship of trying to improve new business sales. Notice that each item contains the number of inbound and outbound arrows documented in the lower portion of the box. Those items with the highest number of arrows (either outgoing causes or incoming effects) also have bolded outline and assigned person accountable to take action.

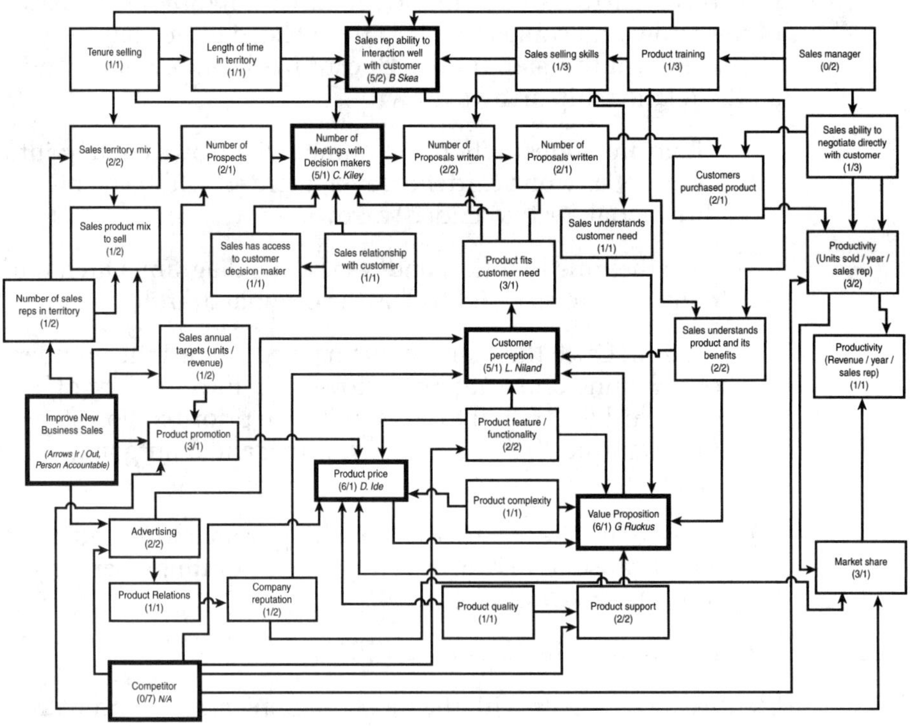

FIGURE I-1: Sample Interrelationship Diagram

A B C D E F G H I J K L M N O P Q R S T U V W X Y Z

Notice that Competition is identified as one of the themes with the highest number of outgoing arrows in Figure I-1. However, since the item is not actionable, no person is assigned to it.

Supporting or Linked Tools

Supporting tools that might provide input when developing an Interrelationship diagram include

- Written report
- Affinity Diagram—7M Tool, **p. 136**
- Tree Diagram—7M Tool, **p. 712**
- Fishbone or Cause-and-Effect diagram (**See Also** "Cause-and-Effect Diagram," **p. 173**)
- Brainstorming (**See Also** "Brainstorming Technique," **p. 168**)

A completed Interrelationship diagram provides input to tools such as

- Brainstorming (**See Also** "Brainstorming Technique," **p. 168**)
- QFD (**See Also** "Quality Function Deployment (QFD)," **p. 543**)
- Root cause analysis techniques (such as statistical Hypothesis testing and Regression) (**See Also** "Hypothesis Testing," **p. 335**, and "Regression Analysis," **p. 571**)
- Concept generation techniques
- FMEA (**See Also** "FMEA," **p. 287**)

Figure I-2 illustrates the link between an Interrelationship diagram and its related tools and techniques, showing how some provide input to, while others utilize the information from a completed Interrelationship diagram.

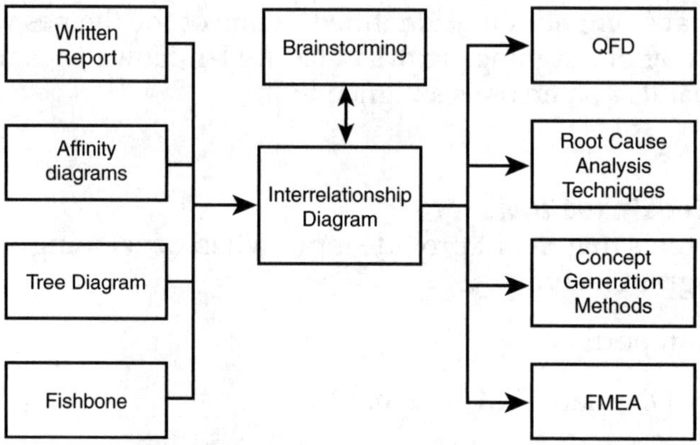

Figure I-2: Interrelationship Diagram Tool Linkage

K

KJ Analysis

What Question(s) Does the Tool or Technique Answer?

What matters to your customers? What are the natural groupings, categories, or affinities of a large number of topics, ideas, and quotations, and how can you best translate the verbal input into requirements?

A KJ Analysis helps you to

- Discover and clarify customer requirements
- Organize and categorize large volumes of language-based data (that is, text or verbal input)
- Show a "parent-child" relationship between more detailed ideas to their higher-order theme, also called Affinity groupings
- Translate the input into customer requirements

Alternative Names and Variations

This tool is also known as

- Affinity diagram or chart

Variations on the tool include

- House of Quality (HOQ) or Quality Function Deployment (QFD)
- Fishbone diagram (rarely)
- Tree diagram or matrix (rarely)

When Best to Use the Tool or Technique

Successful improvement efforts start with a focus on customer needs. The KJ Analysis starts with Voice of the Customer (VOC) input, which is gathered for the purpose of either developing an offering concept for the strategic Portfolio Renewal process or at the early stages of the tactical Offering Development process. (**See Also** "VOC Gathering Techniques," p. **737**, "Six Sigma for Marketing (SSFM)," in Part I , **p. 67**, and "Listening to the Customer First-hand; Engineers Too," in Part III, **p. PDF:851**.)

A KJ Analysis organizes a large number of ideas, quotations, and facts into themes. This categorization allows the content to be translated into customer requirements. This approach often times translates the implicit into explicit needs. It helps reduce complexity and connect related topics. This tool analyzes customer interview data or survey results and summarizes it for purposes of communication and devising responding actions, such as specifications, root cause analysis, and solution generation techniques.

Brief Description

The tool was developed in the early 1960s, by a Japanese anthropologist, Jiro Kawakita, which is where its name comes from, "KJ." [Recall that a Japanese name reverses the order of the given and sir name relative to an English name.] Dr. Kawakita was the founder of the Kawayoshida Research Center.

Kano Model: NUD Versus ECO

The KJ Analysis translates customer input into NUD requirements [pronounced *"nŭd"*]—what is **N**ew, **U**nique, and **D**ifficult. NUD requirements add value to the customer and differentiate an offering from its competition. Within this context, the NUD customer requirements are defined as

- **New**—Something the customer has not asked for or thought of before; it is completely new.
- **Unique**—Something different that is not currently being provided in the marketplace today (either by your organization and/or by a competitor). This could be a substitution product or service.
- **Difficult**—Something that is very challenging to offer (not necessarily new or unique but often is).

NUD requirements delight customers. If you do not satisfy your customers' needs, your competition will. NUD requirements put pressure on competition because these attributes require more lead-time, know-how, and investment to replicate or mirror. Per the Kano model, *customer delighters* satisfy customer needs better than those offerings that provide expected or basic quality, features, and functionality. The Kano model defines three types of product or service quality that provide customer satisfaction:

1. **Fitness to Latent Expectations** and **delighting quality**—Where a neutral effect if not present, but extremely powerful if present.
2. **Fitness to Use** and **performance quality**—Where more is better, which is often the focus of VOC efforts.
3. **Fitness to Standards** and **basic quality**—These items must be present, yet have a neutral effect from the customer. They may be assumed and often are not clearly documented.

The attributes of new, unique, and difficult describe how best to produce a delighter that excites and stimulates customer buying behavior. Figure K-1 illustrates how the Kano model of customer satisfaction relates the three levels of quality—delighting, performance, and basic.

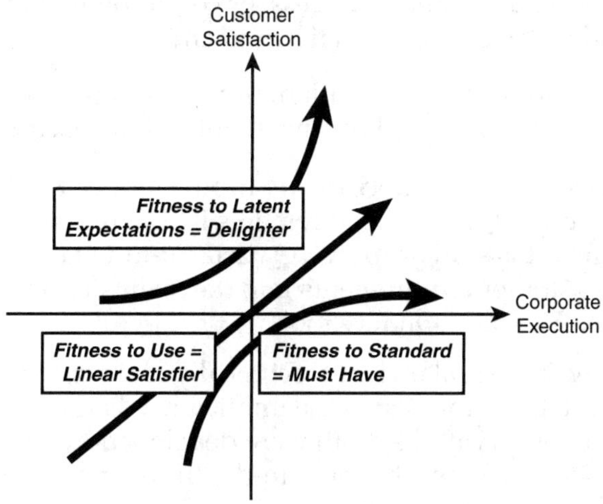

FIGURE K-1: Kano Model of Customer Satisfaction

The idea is to avoid working on *ECO* customer requirements [pronounced *"ě?k-ȳ"*]—those that are **E**asy, **C**ommon and **O**ld requirements. These attributes describe the basic quality curve on the Kano model—the mere minimum expected by the customer. The ECO requirements provide little to no value to the customer but must to be present in a product or service. Plus, competitors can easily replicate or respond in-kind. The ECO requirements are defined as

- **Easy**—Simple to provide with low effort, low cost, and low risk.
- **Common**—It has been offered before with a repeatable history of success; then it is not unique.
- **Old**—Something we have experience fulfilling, and it does not represent a critical requirement.

A good source of NUD requirements stems from the context within which a customer currently works. This data often comes from images in the form of descriptive sentences or observations. An important aspect and distinguishing feature of a KJ Analysis is the processing of customer input as "images." Images or pictures that describe the context provide the best source of data. This image data provides insight into the customer's "latent" needs (intangible or implicit). From a customer environment perspective, there are three kinds of images:

1. A picture of some event, situation, process, or condition that the **customer describes** during an interview discussion.
2. An actual event, situation, process, or condition that the **interviewer witnessed** in the customer's environment.
3. A picture of some event, situation, process, or condition that **forms in the interviewer's mind** during an interview discussion.

The KJ produces three main document outputs: Image KJ Document, Translation Document, and a Requirements KJ Document. The Requirements KJ document is similar to a CTQ (Critical-to-Quality) in that it translates the customer requirements into the team's understanding and places them in a House of Quality tool.

The KJ Analysis is similar to but technically more complex than an Affinity diagram. It comprehends the implicit needs revealed through story-telling and observations. Both were developed by Jiro Kawakita. Hence, this article discusses the more in-depth KJ Analysis technique. (**See Also** "Affinity Diagram," **p. 136**)

How to Use the Tool or Technique

This technique is performed usually by a small group made up of members of a product development team assigned to work on analyzing the VOC data. The team produces three main document outputs when conducting a KJ Analysis:

1. Image KJ Document
2. Translation Document
3. Requirements KJ Document

The KJ Analysis procedural steps to build these three documents are as follows:

Image KJ Analysis Procedure

This procedure describes the development of an Image KJ Document.

Step 1. **Gather VOC**. Identify the topic and then collect and/or capture the text or verbal data. Examples of different means of gathering such input are interviews or surveys.

   a. Identifying the "topic" may simply involve agreeing what aspect of the VOC Interview requires the team to dissect and analyze.

b. Ideally seek "imagery" input from the customers because the verbal data is richer. Imagery data paints a scenario, sets the context within which the customers are doing something, and uses something to describe their requirements.

c. VOC Interview gathering techniques are best done in the customers' own environment. Observations of the work setting assist the customers in recalling the current state and visualizing what they would like to keep the same; start doing differently or stop doing all together.

d. Inquiry should focus on "in-context" analysis to gain additional insights on how and why things are done. This inquiry can effectively be achieved through both further verbal probing or simply observing the surrounding environment in total.

Step 2. **Prepare the data** by putting one idea, thought, or suggestion (either written or verbal input) on either a single piece of paper, index card, or sticky note. (Only one idea per piece of paper, index card or sticky note.) This one idea per card allows ideas to be moved around, mixed-and-matched, and grouped more easily, rather than placing several ideas on one piece of paper and deciding later that one of the ideas may be grouped better with another set of ideas.

Note

If the data already exists in a text document, a shortcut may be to simply cut the document into strips, such that one strip of paper contains only one written idea.

Step 3. **Sort the input by content into themes or categories**. Review each individual piece of paper and begin to place them in piles. Those ideas in the same pile relate to one another (or have an "affinity" for one another).

If using sticky notes, follow the same procedure, but place the ideas with a common theme on a single flip chart sheet or section of the wall. Continue this step until all the individual ideas are placed in a group.

a. Seek clarification on an idea if unclear by discussing among team members and reviewing additional customer input. Edit the idea, as needed.

b. Eliminate redundancy by combining an idea onto one piece of paper or sticky, such that within a group, one piece of paper represents one diverse idea.

c. If an idea relates to more than one group, create a duplicate piece of paper and place it in all the appropriate groups.

d. A few ideas may not fit well into the other groupings or themes; they may be independent ideas or loners. If so, place them in a "Miscellaneous" grouping.

Step 4. **Identify important ideas using red dots**. With either a red pen or red dot sticker, have the team indicate which **ideas seem important**. Only one red dot per statement; if someone has already put a red dot on a note, it has been identified as important. This helps identify the idea easily in later steps.

a. Remove all slips of paper that are absent of a red dot.

b. Count the red dot slips. If 20 to 30 remain, proceed to the next step. If more than 30 remain, then repeat this step by adding a second red dot to a slip, keeping only those with two red dots.

Step 5. **Organize the group and structure it all into a *Ladder of Abstraction*.** Start by reviewing each idea within a grouping or affinity then again sort them into further subsets. Identify those that are similar and hierarchically group them.

Figure K-2 illustrates a generic Ladder of Abstraction structure.

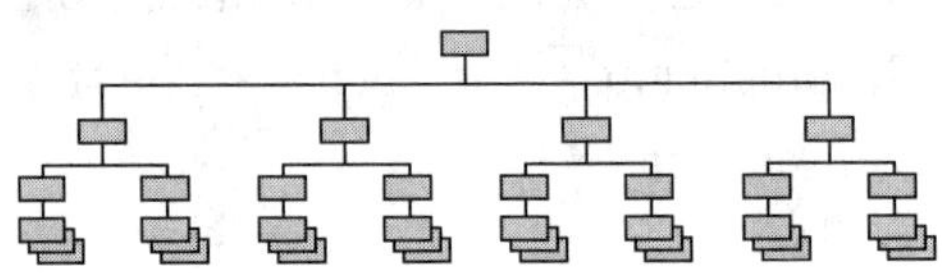

Figure K-2: Ladder of Abstraction

a. Double-check for redundancy and eliminate an unnecessary duplication.

b. A sub-set classification example includes images and statements (verbatim comments). Use the image ideas; hold the statements for the Requirements KJ document described next.

i. **Images**—Describe the customer needs using imagery by painting a scenario or story about tangible items.

ii. **Statements (verbatim)**—Describe the customer needs using less tangible language.

c. Ladder of Abstraction. Arrange the input within each group into hierarchical clusters and arrange it into a rational Tree diagram, wherein the content relates to one another. The hierarchy is determined by sorting or ranking the data within each group by importance or the ability to meet the objective of the study.

i. The first level of (or lowest level of detail in) the Ladder of Abstraction is called the black-level, named so for easy reference in future steps.

d. Look across groups and determine if any groups are related to one another. Determine if there is a strong connection; if so, draw a dotted-line between the two main groups. Determine if any inputs are misplaced and rearrange.

Step 6. **Name the themes** by first identifying a category heading or name for the highest-level of commonality in a given group, and record it on another slip of paper (index card or sticky note). Start at the lowest-level of detail and continue this until all the groups have a label, except for the Miscellaneous group.

a. Working from the lowest level of detail up, create a category name or sentence that states a high-level image that represents what is contained in each group. These statements should be only one level of abstraction higher. They are abstractions, not summary statements.

i. Separating out the inherent qualities or properties.

ii. Do not add new facts during this activity.

iii. The name or theme title slips are referred to as the red-level in the Ladder of Abstraction.

b. If the Ladder of Abstraction has multiple levels of subgroups within a big category, continue the naming process by labeling those higher-level groupings.

i. This level constitutes combining lower-level (or red-level) subgroups.

ii. The highest-level of title statements in a Ladder of Abstraction is referred to as the *blue-level*.

c. Reexamine if the Miscellaneous ideas relate to any of the named groupings; if so, move them to that category. These statements do not require a category name or statement.

Step 7. **Diagram the relationships**. Start with the Ladder of Abstraction structured in a hierarchical flow, similar to a Tree diagram, with the blue-level themes at the upper-most level and the corresponding red-level flowing from the higher-level, followed by the black-level. Using arrows, indicate which blue-level groups or themes support or conflict with another group by placing the appropriate symbol between the two blue groups.

Note

Convention calls for the relationship arrows to be drawn in blue.

a. Directional arrows (→) indicate a supportive relationship.

b. Crossed-lines (an "X") indicate a conflicting relationship.

Figure K-3 illustrates at a high-level the flow of steps Two through Seven in conducting a KJ Analysis.

Step 8. **Draw conclusions to complete the Image KJ document**. Summarize the boundaries of the groups and draw conclusions about the structure.

a. Draw boundaries around each blue grouping in its entirety.

b. Write a conclusion statement about the relationships of the major (blue) themes, place them above the Relationship Arrows of the Image KJ document, and have the team sign their names on a bottom corner.

At this point, the Image KJ document has been completed and represents the scenes or images described by the customer or observed by the team.

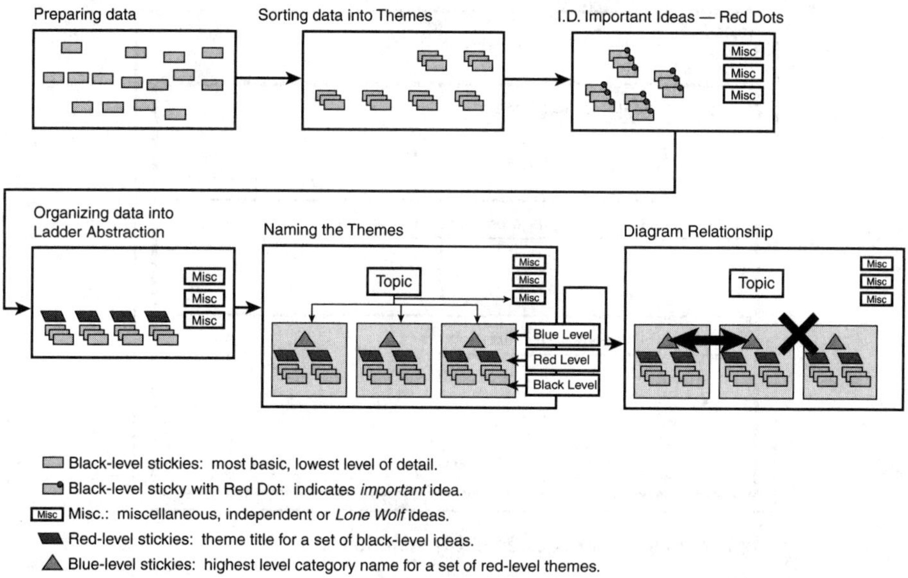

FIGURE K-3: KJ Analysis Flow of First Few Steps

A B C D E F G H I J K L M N O P Q R S T U V W X Y Z

Requirements KJ Analysis Procedure (and Translation Document)
This procedure describes the development of both the KJ Requirements document and a KJ Translation document.

Step 1. **Prepare to conduct the KJ Analysis and develop the KJ Translation document.** Use the held statements (verbatim) that were sorted in Step 5 of the previous Image KJ procedure.

a. Link each statement to at least one image. (A single statement may be liked to numerous images.) As explained earlier, KJ statements refer to those written items that describe the customer needs using less tangible language.

b. Create several blank *Customer Requirement Translation Worksheets,* one per statement. The template should feature three sections: 1) Customer Statements (Voice), 2) KJ Images (from the KJ Image document), and 3) Translated Customer Requirements, as illustrated in Figure K-4.

> **Note**
>
> If working in a team setting, use flip chart sheets for all to view.

Project:
Reqt. #

Customer Statements (Voice):

KJ Images (from KJ Image document):

Translated Customer Requirements:

FIGURE K-4: Sample KJ Customer Requirement Translation Worksheet

> **Note**
>
> Steps 2 through 5 mirror the same procedure as for the earlier Image KJ procedure described at the beginning of this section.

Step 2. **Organize the group and structure into a Ladder of Abstraction**. Start by reviewing each "held" statement within a grouping or affinity and then sort them again into further subsets. Identify those that are similar and hierarchically group them.

a. Double-check for redundancy and eliminate any unnecessary duplication.

b. Sub-set classification example includes images and statements (verbatim comments). Use the image ideas; hold the "statements" for the Requirements KJ document described next.

Step 3. **Ladder of Abstraction**. Arrange the input within each group into hierarchical clusters and arrange it into a rational Tree diagram, wherein the content relate to one another. The hierarchy is determined by sorting or ranking the data within each group by importance or ability to meet the objective of the study.

a. The first level of (or lowest level of detail in) the Ladder of Abstraction is called the black-level.

b. Look across groups and determine if any groups are related to one another. Determine if there is a strong connection; if so, draw a dotted line between the two main groups. Determine if any inputs are misplaced and rearrange.

Step 4. **Name the themes** by first identifying a category heading or name for the highest-level of commonality in a given group and record it on another slip of paper (index card or sticky note). Start at the lowest-level of detail and continue this until all the groups have a label, except for the Miscellaneous group.

a. Working from the lowest level of detail up, create a category name or sentence that states a high-level image that represents what is contained in each group. These statements should be only one level of abstraction higher. They are abstractions, **not** summary statements.

i. Separate out the inherent qualities or properties.

ii. Do not add new facts during this activity.

iii. The name or theme title slips are referred to as the red-level in the Ladder of Abstraction.

b. If the Ladder of Abstraction has multiple levels of subgroups within a big category, continue the naming process by labeling those higher-level groupings.

i. This level constitutes a combining of lower-level (or red-level) subgroups.

ii. The highest-level of title statements in a Ladder of Abstraction is referred to as the blue-level.

c. Reexamine if the Miscellaneous ideas relate to any of the named groupings; if so, move them to the appropriate category. These statements do not require category names or statements.

Step 5. **Diagram the relationships**. Start with the Ladder of Abstraction structured in a hierarchical flow, similar to a Tree diagram, with the "blue-level" themes at the upper-most level and the corresponding red-level flowing from the higher-level, followed by the black-level. Using arrows, indicate which blue-level groups or themes support or conflict with other groups by placing the appropriate symbol between the two blue groups.

> **Note**
>
> Convention calls for the relationship arrows to be drawn in blue.

a. Directional arrows (→) indicate a supportive relationship.

b. Crossed-lines (an "X") indicate a conflicting relationship.

> **Note**
>
> After this step, the procedure involves unique steps from the earlier Image KJ procedure.

Step 6. **Rank the top three red-level customer requirements.** Using a voting system, identify the top three red-level requirements. This ranking system is the team's opinion on rank and importance. The team should validate the results at a later time with the customer (via a survey, for example). Select one of the two approaches that follow to rank the red-level ideas:

a. **Three Color Dot Stickers**—Use red, blue and green sticky dots (or any other three colors) to vote for the most important red-level customer requirements. Provide one of each color sticky dot to each team member. Everyone votes using the dot-stickers, whereby a red dot equals three points, a blue dot equals two points, and a green dot equals one point. Sum up the total votes for each statement to determine the top three vote getters.

b. **Weighed Voting**—Each team member receives a total of six points to distribute among the statements any way he/she deems appropriate. Sum up the total votes for each statement to determine the top three vote getters.

Step 7. **Draw conclusions to complete the Requirements KJ document.** Summarize the boundaries of the groups and draw conclusions about the structure.

a. Draw boundaries around each blue grouping in its entirety.

b. Write a conclusion statement about the relationships of the major (blue) themes and place them above the Relationship Arrows of the Image KJ document and have the team sign their names on a bottom corner.

A B C D E F G H I J K L M N O P Q R S T U V W X Y Z

Note

At this point, the Requirements KJ document has been completed. The last step is to complete the Translation document.

Step 8. **Fill in the Translation Worksheet** for each of the top three red-level customer requirement statements. Using **one customer statement per form**, write a clear and non-prescriptive customer requirement statement, characterized as a "solution-free statement." Use factual language.

Whenever possible, use continuous variables language (for example, use "75 degrees," which is less ambiguous than "hot.") This more precise language enables a crisper translation into actionable terms that an offering designer finds meaningful.

a. Customer Statements describe the customer's voice articulating their desires. Typically these statements start with "*I want…*" or "*I need…*"

b. KJ Images are the references to the Image KJ document. They are the stories and observed behaviors that describe the context, the setting, and the scenario and lead to the latent expectations of the customers. These ideas are the *innovation stimulators*. (Reference the black-level ideas to enliven the story.)

Step 9. **Screen Customer Requirements**. Create a matrix or a spreadsheet with three columns titled 1) Customer Requirements, 2) NUD Designation, and 3) Kano Designation.

a. **Customer Requirements column**—List the black-level requirements.

b. **NUD Designation column**—Indicate if the requirement is "N" for new; "U" for unique, "D" for difficult, or "SR" for a standard requirement (meaning ECO—easy, common, or old).

c. **Kano Designation column**—Identify what Kano category the requirement falls under—"B" for basic; "LS" for linear satisfier; or "D" for delighter, described earlier in Figure K-1.

d. Transfer this information into the *Translated Customer Requirements* section of the Translation document, thus completing the KJ Translation document.

How to Analyze and Apply the Tool's Output

After the KJ Analysis is complete, the content from its three documents (Image, Requirements document, and the Translation document) can be transferred to a Quality Function Deployment (QFD) or House of Quality (HOQ).

Rarely is this level of detail ported to other tools such as a Fishbone or Tree diagram.

Supporting or Linked Tools

Supporting tools that might provide input when developing a KJ Analysis include

- VOC Data Gathering Tools and Techniques, such as surveys and interviews (**See Also** "Voice of Customer Gathering Techniques," **p. 737**)
- Written reports
- Brainstorming (rarely) (**See Also** "Brainstorming Technique," **p. 168**)

A completed KJ Analysis provides input to tools such as

- House of Quality (HOQ) and/or Quality Function Deployment (QFD) (**See Also** "Quality Function Deployment (QFD)," **p. 543**)
- Tree diagram and matrix (rarely) (**See Also** "Tree Diagram," **p. 712**)
- Fishbone (rarely) (**See Also** "Cause-and-Effect Diagram," **p. 173**)
- Simple matrix (rarely) (**See Also** "Matrix Diagrams," **p. 399**)

Figure K-5 illustrates the link between the KJ Analysis and its related tools and techniques, showing how some provide input **to**, while others utilize the information **from** a completed KJ.

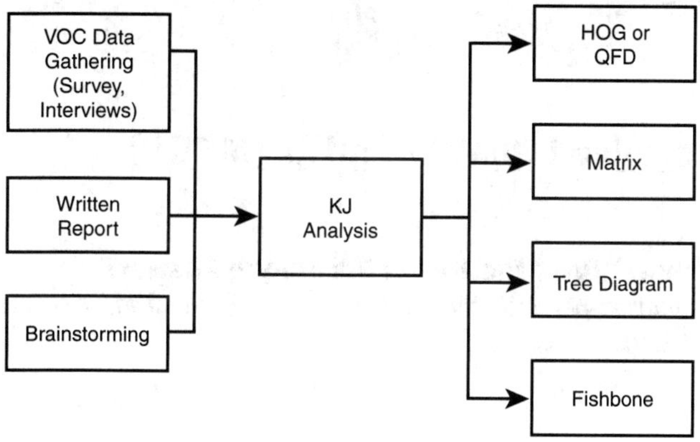

Figure K-5: KJ Analysis Tool Linkage

Variations

Affinity Diagram

The Affinity diagram involves less detail and analysis than what is described in this entry. (**See Also** "Affinity Diagram," **p. 136**)

M

Market Perceived Quality Profile (MPQP)

What Question(s) Does the Tool or Technique Answer?

How does the market perceive the quality of the product and/or services offerings versus competition?

An MPQP helps you to

- Estimate the market's perception of current product and services offerings to evaluate and develop a response and go-forward strategy

When Best to Use the Tool or Technique

Early in the process when gathering requirements from customers, use the MPQP tool to understand current gaps in quality compared to competition's offerings, based upon existing data and internal opinions.

Brief Description

The Market Perceived Quality Profile is a technique that identifies and measures market perceived quality of an offering, relative to meeting customers' requirements and expectations. Quality, as defined by the marketplace, can represent several dimensions—feature/functionality, durability, value, ease of use, and total cost of ownership. MPQP estimates and profiles your company market perception relative to competition. The tool uses data about both the market and market segment to evaluate the value that your company and the key competitors deliver. Presuming that customers prefer to buy on value, the tool calculates value as a function of quality relative to price. It dissects quality into product and/or services performance and features and then contrasts them to their respective market price.

Customers perceive quality by how much an offering conforms to their expectations or requirements. The magnitude of conformance of both implicit and explicit expectations determines how much they are willing to pay for it. To start, MPQP utilizes a set of default customer quality criteria to determine value as surrogate Voice of the Customer (VOC) requirements as to the critical quality parameters. However, it should be adapted, expanded, and refined to meet individual market needs and reflect actual VOC input, if available. The default set of quality characteristics varies depending on whether the offering is a product, service, or both.

Dimensions of Quality

For **product quality**, MPQP utilizes *David Garvin's eight dimensions* of quality to represent customer's criteria. They are as follows:

- **Performance**—Primary functional characteristic of a product or service
- **Features**—Secondary characteristics of a product or service
- **Reliability**—Frequency with which a product or service fails
- **Conformance or consistency**—Ability to fulfill specifications or comply with standards
- **Durability**—Product resilience and life
- **Serviceability**—Speed, courtesy, and competence of repair or adjustment
- **Aesthetics**—Form, fit, and finish
- **Perceived** Quality—Reputation

For **service quality**, MPQP utilizes *Berry's ten characteristics* to represent customer's criteria. They are as follows:

- **Reliability**—Consistency of performance
- **Responsiveness**—Timeliness of service
- **Competence**—Possession of required skills and knowledge
- **Access**—Approachability and ease of contact
- **Courtesy**—Politeness, respect, consideration, and friendliness
- **Communication**—Listening to customers and keeping them informed in a language they can understand
- **Credibility**—Trustworthiness, believability, honesty
- **Security**—Freedom from danger, risk, or doubt
- **Understanding the customer**—Efforts to understand the customer's needs
- **Tangibles**—Physical evidence of the service

The MPQP calculates a gap analysis to **estimate** where your company's score is relative to its competitors in the market perceived quality. The MPQP is used to help identify opportunities for growth and threats that present a potential risk to your current and future near-term offerings. The tool starts with internal VOC data (actually it is the voice of your marketing experts plus secondary market research data). Then the estimated-MPQP profile is validated through actual customer interviews. This **market validation** process is essential to confirm your competitive advantage and customer requirements.

Competitive Summary

Market validation confirms the competitive marketplace drivers as quality-based or price-based. The quality-based markets favor dominant traits in technology and performance leadership. The price-based markets characteristically prefer the low cost production of commodity products. The MPQP tool to summarize the competitive nature of the marketplace is the *Competitive Position Matrix*, illustrated in Table M-1.

In general, the competitive summary determines which market player possesses the dominant theme relative to quality or performance sensitivity. Once the competitive landscape is defined, the question is raised as to what gaps need to be closed with respect to quality performance, based on the customer expectations.

Customer Requirements

Market validation also confirms the VOC requirements as to the critical quality parameters and their relative rank order. During this phase, any customer's new, unique, and difficult (NUD) requirements should be solicited as input to the design of future offerings. In turn, these NUD requirements and the MPQP customer quality criteria feed into the future offering's Quality Function Deployment (QFD). (Moreover, the MPQP gap can be used in the competitive column of the QFD tool.) Responding to the NUD requirements will increase the likelihood of reaching the full entitlement of that offering's business case. (**See Also** "Kano Model: NUD versus ECO," section of "KJ Analysis," **p. 375** and Quality Function Deployment (QFD), **p. 543**, for more detail.)

Ensure that the customer perspective is broad enough to cover not only the current customer base, but prospective customers. Data also should be collected from non-customers. Within the served market, identify key product and service attributes that affect the purchase decision (purchase criteria, dimensions of performance, needs). As previously mentioned, start by using the default Dimensions of Quality (either Garvin's eight or Berry's ten).

The sample customer data needs to reflect the market segmentation within the served market. For these market segments, identify the quality attributes (dynamics) that help characterize the served market.

In addition, the sample customer data need to reflect the functional disciplines representing the target audience. These functions drive the actual purchasing decision within the customer account. Identify the different functions that consume or appreciate the quality attribute(s) represented in the customer's purchasing decision group (for example, purchasing, operations, and design). Determine the unique attributes of quality that this group values.

Completed MPQP

A completed MPQP helps define opportunities to improve your competitive position. As a result, subsequent strategies should focus on improving your company's performance ratings, shifting focus to the criteria where your company has a sustainable competitive advantage, and focusing on market segments representing a new or sustainable competitive advantage.

A completed MPQP (as shown in Table M-2) helps a company answer the following key business questions:

- What are the major product and service attributes that affect the customers' decisions to purchase from your company rather than a competitor?
- Who are your major competitors?
- How do customers perceive your company's product/service offering versus your competitors'?
- Can the *Key Customer Criteria* suggest meaningful ways to segment the market? Are market dynamics active in these areas of quality? (You may use the appropriate default set—Garvin's eight product or Berry's ten service dimensions of quality.)
- Can you identify areas where your company can improve its estimated market perceived quality profile?
- Which quality attributes represent NUD opportunities? These attributes could come from one of the following sources:
    - **Baseline**—All providers perform well; no competitive edge.
    - **Competitive**—Differences in performance determine competitiveness and thus opportunities or threats.
    - **Secondary**—Currently a competitive attribute, but catch up efforts and/or declining weight might take away the top performer's competitive edge.
    - **Dynamic Opportunities**—Will become a Competitive attribute when some provider pulls ahead and/or more weight shifts to this quality attribute.

How to Use the Tool or Technique

The MPQP process is comprised of two parts—the Competitive Position Matrix and the actual Market Perceived Quality Profile (MPQP) Matrix.

The first component summarizes the competitive landscape using a *Competitive Position Matrix*, to assess a company's relative position on quality and value to its competitors. This matrix provides input into the MPQP.

The second component is the actual *Market Perceived Quality Profile (MPQP)*. To simplify the calculations, use a software application such as Microsoft Excel to build the matrix and embed the various formulas.

Step 1. Complete a competitive summary by using the Competitive Position Matrix.

a. Select your company's leading competitors relative to a particular offering (product and/or service).

b. Enter your company's percent of market share and that of your leading competitors in the second row of the matrix—labeled "Market Share."

c. Enter the price index of your leading competitors' products and/or services as a percentage relative to your company's price in the third row of the matrix—labeled "Price Index."

The relative price index is expressed as a percentage compared to your company's priced offering (that is, Competitor A's price is 110%, or 90% of your price).

d. Enter the direct cost of the leading competitor's offering as a percentage relative to your company's direct costs in the matrix's fourth row—labeled "Direct Costs."

This cost index will be an estimate based on benchmarking research or marketing data.

e. Identify whether your competitors' technology and/or performance quality positions are *ahead*, *equal*, or *behind* relative to your company's offering and record that position in row five—labeled "Technology/Performance Quality."

Table M-1 provides a snapshot of the competitive landscape comparing four dimensions: 1) market share, 2) price, 3) direct cost, and 4) quality and/or technology performance.

TABLE M-1: Competitive Position Matrix

	Our Company	Competitor A	Competitor B	Competitor C
Market Share				
Price Index (% relative to ours)	100%			
Direct Costs Index (% relative to ours)	100%			
Technology / Performance Quality (relative to ours)	Equal			

Step 2. Define your estimated competitive gaps by completing the Market Perceived Quality Profile (MPQP) matrix, as illustrated in Table M-2.

a. Create an eleven column L-shaped matrix, preferably in Microsoft Excel. Label the columns as follows:

i. Key Customer Criteria

ii. Relative Importance, Criteria Weight

iii. Relative Importance, Share of Importance

iv. Quality Performance Rating, Our Company

v. Quality Performance Rating, Competitor A (per the leading competitors selected in Step 1.a.)

vi. Quality Performance Rating, Competitor B (per the leading competitors selected in Step 1.a.)

vii. Quality Performance Rating, Competitor C (per the leading competitors selected in Step 1.a.)

viii. Weighted Ratings, Our Company

ix. Weighted Ratings, Quality Performance Leader

x. Weighted Ratings, Share Leader

xi. Gap (Our Company versus Quality Performance Leader)

(**See Also** "Matrix Diagrams—7M Tool," **p. 399**, for details about different shaped matrices.)

b. Select the set of appropriate Key Customer Criteria and list them down the first column of the MPQP matrix.

You may use actual VOC data or start with and/or modify the appropriate Garvin's eight dimensions for product quality or Berry's ten dimensions for service quality, which were previously mentioned.

In the last row of this list, record the name "Total" in the final cell to serve as the row heading for the total scores that will be calculated.

c. Estimate the Customer Criteria stack rank order as relative weights from 0 to 100 points, with **1 as low** and **100 as high**, for each and document them in the second column—labeled "Relative Importance, Criteria Weight."

   Complete this task until each Customer Criteria is stack ranked. Avoid duplicate (or tie) scores to better differentiate one from another.

   Add the scores in the column, and document it in the final row of that column, labeled "Total."

d. Calculate the percent importance contribution of each Customer Criteria relative to the total and record it in the third column—labeled "Relative Importance, Share of Importance."

   The calculation formula is: the individual Key Customer Criteria Weight divided by the Total Criteria Weight, times 100%. Complete this calculation for each cell in the third column containing a Key Customer Criteria Weight.

e. Using the same leading competitors selected in Step 1.a., determine the **relative** Performance Rating for them and your company for each Key Customer Criteria, using the scale of 1 to 10, with 1 as low and 10 as high. Record the rating score in the appropriate columns four through seven—labeled "Quality Performance Rating for Our Company or Competitor A, B, or C."

   Avoid duplicate (or tie) scores for a given Key Customer Criteria to better differentiate one from another across the market players. Complete this relative comparison rating for each cell in the columns four through seven aligned with a Key Customer Criteria Weight.

f. Determine your company's *Weighted Ratings* for each Key Customer Criteria by multiplying its performance rating by the share of importance percentage and recording it in the eighth column—labeled "Weighted Ratings, Our Company."

The calculation formula is: the individual Share of Importance times the Quality Performance Rating, Our Company. Complete this calculation for each cell in the eighth column aligned with a Key Customer Criteria Weight.

g. Determine the Quality Performance Leader's Weighted Ratings for each Key Customer Criteria by selecting the market player with the **highest** Performance Rating (from columns four through seven); multiplying it by the share of importance percentage; and recording it in the ninth column—labeled "Weighted Ratings, Quality Performance Leader."

   The calculation formula is: the individual Share of Importance times the highest Quality Performance Rating found across columns four through seven. Complete this calculation for each cell in the ninth column aligned with a Key Customer Criteria Weight.

h. Identify the market share leader from the data gathered in Step 1. Determine the Weighted Ratings Share Leader scores for each Key Customer Criteria by using the **market share percentage as a constant** and multiplying it by the Share of Importance, and recording the calculation in the tenth column—labeled "Weighted Ratings, Share Leader."

   The calculation formula is: the individual Share of Importance times the market share percentage as a constant, times 100%. Complete this calculation for each cell in the tenth column aligned with a Key Customer Criteria Weight.

i. Calculate the Offering Quality Gaps by subtracting the Quality Performance Leader's Weighted Rating score from your company's score and recording it in the appropriate cell of the eleventh column—labeled "Gap."

   A rating of either a positive or negative score is possible. If your company is the Quality Performance Leader for a given Key Customer Criteria, the gap score will be zero.

   The calculation formula is: the Individual Weighted Ratings, Our Company minus the corresponding Weighted Ratings, Quality Performance Leader for the appropriate Key Customer Criteria. Complete this calculation for each cell in the eleventh column aligned with a Key Customer Criteria Weight.

TABLE M-2: Market Perceived Quality Profile (MPQP) Sample Matrix

Key Customer Criteria	Relative Importance (Avoid ties)		Quality Performance Rating (Rating scale 1 to 10; avoid ties or duplicates)				Weighted Ratings			Gap (Our Company versus Quality Perf. Leader)
	Criteria Weight (1-100)	Share of Importance	Our Company	Competitor A	Competitor B	Competitor C	Our Company	Quality Perf. Leader	Share Leader	
Product or Service Performance	90	30.0%	8	9	6	7	2.4	2.7	13.5	-0.3
Proprietary Features	45	15.0%	9	6	5	4	1.4	1.4	6.8	0.0
Reliabilility	60	20.0%	5	6	4	7	1.0	1.4	9.0	-0.4
Conformance or consistancy	15	5.0%	6	7	8	9	0.3	0.5	2.3	-0.2
Durability	30	10.0%	3	5	4	8	0.3	0.8	4.5	-0.5
Serviceability	25	8.3%	7	8	5	6	0.6	0.7	3.8	-0.1
Appearance or aesthetics	20	6.7%	8	9	6	7	0.5	0.6	3.0	-0.1
Quality Reputation & intangibles	15	5.0%	6	9	4	8	0.3	0.5	2.3	-0.2
Total:	300	100.0%								

Table M-2 uses an example wherein the market share leader possesses a 45% share. This percentage is used to calculate the Weighted Ratings, Share Leader scores found in column ten.

Supporting or Linked Tools

Supporting tools that might provide input when developing an MPQP matrix include

- Benchmarking (**See Also** "Benchmarking," **p. 160**)
- KJ Analysis (**See Also** "KJ Analysis," **p. 375**)
- Porter's 5 Forces (**See Also** "Porter's 5 Forces," **p. 464**)
- SWOT (**See Also** SWOT (Strenghs-Weaknesses Opportunities-Threats), **p. 699**)
- VOC gathering techniques (**See Also** "Voice of Customer Gathering Techniques," **p. 737**)

A completed MPQP matrix provides input to tools such as

- Fishbone (**See Also** "Cause-and-Effect Diagram—7QC Tool," **p. 173**)
- FMEA (**See Also** Failure Modes and Effects Analysis (FMEA), **p. 287**)
- GOSPA (**See Also** Goals, Objectives, Strategies, Plans, and Actions (GOSPA), **p. 320**)
- QFD (**See Also** Quality Function Deployment (QFD), **p. 543**)
- Solution Selection techniques (**See Also** "Solution Selection Matrix," **p. 672** and "Pugh Concept Evaluation," **p. 534**)

Figure M-1 illustrates the link between the MPQP matrix and its related tools and techniques.

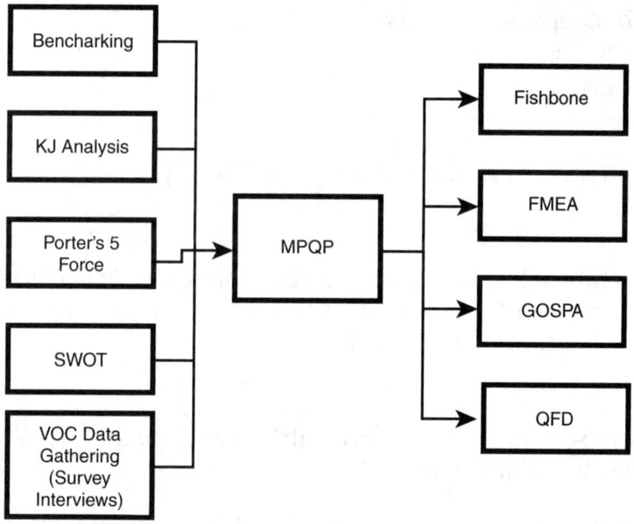

FIGURE M-1: MPQP Tool Linkage

Additional Resources or References

- Gale, B. T., ***Managing Customer Value: Creating Quality and Service That Customers Can See,*** New York: The Free Press, 1994; ISBN: 0-02-911045-9.
- Garvin, D. A., *Managing Quality*, New York: The Free Press, 1988; ISBN: 0-02-911380-6.

Matrix Diagrams—7M Tool

What Question(s) Does the Tool or Technique Answer?

How do these two (or more) groups relate to one another?

Matrix diagrams help you to

- Communicate how groups relate to one another and often the strength of that relationship
- Collect, organize, plan, manage, prioritize, focus on, and decide about a specific topic

Alternative Names and Variations

This tool is also known as

- Chart
- Checklist (**See Also** "Checklists—7QC Tool," **p. 204**)
- Table

Variations on the tool include a specialized matrix used to collect, organize, plan, manage, prioritize, and focus on a specific topic. Often these specialized matrices are known by a unique name.

Examples of such specialized matrices include

- Cause-and-Effect Prioritization matrix (**See Also** the "Cause-and-Effect Prioritization Matrix," **p. 188**)
- Communication plan (**See Also** the Communication Plan "Variation," section, **p. 405** for an example.)
- Control plan (**See Also** the Control Plan section, **p. 406** for an example.)
- CTQ matrix (**See Also** "Critical to Quality (CTQ)," **p. 242**)
- Decision Authority Matrix (**See Also** the Decision Authority Matrix section, **p. 408** for an example.)
- FMEA (**See Also** Failure Modes and Effects Analysis (FMEA), **p. 287**)
- Launch plan (**See Also** the Launch Plan (or Implementation Plan, Pilot Plan) section, **p. 409** for an example.)
- Prioritization matrices (**See Also** "Prioritization Matrices—7M Tool," **p. 470**)
- Pugh Concept Evaluation (**See Also** "Pugh Concept Evaluation," **p. 534**)
- RACI (**See Also** "RACI Matrix (Responsible, Accountable, Consulted Informed)," **p. 554**)
- Real Win Worth (RWW), uses three set of specific matrices. (**See Also** "Real Win Worth (RWW) Analysis," **p. 560**)
- Quality Function Deployment (QFD) or House of Quality (HOQ) (**See Also** "Quality Function Deployment (QFD)," **p. 543**)
- SIPOC (**See Also** "SIPOC (Supplier-Input-Process-Output-Customer)," **p. 663**)
- Solution Selection Matrix (**See Also** "Solution Selection Matrix," **p. 672**)
- Transition plan (**See Also** the Transition Plan section, **p. 410** for an example.)

When Best to Use the Tool or Technique

The matrix is such a flexible tool that it can be used at any time during a project or process. If it is to be used to communicate the current (prior to any improvements) state, then it should document the improved state as well.

Brief Description

The *matrix* is a management and planning tool that evolves depending on the complexity of the situation. It organizes data or knowledge to examine the relationship between two or more groups of topics. Fundamentally, it is a grid-structured tool comprised of rows and columns.

The simplest matrix evaluates two groups, where one is plotted horizontally as column headings (for example, the X-axis), and the second is plotted vertically as row headings (for example, the Y-axis). The strength of relationship is indicated at the intersection of a row and column, in the grid's cell. The cell data typically is comprised of numbers, letters, or symbols. Generally, the associations being explored in a matrix can include: objectives and methods (for example, current approach, results, or possible solution), causes and effects, categories and respective results, and tasks and people. The preceding list of matrix variations align to these association categories as follows:

- Objective and Method:
    - Communication Plan
    - Control plan
    - CTQ matrix
    - Launch plan
    - Prioritization matrices
    - Pugh Concept Evaluation
    - Quality Function Deployment (QFD) or House of Quality (HOQ)
    - Solution Selection matrix
    - Transition plan
- Cause-and-Effect:
    - Cause-and-Effect Prioritization matrix
    - FMEA

- Category and Result:
    - Checklists
    - Real-Win-Worth (RWW)
    - SIPOC
- Tasks and People:
    - Decision Authority Matrix
    - RACI

The shape of a matrix depends on the number of groups or topics being compared. As a comparison category gets added, the dimension of the matrix changes from a simple two-dimension framework comparing two groups to comparisons between and among multiple groups. Hence, there are six different types of matrices:

- **L-shape**—Comparing two groups. Simplest of matrices, showing a one-for-one relationship of two topics, where the rows and columns of a grid form an sideways "L." This is the most common matrix shape and depicts how Topic A relates to Topic B.
- **T-shape**—Comparing three groups, one topic relative to two dependent topics. This matrix compares two topics to a common third topic. The common topic is plotted along the X-axis that divides the other two related topics—one plotted on the upper Y-axis, and the second plotted on the lower Y-axis, thereby forming a sideways "T."

> **Note**
>
> The common topic relates to two dependent topics; however, the two dependent topics are not compared to one another. Topic A relates to Topics B and C.

- **Y-shape**—Comparing three groups, showing a **circular** relationship among the three. This matrix combines two L-shaped matrices along the Y-axis to create a three-dimension grid, shaped similar to a house (with a roof). Topic A relates to Topic B, which relates to Topic C, which relates to Topic A.

- **C-shape**—Comparing three groups, showing a **criss-crossed** relationship among the three topics. The matrix combines two L-shaped matrices along the Y-axis to represent two topics, and a third topic is plotted along the Z-axis, as the third dimension, to form a **cubed** structure. (Often one of the L-shaped matrices may be at a 45-degree angle to the other.) This relationship is often difficult to draw, thus often left to computer software to create. (Software packages tend to be specific to the context, such as engineering or healthcare.) Topics A, B, and C **simultaneously** relate to one another.

- **X-shape**—Comparing four groups—in which two sets of topics are compared. This matrix essentially combines two T-shaped matrices with a common Y-axis but different X-axes. Topic A relates to Topics B and C, and Topic D relates to Topics B and C, but it does **not** examine the relationship between Topics B and C.

- **Roof-shaped**—Compares one set of topics internally to one another. This matrix forms a triangle that indicates the interrelationship of related topics to one another, such as potential features (or components) within a product (or service). Matched topics with a relationship may indicate a positive reinforcing and supportive, or negative, counter-balancing effect. These cells contain a symbol to indicate the strength of the relationship [that is, circle, concentric circles (or doughnut), triangle, square, arrow]. Blank cells indicated no relative impact on one another. It usually accompanies an L-shaped or T-shaped matrix and commonly caps a Quality Function Deployment (QFD) or House of Quality (HOQ) matrix. Topic A relates to Topic B within the same group.

Figure M-2 illustrates the different shaped matrices.

Matrix diagrams are a member of the 7M Tools, attributed in part to one of the first quality thought-leaders, Dr. Shewhart, as seven "management" tools, or sometimes referred to as the "7MP" or *seven management and planning tools*. These 7M Tools make up the set of traditional quality tools used to analyze quantitative data. The 7M Toolset includes: 1) Activity network diagrams or Arrow diagrams; 2) Affinity diagrams; 3) Interrelationship digraphs or Relations diagrams; 4) Matrix diagrams; 5) Prioritization matrices, often replacing the more complex Matrix data analysis; 6) Process decision program charts (PDPC); and 7) Tree diagrams. *The Quality Toolbox*, by Nancy Tague, presents the 7M Tools ranked from those used for abstract analysis to detailed planning: Affinity diagram, Relations diagram, Tree diagram, Matrix diagram, Matrix data analysis (commonly replaced by a more simple Prioritization Matrix), Arrow diagram, and Process Decision Program Chart (PDPC).

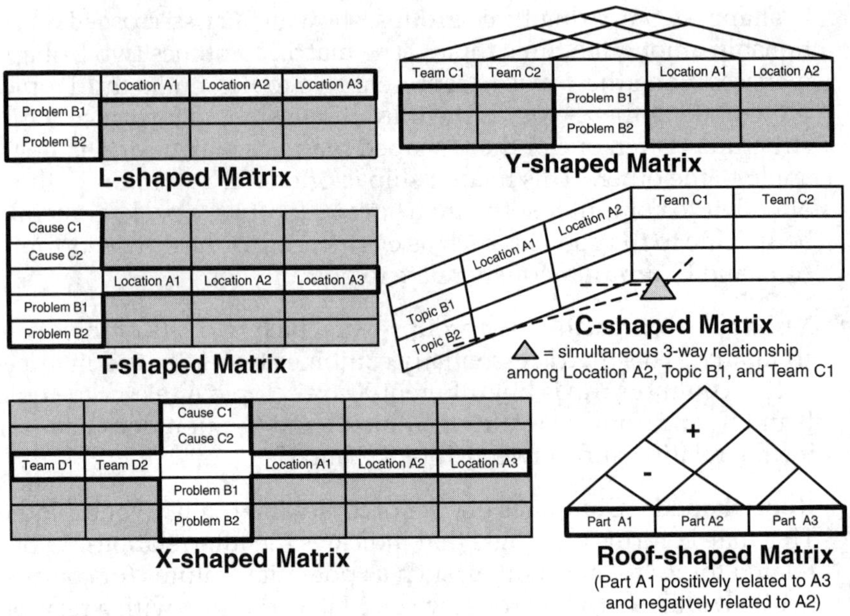

FIGURE M-2: Different Types of Matrices

How to Use the Tool or Technique

Matrices that serve a specific function and are so named for that purpose have a unique set of guidelines or procedures, for example one that prioritizes potential root causes is called a *Cause-and-Effect Matrix*. A separate entry in this reference book documents many of those unique matrix tools (Please see "Cause-and-Effect Prioritization Matrix," "CTQ Matrix," "RACI," "SIPOC," and so on). A list of such matrices with the exact entry name and page numbers to reference is documented in the previous "Alternative Names and Variations" section.

When creating a generic matrix, the procedure is as follows:

Step 1. Define the purpose of the matrix—its objective/what relationship is to be studied.

Step 2. Identify the topics or groups to be compared.

Step 3. Select the appropriate matrix format and symbols to be used.

Step 4. Complete the matrix diagram.

Additional Resources or References

- U.S. Government web sites with instructions and diagramming help: http://www.usbr.gov/pmts/guide/toolbox/matrixdi.html, and http://www.ed.gov/inits/americareads/resourcekit/MakingInfo/title2.html
- Visual Matrix, http://www.visualmatrixpms.com/?gclid=CJXz_9mj6okCFSIOgQodjinUGw.

Variations

Communication Plan

To plan and organize the execution of a project or process communication plan, a customized matrix tool can identify the key message, target audience, type of communication vehicle, timing of delivery, and who should develop and distribute the final communiqué.

A communication plan can take on several dimensions. Its objective can show the relationship between the sender (or deliverer) of a message and the receiver. Another objective could demonstrate the relationship between the message and different recipients. A more complex plan could examine the relationship among what needs to be communicated to whom; why is it important; how the message will be communicated (in what format and venue) and when; who needs to develop and deliver the message and by when; and how much the development and delivery will cost. A communication plan also may consider how the sender can test for understanding with the receiver (or a question-and-answer venue) or how the receiver can notify the sender that the message was received.

These multiple dimensions may dictate multiple formats, such as a written report, presentation, or matrix. Sometimes a matrix can suffice alone or be incorporated within a report as a planning and management tool. Figure M-3 shows a sample communication plan template for a marketing department to develop and deliver key customer messages.

Figure M-4 shows a sample of a project team's communication plan to show how they will work together.

A B C D E F G H I J K L M N O P Q R S T U V W X Y Z

						Development					Delivery				
Communication Message (What; Topic)	Target Audience (To Whom)	Objective (Why they care)	Method (How)	Frequency (When)	Project Manager	Team Member 1	Team Member 2	Team Member 3	Team Member 4	Completion Date *(Target); Actual*	Team Member 5	Team Member 6	Team Member 7	Team Member 8	Completion Date *(Target); Actual*

Objective: (**D**)irective; (**I**)nformative; (**O**)ther
Method: (**E**)Mail; (**R**)eport; (**M**)emo; (**V**)oicemail; (**CC**) Conference Call; (**Mtg**) Meeting ; (**B**)rochure
Frequency: (**D**)aily; (**W**)eekly; (**M**)onthly; (**Qtr**) Quarterly; (**A**)nnually

FIGURE M-3: Example of a Communication Plan Template for Messages

Communications	Method	Frequency	Project Customer	Project Manager	Team Member 1	Team Member 2	Team Member 3	Team Member 4	Team Member 5	Team Member 6
Meetings (Oral)										
Sub-team										
Project Team										
Status										
Management Review										
Interim Project Review										
Post Implementation										
Customer/Client										
Reports (Written)										
Meeting Agenda										
Meeting Minutes										
Action Items										
Project Schedule										
Variance Reports										
Project Status Report										

Method: (**E**)Mail; (**R**)eport; (**M**)emo; (**V**)oicemail; (**CC**) Conference Call; (**Mtg**) Meeting
Frequency: (**D**)aily; (**W**)eekly; (**M**)onthly; (**Qtr**) Quarterly; (**A**)nnually

FIGURE M-4: Example of a Communication Plan Template for a Project Team

Control Plan

To prepare for transition from an improvement project environment to the process players or to simply manage a process, a customized matrix tool can identify key control plan elements. The document might include the key process parameters to be monitored, their respective metrics, target values, source of the data, and who is accountable to gather, monitor, control, and provide status on each factor.

Similar to a communication plan, a control plan can take on multiple dimensions. It should document how to sustain any improvements, plus manage and adapt the ongoing operations of a process. It defines the technical method of control, and documents the response plan to adjust to any changes in either a written report, standard operating procedure, and/or matrix. The control methods may include a data collection matrix, statistical Control charts, dashboards, or operations review meetings. The response plan involves risk identification and response. Similar to a communication plan, a control plan may contain all these elements in a written report and/or be referenced in a single matrix or several smaller matrices. This type of matrix should be specifically customized for the organization's purpose.

Figure M-5 provides an example of a Control plan using a matrix structure.

Control Plan

Process Name: ____________		Prepared by: ____________	Page: ____ of ____
Process Owner: ____________	Int/Ext ______	Approved by: ____________	Date: ____________

Process Management Components

Component	Accountable	Key Metrics	Frequency Cycle	Key Customer
Standard Work Policies and Procedures				
Critical Process Parameter Data Gathering Coordination				
Planning and Preparation for Operations Review of process performance *(calendar, agenda, logistics)*				
Conduct Operations Reviews				
Integration of Risk Mitigation Plan *(Data and Action Plan integration)*				
Communication Plan *(within Process, in-company, external)*				
Integration of Risk Mitigation Plan				
Process Player Training Plan *(new members and ongoing)*				
Communication Plan *(within Process, in-company, external)*				

Critical Process Parameters

Process Step	CTQ Characteristic	CTQ Metric		Target +/or Control Limit		Measurement Method	Sample Size	Frequency	Who Measures	When Recorded	Decision Rule/ Corrective Action	SOP Reference
		KPOV	KPIV	USL	USL							

FIGURE M-5: Example of a Control Plan Using a Matrix Format

(**See Also** "Data Collection Matrix," **p. 248**; "Control Charts—7QC Tool," **p. 217**; Failure Modes and Effects Analysis (FMEA), **p. 287**; and "Poka-Yoke," **p. 462**)

Decision Authority Matrix

To resolve any ambiguity, gaps, or redundancy for a given decision-making process, a customized matrix tool can define the standard procedure. The tool should identify the key players involved in the process, their deliverables, and ultimately who has the final vote. The matrix defines the type of decision-making process pertinent to a given topic, such as authoritative, collaborative, participative, or consensus.

This matrix tool identifies the roles or people accountable for making decisions in a process or organization. The Decision Authority Matrix reflects and reinforces the type of decision-making style or culture of the organization—consensus, democratic, participative, or authoritarian. This kind of documentation is helpful on complex, new, or cross-organizational processes.

As its name states, a Decision Authority Matrix identifies one, **and only one**, decision-maker who is accountable for making the decision and its consequences. For complex scenarios, it also defines the decision-making process and who else is providing appropriate background information and context, including those responsible for any analytical activities. Complex processes may identify a person accountable for integrating the relevant data, synthesizing it, and making a recommendation to the decision-maker. At times, decision-making may be delegated or escalated to another role or person, depending on the level of risk and financial impact.

Figure M-6 provides an example of a Decision Authority Matrix template.

Decision Authority Matrix
for_______________process

D = Decision-maker (only 1 entry)
R = Recommend C = Concur
P = Perform E = Explain

Deliverable/Task Description	Member A	Member B	Member C	Member D	Member E	Member F	Member G	Member H		
Decision 1	D	R	P	E	P	P	P	C		
Decision 2		D		E	C					
Decision 3	C			D	R	P	P	R		
Decision 4	D	R	P		P					

FIGURE M-6: Example of a Decision Authority Matrix Template

This tool is often compared with an *RACI Matrix*, since both tools define the responsibilities for those involved in a process.

Although the names are similar, the *Decision Authority Matrix* distinguishes itself from two other kinds of matrices—a *Decision Matrix* and a *Prioritization Matrix*. A Decision Matrix may refer to a specific tool or a category of matrices used to evaluate and prioritize a list of options, often called by a specialized name—Pugh Matrix or Solution Selection Matrix.

A Prioritization Matrix is a category of three different types of matrices used to narrow options by comparing groups.

(**See Also** "RACI Matrix (Responsible, Accountable, Consulted, Informed)," **p. 554**; "Pugh Concept Evaluation," **p. 534**; "Solution Selection Matrix;" **p. 672**; and "Prioritization Matrices—7M Tool," **p. 470**)

Launch Plan (or Implementation Plan, Pilot Plan)

To plan and organize the execution of a product and/or services offering launch to the marketplace, a customized matrix tool can identify the key elements. It should identify the key deliverables, respective metrics, respective target audience (customer), timing of delivery, and who should develop and distribute the final output.

In preparation to launch a product or services offering, a launch team uses a *launch implementation plan* to introduce the offering to the marketplace and the customer value chain. Such a plan can apply to both a pilot, the implementation of an improvement, or the launch of a new offering. This plan documents the readiness requirements, deliverables and process to plan and manage the launch. The plan components may include the following:

- Schedule or timeline of activities, events, deliverables, and milestones.
- Work Breakdown Structure (WBS); a Tree diagram or matrix defining the key tasks, deliverables, task owners, task customer, timeframe to complete the task.
- Control plan documenting the key process and performance metrics, the score card (or dashboard) to monitor the metrics.
- Standard operating procedures, as applicable.
- Communication plan defining what will be communicated to whom, when, and how.
- Process map or flow (the former and enhanced process). (**See Also** "Process Map (or Flowchart)—7QC Tool," **p. 522**)
- SIPOC to changes in any one of the following areas: suppliers, inputs, high-level process, outputs, and/or customers. (**See Also** "SIPOC (Supplier-Input-Process-Output-Customer)," **p. 663**)
- RACI matrix to highlight those involved in the launch assigned tasks, deliverables, and/or responsibilities. (**See Also** "RACI Matrix (Responsible, Accountable, Consulted, Informed)," **p. 554**)

- Training plan of the initial and ongoing training and potentially any other human resource concerns (such as compensation or bonus impacts).
- Risk mitigation plan identifying potential risks and the corresponding contingency and response plan to each prioritized risk. This plan should include who is accountable for taking action if the risk occurs and any key triggers to detect if the risk occurs. (**See Also** Failure Modes and Effects Analysis (FMEA), **p. 287**)
- Budget allocated to the launch activities.

Figure M-7 illustrates a Launch plan matrix set up similar to a status report as an example. However, this type of matrix should be specifically customized for the organization's purpose.

XXX Launch Plan
Key Deliverables

Task #			Deliverable	Target Completion	Status	Accountable	Start Date	Completed Date	Responsible Individuals
Objective:			Launch XX in Direct Sales + Dealer Channel	12/30/20XX					
1.00			Marketing						
	1.10		**Features & Functionality Documented**	**2/18/20XX**	Completed	G. Testa	1/24/20XX	2/28/20XX	N, Minocha, M. McCandless
		1.10.1	Draft 1.0 Completed & Reviewed w/C. Magee	1/31/20XX	Completed	G. Daniel	1/24/20XX	1/31/20XX	N, Minocha, M. McCandless
		1.10.2	Draft 2.0 Completed & Reviewed w/C. Magee	2/2/20XX	Completed	G. Daniel	1/31/20XX	2/2/20XX	
		1.10.3	Draft Reviewed w/M. George	2/3/20XX	Completed	G. Testa	2/2/20XX	2/3/20XX	
		1.10.4	Draft Reviewed w/T. Garrett	2/17/20XX	Completed	G. Testa	2/2/20XX	2/28/20XX	
		1.10.5	Draft Reviewed w/D. Washington	2/17/20XX	Completed	G. Testa	2/2/20XX	2/28/20XX	
		1.10.6	Field Service level of involvement determined		Completed	G. Testa	2/2/20XX	2/28/20XX	J. King
	1.30		**Value Proposition Defined by Target Audience**	**2/28/20XX**	In-process	D.Garcia			W. Jewett, S. Jones
		1.30.1	Target Audience Identified (by market segment)	2/17/20XX	Completed	D.Garcia	2/2/20XX	2/28/20XX	W. Jewett, S. Jones
		1.30.2	Value Propositions drafted	3/30/20XX	Completed	D.Garcia	2/20/20XX	1/30/20XX	W. Jewett, S. Jones
		1.30.3	Value Propositions tested with market segment	5/30/20XX	In-process	D.Garcia			W. Jewett, S. Jones

FIGURE M-7: Example of a Launch Plan Template

A well-executed launch implementation plan of a poor idea is more likely to be successful than the other way around. And a well-developed launch implementation plan becomes the control plan. Similar to a communication plan, a launch plan may contain any or all of these elements in a written report and/or referenced in a single matrix or several smaller matrices.

Transition Plan

To plan and organize the hand-off of the project work to the ongoing process owner and players, a customized matrix tool can identify the key elements of a transition plan. Similar to a Launch plan, a Transition plan prepares the final project recommendations for ongoing execution by those working in the process, and allows the project team to disband while maintaining or sustaining the desired project results.

In preparation to closing out a project, a project team uses a transition plan to hand off its improvements to the ongoing process players in the organization. This plan documents the changes made to the current process and may include the following:

- Training plan of the initial and ongoing training
- Process map or flow (the former and enhanced process) (**See Also** "Process Map (or Flowchart)—7QC Tool" **p. 522**)
- SIPOC to changes in any one of the following areas: suppliers, inputs, high-level process, outputs, and/or customers (**See Also** "SIPOC (Supplier-Input-Process-Output-Customer)," **p. 663**)
- RACI matrix to highlight changes in the process players' assigned tasks, deliverables, and/or responsibilities (**See Also** "RACI Matrix (Responsible, Accountable, Consulted, Informed)," **p. 554**)
- Financial plan or budget to document any initial or ongoing funding requirements of the improvements

Similar to a communication plan, a transition plan may contain all these elements in a written report and/or referenced in a single matrix or several smaller matrices. This type of matrix should be specifically customized for the organization's purpose.

Figure M-8 provides an example of a Transition plan using a matrix structure.

Transition Plan

Process Name: ______________ Prepared by: ______________ Page: ____ of ____
Process Owner: ______________ Int/Ext ______ Approved by: ______________ Date: ______

Process Management Components

Component	Accountable	Key Metrics	Frequency Cycle	Key Customer
Standard Work Policies and Procedures *(revised as needed)*				
Process Map *(revised as necessary)*				
RACI Document *(responsibilities revised as necessary)*				
Critical Process Parameter metrics *(revised as necessary)*				
Other Improvement Revisions				
Improvement Implementation Risk Mitigation Plan *(Integrated with ongoing process)*				
Change Management Plan *(integrated with ongoing process)*				
Communication Plan of Improvements *(within Process, in-company external)*				
Process Player Training Plan *(current process players, and added to new members training)*				
Operations Review of Hand-off *(after XXXX $ of observed cycles)*				
Project Close-Out document *(after transition completed and approvals gained)*				

Revised Critical Process Parameters

Process Step	CTQ Characteristic	CTQ Metric		Target +/or Control Limit		Measurement Method	Sample Size	Frequency	Who Measures	When Recorded	Decision Rule/ Corrective Action	SOP Reference
		KPOV	KPIV	USL	USL							

FIGURE M-8: Example of a Transition Plan Template

Measurement System Analysis (MSA)

What Question(s) Does the Tool or Technique Answer?

How accurate is the measurement system? Is the process truly performing the way the data seems to be reporting, or is the measurement system inaccurate?

MSA helps you to

- Determine the reliability of the measurements
- Understand the integrity of the data

Alternative Names and Variations

This tool is also known as

- This tool is also known as theMeasurement System Evaluation method

When Best to Use the Tool or Technique

Before any data is collected, evaluate the measurement system (the calibration of device(s) and utilization procedure) for accuracy to ensure that it is not a source of variability and that in fact any observed variation is due to the product or process of interest—ensuring data integrity.

Brief Description

In Six Sigma, variation is the enemy, particularly if it is discernable by the customer. Often times unknowingly, the measurement system can introduce variability (noise) in the measurement data. Thus, observed variation may contain two sources of dispersion—the product (or part) itself and how it is measured. The objective is to minimize those controllable factors that exacerbate variation in data.

The total variability observed equals the sum of product (or part variability) plus the variability in the measurement system of the product (or part). The equation is written as: ***Total standard deviation*** **(σ_{Total}^2) = *Product (or Process) standard deviation*** **($\sigma_{Product}^2$) + *Measurement System standard deviation*** **($\sigma_{Measurement\ System}^2$)**. A measurement system is comprised of the measurement device (or tool) and its calibration, the procedure to take a measurement with the device, and the person or machine performing the process with the device. The utmost data integrity represents isolating any detected variation as attributed to the product (or part) of interest and ***not*** the measurement system.

The Measurement System Analysis (MSA) is a set of tests designed to evaluate how much variability is in the measuring device. The objective is to ensure that the measurements reflect the true nature of what is being measured and avoid a poor (noisy) measurement system. The measurement results equals the true value plus the error of a measuring instrument, represented by the equation: ***Measurement standard deviation*** ($\sigma_{Measurement}^2$) **= *True standard deviation*** (σ_{True}^2) **+ *Error*** (σ_{Error}^2). However, measurement systems have the potential to add a large amount of variability to the data of interest.

A poor measurement system could cause an organization to reject products that are actually good, simply due to inaccurate and unreliable measurements—thereby potentially rendering the process incapable of meeting target. If a poor measurement system exits, it may be unable to detect subtle differences that are important. The measurement system needs to be *discriminating* enough (have enough resolution) to detect differences of interest. It is important to conduct an MSA to validate the accuracy of the measurement system before determining the baseline capability of the process. Ideally, measurements are accurate and reliable and do not introduce any variability; if not, calibrate the measurement system before proceeding with the baseline capability test. The MSA tries to avoid the results of the old adage of "garbage in, garbage out" from occurring—so that good output is properly accepted and bad output is properly rejected. (**See Also** "Process Capability Analysis," **p. 486**)

MSA applies to any kind of data—continuous (variable) or discrete (attribute). As the data type changes, so does the measurement instrument. A tangible item can be evaluated for its size (inches or metrics), weight (pound or grams), or temperature (Celsius or Fahrenheit)—continuous measures. The same item also can also be evaluated for its attributes, such as color, type or name category, or customer satisfaction—discrete measures (count, rank, yes/no, and so on). The appropriate MSA test changes to align with the data type, which is no surprise, given that the measurement instrument differs with the data type as well. (**See Also** "Graphical Methods," **p. 323** for a more detailed discussion on different types of data.)

Accuracy and Precision

Two terms characterize variability—accuracy and precision. *Accuracy* describes how close to the target; and *precision* measures the degree of repeatability and reproducibility.

A B C D E F G H I J K L M N O P Q R S T U V W X Y Z

Which scenario is preferable—one in which the output is precise but off-target or one that occasionally hits the target (occasionally accurate) but contains a high amount of variability? The answer is precision, but off-target is easier to correct than highly variable, occasionally accurate output, as shown in Figure M-9.

Figure M-9 illustrates a target (a bulls-eye in the center of the circle) that both Lori and George are trying to hit. George's work may be viewed as more accurate than Lori's, considering he has hit the target once, but also the output contains a lot of variability. George may need more training or change how he is using the equipment to hit the target; multiple components of his work may need to improve. Lori's results have missed the target but collectively are more tightly clustered than George's. Hence, her results are more precise but not accurate. Lori's work is easier to correct. Lori's may simply need to aim differently but keep the process that she is using the same. Accuracy is described relative to hitting a target. Precision describes the amount of variability in a set of data. Both an accurate and precise outcome would be output centered directly on the target.

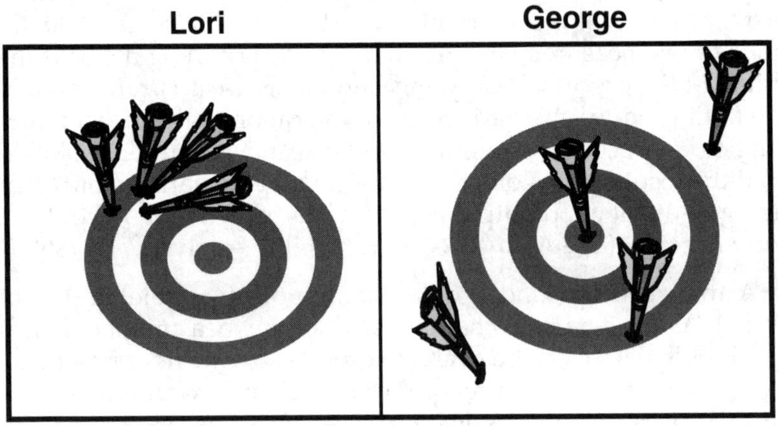

FIGURE M-9: Accuracy Versus Variability Illustration

Accuracy defines the precision with which a target is reached and is measured as the difference between the outcome and a target or standard value. If a sample of items were taken, and each were measured for accuracy, the average of those accuracy measurements for each sample item is known as *bias*. Figure M-10 illustrates the bias for two distributions.

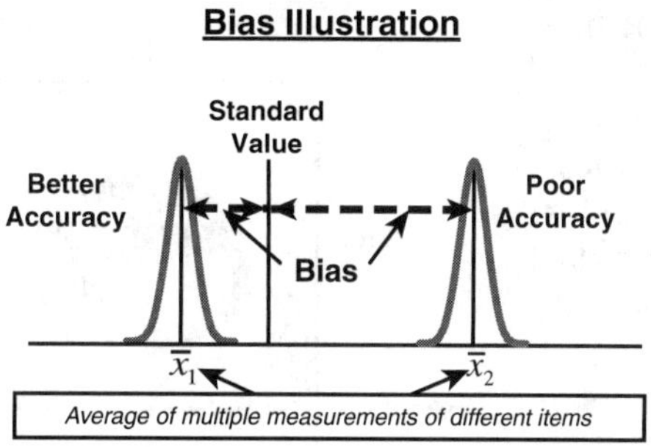

FIGURE M-10: Bias Illustration

Precision defines the variation in repeated measurements of the same item. There are two major ways to measure precision—repeatability and reproducibility. *Repeatability* is calculated by repeated measurements of the same item, by the same person, with the same measuring device. A measurement system is repeatable if the variation of the repeated measurement is tight. *Reproducibility* is when the variation source is the averages from repeated measurements (or factors) *made by different people (or machines) on the same item.* This variation between factors is also called *operator* or *technician error*. Figure M-11 depicts both repeatability and reproducibility.

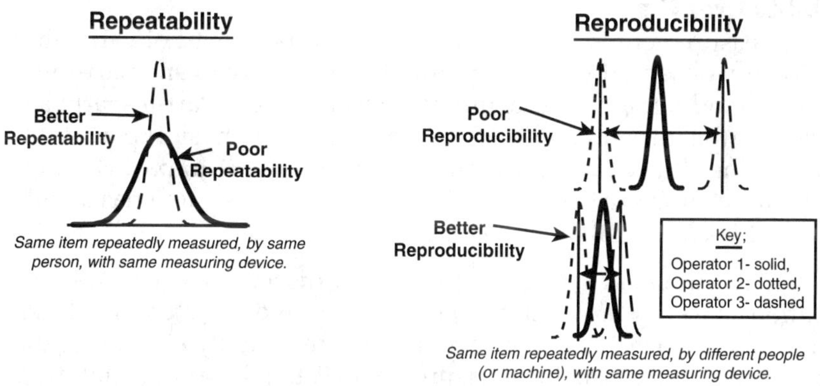

FIGURE M-11: Repeatability and Reproducibility Illustration

The operator error depicted in Figure M-11 represents two sources of variation—the differences from one operator's technique to the next and the interaction between the operator and the part. The latter variation source occurs when an operator is inconsistently handling the part when measuring it, resulting in different measurements, perhaps from different sections of the part or improper measurement of the part. Often times operator error is corrected with training and clarification of the standard operating procedure (SOP). (**See Also** "Attribute MSA," **p. 425**)

In summary, continuing with the dart board analogy, it integrates well the concepts of accuracy and precision. It is easier to adjust the aim of a dart thrower who consistently hits the same spot than it is to correct someone's throw whose darts hit all over the place. Thus, it is easier to correct poor accuracy than poor precision. Relative to a target, accuracy describes centering (how tight, consistent), and precision refers to the spread. Figure M-12 examines the relationship between good versus poor accuracy and good versus poor precision.

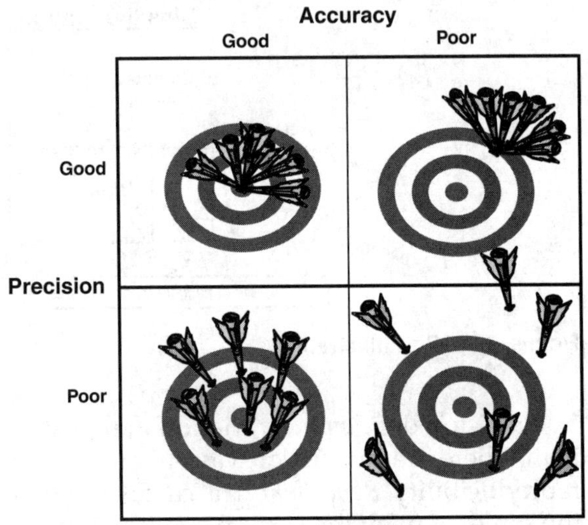

FIGURE M-12: Accuracy Versus Precision

Stability Over Time

Every measurement is a combination of various sources of variability. Ideally, any observed variation would be due solely to any part-to-part variation, and not to the measurement system itself. Part-to-part variation represents the measurement differences from one physical part to another. The ability to discern minor changes in part-to-part variation, or small part characteristics, means the measurement system has a high-degree of *resolution* or *discrimination*.

The measurement system, as previously discussed, can be a source of variation as well. *Stability* describes a measurement system that lacks change in bias or precision over time when repeatedly measuring the same items. Hence, a stable measurement system is said to "not drift." The measurement device may have an *inherent bias across its operating range*, which is defined by its measurement scale. For example, a device may introduce variability at either extreme of its scale—bias relative to a true value. This type of variability is called *linearity*. Figure M-13 illustrates both stability and linearity.

MSA Study

A Measurement System Analysis (MSA) looks to reduce or eliminate any variability coming from the measurement system to obtain stability.

Figure M-14 summarizes the different variation components in a Tree diagram. The MSA focuses on the entire right-half of the Tree diagram, under the Measurement System Variability branch, indicated by the rectangles with rounded corners.

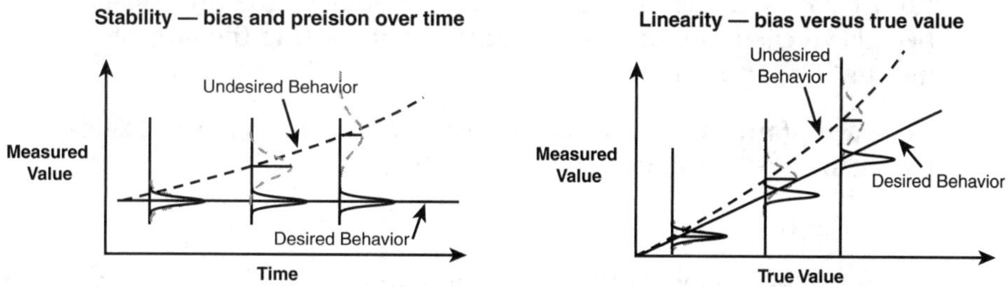

FIGURE M-13: Illustration of Stability and Linearity

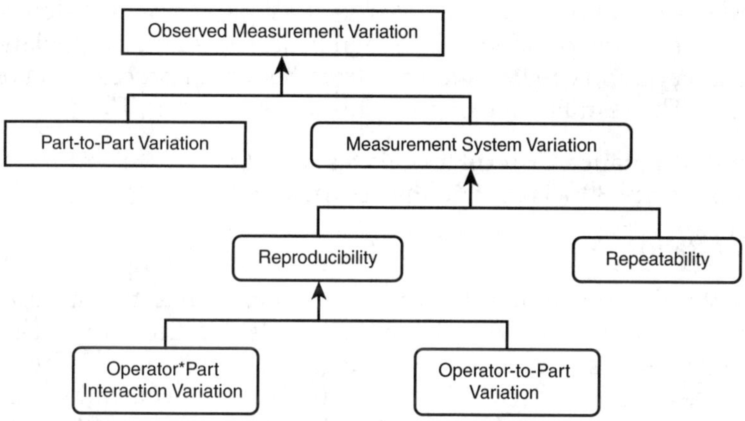

FIGURE M-14: Components of Variation

MSA for Variable Data—Gage R&R

Recall that the Measurement System Analysis (MSA) is a set of tests designed to evaluate how much variability is in the measuring device. One test focuses on variable data, known as the *Gage Repeatability & Reproducibility* (**Gage R&R, Gauge R&R, or GR&R**). This test examines the total variation (TV), the product variability, and the measurement system variability[also known as measurement system error or precision (P)]. Total variation (TV) equals the product variation plus the measurement variation. The Gage R&R contains three tests—% Contribution, P/TV, and P/T:

A B C D E F G H I J K L M N O P Q R S T U V W X Y Z

- % Contribution

  The ratio of variances of the measurement system standard deviation ($\sigma_{Measurement}^2$) to the total standard deviation (σ_{Total}^2) times 100% defines the %Contribution test for each component of variation. This test is the most common or popular of the three tests because it easily identifies the culprit contributing the largest amount of variation.

  The **Acceptance Criteria** for this test is <1% is Excellent; <4% is Good; and <9% is marginally acceptable.

- P/TV Ratio

  A second test compares the *precision (P or* $\sigma_{Measurement}$*)* against the total variation (TV or σ_{Total}), known as the **P/TV Ratio**. Reducing the P/TV ratio minimizes the effect of measurement variation on the process variation. It is calculated as the measurement system standard deviation divided by the total observed standard deviation. Total variation can be estimated from historical process data or sampling. The results often express this statistic as %P/TV.

  The **Acceptance Criteria** for this test is <10% is Excellent; <20% is Good; and <30% is marginally acceptable.

- P/T Ratio

  Another test examines the ratio of the precision to tolerance, called the P/T Ratio. The P/T calculation equals 5.15 times the measurement system standard deviation ($\sigma_{Measurement}^2$) divided by the tolerance (USL-LSL); wherein USL is the Upper Specification Limit, and LSL is the Lower Specification Limit. The 5.15 constant represents the number of standard deviations needed to obtain 99% of the variation.

  This ratio represents the estimated measurement error (precision) and the tolerance of the characteristic being measured (the part or product). The results often express this statistic as %P/T.

  The **Acceptance Criteria** for this test is <10% is Excellent; <20% is Good; and <30% is marginally acceptable.

The %Contribution, P/TV Ratio, and P/T Ratio evaluate the acceptability of the measurement system. As their calculated value increases, the ability to discriminate a change in the processes decreases. If the measurement system were inadequate, it would be unable to detect any part-to-part variation. These tests assume that the measurement errors are independent and normally distributed and independent of the magnitude of the measurement itself.

MSA Versus Attribute MSA
An MSA for variable data is a statistical test that looks at measurement variability over time (*stability*) and within the measuring device's operating range (*linearity*). The continuous data measurement devices (or gauges) include weights, scales, thermometers, and rulers, wherein the units of measure are variable. Each type of gauge offers a version with varying degree of discrimination, such as how rulers or scales found in a school science lab measure at a more gross level of resolution than those found in a pharmaceutical laboratory.

An MSA for attribute data is less statistically powerful than its variable kin, but often that is all that is available. Thus, an Attribute MSA requires a relatively large sample size of **at least 30 to 50 samples** across the span of extremes. An Attribute MSA does not have to be very complex, and it can yield quick and useable information.

Attribute MSA devices require similar information as does an MSA for variable data, but using a different approach, for example: *Repeatability* (that is, the % Agreement Within Appraisers), *Reproducibility* (that is, the % Agreement Between Appraisers), and *Bias* (that is, the % Agreement relative to a known standard).

The operators, or the appraisers themselves are often the measuring devices for attribute data. They inspect an item and determine if it has passed or failed relative to a standard. The standard, and different operators' consistent understanding of that standard is referred to as an *Operational Definition*. A challenge for an Attribute MSA is one of human error. To ensure common understanding of a standard and consistently apply it involves an operational definition, which defines what good looks like (and defines a defect within a unit or a defective unit). If the operational definition is written text, it needs language that is precise and easily understood across the full range of its users. It might include not only prose (or text) descriptions, but also pictures, photographs, models, or samples. A well-constructed and well-deployed operational definition addresses repeatability (within operator variation), reproducibility (consistency among multiple operators), and bias (the accuracy of a single operating matching a standard, and of multiple operators matching a standard).

How to Use the Tool or Technique
When conducting an MSA, it is best to use statistical software, such as MINITAB. The following are MSA procedures for both an MSA for variable data and an Attribute MSA.

MSA for Variable Data

Use the following procedure to develop an MSA using MINITAB: Enter the data into the MINITAB Worksheet reflecting the following columns of data—the parts (or items) measured, the operators (or machine) conducting the measurements, the sequence in which these measurements were taken, and finally the measurement results, themselves (the response variable).

For example purposes, paper helicopters were used. Four unique airplane models were flown for three separate runs. The flight times were measured and recorded in a MINITAB Worksheet by two different flight teams. Conduct an MSA on the flight time measurement system.

Gage Run Chart

Plot the data to look for unusual observations and trends. Do this by using a Gage Run Chart.

Select the following commands from its drop-down menu: ***Stat > Quality Tools > Gage Study > Gage Run Chart…***.

Select the column containing the appropriate data for the following fields on the Gage Run Chart main screen:

- **Part numbers**—Select the column that identifies the different models or parts measured.
- **Operators**—Select the column listing the different names or categories of operators involved in the measurement system.
- **Measurement data**—Select the column containing the response variable, the result of the measurement.
- **Trial numbers**—Select the column corresponding with the run sequence of the various measurements taken.

Figure M-15 displays a sample MINITAB screen where the appropriate data are selected to produce the final sample Gage Run Chart found in Figure M-16.

Based on the graph in Figure M-16, the part-to-part variation appears to lack flight time discrimination across the four airplane models. The key part-to-part variation question is, if given a flight time, would it indicate the airplane model? The parts (in this case, airplane models) are displayed in the columns WS, WL, NS, NL—indicating Wide-Short (WS), Wide-Long (WL), Narrow-Short (NS), Narrow-Long (NL) winged models. Looking horizontally across the columns, the NL and WS models can be distinguished based on their flight times, but it appears difficult to distinguish the NL from the WL models because they both have measures of a 2.0 flight time.

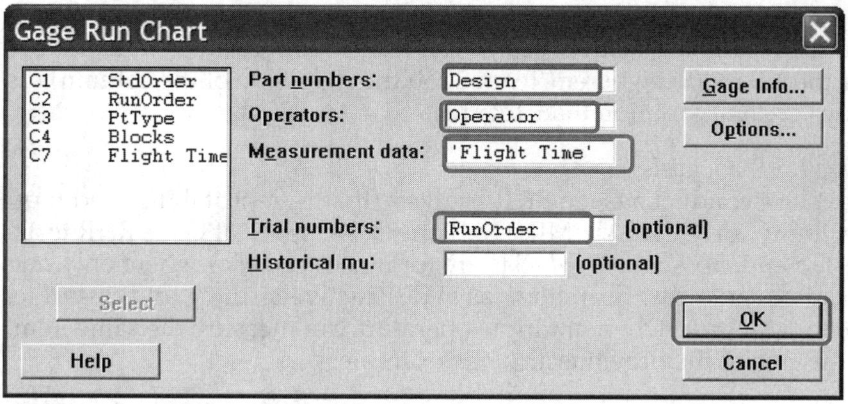

FIGURE M-15: Example MINITAB Gage Run Chart Main Screen

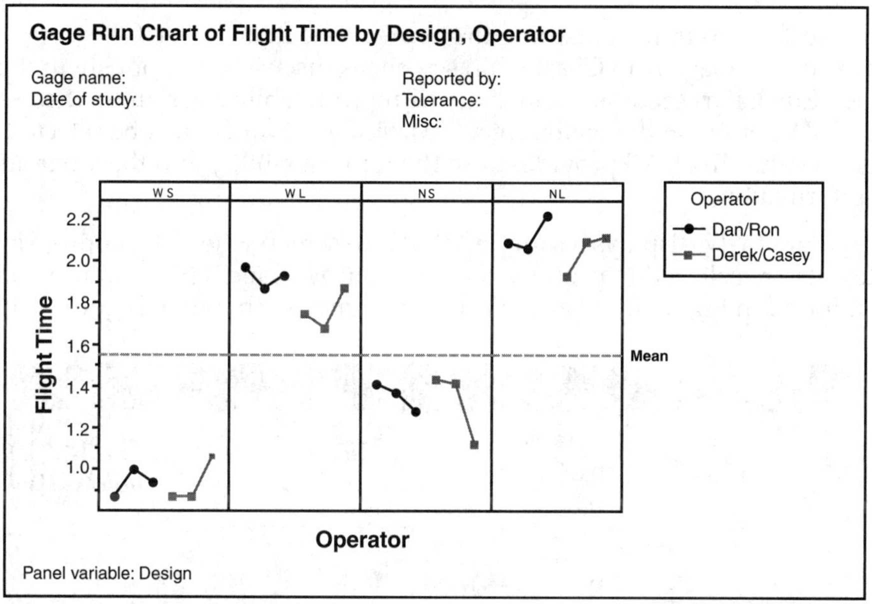

FIGURE M-16: Example MINITAB Gage Run Chart

For repeatability, look at a single three-dot cluster within a column in Figure M-16. In general there appears to be fairly little variation within the first team indicated on the far left of each airplane model column since the three-dot cluster appears tight. However, the second team displayed on the right side of each column shows greater variation within the team, particularly for the NS model.

For reproducibility, examine the three-dot operator clusters as pairs by model to compare team-to-team. The teams seem relatively similar except for the NS model, wherein Team 2 (on the right side of the column) displays greater variation than Team 1.

Gage R&R Analysis

Next let's conduct a Gage R&R analysis (that is, repeatability and reproducibility) in MINITAB. MINITAB offers two types of Gage R&R tests—nested and crossed. *Nested* tests are for those scenarios where only one operator can measure an item (as in destructive testing), and *crossed* tests are for scenarios where multiple operators can measure the same item, as is the case in the previous airplane example.

Using the same data as in the previous airplane example, select the following commands from its drop-down menu: ***Stat > Quality Tools > Gage Study > Gage R&R Study (Crossed)....***

Select the column containing the appropriate data for the following fields on the Gage Run Chart main screen, as discussed previously in the Gage Run Chart section. To analyze the reproducibility variation, the ANOVA option on the main screen's *Method of Analysis* must be selected. As a result, MINITAB provides both the reproducibility and the x-bar and R information.

Figure M-17 displays a sample MINITAB screen where the appropriate data are selected to produce the final sample Gage R&R graphical output found in Figure M-18 and the session window shown in Figure M-19.

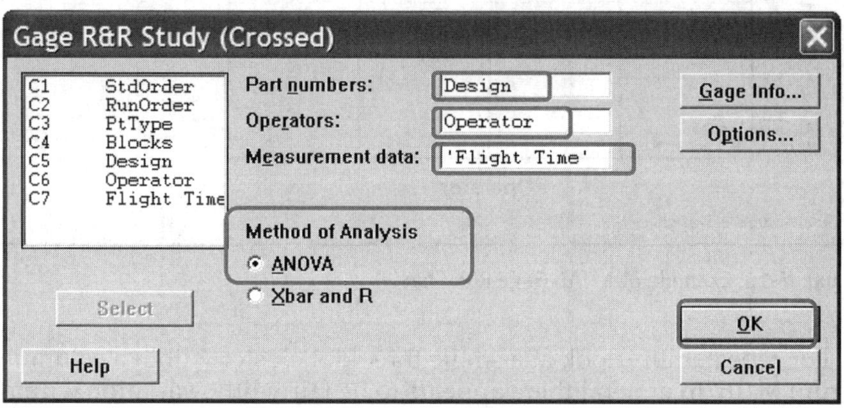

FIGURE M-17: Example MINITAB Gage R&R Study (Crossed) Main Screen

A B C D E F G H I J K L M N O P Q R S T U V W X Y Z

Figure M-18 shows the Components of Variation in the upper-left corner as having the part-to-part variation representing the largest portion of overall variation, which is good. The next three graphs reinforce this point. Going clock-wise, the Flight Time by Design in the upper-right corner shows non-zero slopes indicating that the different types of airplane designs are distinguishable. Below that graph, the Flight Time by Operator displays no slope between teams, which is good reproducibility. In the Operator*Design Interaction graph in the lower-right corner, the parallel lines indicate no interaction between the operator and the part (or model-type), which is good.

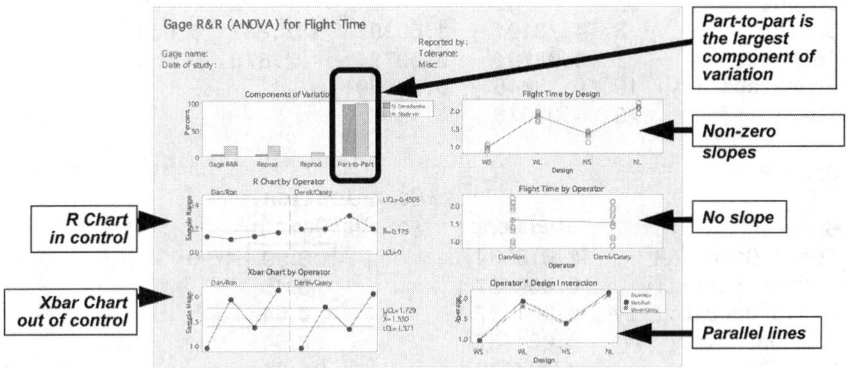

FIGURE M-18: Example MINITAB Gage R&R Graphical Output

The graph in the lower-left corner, the x-bar Chart by Operator, displays the part-to-part variation between operators. The points out of control (beyond the control limits) indicate that for both operator teams, the four airplane models (the four points) are distinguishable from one another (part-to-part variation), which is good. Last, the R-Chart by Operator shows a graph in control, which indicates stability over time and within the control limits—hence, no special cause variation, which is good. If the R-Chart indicated a special cause event, it would have to be corrected and the MSA run again.

All in all, the six MSA graphs display an adequate measurement system. These findings are reinforced by the analytical results found in MINITAB's session window; for this example, see Figure M-19.

Figure M-19 shows an overall "%Study Var" for the Total Gage R&R at 19.98, and the Number of Distinct Categories at 6. A rule of thumb is important to judge if those are adequate results, provided in Table M-3.

Based on the decision criteria in Table M-3, the example's airplane measurement system variation (Total Gage R&R) of 3.99% is adequate [looking at the % Contribution (of VarComp)]. Continuing to examine the numbers in this column, the biggest contributor to the measurement system variation is repeatability at 3.45%, and the % contribution for part-to-part variation is 96.01%. It's total Gage R&R %Study Variability at 19.98% (falling between 10 and 30% decision criteria) is a **marginal.** The Number of Distinct Categories of 6 is **excellent**. The latter number defines the granularity or resolution that the measurement system is capable of distinguishing.

Two-Way ANOVA Table Without Interaction

Source	DF	SS	MS	F	P
Design	3	4.78191	1.59397	167.837	0.000
Operator	1	0.02734	0.02734	2.878	0.106
Repeatability	19	0.18045	0.00950		
Total	23	4.98970			

Gage R&R

Source	VarComp	%Contribution (of VarComp)
Total Gage R&R	0.010984	3.99
Repeatability	0.009497	3.45
Reproducibility	0.001487	0.54
Operator	0.001487	0.54
Part-To-Part	0.264079	96.01
Total Variation	0.275063	100.00

Source	StdDev (SD)	Study Var (6 * SD)	%Study Var (%SV)
Total Gage R&R	0.104804	0.62882	19.98
Repeatability	0.097453	0.58472	18.58
Reproducibility	0.038558	0.23135	7.35
Operator	0.038558	0.23135	7.35
Part-To-Part	0.513886	3.08332	97.98
Total Variation	0.524464	3.14679	100.00

Number of Distinct Categories = 6

Figure M-19: Example MINITAB Gage R&R Graphical Output

TABLE M-3: Criteria to Assess Gage R&R Output

Gage R&R Output Type	Decision Criteria	Evaluation
% Contribution (of VarComp)	< 1%	Excellent
	≥ 1% -- < 9%	Acceptable
	≥ 9%	Inadequate
Total Gage R&R % Study Var.	< 10%	Excellent
	10% ≤ Gage R&R% < 30%	Marginal
	≥ 30%	Inadequate
# Distinct Categories	≥ 5	Excellent
	= 4	Marginal
	≤ 3	Inadequate

Attribute MSA

Use the following procedure to develop an Attribute MSA. This type of test evaluates if the operators, as the measurement system for attribute data, can correctly determine a good versus bad item or rank order a characteristic, based on a standard operational definition.

Recall, for attribute measurement studies, a sample size of at least 30 (ideally 50 to 100) is required. This sample data needs to reflect the normal extremes of the process, with a majority of the samples from the "gray" areas where the correct answer is **vague, unclear or uncertain** (that is, 20 items, or 70%) and the remaining samples from the clearly evident good or bad items (that is, 5 items each, or about 15% each). This distinction typically is done by a process expert who is knowledgeable of the customer requirements and corresponding operational definition.

The process for conducting an Attribute MSA is as follows:

Step 1. Create a data collection sheet with three types of columns, the first labeled "Sample" data, the second for the "Reference Standard" attribute data, and the remaining columns for the individual "Operator" data.

Step 2. The process expert rates the sample set and records it in the Reference Standard column.

Ensure the data adequately spans the range of extremes—majority items (70%) representing those with unclear answers with varying defects and 15% from clearly representing either extreme answer (that is, good and bad, or highest and lowest for a rank order).

Step 3. Randomly sort the data set. Each operator (or appraiser) audits and rates the samples and records his findings in an appropriate Operator data column of the data collection sheet. (Trial 1)

Step 4. Repeat Step 3, including randomly resorting the data set and each operator conducting the audit process. (Trial 2)

Step 5. Analyze the measurements data as follows:

a. **Within Operator**—Consistent, successfully repeated measures with the same answer for a given item—**repeatability**.

b. **Operator to Reference Standard**—Where the consistent measures per item (identified in Step 5.a.) match the Reference Standard—**operator bias**.

c. **Across Operators**—Where all the operators' answers agree for an item for both trials—**reproducibility**.

d. **All Operators to Reference Standard**—Those matching answers identified in Step 5.c. also match the Reference Standard—**measurement system bias**.

The All Operators to Reference Standard represents the effectiveness of the measurement system.

Table M-4 summarizes the various Attribute MSA evaluation conditions.

TABLE M-4: Various Attribute MSA Variability Conditions

Sample	Reference Standard	Appraiser 1		Appraiser 2		
		Trial 1	Trial 2	Trial 1	Trial 2	
1	Bad	Good	Bad	Good	Good	*Within Operator*
2	Good	Bad	Good	Good	Good	*Operator to Reference Standard*
3	Bad	Bad	Bad	Bad	Bad	
4	Good	Good	Bad	Good	Good	
5	Good	Bad	Bad	Bad	Bad	*Across Operators*
6	Good	Good	Good	Good	Good	*All Operators to Reference Standard*
...						
50	Bad	Good	Bad	Bad	Bad	

A B C D E F G H I J K L M N O P Q R S T U V W X Y Z

Rule of Thumb

Greater than or equal to 80% successful is **adequate (≥ 80%)**; greater than or equal to 90% is **excellent (≥ 90%)**.

Operator error typically can be improved by tactics such as

- Training
- Improving the Operational Definition
- Changing the process, or environment.
- Sense multipliers (devices to improve human senses)
- Masks/templates (block out unimportant information)
- Checklists
- Automation
- Reorganization of work area
- Visual aids

Attribute MSA Example

Example Scenario: In an administrative department, the measurement system for the invoice process appears to have an increased number of errors, which is causing delays in paying the customers. Randomly collect 30 sample invoices and conduct an Attribute MSA.

Following the Attribute MSA procedure previously described, create a data collection sheet in the MINITAB Worksheet to collect the results of the test. The data can be stacked or unstacked. Figure M-20 shows stacked data, wherein there are two appraisers who evaluate the same invoice twice. (Recall that when this test is being conducted, that the sequence of invoices must be random.) Figure M-20 shows the appraiser results in the Result column, C4-T (meaning Column 4-Text), and the Reference Standard in the Expert column, C5-T.

MINITAB needs the result data (provided in text) translated to numeric data (a.k.a. coded data) to run the analysis. Label the next column—for this example, label column C6 as `Coded Results`. To do so, select ***Data > Code > Text to Numeric*** and then enter which column contains the result text data that requires coding (`Area 1`, Code data from columns), which column the newly coded numeric data will be recorded (Area 2, Into columns), and the original to new values, one for one [Area 3, Original values (that is, red light blue)]. In this case, those invoices determined as good will be coded as 1, and the bad as 2. Click OK, and MINITAB will populate the coded data in column C6, as illustrated on the right-half of Figure M-20.

Now the data is ready to conduct an *Attribute Agreement Analysis* among the appraisers. From the MINITAB main screen, select the following commands from its drop-down menu: ***Stat > Quality Tools > Attribute Agreement Analysis...***

Figure M-21 displays a sample MINITAB main screen where the appropriate attribute test data is selected to produce the final sample graphical display showing the Within Appraiser and Appraiser versus Standard results.

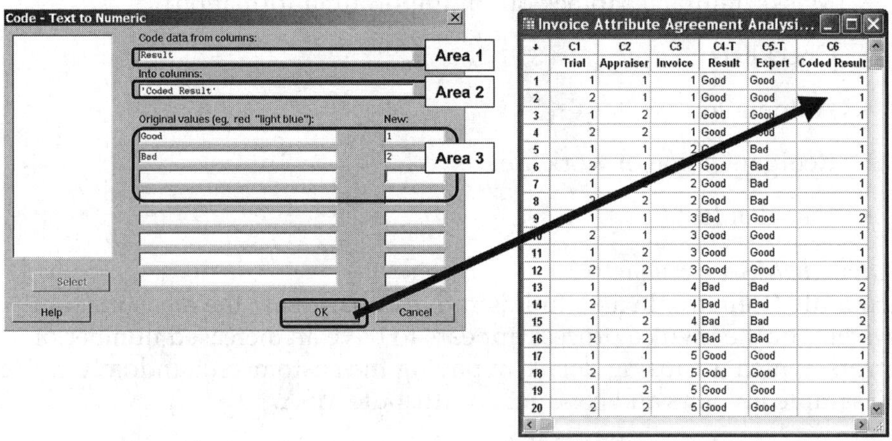

Figure M-20: Example of MINITAB's Code—Text to Numeric Main Window and Resulting Worksheet

The overlapping confidence intervals, shown in the graph of Figure M-21, indicate *no* significant difference between appraisers in repeatability (Within Appraisers) and between appraisers in bias (Appraiser vs. Standard). Hence, there is no consistent bias, which is good.

Now examine whether there is any specific discrepancy between and within the appraisers' judgments. To do this, conduct a Gage Run Chart by selecting the following commands from its drop-down menu: ***Stat > Quality Tools > Gage Study > Gage Run Chart...***

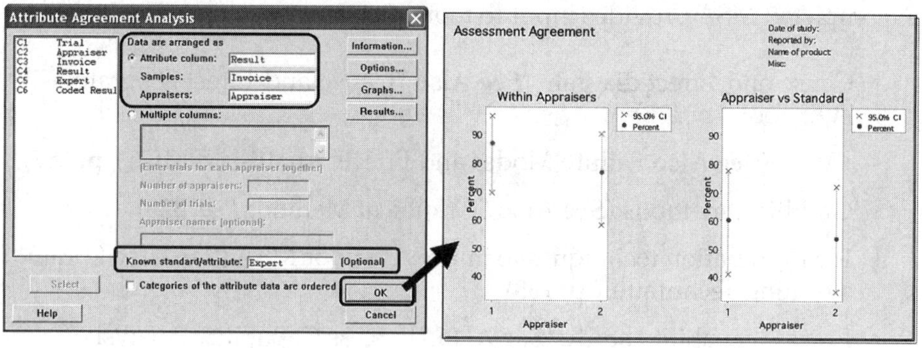

Figure M-21: Example MINITAB Attribute Agreement Analysis Main Screen and Graphical Output

Figure M-22 displays a sample MINITAB main screen where the appropriate attribute test data is selected. This test requires the coded (numeric) results versus the text attribute data. Figure M-22 also exhibits the final graphical display of the "Gage Run Chart of Coded Results by Invoice, Appraiser" results for each of the 30 invoice samples in this example. In the Gage Run Chart graph, notice the three highlighted scenarios for the different disagreement combinations—for just one appraiser, between both, and within both.

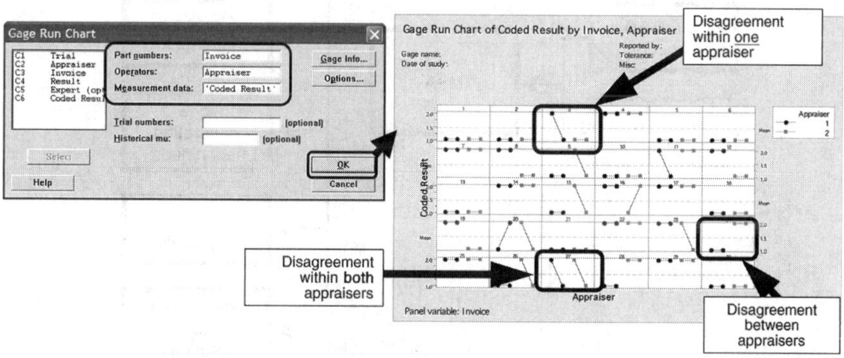

FIGURE M-22: Example MINITAB Gage Run Chart Main Screen and Graphical Output

Supporting or Linked Tools

Supporting tools that might provide input when conducting an MSA include

- Data collection sheet (**See Also** "Data Collection Matrix," **p. 248**)
- Process map (**See Also** "Process Map (or Flowchart)—7QC Tool," **p. 522**)

A B C D E F G H I J K L M N O P Q R S T U V W X Y Z

A completed MSA provides input to tools such as

- Cause-and-Effect diagram (**See Also** "Cause-and-Effect Diagram—7QC Tool," **p. 173**)
- FMEA (**See Also** Failure Modes and Effects Analysis (FMEA), **p. 287**)
- Graphical methods (**See Also** "Graphical Methods," **p. 323**)
- Idea generation techniques to improve a poor MSA (**See Also** "Brainstorming Technique," **p. 168**)
- Process capability study (**See Also** "Process Capability Analysis," **p. 486**)
- Statistical analytical tools (**See Also** "Statistical Tools," **p. 684**)

Figure M-23 illustrates the link between an MSA and its related tools and techniques.

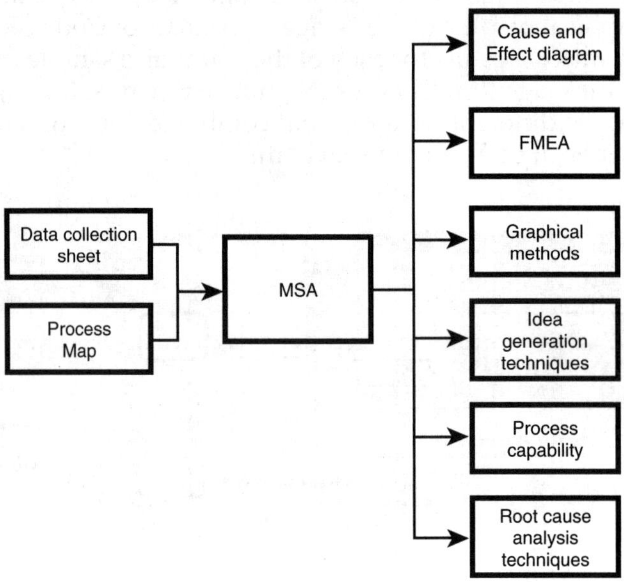

FIGURE M-23: MSA Tool Linkage

Monte Carlo Simulation

What Question(s) Does the Tool or Technique Answer?

What are the probabilities and risks associated with several possibilities (how about scenarios, outcomes, or decisions)?

The Monte Carlo simulation helps you to

- Simulate the range of outcome possibilities for scenarios to aid in the decision-making process
    - For example: scheduling of a project, program, or process; forecasting financials; and managing an offering portfolio
- Understand the variability in a process or system
- Identify problems within a process, system, or system design
- Manage risk by understanding the cost/benefit relationship

(**See Also** "Selecting Project Portfolios using Monte Carlo Simulation and Optimization," in Part III of this book, **p. PDF: 921**)

When Best to Use the Tool or Technique

Construct a Monte Carlo simulation at the beginning of a project or program to explore a range of possibilities and their associated risks and to inform the selection of a path forward with a degree of certainty. This tool is useful for any problem where variation and uncertainty cause risk or defects. It also can be used to solve problems that are computationally intensive by sampling on the computer.

Common applications for a Monte Carlo simulation include models for

- Business process
- Engineering design
- Financial planning and forecasting
- Manufacturing production system
- Offering Portfolio management (**See Also** "Selecting Project Portfolios using Monte Carlo Simulation and Optimization," in Part III, **p. PDF: 921**)
- Sales planning and forecasting

Brief Description

A Monte Carlo simulation produces different combinations of possible outcomes for different key parameter distributions. It models variation and uncertainty to help make better decisions. A simulation is the process of experimenting with a model to measure performance and behavior of a system's inputs. Modeling is an iterative process that tries to replicate or represent a real system. Often times the model is a simple representation of a complex system or process.

A model built via a spreadsheet is easy to use and is flexible; however, it is limited to only a one number answer (deterministic), and it is difficult to audit the model, its data and assumptions. The single-point estimates (even if they include the most-likely, best- and worst-case scenarios) lack probability information. In contrast, a simulation model uses random numbers to measure the effects of uncertainty.

Monte Carlo gets its name from the city in Monaco that is famous for its roulette wheel as a kind of random number generator. In the 1940s, the Los Alamos National Labs used Monte Carlo to simulate the impact of a nuclear bomb. Interest in generating random numbers dates back to the 1800s when mathematicians were attempting to approximate pi. However, the real power behind simulation modeling came about with increasing computer capabilities. With today's technological capabilities, computer software packages such as Decisioneering®, Inc.'s Crystal Ball® software, which works on top of Microsoft Excel, have simplified the user interface and run the calculations quickly in the background.

Monte Carlo simulation recognizes that independent variables (Xs) in a math model can vary over a range. The Y = f(X) equation describes a dependent variable (Y) as a function of independent variables (Xs). The independent variables can have a range of values. This range can be characterized by a probability distribution. By sampling the values of X parameters and making repeated calculations of the result, the model develops probabilities of the variability to forecast the impact on the dependent Y variables and the resulting risk. (**See Also** "Y = f(X)," and "Regression Analysis," **p. 758** and **p. 571**, respectively.)

Simulation is an inexpensive means to evaluate decisions prior to action and reveals critical sensitivity points of a system. The results are sensitive to the accuracy of the input data. Simulation does not solve the problem; it simply aides in the decision making process.

How to Use the Tool or Technique

For purposes of demonstration, the following procedure develops a Monte Carlo simulation model using Crystal Ball software from Decisioneering. Crystal Ball software runs as an add-in to Microsoft Excel, so both software application packages are required.

The procedures to run a Monte Carlo simulation are as follows:

Step 1. Develop a system flow diagram or algorithm.

a. Collect the input. Establish the functional relationships (the math model) that exist between the various input assumption terms to produce the forecasted output response.

Step 2. Record the data in a Microsoft Excel spreadsheet with Crystal Ball software open. (For example, list each project task and record the time associated with the minimum, maximum, and median (or most likely) estimates.) The spreadsheet should have tasks listed down the first column, as the row headings, as shown in Figure M-24.

	A	B	C	D	E
1	**Project Mgt. Example**	Median	Minimum	Maximum	Type
2		Task Time	Time	Time	of
3	**Phase 1 Concept Development Tasks:**	(days)	(days)	(days)	Distribution
4	Gather the Voice of the Customer	12	7	20	Triangular
5	Refine & Rank the Voice of the Customer using KJ Analysis	3	2	4	Triangular
6	Create Product or System Level House of Quality	5	4	6	Triangular
7	Conduct Competitive Product Bench marking	9	7	10	Triangular
8	Generate Product or System Level Requirements Document	14	12	16	Triangular
9	Define the Functions that Fulfill the System Requirements	2	1	2.5	Triangular
10	Generate System Concept Evaluation Criteria	1.5	1	2	Triangular
11	Generate System Concepts that Fulfill the Functions	25	20	30	Triangular
12	Evaluate System Concepts	6	4	7	Triangular
13	Select Superior System Concept	3	2	4	Triangular
14	Analyze, Characterize, Model & Predict Nominal Performance	60	54	66	Triangular
15	Develop System Reliability Reqts, Initial Reliability Model & FMEA	20	18	22	Triangular
16					
17	**Total Phase 1 Development Time**	160.5			

Figure M-24: Sample Crystal Ball Software Spreadsheet

Step 3. Use Crystal Ball software to model assumptions and forecast probability distributions.

a. Set up the problem by defining the input assumptions.

i. Define critical X variable assumptions, by clicking and selecting a cell with the data, and going to the Define drop-down menu and select ***Define > Define Assumptions***. Enter the appropriate assumptions, as shown in M-25.

ii. Select and enter the distribution for your assumption. **Repeat** this step for each component or element in your model.

A B C D E F G H I J K L M N O P Q R S T U V W X Y Z

b. Select Define Forecast (Y variable) from the Define drop-down menu (*Define > Define Forecast*).

  i. Specify the items being calculated in the forecast formula. Repeat this step for each forecast formula in your model.

  ii. **Option in the Define Forecast**: highlight the cell containing the total median time estimate, and enter any of the conditions of interest—LSL (Lower Spec Limit), USL (Upper Spec Limit), and/or Target, as illustrated in Figure M-25.

  iii. More than one forecast can be generated by defining several forecast formulas within the same spreadsheet.

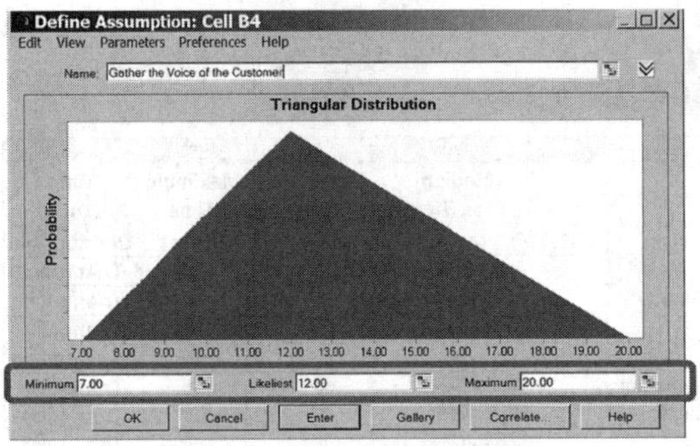

Define > Define Assumptions

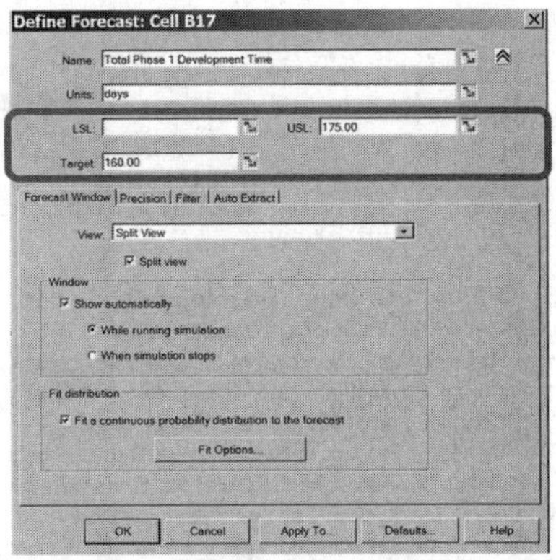

Define > Define Forecast

FIGURE M-25: Sample Crystal Ball Software Define Assumptions and Forecast Windows

Step 4. Run the simulation and analyze the results.

a. Choose Run Preferences from the Run drop-down menu (***Run > Run Preferences***).

b. Under the Trials tab, type in the number of trials you would like to run—must be a 200 minimum.

c. Run the model by selecting ***Run > Start Simulation***.

Step 5. Improve the model and/or make decisions.

a. Interpret the results using the Monte Carlo outputs—descriptive statistics, frequency chart, and sensitivity chart for the X and Y variables evaluated in the model.

Figure M-26 illustrates the various outputs from a Crystal Ball software simulation and includes a cumulative frequency chart just underneath the histogram. Both graphical charts contain the target indicated by a line and a text box containing the value, if defined in the Define Forecast drop-down menu. It identifies the optimal statistical test and provides the common statistical data associated with the simulation to the right of the resultant histogram. Place the cursor in the **upper-right box** of the output screen with the statistical tests and then hit the space bar on your computer's keyboard to open a second table of statistics containing the Normality test at the top, followed by the various metrics, including process capability. Repeat tapping the space bar to open a third window that contains the percentile distribution. (**See Also** "Normal Probability Plots, Normal versus Non-normal Data," **p. 227** for a discussion on Normality test; and "Process Capability Analysis," **p. 486**)

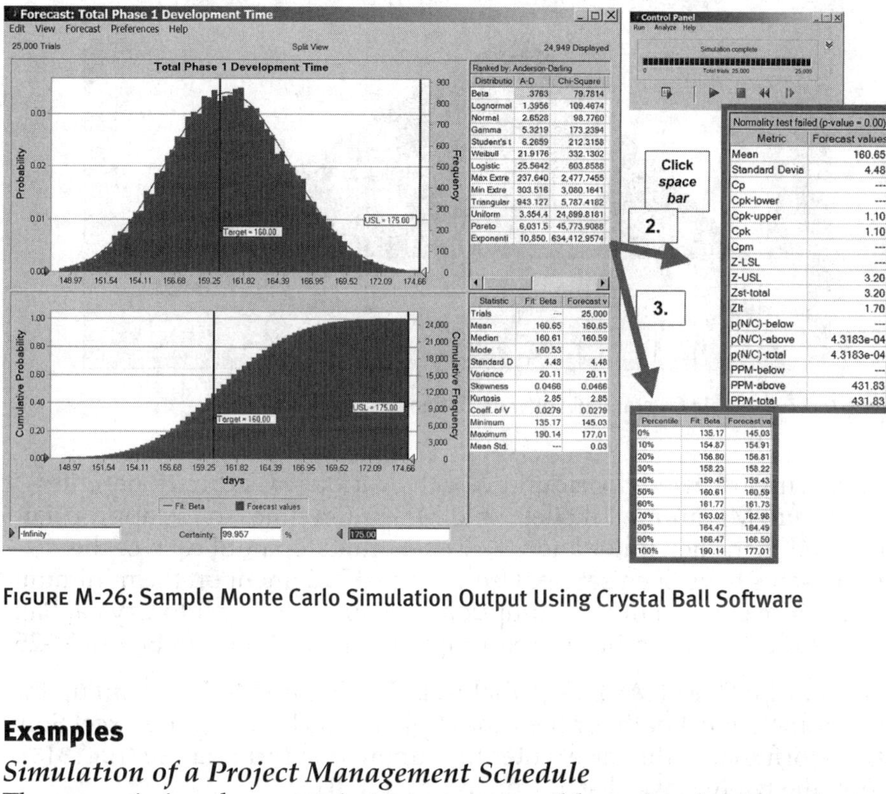

FIGURE M-26: Sample Monte Carlo Simulation Output Using Crystal Ball Software

Examples

Simulation of a Project Management Schedule

The scenario involves a project team responsible to develop a product concept. The process starts with gathering the voice of the customer up through the development of specific product requirements. In such a complex, cross-functional project, the schedule could contain some competing critical paths, as the team works toward driving a launch schedule. Traditional PERT analysis looks at the variation of tasks on the critical path but fails to recognize that the critical path can change. Crystal Ball software looks at all paths through the activity network and understands how other paths can influence total time. (**See Also** "Activity Network Diagram (AND)—7M Tool," and PERT (Program Evaluation and Review Technique) Chart, **p. 127** and **p. 453**, respectively.)

A B C D E F G H I J K L M N O P Q R S T U V W X Y Z

The team mapped out a 12-step project plan and listed the activity steps down the first column in their Excel spreadsheet, as shown in Figure M-24.

Click inside the first cell with the median task assumptions (that is, cell B4), and from the Crystal Ball software main toolbar or the Excel Define menu, select ***Define Assumption…*** from the drop-down menu. The screen to define that cell's distribution assumptions will open, as displayed in Figure M-27.

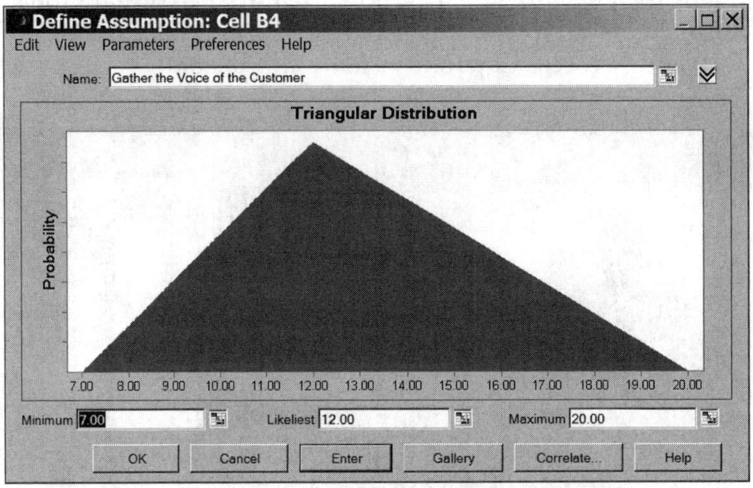

FIGURE M-27: Example Crystal Ball software Define Assumption Dialog

To set up a skewed distribution for the various task durations, enter the different Minimum, Likeliest, and Maximum times in the appropriate boxes and then click Enter and OK. Repeat this step until each of the activity steps have been set up. Use your best judgment on the minimum, likeliest & maximum time assumptions you are entering into Crystal Ball software. For this example, the assumptions can be found in Figure M-25.

Define the FORECAST Cell (that is, B17 in Figure M-25) to sum up the time for the critical path for the project plan tasks by using the Excel Summation formula. In this case, enter the summation formula `=SUM(B4:B15)` to sum the twelve tasks listed from cells B4 to B15.

Define the Forecast by selecting ***Define Forecast…*** from the Excel Define menu or using the Crystal Ball toolbar icon. Fill in the Forecast Name & Units and then click OK, as illustrated in Figure M-28.

Now Crystal Ball software is ready to run the simulation. Select ***Run Preferences…*** from the Excel Run menu or by using the Crystal Ball toolbar icon. Select the Trials tab in the upper-left corner of the Run Preferences screen to enter the number of trials to simulate. In this case, enter 10,000 and keep the default of 95% for the confidence level, as shown in Figure M-29. Then select OK.

From the Excel Run menu, select ***Start Simulation*** to run the simulation or just click the Run icon on the Crystal Ball toolbar. The Forecast window will fill with the sample data being produced during the 10,000 trials being run and eventually will produce a distribution graph similar to the one found in Figure M-26.

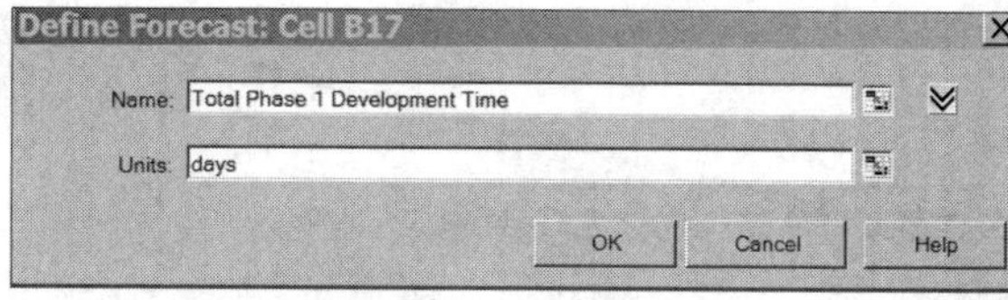

FIGURE M-28: Example Crystal Ball Software Define Forecast Screen

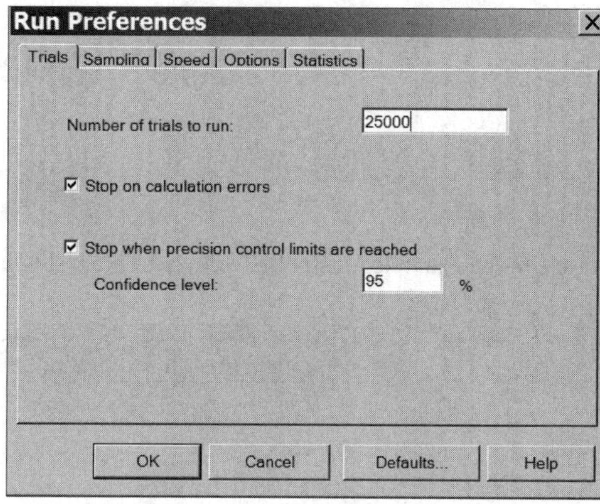

FIGURE M-29: Example Crystal Ball Software Run Preferences Screen

To view the statistics associated with the *Forecast* window, select ***View > Statistics*** to get the window displayed in the upper-right of Figure M-26, or Figure M-31 (or hit the space bar on your keyboard). It provides information about the output distribution, such as the mean, median, mode, and standard deviation. Given the forecasted standard deviation, the process capability can be calculated.

Also in the Forecast window, select ***Forecast > Open Sensitivity*** to see the various sensitivities in the model and which tasks contribute the largest variance, as illustrated in Figure M-30. This information may help identify steps requiring corrective action and specify specific failure modes in an FMEA. (**See Also** Failure Modes and Effects Analysis (FMEA), **p. 287**)

In this example, the simulation indicates that the project plan can deliver its requirements within the upper specification limit of 175 days (a 95% confidence level). However, as shown in Figure M-34, the process capability of about 1.0 indicates the project is barely capable of completing by the target of 160 days (with a mean of 160.64). The project plan is tight, and all the task estimates need to hit the "most likely" estimate.

> **Note**
>
> Figure M-31 represents only the statistical portion of the simulation Forecast output and zooms in on the upper-right corner of Figure M-26 after hitting the space bar once.

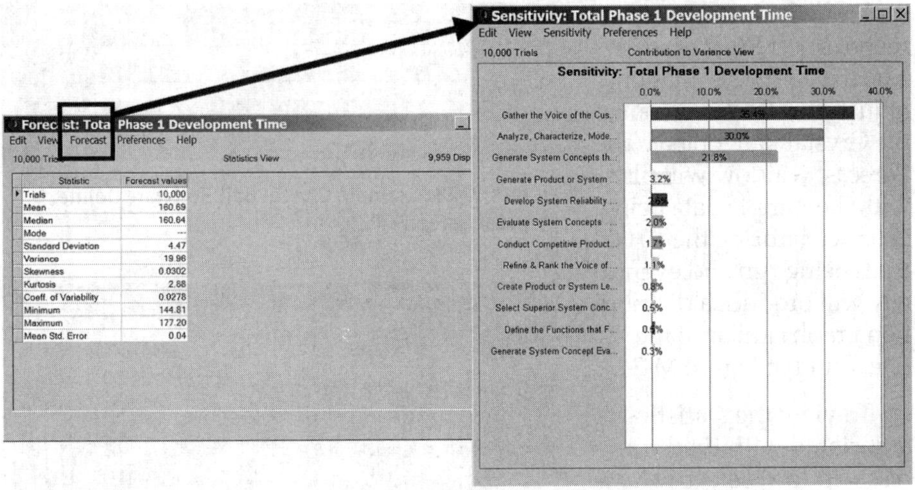

FIGURE M-30: Example Crystal Ball Software Forecast Window and Sensitivity Chart

Normality test failed (p-value = 0.003)	
Metric	**Forecast values**
Mean	160.64
Standard Devia	4.50
Cp	---
Cpk-lower	---
Cpk-upper	1.09
Cpk	1.09
Cpm	---
Z-LSL	---
Z-USL	3.19
Zst-total	3.19
Zlt	1.69
p(N/C)-below	---
p(N/C)-above	4.6769e-04
p(N/C)-total	4.6769e-04
PPM-below	---
PPM-above	467.69
PPM-total	467.69

FIGURE M-31: Example Crystal Ball Software Statistical Portion of the Final Output

As a result, the project team might examine the critical path of its project plan to explore opportunities for change based on the simulation's results. Perhaps some activities can occur in parallel. Moreover, an expectation setting discussion with the project sponsor would also be appropriate.

Multi-vari Chart

What Question(s) Does the Tool or Technique Answer?

Across multiple sources of variability, which one contributes the most?

A Multi-vari chart helps you to

- Visually analyze variance from multiple sources
- Understand the differences in the means and variability together in one graphical analysis tool
- Visually stratify multiple sources of variation to examine possible patterns and largest contributors

Alternative Names and Variations

This tool is also known as

- Multi-vari plot
- Multi-vari study

When Best to Use the Tool or Technique

Before conducting a full analysis of variance, conduct a Multi-vari study to visualize which variable contributes the most variation. Look for patterns, trends, or interactions among the multiple sources to determine where to focus further attention to minimize the variation or identify areas not needing further investigation. (**See Also** Analysis of Variance (ANOVA)—7M Tool, **p. 142**)

Brief Description

The Multi-vari chart is a good executive communication tool of a relatively complicated comparative analysis. It is a powerful graphical tool that draws the relationship between one variable with many other variables, to examine the differences in their **means** and **variability** on a single plot. The variation can either be controllable or uncontrollable (or noise). The tool uses continuous and discrete data, with an average of 15 items per sample variable.

The variation may come from 1) within an item, 2) part-to-part, or 3) time-to-time (over-time). The Multi-vari chart examines all three of these types of variation.

There are two types of Multi-vari studies—nested and crossed. The *nested* design looks at variation of one item within another item. A nested design is plotted over time; therefore the data adheres to the order within which it was produced. The tool can be used to examine the stability of a process if used along a time scale, and the length of each line plotted represents the range for that sample data. It exhibits three types of "within variation" for a product or process:

- **Positional**—Variation within a part (or batch). Positional variation has three components:
    - **Cylinder**—End-to-end.
    - **Batch**—Top-to-bottom, or side-to-side.
    - **Flat piece**—Across width, or front-to-back.
- **Cyclical**—Variation from consecutive piece-to-piece (or batch-to-batch).
- **Temporal**—Variation occurring over time (that is, shift-to-shift or day-to-day).

A *crossed design* manipulates the three sources of variation as independent variables. This design analyzes the two-way interaction of each of them to an output variable—the impact of the independent factor on the dependent one. The chart diagrams these multiple variables, known as input (Xs) against a single output variable (Y) to help identify which might be attributed to the largest source of variation. The independent variables typically are tested at two levels (that is, high/low or on/off). And the design set-up should be orthogonally balanced, as in a Design of Experiment (DOE). Different from the nested design, the crossed-design charts are not time-based. (**See Also** "Y=f(X)," and "Design of Experiment (DOE)," **p. 758** and **p. 250**)

How to Use the Tool or Technique

Using MINITAB, the procedure to conduct a nested or crossed design is the same; the data determines which test MINITAB runs. For purposes of an example, data for a crossed design will be used.

Enter the data into the MINITAB Worksheet. Select the following commands from its drop-down menu: ***Stat > Quality Tools > Multi-Vari Chart....*** Within the main screen of MINITAB's *Multi-Vari Chart*, select the column containing the appropriate data for the response (output or dependent variable) and any of the independent factors of interest, shown in Figure M-32.

Select the Options button and notice that MINITAB's default includes connecting the means for as many factors identified as of interest. To see the granularity of the variation, sometimes it is helpful to see the individual data points; therefore, select Display individual data points in the Options screen, shown in M-32.

Click OK in both the Options and main screens to produce the Multi-vari graph, shown in Figure M-33.

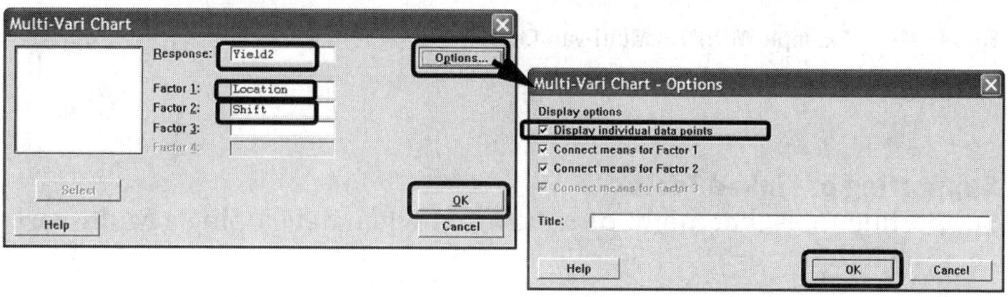

FIGURE M-32: Example MINITAB Multi-vari Chart Main and Options Screens

Figure M-33 plots three work shifts along the horizontal axis. The vertical axis (Y-axis) displays the yield that the shifts produce as the response. The data points represent two different locations—Plant A (indicated by an open circle) and Plant B (indicated by a circle with an embedded cross). The spread in the data is similar between plants and across shifts. There is very little difference among the shifts within a location. There is a consistent difference of about 10% in Yield between the two locations, as indicated by the slope of the three solid lines connecting the means of the two locations by shift. Based on the chart, further exploration seems warranted as to why Plant A consistently produces a higher yield than Plant B, regardless of shift.

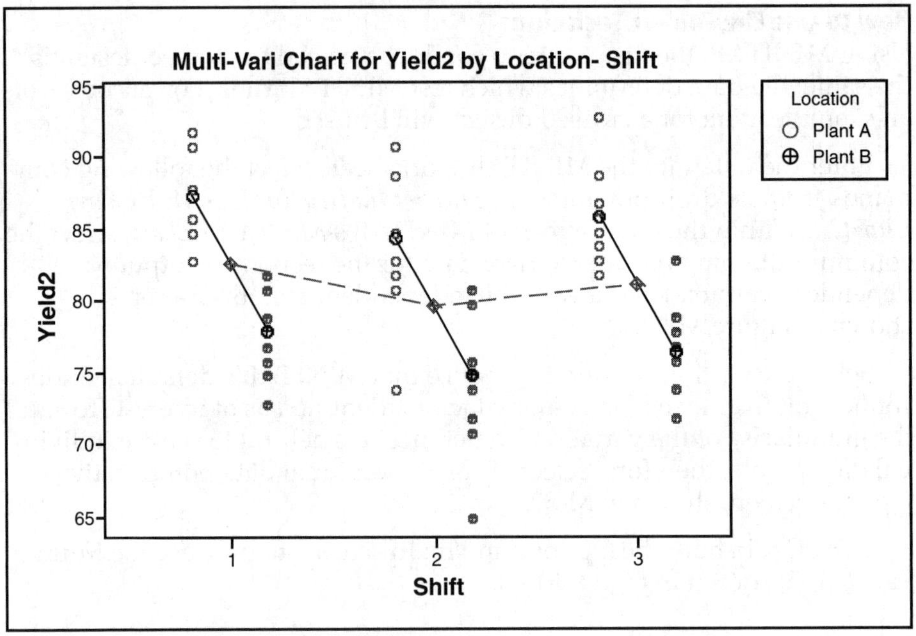

FIGURE M-33: Example MINITAB Multi-vari Chart

Supporting or Linked Tools

Supporting tools that might provide input when developing a Multi-vari chart include

- Data gathering and data collection sheets (**See Also** "Data Collection Matrix," **p. 248**)
- Performance charts and dashboards

A completed Multi-vari chart provides input to tools such as

- Cause-and-Effect diagram (**See Also** "Cause-and-Effect Diagram—7QC Tool," **p. 173**)
- FMEA (**See Also** Failure Modes and Effects Analysis (FMEA), **p. 287**)
- QFD (**See Also** "Quality Function Deployment (QFD)," **p. 543**)
- Statistical analysis tools (**See Also** "Statistical Tools," **p. 684**)

Figure M-34 illustrates the link between a Multi-vari chart and its related tools and techniques.

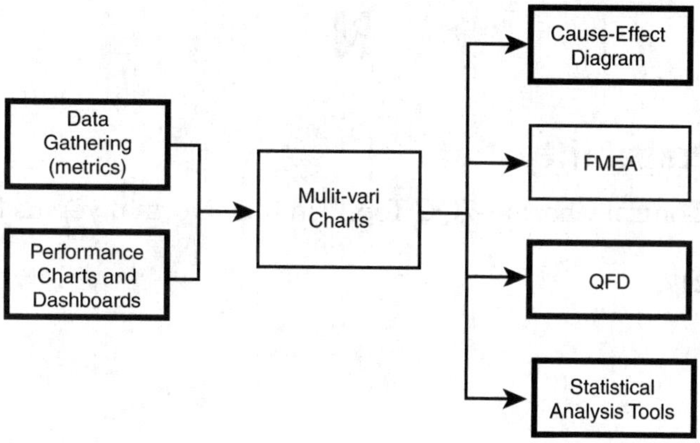

Figure M-34: Multi-vari Chart Tool Linkage

N

Normal Probability Plot

See Also "Control Charts—7QC Tool," in the "Normal versus Non-normal Data" section, **p. 217**.

P

Pareto Chart—7QC Tool

What Question(s) Does the Tool or Technique Answer?

What are the vital few items with the biggest impact? Which 20% of items produce 80% of the impact (the 80/20 rule)?

A Pareto chart helps you to

- Prioritize and select the biggest problem areas or largest areas of opportunities
- Analyze the frequency of an event (occurrences or number of items) in a process and identify the biggest contributors
- Communicate a snapshot summary of how 80% of the problem comes from 20% of the causes

Alternative Names and Variations

- This tool is also known as a Pareto diagram
- Variations on the tool include Weighted Pareto Analysis

When Best to Use the Tool or Technique

A Pareto chart is a tool to focus attention on priorities when trying to make decisions. It is a good communication tool describing the data in a simple and easy to read bar chart. It highlights the vital few major contributors with the largest impact.

Brief Description

The Pareto chart is used to prioritize which contributors make the biggest impact on a problem, or which represent the largest areas of opportunity. These diagrams communicate the *80/20 rule*, which states that 80% of an effect comes from 20% of the causes. For example, 80% of the revenue comes from 20% of the customers; or alternatively 80% of the customers contribute only 20% of the revenue. Another example, 80% of the customer complaints come from only these few causes (20%), or 80% of the repair time is spent on these few problem areas (20%). Essentially the 80/20 rule means that focusing on the **vital few** yields larger gains than the trivial many.

The term "Pareto" stems from an economist, named Vilfredo Pareto (1848–1923). He was born in Paris but fled to Italy in search of political freedom. Pareto studied the distribution of wealth and devised mathematical models representing his findings that 80% of wealth was distributed among only 20% of the population. His publications supposedly were attributed to initiating Fascism in Italy.

In the 1940s, Dr. Joe Juran, a renowned quality thought-leader, used a cumulative frequency chart to illustrate the concept of the vital few versus the trivial many in his *Quality Control Handbook.* To describe these curves, he wrote a caption underneath them stating, "Pareto's principle of unequal distribution…." Juran applied Pareto's economic analysis of unequal income and wealth distribution to a broader context. Interestingly, the actual concept of these cumulative curve diagrams first was attributed to M.O. Lorenz in 1904, but Juran is recognized as having popularized not only the use of these charts, but also the more universal application of the Pareto principle.

The Pareto chart serves as a good executive management communication tool and is easy to construct. It is a specialized bar chart displaying the frequency and accumulation of nominal (count or attribute) data. The horizontal axis lists the categories of interest. Structurally, the Pareto plots are from left-to-right organized by height from tallest to shortest. The bars' height indicates the frequency or count, as indicated along the vertical axis. By convention, a connecting line from the center-top of each bar depicts the cumulative percent portion of each bar's count, adding from left-to-right until 100% total is reached. Some Pareto charts lack this connecting line. However, if a cumulative line exists, the left vertical axis represents the count, and a right-axis shows the cumulative percentage. Typical sample size used to construct a Pareto is 30 or more.

The Pareto chart is a member of the 7QC Tools, (or seven *Quality Control* tools), attributed to Dr. Kaoru Ishikawa. The 7QC Tools sometimes are called the *seven basic tools* because they were the first set of tools identified as the *core* quality improvement tools. Ishikawa's original 7QC toolset includes: 1) Cause-and-Effect diagram; 2) Check sheet (or checklist); 3) Control charts; 4) Histogram; 5) Pareto chart; 6) Scatter diagram; and 7) Stratification. More recently, the 7QC toolset is modified by substituting the Stratification technique with either a flowchart (or Process map) or a run chart (or Time Series plot).

How to Use the Tool or Technique

Pareto charts can be constructed manually or by using various software packages such as Microsoft Excel or MINITAB.

A B C D E F G H I J K L M N O P Q R S T U V W X Y Z

For illustration purposes, the example scenario examines defects of painting metal components as part of an outdoor art exhibit produced in two locations: Bath and Neenah.

When creating a Pareto chart, MINITAB allows data to be entered into its Worksheet one of two different ways, as shown in Figure P-1:

- **Detailed information**—The raw data is entered by column, and each column is labeled by the category name (that is, type of defect).
- **Summary information**—A column lists the category names (that is, type of defect) and the adjacent column of summary frequency (or count) data alongside the appropriate category.

Figure P-1 displays a partial snapshot of the same set of data, formatted in the two different MINITAB Worksheet structures. Notice the asterisk (*) in the Frequency (Freq) column of the Summary Structure image of the MINITAB Worksheet. That asterisk indicates the absence of data; hence, no counts of chipping existed in the sample set.

Detailed Structure

C1	C2-T	C3-T
ID	Plant	Defect Category
1	Bath	Peeling
2	Bath	Missing (Thin)
3	Neenah	Nonlevel Layers
4	Bath	Peeling
5	Bath	Peeling
6	Neenah	Missing (Thin)
7	Neenah	Interfacial Cracking
8	Bath	Peeling
9	Neenah	Missing (Thin)
10	Bath	Missing (Thin)
11	Neenah	Interfacial Cracking
12	Bath	Missing (Thin)
13	Bath	Peeling
14	Bath	Missing (Thin)
15	Bath	Chipping
16	Bath	Peeling
17	Bath	Peeling
18	Neenah	Missing (Thin)
19	Bath	Peeling
20	Bath	Interfacial Cracking
21	Bath	Missing (Thin)
22	Neenah	Missing (Thin)
23	Neenah	Interfacial Cracking
24	Bath	Mottled Layers

Summary Structure

C5-T	C6-T	C7
Plant Code	Defect	Freq
Neenah	Missing (Thin)	22
Neenah	Chipping	*
Neenah	Peeling	6
Neenah	Interfacial Cracking	10
Neenah	Mottled Layers	1
Neenah	Nonlevel Layers	3
Bath	Missing (Thin)	26
Bath	Chipping	5
Bath	Peeling	20
Bath	Interfacial Cracking	6
Bath	Mottled Layers	1
Bath	Nonlevel Layers	3

FIGURE P-1: Example of MINITAB Worksheet Data Structure for Pareto Chart

Use the following procedure to develop a Pareto chart using MINITAB: Given that numeric data has been entered into the MINITAB Worksheet, select the following commands from its drop-down menu: ***Stat > Quality Tools > Pareto Chart....***

(Given that the Pareto is recognized as an early quality tool, it is found under MINITAB's Quality Tool category of its main menu, rather than under Graphing.)

If the data were entered into the Worksheet using the Summary Structure, the procedure to construct a Pareto chart would involve the lower half of the Pareto Chart main screen. Select the Chart defects table and enter the appropriate column of data in the two dialog boxes—"Labels in" (indicating the column containing the category names), and "Frequencies in" (referring to the column containing the summary counts by category), as shown in Figure P-2. Keep the MINITAB default "Combine remaining defects into one category after this percent—95" as is. Click OK to generate the Pareto chart, as displayed in Figure P-3.

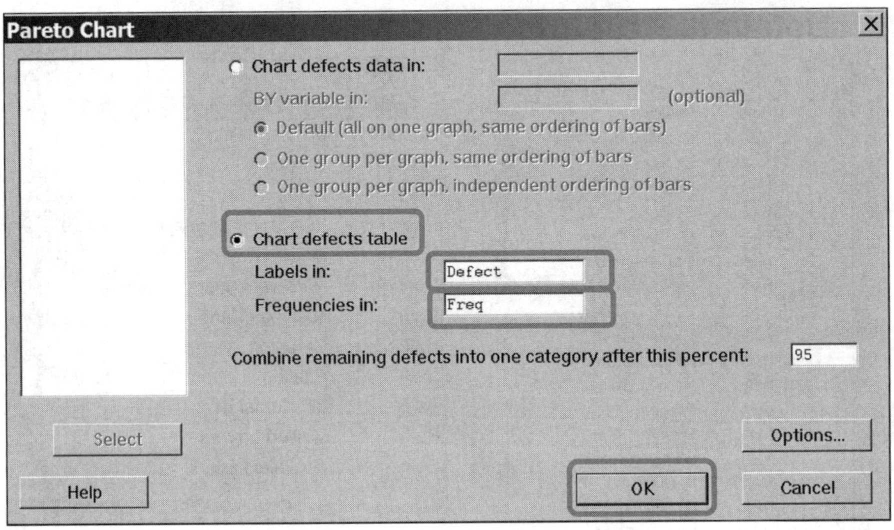

FIGURE P-2: Example of MINITAB Pareto Chart Main Screen Using Data in a Summary Structure

A B C D E F G H I J K L M N O P Q R S T U V W X Y Z

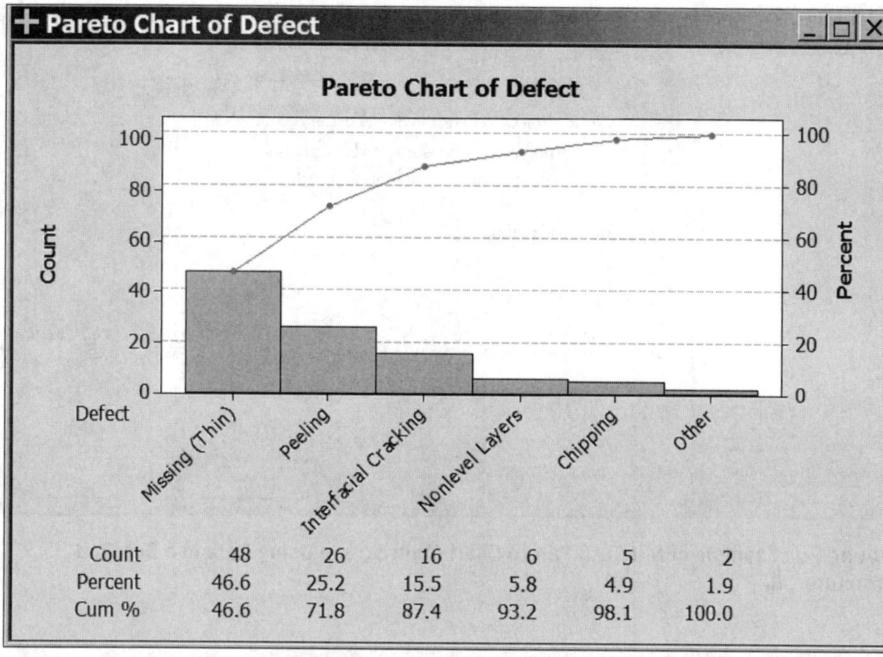

FIGURE P-3: Example of MINITAB Pareto Chart

Combining both the locations, Bath and Neenah, shows that about 75% of the defects come from paint that is either *Missing (Thin)* or *Peeling*. Hence, if a process improvement team were evaluating this data, those would be the two defects they would address first.

Look at the data by location to see if the defect prioritization is the same in the two locations. Create a stratified Pareto chart by using the data found in the Detailed Structure and select the top portion of the Pareto chart screen labeled "Chart defects data in," as displayed in Figure P-4. Select the column containing the defects listed as attribute data and enter it in the first dialog box, which is placed along this screen portion's title. Enter the optional variable data column containing the segmented information in the By variable in dialog box, (in this case, the column labeled "Location"). Keep the MINITAB defaults—both the highlighted for the Default (all on one graph, same ordering of bars) and the combine remaining defects into one category after this percent—95. Click OK to produce the final segmented Pareto chart, shown in Figure P-5.

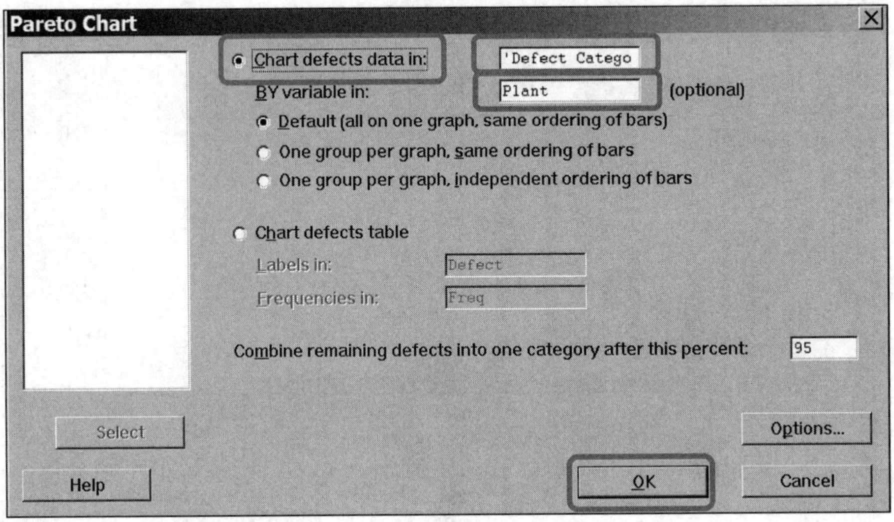

FIGURE P-4: Example of MINITAB Pareto Chart Main Screen Using Data in a Detailed Structure

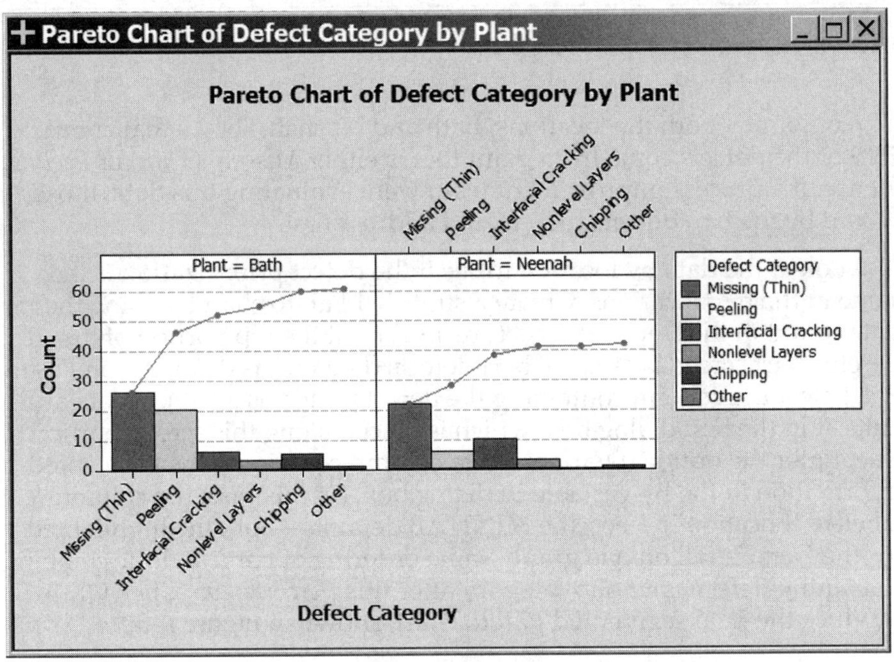

FIGURE P-5: Example of MINITAB segmented Pareto Chart by Location

For this example scenario, when stratified by location, the *Missing (Thin)* defect is the largest defect category in both locations. However, *Peeling* appears to be of interest at the Bath location as the second largest contributor, but not in Neenah. Neenah's second biggest defect is *Interfacial Cracking*.

Supporting or Linked Tools

Supporting tools that might provide input when developing a Pareto chart include

- Data Gathering (metrics) (**See Also** "Data Collection Matrix," **p. 248**)
- Performance Charts and Dashboards

A completed Pareto chart provides input to tools such as

- Cause-and-Effect diagram (**See Also** "Cause-and-Effect Diagram—7QC Tool," **p. 173**)
- FMEA (**See Also** "Cause-and-Effect Diagram—7QC Tool," **p. 173**)
- QFD (**See Also** "Quality Function Deployment," **p. 543**)
- Statistical Analysis Tools (**See Also** "Hypothesis Testing" and "Regression Analysis," **p. 335** and **p. 571**, respectively.)

Figure P-6 illustrates the link between a Pareto chart and its related tools and techniques.

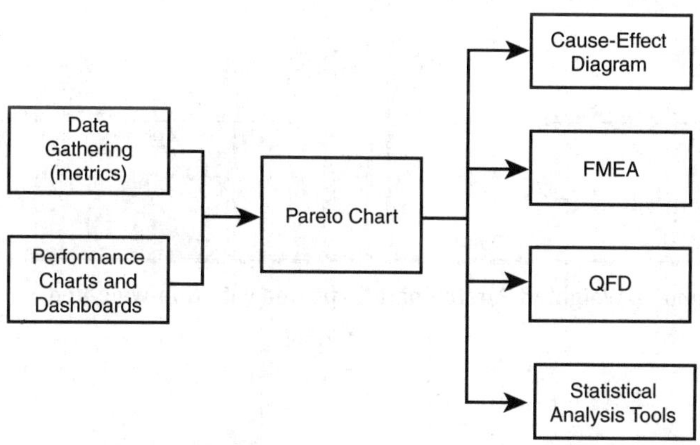

FIGURE P-6: Pareto Chart Tool Linkage

Variations

Weighted Pareto Analysis

A *weighted Pareto* is used when a Pareto chart is needed, but the categories represent unequal magnitude of importance. Hence, a conventional Pareto with each frequency weighted equally might misrepresent the prioritization. A category weighting system often reflects either unequal issues of importance (that is, health and safety) or cost concerns.

Constructing a weighted Pareto chart requires that a weight of importance be assigned to each category, typically 1 to 100. It gets applied to all the counts in that category. Multiply the frequency counts by its given category weights and record the weighted-count in the data collection sheet or MINITAB Worksheet. To construct a weighted Pareto chart, use the same procedure as constructing a Pareto chart, but reference the new weighted-count data.

Figure P-7 shows the preceding example of painted metal components from both Bath and Neenah, with the defect categories weighted based on repair and maintenance costs. The resulting weighted Pareto chart emphasizes a different set of vital few defects compared with the conventional non-weighted Pareto. Notice that *Peeling* has increased in overall magnitude of impact in the weighted Pareto, versus *Missing (Thin)* in the conventional approach. Also notice in Figure P-7 that the vertical axis scale of the weighted Pareto chart reflects the new weighted-counts.

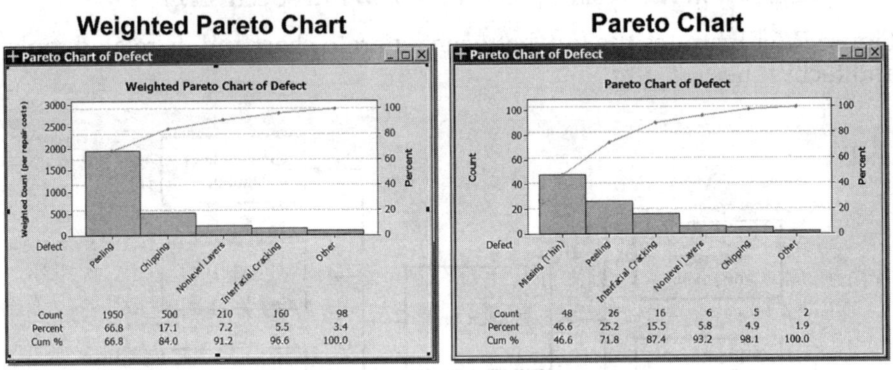

FIGURE P-7: Example of Weighted Pareto Chart Compared with Non-weighted

PERT (Program Evaluation and Review Technique) Chart

What Question(s) Does the Tool or Technique Answer?

What is the most efficient way to complete this process or project?

A PERT chart helps you to

- Plan and manage the timing of activities and overall completion of a project or process.
- Graphically organize process steps into the most efficient sequence.
- Show the most critical path and any parallel paths with time to completion estimates.
- Evaluate and reorganize the step sequence. It identifies any simultaneous tasks and tasks that will take the longest to complete.
- Identify any slack time—that is the amount of time a non-critical path task can be delayed without delaying the project.
- Manage resources and understand upstream and downstream dependencies within the process.

Alternative Names and Variations

- Variations on the tool include Activity Network Diagram (AND), Arrow diagram, or critical path (**See Also** "Activity Network Diagram (AND)—7M Tool," **p. 127**)

When Best to Use the Tool or Technique

The Program Evaluation and Review Technique (PERT) is a project planning and decision-making tool to evaluate project timelines.

Brief Description

The Program Evaluation and Review Technique, a variation on the Activity Network Diagram (AND) technique, accounts for uncertainty in projected estimates of activities in the time management planning of programs and projects. PERT follows the same approach as the AND technique, up until the very last step in the procedure, wherein the three estimates are calculated (minimum, maximum, and most likely), to bound the range of timing possibilities. (**See Also** "Activity Network Diagram (AND)—7M Tool," **p. 127**)

To construct a PERT chart, abide by five general network planning tool guidelines, which include

- Activities should be numbered, and each must have a unique number.
- The network diagram starts only with a single activity and ends with only a single activity.
- Before an activity begins, the preceding activity is completed.
- Arrows represent the sequencing of activities; however, the length of the arrow between different activities does **not** represent the amount of time between each, as found when one arrow is longer than another.
- Two events may be connected by only one activity.

Moreover, the PERT chart distinguishes itself as a network-planning tool by also following three additional requirements needed to complete it:

- Include all project individual activities.
- Calculate three time estimates for each activity—minimum (or pessimistic), maximum (or optimistic), and most likely.
- Determine the critical path and slack times.

The *critical path* is defined as the sequence of activities that requires the longest amount of expected time; therein determining the shortest amount of time to complete the project. *Slack time* (denoted as S) is the amount of time a task can be delayed without delaying a project. It is the latest date that an activity can be completed without extending the project. Hence, events on the critical path have zero slack time. In contrast, *parallel paths* contain slack time or a shorter duration than the critical path.

The PERT method may complicate the project planning process because it requires more data to calculate the activity network. However, this extra work pays off given that the PERT chart works well for unique, non-repetitive projects where their historical data can provide (at best) estimates as to the project schedule. The PERT method helps to define interrelationships between and among activities, which can identify problem areas. The probability of achieving the project deadline may be better estimated, and the work spawns alternative planning to improve the likelihood of on-time completion. Moreover, the impact of any project changes can be quickly evaluated. And finally, it is a good communication tool to illustrate the critical path and make decisions about streamlining it or how best to execute parallel paths while conserving resources (considering the PERT chart organizes a large amount of data into a diagram format). If the project complexity and uncertainty impact the risk of a significant project, consider using Monte Carlo simulation to better estimate the probabilities to complete activities and the overall project. (**See Also** "Monte Carlo Simulation," **p. 431**)

How to Use the Tool or Technique

The procedure to develop a PERT chart follows the similar steps as an Activity Network Diagram (AND). (**See Also** "Activity Network Diagram (AND)—7M Tool," **p. 127**)

The procedural steps are as follows:

Step 1. Identify the topic, and then collect any data about the task.

Examples include, Brainstorming, Process maps, Procedure manuals

Step 2. List the necessary tasks.

One technique used is to create a master list and then put one activity on either an index card or sticky note to make it easy to sequence the tasks. *(Only one activity step per index card or sticky note.)*

Step 3. Sequence the tasks into a logical order.

Identify which tasks must precede another task and arrange them accordingly. If some tasks can occur simultaneously, then put them on the same vertical plane (parallel to one another).

If the flow involves loop-back tasks, identify the determining (or evaluation) task and place in the sequence, similar to a process map's decision diamond that poses a binary question (that is, yes/no; pass/fail). (**See Also** "Process Map (or Flowchart)—7QC Tool," **p. 522**)

If using sticky notes, follow the same procedure and place the activities on a flip-chart sheet or section of the wall.

Continue this step until all the individual steps are placed in a logical flow. Be sure to leave space in-between each step to allow additional notes to be documented.

Step 4. Calculate and document the three time estimates

a. **Pessimistic (P)**—Represents the timing if everything were a disaster; all that could go wrong did (minimum time, denoted as P).

b. **Most Likely (ML)**—Represents the average or most expected timeframe, which may or may not be halfway between the optimistic and pessimistic timing (denoted as ML).

c. **Optimistic (O)**—Represents the timing if everything happened in the ideal state—smoothly, without any delays (maximum time, denoted as O).

d. Document all three times on the network diagram in sequence from lowest to highest time, separated by dashes: Pessimistic-Most Likely-Optimistic (or P-ML-O).

e. Calculate the Expected time and Variance for each activity using the following formulas:

i. **Expected Time (T_E):**

$$T_E = (P + 4ML + O)/6$$

ii. **Variance (V or σ^2):**

$$V = ((O–P)/6)^2$$

Step 5. Determine the critical path, using the Estimated Times (T_E).

a. Identify all the possible path combinations from start to finish, using the estimated times (T_E) for each task or activity.

b. Sum the estimated times (T_E) of each activity in each of the possible path combinations and record the path and its respective duration.

c. Identify the longest path and highlight by bolding (highlighting or darkening) the path arrows—this is the critical path.

d. Calculate the *Earliest Start* (ES) times and *Earliest Finish* (EF) times based on how long the preceding task(s) take. Hence, the first activity has zero ES time, and its EF time equals the total duration for the task. Preceding tasks have a compounding (or cumulative) effect on subsequent tasks' ES and EF times.

i. ES = the latest EF from the preceding task.

ii. EF = the ES time–the actual task time.

e. Calculate the *Latest Start* (LS) times and *Latest Finish* (LF) times by starting at the end of the network (project or process completion) and working backwards toward the start activity.

   i. LF = the shortest LS of the subsequent task.

   ii. LS = the LF time–the actual task time.

f. Calculate the slack times for each task and for the project. *Total slack* is the estimated time (T_E) an activity can be delayed without affecting the project schedule. *Free slack* is the time an activity can be delayed without affecting the early start of any subsequent task.

   i. Total slack = LS–ES = LF–EF

   ii. Free slack = the earliest ES of all subsequent tasks—EF

Step 6. Determine the probability (P) that the project will be finished by the deadline (T_D).

a. Calculate the Total Time Expected time and Total Variance for the project using the following formulas:

   i. **Total Time Expected (TT_E):**

   TT_E = Sum of the expected times on the critical path

   or

   TTE = Σ (all the T_E on the Critical Path)

   ii. **Total Variance (TV or σ_T^2):**

   TV or σ_T^2 =Sum of the variances (for example, $\sigma^2 = ((O—P)/6)^2$) on the critical path

   or

   TV = Σ (all the V or σ^2on the Critical Path)

   iii. **Z value** for the Deadline less the Total Time Expected, divided by the standard deviation:

   $Z = ((T_D–TT_E)/(\text{Square Root of } \sigma_T^2))$

iv. Look up the Z value on the Standard Normal Table to find the area under the curve. The table value represents the probability that the project is completed by the deadline.

Note

If the Standard Normal Table displays the area for the tail (as in the Appendix B table), and the area for the main body of the curve is what is needed, then subtract the tail area from 1. (**See Also** "Statistical Tools," for a discussion on Standard Normal distribution, **p. 684**; and "Standard Normal Distribution," in Appendix B, **p. 975**)

Step 7. Examine and adjust the diagram as needed.

A B C D E F G H I J K L M N O P Q R S T U V W X Y Z

Examples

The following example depicts a PERT Chart for six activities using the Finish-Start dependency. The list of activities includes:

- Activity A @ 3-day most likely duration; with 2 for pessimistic, and 5 for optimistic
- Activity B @ 5-day most likely duration; with 2 for pessimistic, and 8 for optimistic
- Activity C @ 7-day most likely duration; with 5 for pessimistic, and 10 for optimistic
- Activity D @ 4-day most likely duration; with 3 for pessimistic, and 7 for optimistic
- Activity E @ 9-day most likely duration; with 7 for pessimistic, and 12 for optimistic
- Activity F @ 3-day most likely duration; with 1 for pessimistic, and 4 for optimistic

The project's deadline is 25 days.

Table P-1 provides the necessary calculations of the time estimates to determine the critical path, Total Time Estimates (TT_E), Total Variance (TV), Z value and the Probability.

TABLE P-1: Calculations for Six-step PERT Example

Activity	P-ML-O	T_E TE = (P + 4ML + O) / 6	V V = ((O – P) / 6)^2
A	2-3-5	**3.17**	**0.25**
B	2-5-8	**5**	**1**
C	5-7-10	**7.17**	**0.69**
D	3-4-7	4.33	0.44
E	7-9-12	**9.17**	**0.69**
F	1-3-4	**2.83**	**0.25**
Path Options			
A-B-D-F		(3.17 + 5 + 4.33 + 2.83) = 15.33	
A-B-C-E-F		(3.17 + 5 + 7.17 + 9.17 + 2.83) = **27.33**	(0.25+ 1+ 0.69 + 0.69 + 0.25) = **2.88**
A-B-C-F		(3.17 + 5 + 7.17 + 2.83) = 18.17	
Total for the Critical Path		**TT_E = 27.33**	TV ($\sigma_T$$^{2)}$ = **2.88**
For a Time Deadline @ 25 (a given)		**Z** = **(T_D-TT_E) / (Sq.Rt. TV)** (25 – 27.33) / (Sq. Rt. 2.88) = (-2.33) / 1.62 = **-1,373**	Look up -1.37 in Z table; Find Z value @ 0.0853; Probability = 1 - 0.0853) = 0.9147 or about 92% probability

Notice that Table P-1 shows a 92% probability of finishing on time for this example project to complete within the 25-day deadline. Hence, the project planning probably requires very few adjustments, if any. Table P-1 shows the example's path calculations as

- A-B-D-F = 15.33 days
- A-B-C-F = 18.17 days
- A-B-C-E-F = 27.33 days

The longest path is A-B-C-E-F, at 27.33 days; therefore, it is the critical path.

The formulas for the Earliest and Latest times are as follows, and this example's calculated results can be found in Figure P-8:

- ES = the latest EF from the preceding task.
- EF = the ES time–the actual task time.
- LF = the shortest LS of the subsequent task.
- LS = the LF time–the actual task time.

A B C D E F G H I J K L M N O P Q R S T U V W X Y Z

The formulas for the slack times are as follows, and this example's calculated results can be found in Figure P-8:

- **Total slack** = LS–ES = LF–EF;
- **Free slack** = the earliest ES of all subsequent tasks—EF.

Notice, there is **zero slack time along the critical path.**

The PERT chart showing the example's six-step scenario, its critical path and slack time calculations displayed on a network diagram can be found in Figure P-8:

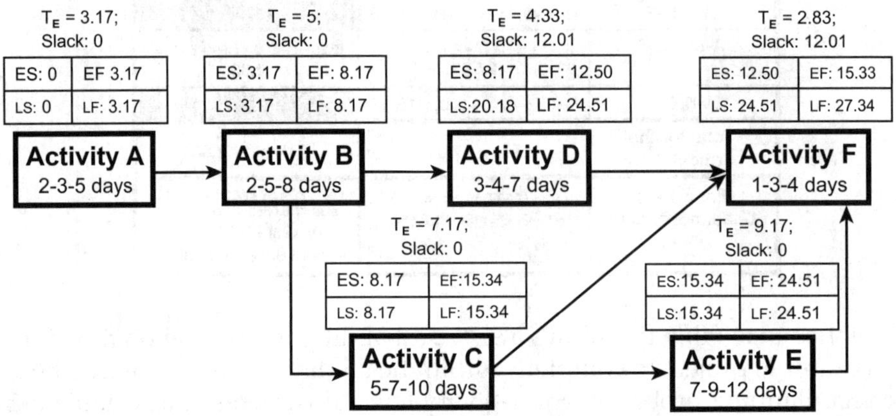

FIGURE P-8: Six-step PERT Chart Example

Hints and Tips

- Building a PERT chart by hand can be tedious. For complex projects, use computer software packages to calculate the critical path and probabilities.
- If the project complexity and uncertainty impact the risk of a significant project, consider using Monte Carlo simulation to better estimate the probabilities to complete activities and the overall project. (**See Also** "Monte Carlo Simulation," **p. 431**)

Supporting or Linked Tools

Supporting tools that might provide input when developing a PERT chart include

- Activity Lists
- Brainstorming sessions (**See Also** "Brainstorming Technique," **p. 168**)
- Process map (**See Also** "Process Map (or Flowchart)—7QC Tool," **p. 522**)

A completed PERT chart provides input to tools such as

- Brainstorming sessions (**See Also** "Brainstorming Technique," **p. 168**)
- Process map (**See Also** "Process Map (or Flowchart)—7QC Tool," **p. 522**)
- FMEA (**See Also** "Failure Modes and Effects Analysis (FMEA)," **p. 287**)
- Cause-and-Effect diagram (**See Also** "Cause-and-Effect Diagram—7QC Tool," **p. 173**)
- Value Stream Activity Matrix (**See Also** "Value Stream Analysis," **p. 727**)

Figure P-9 illustrates the link between a PERT chart and its related tools and techniques.

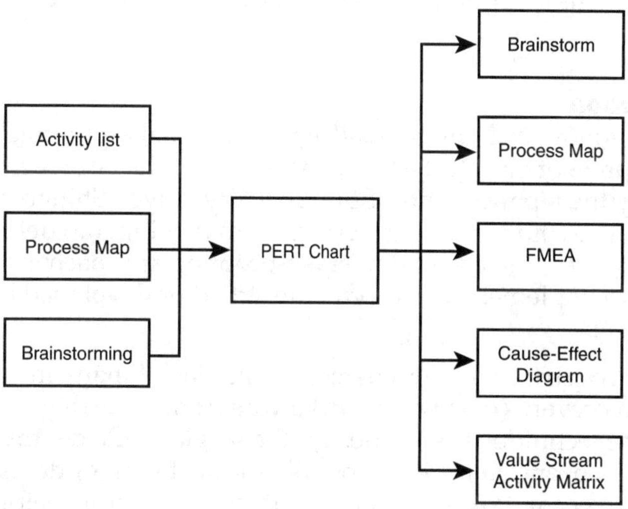

FIGURE P-9: PERT Chart Tool Linkage

A B C D E F G H I J K L M N O P Q R S T U V W X Y Z

Poka-Yoke

What Question(s) Does the Tool or Technique Answer?

How best to prevent or correct in-process errors (often human mistakes)?

Poka-yoke helps you to

- Detect errors before they occur, and prevent them
- Respond and adjust to in-process errors in real time

Alternative Names and Variations

This tool is also known as

- Mistake-proofing
- Fail-safing

When Best to Use the Tool or Technique

Ideally, the best time to implement Poka-yoke strategies is at the onset of planning and designing a new process, system, product, or service to prevent errors from occurring. However, Murphy's Law is ever present; errors will occur at some point within a process involving humans. Hence, any time is an appropriate time to deploy Poka-yoke.

Brief Description

Poka-yoke [pronounced "*POKA-yolk-ey*"] is a mistake-proofing method or device to prevent or detect an in-process error. The approach was developed by the Japanese and popularized by Shigeo Shingo in the mid- to late-1980s, to avoid inadvertent errors from turning into defects or defective products. A successful mistake-proofing approach is relatively inexpensive to implement and is often designed or developed by the process players.

The objective strives for zero defects using one of many in-process techniques to prevent (mostly human) errors from occurring; without sampling. The technique tries to design the single method—**the right way**—to do a job, and make it impossible for mistakes (or deviations on the method) to occur. Poka-yoke also aims to detect errors before they become defects by using shutdown, control, or warning techniques. For example, it prevents incorrect parts from being made or assembled or easily identifies a flaw or error by providing a visual or other signal to indicate a characteristic state. Another example is in a control system, where operations halt when a defect is detected; immediately root cause analysis begins, and the problem is fixed before resuming operations. This avoids scrap from increasing.

Shingo understood that human errors take on many forms and can occur at any time in a process that involves people. People include not only the process players, but also the partners and customers. The concept of concentrating on the task 100% of the time still fails to guarantee zero defects. Thus, if an error occurs, Poka-yoke minimizes the time between the error occurring and the detection warning.

Poka-yoke purports five basic principles with varying degrees of effectiveness. The intent is to utilize the best most effective approach possible in a given situation. The following lists the principles in descending order of effectiveness:

- **Elimination**—The preferred and most effective method, which removes the possibility of an error.

  Example: redesign the process (or product) so that the task is no longer necessary; fewer component parts

- **Replacement**—A better method than the subsequent approaches; it substitutes a procedure with a more reliable one.

  Example: use robotics or automation; bar-coding prevents data entry errors

- **Facilitation**—A slightly less effective method than those previously mentioned, but still good since it makes the work easier to perform.

  Example: color-coding; combining steps; checklist; and so on

- **Detection**—A less desirable method because it failed to prevent the error from occurring, but it quickly identifies the mistake before further processing.

  Example: develop computer software that notifies a worker when a wrong input is made (an error message); fail-safe cut-off mechanism that shuts-down the process; flashing light or buzzer

- **Mitigation**—The least effective, but good strategy that minimizes the effect of the error.

  Example: utilize electrical fuses for overloaded circuits; a computer back-up system (redundancy)

Poka-yoke devices can be coupled with other inspection systems to enhance their effectiveness. However, the best mistake-proofing is designing the quality into the process or product from the beginning, as in Design for Six Sigma (DFSS).

Porter's 5 Forces

What Question(s) Does the Tool or Technique Answer?

How are we performing relative to the competitive business threats?

Porter's 5 Forces analysis helps you to

- Develop a competitive business strategy to gain or maintain an advantageous market position
- Communicate a high-level summary of the competitive landscape

When Best to Use the Tool or Technique

A planning tool used to analyze the competitive marketplace at an industry level to identify capabilities needed to enhance competitive advantage and to provide useful information to a SWOT Analysis. (**See Also** "SWOT (Strengths-Weaknesses-Opportunities-Threats)," **p. 699**)

Brief Description

The Porter's 5 Forces analysis is a method to determine your firm's position relative to the competitive forces active in the markets you are currently serving or intend to serve. Michael Porter, a Harvard Business School professor, developed this technique of analyzing industries and competitors for organizations to use an outside-in perspective as input to their competitive business strategy. In his book, *Competitive Strategy: Techniques for Analyzing Industries and Competitors*, Porter introduces this structural analysis of industries with a five competitive forces model. Figure P-10 illustrates the model and its five components.

This model analyzes the attractiveness of an industry opportunity by examining its five competitive forces in aggregate, which are

- **Intensity of Competitive Rivalry**—Determine the rivalry among the current market players. Is the competition strong among the existing organizations? Are they relatively equally balanced in size, strength, offerings, market share, or is there a dominant company?

  The intensity of rivalry depends on several factors including: size of market (regional, domestic, global); degree of offering differentiation (commodity versus specialty); leadership costs and market maturity.

A B C D E F G H I J K L M N O P Q R S T U V W X Y Z

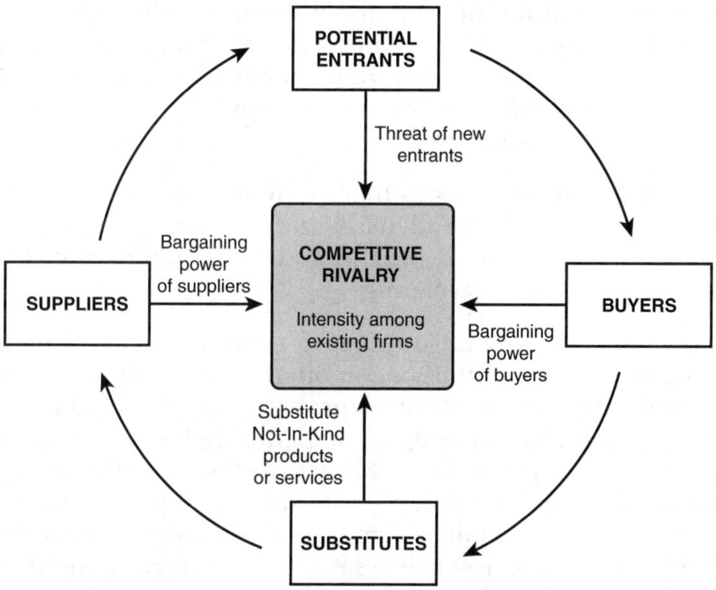

FIGURE P-10: Porter's 5 Forces Model

- **New Entrant**—Ascertain the likelihood of a new entrant to start competing in the market. Do barriers exist to block entrance, or is it easy for a new entrant to offer a new, unique, or difficult (NUD) offering to meet current customer demands? (For example, the entrance of the Japanese firms, such as Toyota and Honda, into the U.S. auto market.) (**See Also** the "Kano Model: NUD versus ECO" section within the "KJ Analysis," **p. 376**)

  Threats to new entrants depend on several factors including: economies of scale; capital investment requirements; technological capabilities and reliability; brand loyalty; customer switching costs (such as with information technology (IT) platform vendors); competitive response from current market players; and any government assistance and/or regulations.

- **Substitute**—Evaluate how easily a substitute offering (product or service) can be made; often times it is an offering intended for one purpose but expanded to address a new application, and sometimes will be faster, cheaper, or better. (For example, some cell phones replace a PDA (personal digital assistant), such as the Palm Pilot, or the Hummer substituting for SUV (Sports Utility Vehicles), versus its original military purpose.)

Threats to substitutes depend on several factors including: quality (for example, reliability, durability); feature/functionality; buyer's willingness to switch; relative value of new versus current offering (price to feature, functionality and performance, and total cost of ownership) and switching costs.

- **Power of Suppliers**—Assess the bargaining power of the suppliers relative to the sellers. Are there several suppliers to choose from or only a few? Is their offering a commodity, or do they have a unique, specialized skill and/or offering?

  The bargaining power of suppliers depends on several factors including: their profitability; their offering as a commodity versus specialization; number of suppliers; brand loyalty; quality and switching costs. In addition, as companies right-size, mergers and acquisitions take place or capability expansion or retrenching occurs, these trends introduce threats—for example, suppliers expanding to establish their own retail outlets and sell directly to consumers, rather than business-to-business. Conversely, buyers could threaten to integrate backward into the supply chain.

- **Power of Buyers**—Evaluate the bargaining power of the buyers. Can they form alliances (that is, associations, such as AARP) and in aggregate exert power to make demands, including ordering volumes, availability, pricing discounts, quality, and regulations?

  The bargaining power of buyers depends on several factors including: the concentration of buyers; disposable income and the relative need versus want and switching costs.

The model sometimes adds a sixth competitive force if appropriate—the government, or a regulatory agency. This dimension is industry-specific and may or may not be applicable on a steady-state basis.

How to Use the Tool or Technique

The following procedure provides a guideline to applying the Porter's 5 Forces model to an organizations business strategy.

Step 1. Rate each of the five competitive forces.

  a. Determine the list of appropriate assumptions behind each of the five competitive forces.

  b. By dimension, rate each assumption. Determine the rating scale to score the assumptions. Options to score the impact or magnitude include

i. High or Low

ii. High, Medium, or Low

iii. 9 (High), 3 (Medium), 1 (Low)

Figure P-11 illustrates an example of rating of the forces' assumptions.

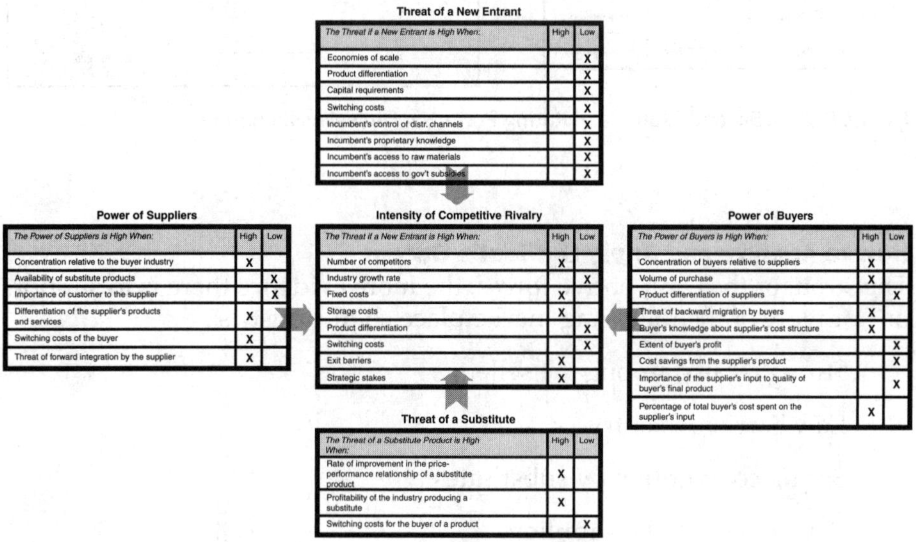

FIGURE P-11: Example of Rating Porter's 5 Forces' Assumptions

c. Evaluate the rating for the overall competitive force, using the same scale selected in Step 1.b.

Step 2. Summarize the overall competitive force ratings in a spider chart.

a. Using Microsoft Excel's spider chart format, enter the summary ratings for each of the five competitive forces.

Figure P-12 illustrates an example of the overall rating of the five forces' and a spider chart template.

Overall Assessment

Competitive Forces	Industry
Intensity of Competitive Rivalry	Low, Medium, or High
Threat of a New Entrant	Low, Medium, or High
Threat of Substitute Products	Low, Medium, or High
Power of Buyers	Low, Medium, or High
Power of Suppliers	Low, Medium, or High
Overall Competitive Assessment	Low, Medium, or High

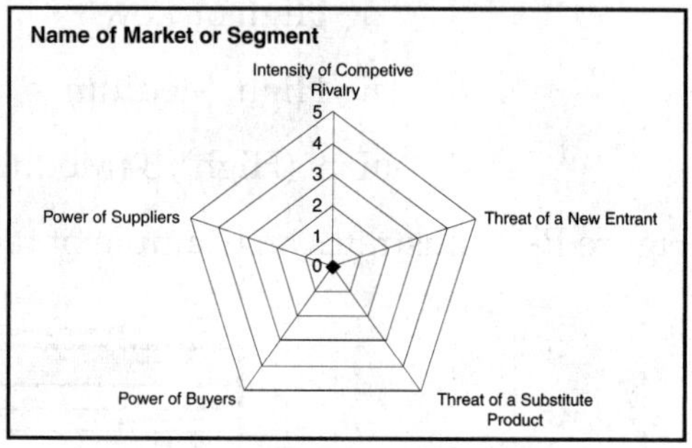

FIGURE P-12: A Second Example of Rating Porter's 5 Forces' Assumptions

How to Analyze and Apply the Tool's Output

The stronger the competitive forces, the more evidence there is to support one or more of the following marketplace characteristics:

- Strong rivalry among the sellers
- Low barriers to entry
- Strong competition by substitutes
- Power held by the suppliers
- Power held by the buyers

As a result, the stronger the competitive forces, the higher the score. The large scores indicate less profitable market opportunities and purportedly a less attractive industry. Conversely, the smaller the score, the less competitive forces exist.

Figure P-13 shows a completed Porter's 5 Forces diagram as a spider chart for the cell phone industry, indicating a strong threat with the entrance of MP3 players from companies such as Apple's iPhone and the Asia-Pacific manufacturers. The MP3 devices may serve as one of the cell phone's threats as a substituted product.

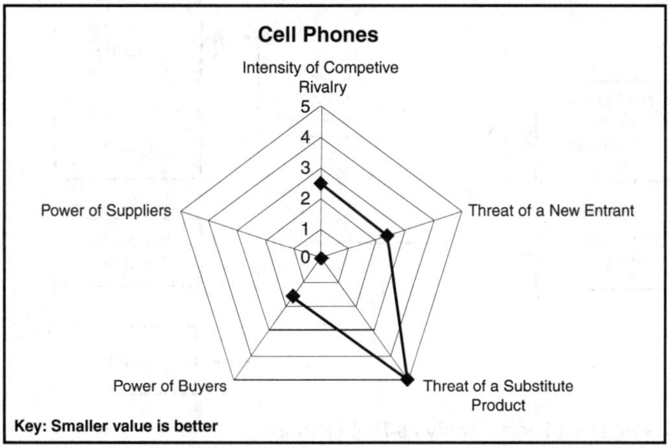

FIGURE P-13: Example Porter's 5 Forces Illustrated as a Spider Diagram

Supporting or Linked Tools

Supporting tools that might provide input when developing a Porter's 5 Forces Analysis include

- Market and Segment characteristic data
- Buyer, supplier, and competitive intelligence data

A completed Porter's 5 Forces Analysis provides input to tools such as

- Cause-and-Effect diagram (**See Also** "Cause-and-Effect Diagram—7QC Tool," **p. 173**)
- FMEA (**See Also** "Failure Modes and Effects Analysis (FMEA)," **p. 287**)
- Prioritization Matrix (**See Also** "Prioritization Matrices—7M Tool," **p. 470**)
- SWOT Analysis (**See Also** "SWOT (Strengths-Weaknesses-Opportunities-Threats)," **p. 699**)

Figure P-14 illustrates the link between a Porter's 5 Forces Analysis and its related tools and techniques.

A B C D E F G H I J K L M N O P Q R S T U V W X Y Z

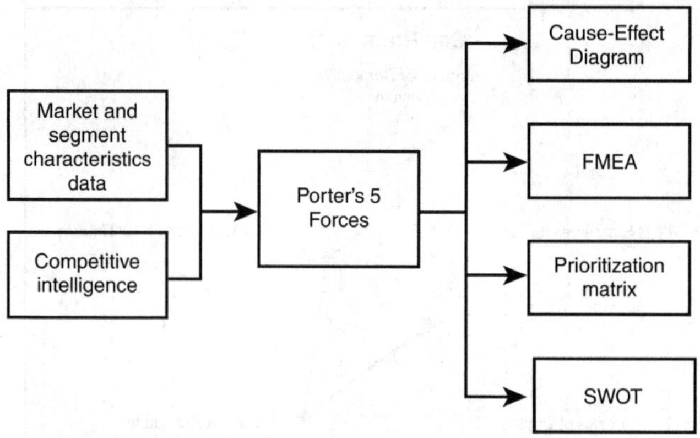

FIGURE P-14: Porter's 5 Forces Analysis Tool Linkage

Prioritization Matrices—7M Tool

What Question(s) Does the Tool or Technique Answer?

What is the best option among the several possibilities for a crucial (often mission-critical) decision that carries risk of significant consequences if wrong?

- Prioritization Matrices help you to Narrow down options for a weighty decision to select the best one using several subjective criteria

Alternative Names and Variations

This tool is also known as

- Analytical Criteria method or Full Analytical Criteria method
- Consensus Criteria method
- Combination I.D./Matrix method, where I.D. refers to Interrelationship Diagraph (**See Also** "Interrelationship Diagram—7M Tool," **p. 369**)

Variations on the tool include

- Decision matrix (**See Also** Matrix Diagrams—7M Tool," **p. 399**)
- Pugh Concept Evaluation (**See Also** "Pugh Concept Evaluation," **p. 534**)
- Solution Selection Matrix (**See Also** "Solution Selection Matrix," **p. 672**)

When Best to Use the Tool or Technique

When the decision is of high importance to the organization, the stakes are high, and grave consequences threaten the business if wrong. The set of Prioritization Matrices are best used to narrow a list of available options, by comparing the options using three or more essential (often subjective) criteria to select the best one.

Brief Description

Prioritization Matrices represent a set of three different matrices. These set of tools were introduced early in the history of quality and Six Sigma to **balance** the heavy quantitative emphasis of the matrix data analysis introduced by the Japanese with some *qualitative-oriented* tools. This set of tools is designed for crucial organizational decisions that carry significant consequences if wrong. Depending on the size, scale, and scope of the organization, such weighty decisions may include selecting an acquisition candidate, making key personnel decisions, or choosing a new enterprise-wide IT platform. These matrices can highlight any disagreement among a project team and help to focus them on the best alternative by building consensus around the unbiased decision-criteria.

Typically, these matrices use an L-shaped structure to compare a set of options with a set of decision-criteria. The Prioritization Matrices entail a systematic approach to select criteria, determine criteria weights, and compare choices. Of all the matrix diagram tools, the Prioritization Matrices are considered the most rigorous and time-consuming to use. (**See Also** "Matrix Diagrams—7M Tool," **p. 399**, for a discussion on other matrices and their shapes.)

Although the names are similar, the Prioritization Matrices distinguish themselves from two others—the *Decision Matrix* and *Decision Authority Matrix*. A Decision Matrix may refer to a specific tool or a category of matrices used to evaluate and prioritize a list of options, often called by a specialized name—Pugh Matrix and Solution Selection Matrix. A Decision Authority Matrix identifies the roles or people accountable for making decisions in a process or organization.

Prioritization Matrices are a member of the 7M Tools, attributed in part to Dr. Shewhart, as seven "management" tools, or sometimes referred to as the 7MP or *seven management and planning tools*. These 7M Tools make up the set of traditional quality tools used to analyze quantitative data. The 7M Toolset includes: 1) Activity network diagrams or Arrow diagrams; 2) Affinity diagrams; 3) Interrelationship digraphs or Relations diagrams; 4) Matrix diagrams; 5) Prioritization Matrices, often replacing the more complex matrix data analysis; 6) Process decision program charts (PDPC); and 7) Tree diagrams. The book, *The Quality Toolbox* by Nancy Tague, presents the 7M Tools ranked from those used for abstract analysis to detailed planning: Affinity diagram, Relations diagram, Tree diagram, Matrix diagram, Matrix data analysis (commonly replaced by a more simple Prioritization Matrix), Arrow diagram, and Process Decision Program Chart (PDPC).

The three types of Prioritization Matrices include

- **Full Analytical Criteria method**—The most complex of the three.
- **Consensus Criteria method**—Moderate complexity.
- **Combination I.D./Matrix method**—Simplest of the three.

Full Analytical Criteria Method

The Prioritization Matrix produced from the *Full Analytical Criteria* method should be reserved for those most significant or critical issues. It requires multiple matrices to conduct a pair-wise comparison before completing the final Prioritization Matrix. Though it is the most complex of the three Prioritization Matrices, it is a simplified derivative of the Analytical Hierarchy Process (AHP).

The method's guidelines recommend that a small team of three to eight people share in building this matrix. Otherwise, working through the process and accounting for a large number of perspectives and input can become unwieldy. When the team remains small, reaching a complete consensus for each of the multiple matrices becomes more facile.

A Full Analytical Criteria method prioritizes the decision criteria, weights them, and applies numerical values to the options to indicate the best alternative. A limited number of options and criteria are recommended (about ten or less of each) since the approach is time-consuming and the calculations can become cumbersome.

Consensus Criteria Method

The *Consensus Criteria* method builds a Prioritization Matrix for scenarios where the options appear virtually equal. This Prioritization Matrix essentially is a simplified version of the Full Analytical Criteria method, given that it skips the multiple pair-wise comparisons. Therefore, it takes less time to create. Similar to the previously mentioned method, it requires multiple matrices to complete the final matrix. The Consensus Criteria method uses weighted voting and ranking and applies numerical values to the options to indicate the best alternative. A limited number of options and criteria are recommended (about ten or less of each) to minimize the time to produce the final output.

Combination I.D./Matrix Method

The *Combination ID/Matrix* method creates a Prioritization Matrix for a complex issue that involves a cause-and-effect relationship to determine where to start tackling the problem. Thus, the method focuses on a causal-based approach rather than criteria-based. It prioritizes and selects the root cause to address first. The team assigned to resolve this issue should have a first-hand working knowledge of the process.

This combination method employs a hybrid structure made up of a matrix chart and a Tree diagram. It uses the cause-and-effect relationship information from an Interrelationship diagram (ID) or Tree, rather than decision-criteria. The Tree diagram parses the ideas into further detail and then combines the detail with the criteria in an L-shaped matrix. Within an L-shaped matrix, the root cause options are compared to each other to determine their relative strength. It is considered the simplest of the three Prioritization Matrices because it uses non-numerical values (or symbols and arrows) to identify the best option. (**See Also** "Interrelationship Diagram—7M Tool," **p. 369**)

How to Use the Tool or Technique

The imperative decisions facing an organization are best tackled by a team of the key individuals. Often such a crucial decision elicits varying opinions, all worth consideration. Usually the decision-criteria include qualitative ones that represent different impacts for the different business perspectives. Building a Prioritization Matrix brings this team of key individuals together to gain consensus on the selection criteria and the optimal alternative.

The procedure to develop each of the three Prioritization Matrices is as follows.

Full Analytical Criteria Method

The procedure using the Full Analytical Criteria method involves a small team of the key individuals (ideally three to eight people) to perform the following tasks:

Step 1. Agree on the goal statement.

Step 2. Gain consensus on the final list of decision-criteria.

Step 3. Create the decision-criteria matrix to weigh each of them against one another.

a. Using an L-shaped matrix structure, starting in the second row of the first column, record the final list of decision-criteria by placing one criterion per cell, to serve as the row heading.

b. Starting in the first row, second column, and using the same sequence of decision-criteria as in Step 3.a., record each of them from left to right, placing one topic per cell, to create the column headings.

c. Starting in the upper-left corner of the matrix, shade each cell along the diagonal to the lower-right corner because these cells compare the same decision-criteria to itself, which is unnecessary.

d. Add two additional columns on the far right and title them with the following headings—"Row Total" in the second column from the right and "Relative Value Rating" in the far right column.

e. Add one additional row at the bottom of the matrix and title it "Grand Total."

Figure P-15 provides an illustration of a completed sample decision-criteria matrix.

Step 4. Rate the decision-criteria against one another, using the standard rating scale found in Table P-2:

TABLE P-2: Decision-Criteria Rating Scale for Full Analytical Criteria Method Prioritization Matrix

10	Row Criteria is MUCH MORE IMPORTANT than the Column Criteria
5	Row Criteria is MORE IMPORTANT than the Column Criteria
1	Row Criteria is EQUALLY IMPORTANT to the Column Criteria
1 / 5	Row Criteria is LESS IMPORTANT than the Column Criteria
1 / 10	Row Criteria is MUCH LESS IMPORTANT than the Column Criteria

a. Always starting with the decision-criteria listed along the vertical axis of the matrix—the Row Criteria—rate the row criterion versus the column criterion within each unshaded cell, using the rating scale found in Table P-2.

b. Continue until each un-shaded cell contains a rating.

Step 5. Complete the decision-criteria matrix.

a. Sum the row ratings to compute the Row Total for each row criterion and record the score in the second from the right column, Row Total.

Sum the row totals of this column to compute the Grand Row Total. Record that summed ratings score in the last row's cell of that column, of the row, Grand Total.

b. Compute the Relative Value for each Row Criteria by **dividing** each Row Total score by the Grand Row Total. This calculation determines what percentage a given Row Criteria is of the Grand Total. Each computation should be carried to two decimal positions. Record the calculated relative weighing of each Row Criteria in the adjacent cell of the far right column, Relative Value Rating.

Figure P-15 displays a completed decision-criteria matrix for an acquisition example.

c. If a decision-criteria's Relative Value is extremely small compared with the others on the list, the team may decide to remove it from the list.

Now the decision-criteria list with its Relative Value scores is ready to use to evaluate the options under consideration.

	Criteria				
Criteria	Customer base	Increased Capacity	Technological Advantage	Row Total	Relative Value Rating
Customer base		5	10	15.00 (= 5 + 10)	0.73 (= 15 / 20.50)
Increased Capacity	1/5 (= 0.2)		5	5.20 (= 5 + 0.2)	0.25 (= 5.2 / 20.50)
Technological Advantage	1/10 (= 0.1)	1/5 (= 0.2)		0.30 (= 0.5 + 0.1)	0.02 (= 0.3 / 20.50)
Grand Total				20.50 (= 15+5.2+0.3)	100%

FIGURE P-15: Sample Decision-criteria Matrix for the Full Analytical Criteria Method Prioritization Matrix

> **Note**
>
> The shaded cells along the upper-left-to-lower-right diagonal create a mirror-image of cells on either side. Hence, if the rating for Row Criterion 1 versus Column Criterion 2 were 5 and placed **above** the shaded-diagonal; then the mirror-image cell found **below** the shaded-diagonal formed by Row Criterion 2 versus Column Criterion 1 would have an inverse rating of **1/5**.

Step 6. Create the set of options rating matrices to weigh each of them against one another using one decision-criterion and its relative values.

a. Create additional L-shaped matrices that total **the number of final decision-criteria.** Hence, if there are three final decision-criteria, create four additional matrices.

   Assign each decision-criterion to a matrix and title it as such, one for each decision-criterion. Place each title in the upper-left cell of each matrix.

b. Build each matrix structure similar to the procedure described in Step 3. Start in the second row of the first column and record the list of options by placing one option per cell, to serve as the *row heading*.

c. Starting in the first row, second column, and using the same sequence of options as in Step 6.b., record each of them from left to right, placing one topic per cell, to create the column headings.

d. Starting in the upper-left corner of the matrix, shade each cell along the diagonal to the lower-right corner because comparing the same option to itself is unnecessary.

e. Add two additional columns on the far right and title them "Row Total" in the second column from the right and "Relative Value Rating" in the far right column.

f. Add one additional row at the bottom of the matrix and title the row "Grand Total."

Step 7. Complete each option rating matrix, by rating the options within each option-rating matrix, using the standard rating scale found in Table P-2.

a. Always start with the options listed along the vertical axis of the matrix—the Row Options. Using the pair-wise approach, rate the row option versus the column option within each unshaded cell, using only one criterion and the standard rating scale found in Table P-2.

> **Note**
>
> The shaded cells along the upper-left-to-lower-right diagonal create a mirror-image of cells on either side, as explained in Step 4.a.

b. Continue until each unshaded cell contains a rating.

c. Sum the row ratings to compute the Row Total for each row option and record the score in the second from the right column, Row Total.

Sum the row totals of this column to compute the Grand Row Total. Record that summed ratings score in the last row's cell of that column, of the row, Grand Total.

d. Compute the Relative Value for each Row Option by **dividing** each Row Total score by the Grand Row Total. This calculation determines what percentage a given Row Option is of the Grand Total. Each computation should be carried to two decimal positions. Record the calculated relative weighing of each Row Option in the adjacent cell of the far right column, Relative Value Rating.

e. Continue this sequence of tasks until all the criterion-specific options rating matrices are completed. Hence, the number of completed option-rating matrices equals the number of decision-criteria, as shown in Figure P-16, which continues with the company acquisition scenario illustrated in the previous figure.

Customer-base Criteria	Options			Row Total	Relative Value Rating
	Company A	Company B	Company C		
Company A		5	1	6.00	0.48
Company B	1/5		5	5.20	0.42
Company C	1	1/5		1.20	0.10
Grand Total				12.40	100%

Increased Capacity Criteria	Options			Row Total	Relative Value Rating
	Company A	Company B	Company C		
Company A		10	1	11.00	0.63
Company B	1/10		1/5	0.30	0.02
Company C	1	5		6.00	0.35
Grand Total				17.30	100%

Technological Advantage Criteria	Options			Row Total	Relative Value Rating
	Company A	Company B	Company C		
Company A		1/10	1/5	0.30	0.01
Company B	10		10	20.0	0.79
Company C	5	1/10		5.10	0.20
Grand Total				25.40	100%

FIGURE P-16: Sample Set of Option Rating Matrices for the Full Analytical Criteria Method Prioritization Matrix

> **Note**
>
> At this point only the Summary Prioritization Matrix will be incomplete.

Step 8. Complete the Summary Prioritization Matrix that compares the weighted options to the weighted decision-criteria to identify the optimal alternative.

a. Create the last L-shaped matrix and title it "Summary Prioritization Matrix."

b. Add two additional columns on the far right and title them "Row Total" in the second column from the right, and "Relative Value Rating" in the far right column.

c. Add one additional row at the bottom of the matrix and title it "Grand Total."

d. In the first row, second column, record each decision-criterion from left to right, placing one topic per cell, to create the column headings. In parentheses under the column heading, record each decision-criterion relative value within each appropriate cell. (Recall these values come from the decision-criteria matrix.)

e. Create the row headings by starting in the second row of the first column; record the list of options by placing one option per cell, to serve as the row heading.

f. Gather the option rating matrices. Using one matrix at a time, record the appropriate option's relative value in a given column that corresponds with the decision-criteria used to create that specific option-rating matrix.

Continue until all the columns have been filled with the appropriate option's relative value.

g. By cell, calculate and record the priority rating by multiplying the appropriate decision-criteria's relative value times the option's relative value.

You may want to circle or bold this number to distinguish it from the option's relative value in the cell.

Continue until each cell contains a priority rating.

h. Sum the row priority ratings to compute the Row Total for each row prioritized option and record the score in the second from the right column, Row Total.

Sum the row totals of this column to compute the Grand Row Total. Record that summed ratings score in the last row's cell of that column, of the Grand Total row.

i. Compute the Relative Value for each Row Prioritized Option by dividing each Row Total score by the Grand Row Total. This calculation determines what percentage a given Row Prioritized Option is of the Grand Total. Each computation should be carried to two decimal positions. Record the calculated relative weighing of each Row Prioritized Option in the adjacent cell of the far right column, Relative Value Rating.

The largest Prioritized Option Relative Value indicates the best alternative, as shown in Figure P-17, which continues with the company acquisition scenario illustrated in the previous two figures.

A B C D E F G H I J K L M N O P Q R S T U V W X Y Z

If there are two or more of the top prioritized options that have ratings relatively close together, conduct further exploration.

Figure P-17 displays the final Prioritization Matrix using the Full Analytical Criteria method. For this scenario, Company A appears to be the best option of the three, with a total score of 0.51.

Summary Prioritization Matrix	Criteria			Row Total	Relative Value Rating
	Customer base (CB) (0.73)	Increased Capacity (IC) (0.25)	Technological Advantage (TA) (0.02)		
Company A	0.48 x CB = 0.35	0.63 x IC = 0.16	0.01 x TA = 0.0002	0.51	0.51
Company B	0.42x CB = 0.31	0.02 x IC = 0.01	0.79 x TA = 0.02	0.34	0.34
Company C	0.10x CB = 0.07	0.35 x IC = 0.08	0.20 x TA = 0.004	0.15	0.15
Grand Total	0.73	0.25	0.02	1.00	1.0

FIGURE P-17: Sample Option Rating Matrices for the Full Analytical Criteria Method Prioritization Matrix

Consensus Criteria Method

The procedure using the Consensus Criteria method involves the team of key individuals to perform the following tasks:

Step 1. Agree on the goal statement.

Step 2. Gain consensus on the list of decision-criteria.

Step 3. Create the decision-criteria weightings.

a. Stack rank or sequence each of the criteria from most import to least important.

b. Given that the total weightings must equal 100%, assign a percentage weight to each of the decision-criteria.

Step 4. Create the Consensus Criteria matrix to weigh each option against the weighted decision-criteria.

a. Using an L-shaped matrix structure, in the first row, second column, record each of the prioritized decision-criteria from left to right, starting with the most important, and placing one criterion per cell, to create the column headings.

Underneath each decision-criteria column heading, record the appropriate weighting (as a decimal) for each. Recall, these weightings should sum to 100%.

b. In the second row of the first column, record the options by placing one topic per cell, to serve as the row heading.

c. Add one additional column on the far right and name the column "Row Total."

Step 5. Complete the Consensus Criteria matrix as the final Prioritization Matrix.

a. Count the number of options under consideration; that number becomes the highest possible rank for the options. Create a ranking scale:

i. **1** = worst (or lowest rank), to

ii. "**X**" number of options under consideration = best (or highest rank).

b. Within a column, stack rank and record each option against that column's decision-criteria, from best to worst.

Continue this task until all the decision-criteria columns have been filled with the appropriate stack rankings.

c. By cell, calculate and record the prioritized rankings by multiplying the appropriate decision-criteria's weighting times the option's stacked rank value.

You may want to circle or bold this number to distinguish it from the option's stacked rank value in the cell.

Continue until each cell contains a priority rating.

d. Sum the row prioritized *rankings* to compute the *Row Total* for each row option and record the score in the far right column, Row Total. Each computation should be carried to two decimal positions.

e. The largest Prioritized Option score indicates the best alternative, as shown in Figure P-18.

If there are two or more of the top prioritized options have ratings relatively close together, conduct further exploration.

Figure P-18 shows the final Prioritization Matrix using the Consensus Criteria method. For this scenario, Company B appears to be the best

A B C D E F G H I J K L M N O P Q R S T U V W X Y Z

option of the three, with a total score of 2.6; however, that score is close to that of Company C. Thus, further exploration should be conducted to ensure which option appears to be the clear winner.

Consensus Criteria Prioritization Matrix	Criteria			Row Total
	Customer base (CB) (0.7)	Increased Capacity (IC) (0.2)	Technological Advantage (TA) (0.1)	
Company A	1 x CB = 0.7	1 x IC = 0.2	2 x TA = 0.2	1.1
Company B	3 x CB = 2.1	2 x IC = 0.4	1 x TA = 0.1	2.6
Company C	2 x CB = 1.4	3 x IC = 0.6	3 x TA = 0.3	2.3

FIGURE P-18: Sample Prioritization Matrix Using the Consensus Criteria Method

Combination I.D./Matrix Method

The Prioritization Matrix developed from using the Combination I.D./Matrix method involves the team of key process players to perform the following tasks:

Step 1. Agree on the problem statement.

Step 2. Generate list of root cause options.

a. Gather information and data about the issues its cause-and-effect relationships. Potential input sources include Interrelationship diagram, Fishbone (or Cause-and-Effect diagram), Brainstorming using the 5Ms and P technique and/or 5-Whys, graphical methods, Control charts, and a Process map. (**See Also** "Interrelationship Diagram—7M Tool," **p. 305**; "Cause-and-Effect Diagram—7QC Tool," **p. 173**; "Brainstorming Technique," **p. 168**; "5-Whys," **p. 305**, "Graphical Methods," **p. 323**, "Control Charts—7QC Tool," **p. 217**; "Process Map (or Flowchart)—7QC Tool," **p. 522**)

b. Summarize and document the cause-and-effect relationships in a Tree diagram.

c. Generate a list of the root causes options from the most detailed level of the Tree diagram.

Step 3. Create the Combination ID/Matrix to compare each root cause option against one another, using the cause-and-effect relationship as the evaluation standard.

a. Using an L-shaped matrix structure, in the first row, second column, record each of the options from left to right, placing one per cell, to create the column headings.

b. In the second row of the first column, again record the **options** going down the column, in the same sequence used for the column headings, by placing one topic per cell, to create the row heading.

c. Starting in the upper-left corner of the matrix, shade each cell along the diagonal to the lower-right corner because comparing the same option to itself is unnecessary.

d. Add four additional columns on the far right and name the columns in **sequence from left to right** as follows:

   i. **Total In**—in the column fourth from the right

   ii. **Total Out**—in the column third from the right

   iii. **Total In and Out**—in the column second from the right

   iv. **Strength**—in the far right column

Step 4. Conduct vertical versus horizontal pair-wise comparisons on the root cause options to prioritize and select which to address first.

a. Start with the first option listed on the vertical axis (the Y-axis) in row 3, compare it to the second option listed on the horizontal axis (the X-axis), and find where they intersect in a cell—in the third row, second column.

b. Evaluate the strength of the cause-and-effect relationship between the two options by asking, *"Does the row option effect or influence the column option? If so, what is the strength of the relationship—strong, moderate, or weak-to-none?"*

   Select one of the top three geometric symbols (solid circle, box, or triangle) in following standard rating scale found in Table P-3 to indicate the response. **If no relationship exists, leave the cell blank.** Record the geometric symbol in the appropriate cell of the Combination ID/Matrix.

   Continue this task until all the paired option cells have been filled with the appropriate rating symbols.

TABLE P-3: Cause-and-Effect Rating Scale of Symbols for a Prioritization Matrix Using the Combination ID/Matrix Method

Relationship Strength	●	STRONG cause-effect relationship (9 points)
	□	MODERATE cause-effect relationship (3 points)
	△	WEAK cause-effect relationship (1 point)
Cause Source	⬆	Up (OUT) Arrow indicates Driving Cause of the other option
	⬅	Side (IN) Arrow indicates Effect

Note

Recall: Blank cells indicate **no** relationship. Continue this task until all the paired option cells have been filled iwith the appropriate rating symbols.

c. Indicate which option in a pair-wise comparison drives the cause, by using the arrow symbols (up or side) from the standard rating scale found in Table P-3 to indicate the response. Record the arrow symbol next to the geometric symbol in the appropriate cell of the Combination ID/Matrix.

Step 5. Complete the Combination ID/Matrix as the final Prioritization Matrix.

a. Identify the root cause option associated as driving the most causal relationships—the *biggest Cause Source.*

i. By row, count the number of side (IN) arrows pointing left in that row, and record the total in that row's cell within the ***Total In*** column.

ii. By row, count the number of up (OUT) arrows in that row and record the total in that row's cell within the Total Out column.

iii. By row, add the "Total In" and "Total Out" numbers together and record the sum in the row's cell within the Total In and Out column.

iv. Continue until each row is completed.

b. Identify the root cause option associated with the strongest cause-and-effect relationship.

i. Refer to the point values associated with each cause-and-effect relationship strength rating found in Table P-3.

ii. By row, add up the point values associated with each geometric strength relationship symbol found in that row and record the total in that row's cell within the Strength column.

c. The **top root cause option candidates** are identified as those with the highest strength score and highest number of total in and out arrows, as show in example Prioritization Matrix of Figure P-19.

i. *Highest Strength* indicates a strong relationship with the other options.

ii. *Highest Total In and Out* indicates many connections, with the *Highest Out* characterizing a strong root cause relationship and should be addressed first. The *Highest In* indicates that these options may serve as good leading indicators.

If there are two or more of the top prioritized root cause options that have ratings relatively close together, conduct further exploration.

Figure P-19 shows the final Prioritization Matrix using the Combination ID/Matrix method. For this scenario, an organization wanted to improve employee morale but did not know where to start. The resulting Prioritization Matrix using this combined approach indicates that apparent lack of interest in exercising empowerment and workload issues probably are the two best places to start to address the employee morale problem. Perhaps delving deeper into both of the top candidate topics may uncover a more clear starting point. Further investigation techniques may include using the 5-Whys or interviewing the employees and conducting a KJ analysis on the results. (**See Also** "5-Whys," **p. 305** and "KJ Analysis," **p. 375**)

Combination ID/ Matrix method	Root Cause Options						
	High absenteeism	Unexercised empowerment	Over-worked	Total IN	Total OUT	Total IN & OUT	Strength
High absenteeism		⬆ □	□ ⬅	1	1	2	6
Unexercised empowerment	⬆ □		● ⬅	1	1	2	12
Overworked from down-sizing	⬆ ●	⬆ □		0	2	2	12

FIGURE P-19: Sample Prioritization Matrix Using the Combination ID/Matrix Method

Process Capability Analysis

What Question(s) Does the Tool or Technique Answer?

Is the process able to meet customer requirements?

Process Capability analysis helps you to

- Understand if an existing, steady state, normal process, is **capable** of producing output within specified customer specification limits
- Understand whether the actual process results are acceptable with respect to the customer specifications
- Evaluate the performance of a process
- Compare the performances of two different processes; to select one

Alternative Names and Variations

This tool is also known as

- Capability Analysis or Capability Study
- Process Capabilility Study

When Best to Use the Tool or Technique

Process Capability analysis can be preformed only on processes known to be in "statistical control," meaning that the exhibited process variation is random and steady over time. The term *random* addresses that something occurs as a common event nature, and not a purposeful, special cause event. The term *steady over time* refers to the ability to predict the event within the process' inherent fluctuation. The process data also must be normally distributed. If a process proves incapable of producing customer requirements, then improvements may be in order, if the business can afford the investment and time. Consider the alternatives—redesigning the current process or stopping to use the process all together and designing a new "clean sheet" process. If the process is capable, however, the number of defects or amount of variation miss the target, then process improvements may be in order and probably requires less investment than a process that is incapable. (**See Also** "Control Charts—7QC Tool," **p. 217**)

Brief Description

The concept of process capability describes a predictable pattern of statistically stable behavior wherein the process variation occurring by chance alone is compared to a set of specifications. Ideally, those specifications (upper and lower specification limits) are set by the customer, but often times they are set by the technical engineering design. A capable process is one whose distribution is narrow enough to fit within the specification range.

Process Capability analysis judges whether the process variation fits within the upper and lower specification limits. If the analysis determines the process incapable, the process requires improvements to reduce the variation (as in a Lean Six Sigma project)—**variation is the enemy**. If after improvement initiatives, the process remains incapable, consider either a more aggressive process redesign or clean-sheet design approach (as in DFSS). As a last resort, consider examining the process specifications settings. Perhaps they are out of alignment with both the customer and process requirements, therefore requiring an adjustment.

There are various statistical metrics used to describe process capability, including "Number of Defects, Process Capability Indexes (CPIs), and Yield.

Number of Defects

- **Defects per Unit (DPU)**, which is the total number of defects divided by total number of units.

  Average number of defects found in a sample.

> **Note**
>
> *Defects* are defined as a number of things wrong within a product, though, the product is still considered good or acceptable.

  Defective is defined as a bad part or product, one that is **not** fit for use and must be either scrapped or reworked.

- There are two ways to make a "Bad Part":
  - Process has shifted off target. The center of the process can be measured three ways: mean, median, and mode.
  - Process variation is too wide and extends beyond the customer specification limits. The process spread is measured by the Range, Variance, and Standard Deviation.

- **Defects Per Million Opportunity (DPMO):**
  - A common simple process capability metric.
  - The calculation can be used with either attribute or continuous data and is described as the "Average number of defects per unit divided a million opportunities within which to make a defect." DPMO = DPU x 1,000,000.
  - It can also be expressed as DPMO = [(# of Defects) divided by [(# of Units) times (# of Opportunities)] times 1,000,000].

Table P-4 summarizes the relationship among DPMO, Yield, and the Sigma-level.

TABLE P-4: Translating DPMO into Yield and Sigma Level

DPMO	%Yield	Sigma Level
308,538	69.1	2
66,807	93.32	3
6,210	99.379	4
233	99.9767	5
3.4	99.99966	6

Process Capability Indexes (CPIs)

Process Capability Indexes (CPIs) statistically quantify the ability of a process to produce output that meets customer requirements by calculating variability differently, depending on the duration of the data (short- or long-term).

While there are several CPIs available, the most commonly used indexes are Cp and Cpk. Both of these indexes view process capability through a "short-term" lens. Short-term analysis requires less data to gather by estimating the standard deviation and mean and predicting it onto the population. Over the long-term, a process becomes "less capable" because of changes in its mean over time. Motorola discovered and established that with time, a process tends to drift about 1.5 sigma, as shown in Figure P-20. However, some current industry experts argue against the 1.5 sigma shift as a cover-up and have written several articles on the topic.

The other indexes are defined as follows, but the focus of this section is on the calculation and interpretation of Cp and Cpk. The condition where both Cp $\geq$ 2.0 and Cpk $\geq$ 1.5 exist defines a Six Sigma process.

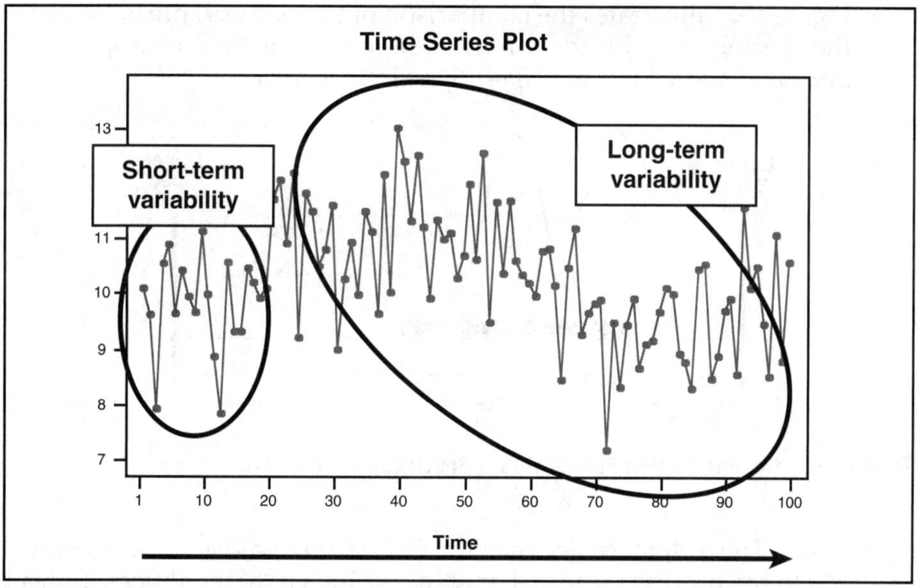

FIGURE P-20: Short-term Versus Long-term Variability

- **Cp:** *Specification Spread (USL—LSL) divided by Process Spread (6s)*; or [(USL—LSL)/(6s)]; known as "short-term" process capability, based on the dispersion of the sample data.
    - Examines whether the process spread fits within the customer specification limits. It examines the process **variation** to the specification limits. This index does not account for increased defect rates due to shifts in the mean.
- USL = Upper Specification Limit.
- LSL = Lower Specification Limit.
    - **s** = Standard Deviation for a *sample*, or an estimated standard deviation.
    - x-bar = the sample mean (or average); often denoted as an "X" with a line or a bar drawn above it.
    - Larger the Cp the better.

> **Rule of Thumb**
>
> Cp ≥ 1.5 to be capable, ideally. Some experts accept with a Cp ≥ 1.0; however, that is the bare minimum.

- Figure P-21 illustrates the comparison of the process spread versus the specification spread, wherein LSL represents the *Lower Specification Limit* and USL is the *Upper Specification Limit*.

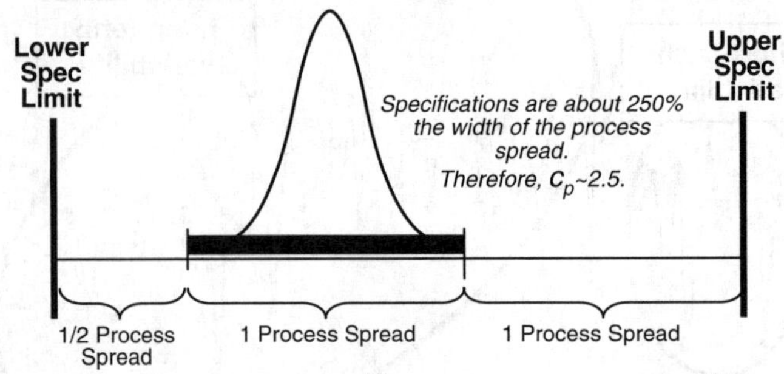

FIGURE P-21: Generic Process Spread Versus Specification Spread (Cp)

- **Cpk**—The difference between the *Closest Specification Limit* (closest SL) to the process mean (x-bar) divided by half of the Process Spread (3s); or [(USL–x-bar)/(3s)] OR [(x-bar–LSL)/(3s)]; whichever is smaller, using the sample mean (centeredness).
    - The most popular calculation of Process Capability because it considers both *variability* and *mean* of process output relative to customer requirements.
    - Examines the proximity of the current process mean to the closest customer specification limit (either the Upper or Lower, whichever is closer). This also is referred to as the *process centeredness*.
    - Therefore, Cpk = 1.0 when the mean of the current process is exactly ½ of the curve away from the nearest specification limit. **One Cpk measures ½ of the process curve.**
        - Cp is bigger than Cpk, given that Cp accounts for the full process spread, and Cpk accounts for half of the spread; Cp>Cpk.
        - Cp can be considered the "upper bounds" of Cpk when the process mean is on target. The difference between Cp and Cpk represents the potential gain in process capability if the process were to be centered.
        - Cpk is largest when the mean is located at the midpoint of the specification spread.

> **Caution**
>
> Always use the smaller Cpk number to reflect the closest specification limit distance to the process mean so as not to overstate the process capability.

> **Rule of Thumb**
>
> Cpk ≥ 1.33, where 1.0 is barely capable, and 1.33 or larger is preferred. For a Six Sigma (6σ) process, Cpk = 2.0.

- Figure P-22 illustrates Cpk relative to the Lower Specification Limit (LSL); representing the following formula [(x-bar–LSL)/(3s)].

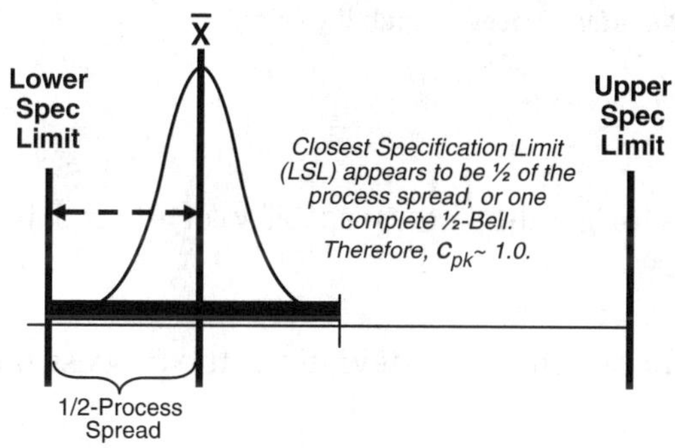

FIGURE P-22: Generic Process Capability (Cpk)

- If the furthest specification limit is used, the process capability is overstated, as shown in Figure P-23.

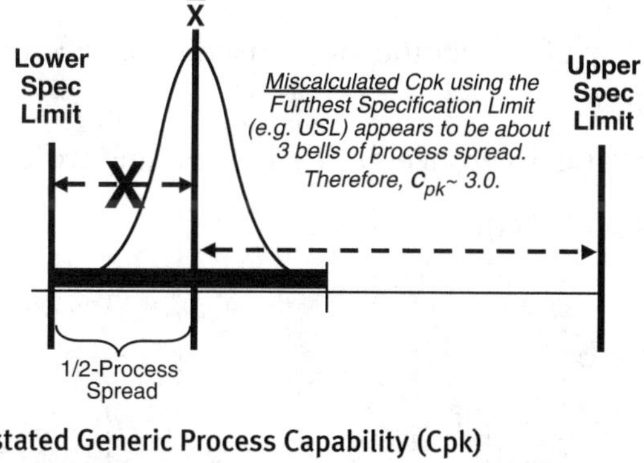

FIGURE P-23: Overstated Generic Process Capability (Cpk)

A
B
C
D
E
F
G
H
I
J
K
L
M
N
O
P
Q
R
S
T
U
V
W
X
Y
Z

Hint

If Cpk is negative, the process mean is outside the closest specification limit as shown in Figure P-24.

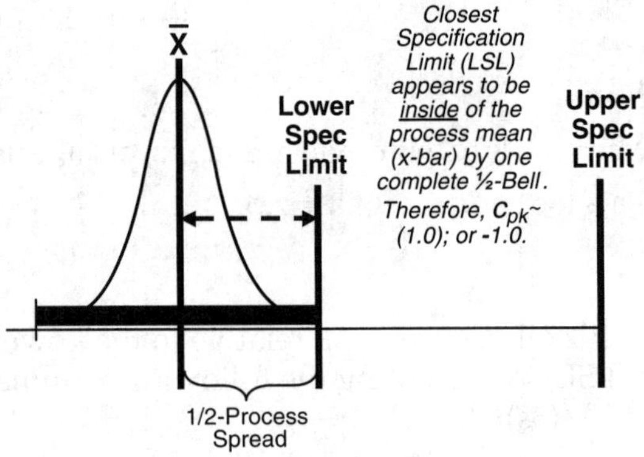

FIGURE P-24: Generic Negative Process Capability (-Cpk)

Hint

Count "½-bells" or ½ of the process capability curve to equal one Cpk unit.

- **Cpm**—Number of standard deviations the process mean is from a target value.
    - Examines the process centeredness relative to a target.
    - If the process were perfectly centered on the target, and if the process were capable, the Cp would equal the Cpm. That is, Cp = Cpm when the process is perfectly centered on its target.
    - Larger Cpm is better.
- **Cm**—Capability of a machine; often used by DFSS (Design for Six Sigma).
    - Engineering Tolerance divided by 8 standard deviations.
    - [(USL–LSL)/(8s)]

Rule of Thumb

Cm $\geq$ 1.33 to be acceptable.

- **Capability Ratio (CR):** ***Process Spread** (6s) **divided by Specification Spread** (USL–LSL);* or [(6s)/(USL–LSL)].
    - Sometimes referred to as the *natural tolerance.*
    - CR is the numerical inverse of Cp.
    - Often expressed as a percentage (%), thereby multiplying the ratio by 100.
    - Smaller the CR the better.

Rule of Thumb

CR $\leq$ 0.5 to be "good" (or 50%).

A B C D E F G H I J K L M N O P Q R S T U V W X Y Z

- **Pp:** ***Specification Spread** (USL—LSL) **divided by Process Spread** (6 σ);* or [(USL—LSL)/(6 σ)]; known as "long-term" process capability based on population data.
    - Pp and Ppk are referred to as *performance metrics* to distinguish from the short-term capability metrics (that is, Cp and Cpk).
    - Defines the long-term process capability, which reflects more inherent variation over time, relative to customer requirements. This index does not account for increased defect rates due to shifts in the mean.
    - Similar to Cp, but using the population data, rather than estimating with sample data.
    - USL = Upper Specification Limit.
    - LSL = Lower Specification Limit.
    - σ = Standard Deviation for a *population.*
    - The larger the Pp the better.
- **Ppk**—The ***difference between the Closest Specification Limit** (closest SL) **to the population mean** μ) **divided by half of the Process Spread** (3 σ);* or [(USL–μ)/(3 σ)] OR [(μ–LSL)/(3 σ)]; whichever is smaller, using the population mean (centeredness).
    - Pp and Ppk are referred to as a *performance metrics* to distinguish from the short-term capability metrics (that is, Cp and Cpk).
    - Defines the long-term process capability, which reflects more inherent variation over time and considers **both** variability and mean of process output relative to customer requirements.

- Similar to Cpk, but using the population data, rather than estimating with sample data.
    - Pp is bigger than Ppk because Pp accounts for the full process spread of the population, and Ppk accounts for half of the spread; Pp>Ppk. Pp can be considered the "upper bounds" of Ppk when the process mean is on target.

- **Z_L**—Upper Specification Limit (USL) minus the mean (x-bar), divided by the standard deviation (s).
    - Measures the percentage area either inside or out of the requirement process by examining the process's location (centeredness of the mean) relative to both the standard deviation and the lower specification limit.
    - Uses the Z statistic and references a Standard Normal Table to determine the area under the process curve.
    - Z_L = [(x-bar–LSL) divided by 1s]
    - The bigger Z_L the better.

Rule of Thumb

$Z_L \geq +3$ to be acceptable, and produce less than 0.01% defects. For a Six Sigma (6 σ) process, $Z_L = +6$.

- **Z_U***—Lower Specification Limit* (LSL) ***minus the mean*** *(x-bar),* ***divided by the standard deviation*** *(s).*
    - Measures the percentage area either inside or out of the requirement process by examining the process's location (centeredness of the mean) relative to both the standard deviation and the upper specification limit.
    - Uses the Z statistic and references a Standard Normal Table to determine the area under the process curve.
    - Z_U = [(USL–x-bar) divided by 1s]
    - Similar to Z_L, the bigger Z_U the better.

> **Rule of Thumb**
>
> $Z_U \geq +3$ to be acceptable and produce less than 0.01% defects. For a Six Sigma (6 σ) process, $Z_U = +6$.

- Z_{MIN}—Simply the *smaller of the two Z_L and Z_U calculations*, determining which is closest to the nearest specification limit to accurately state the process capability. Otherwise, the process capability would be overstated.
  - Used to help compute Cpk, since Cpk = Z_{MIN} divided by 3.
  - Similar to Z_L and Z_U, the bigger Z_{MIN} the better.

> **Rule of Thumb**
>
> $Z_{MIN} \geq +3$ to be acceptable, and produce less than 0.01% defects. For a Six Sigma (6 σ) process, $Z_{MIN} = +6$.

Yield

- **First Pass Yield (FPY)** —Simply divides the number of good units produced by the number of units entering the process.
  - Also known as *First Time Yield.*
  - FPY ignores, or does not account for, the "hidden factory" which consumes resources to do "rework" and scrap products with defects, since the calculation occurs *after* any "inspection" is conducted to determine if a unit is good or not.
  - FPY = [Total Good units produced/Number of Units Entering] $= (1-p)^n = q^n$, where p = the probability of the number of defects, and q = probability that the defect will not occur.
- **Rolled Throughput Yield (RTY)**—Measures the total DPMO for an entire process or product by multiplying the DPMO of each process step.
  - Represents the probability that a single unit can pass through the process free of any defects, thereby decrementing the capability calculation for any rework (and scrap) involved in making a unit good. Hence, RTY is the preferred yield calculation over FPY.

- RTY = [(1-(DPMO$_1$/1,000,000)) x (1-(DPMO$_2$/1,000,000)) x … (1-(DPMO$_n$/1,000,000))], for "n" number of process steps.

- For example, if the process had four steps, and each step produced a 90% yield, the Rolled Throughput Yield would equal:

  RTY = 90% x 90% x 90% x 90% = 66% (not 90%).

  The math calculations breakdown to:

  0.9 x 0.9 = 0.81 (for steps 1 and 2); then 0.81 x 0.9 = 0.73 for steps 1, 2 and 3; then 0.73 x 0.9 = 0.66 for all 4 steps, as shown in Figure P-25.

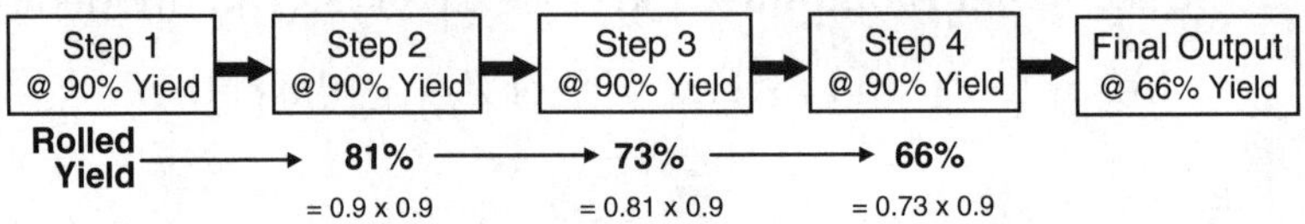

Figure P-25: Rolled Throughput Yield Example

- The RTY is smaller than the lowest yield of any single process step. The RTY becomes exponentially smaller as the number of steps in the process grows.

How to Use the Tool or Technique

Prerequisites to Examining Process Capability

Prior to conducting a Process Capability study, the process in question must pass two tests.

The first test—is *process stability*. This means that the process results remain in control, showing random, and steady state pattern over time. Process stability allows relatively accurate predictions about future yields or outputs, as well as the defect rate. Without process stability, the capability measures would be meaningless.

Process stability can be determined by examining the process variation and whether it is randomly distributed around the mean and within the control limits. Typically three standard deviations on either side of the process mean define the control limits. A steady-state process is **absent of any patterns** such as

- A trend (or drift) up or down
- An oscillating pattern around the mean
- A shift (or jump) in the data over time

Any process spread is explained by common cause variation, rather than special cause variation. *Common cause* variation is the inherent process fluctuations present in all processes. An in-control process with only common cause variation is stable and predictable. In contrast, special cause variation causes the process to be unstable and unpredictable. A special cause occurrence is **not** inherent to, or beyond the natural variation of, the process (or system). Special cause variation is "assignable" to an unusual event, whereby once identified, it should be eliminated. This removal of special cause events from the data should occur prior to conducting a Process Capability analysis.

> **Note**
>
> If common cause is treated as special cause, resulting in process adjustments (or tampering), then more variation is introduced into the process. If special cause is treated as common cause, then the special event is neglected, and the delayed reaction may increase the defect rate.

Tools and techniques determine if a process is in control, include a suite of Statistical Process Control (SPC) charts, or a Time Series plot or Run chart, and show the process data over time. Figure P-26 illustrates a process in control (delineated by a type of Statistical Process Control chart known as an x-bar-R chart). In contrast, Figure P-27 depicts four examples of processes out-of-control. (**See Also** "Control Charts—7QC Tool," **p. 217** for more detail.)

Normal Probability Plot
The second test is the *process normally distributed*. When graphically represented, the frequency distribution follows a bell-shaped curve (that is, symmetrical around the process mean).

> **Note**
>
> Normal distribution is a bell-shaped symmetrical distribution created by data said to be normally distributed.

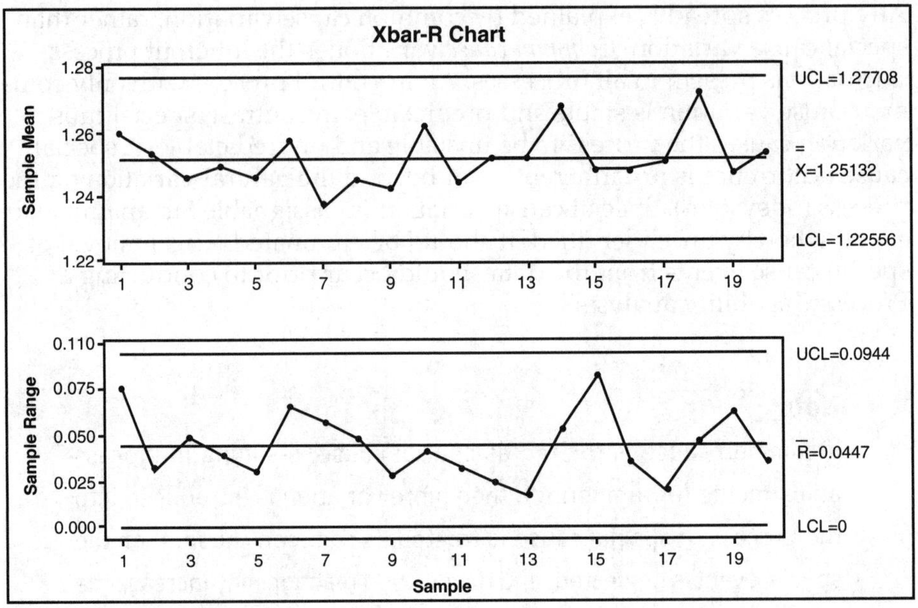

FIGURE P-26: Generic Process in Control

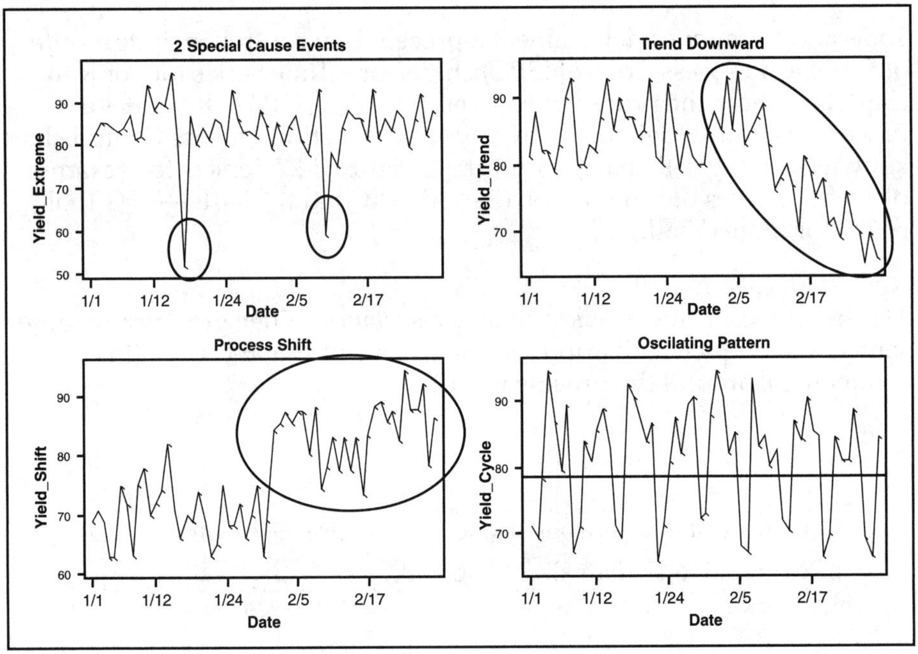

FIGURE P-27: Examples Failing Steady-State, In-Control Process Test

This is commonly displayed as a histogram, as shown in Figure P-28. Another means to test if the data are normal involves using the *normal probability plot* that examines the data around confidence intervals (usually 95%), as displayed in Figure P-29. If the data are normal, 95% of the data points will fall within the confidence interval. Notice how tight the data points are around the mean and within the confidence intervals displayed in Figure P-29. Moreover, the tool provides a statistic that represents the probability that the data falls within that confidence band. Normal data will yield a probability statistic (a.k.a. the *P-value*) greater than 0.05, or P-value > 0.05, again indicating that 95% confidence that the data could be well represented using a normal distribution.

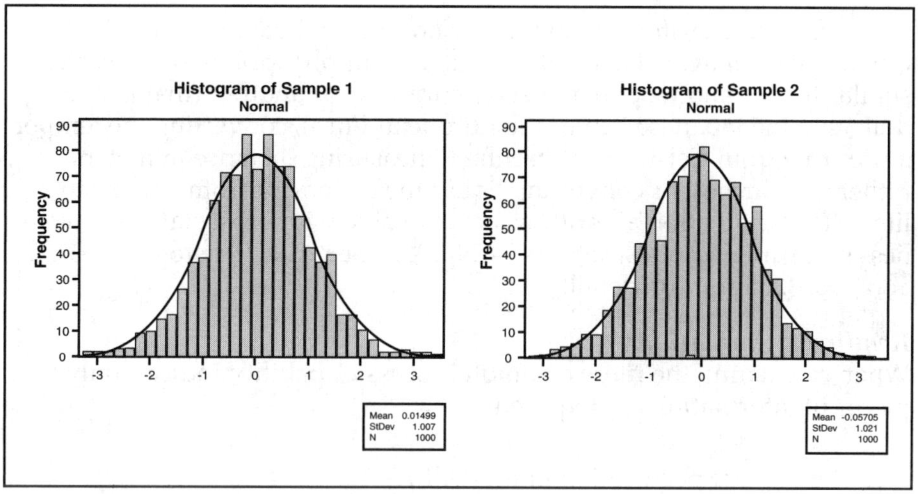

FIGURE P-28: Examples of Two Normal Histograms

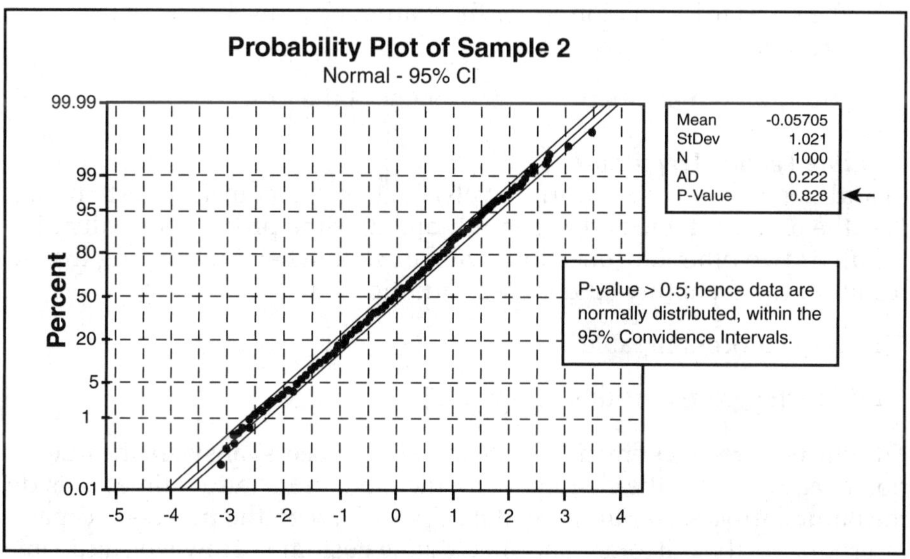

FIGURE P-29: Example of Normal Probability Plot

Transforming Non-Normal Data

If the data are non-normal, then the estimated defect rates and yields will be incorrect. (Note: *non-normal data*, being the opposite of normally distributed data, is defined as being distributed in any shaped-curve other than a symmetrical bell-shaped curve.) The calculated defect rates would be either too high or too low depending on the type of non-normality. Non-normal data can be remedied by transforming it. Using the principle of multiplying by a constant (that is, lambda), data transformation enables tools such as Capability Studies that require normal data to be used. These tools apply a non-linear, shape-changing model.

A common transformation tool for non-normal data is called the *Box-Cox Transformation*. Such a data conversion simply applies a new scale, similar to exchanging one monetary currency for another (that is, U.S. dollars to the Japanese Yen or the European Euro), converting the temperature scale from Celsius to Fahrenheit, measuring distance in meters rather than inches, or converting liquid measurement from gallons to liters. Transformation algorithms are available with most statistical business software packages, such as MINITAB. (**See Also** "Histogram—7QC Tool," **p. 330** for more detail.)

Required Information

When calculating the more common Process Capability Indices, only four pieces of information are required:

1. the Upper Specification Limit (USL)
2. the Lower Specification Limit (LSL)
3. the standard deviation (s); or the "fatness" or dispersion of the process
4. the process mean (x-bar); or the center of the process

Capable Versus Acceptable

In addition to answering affirmatively, the two prerequisite questions are 1) Is the process stable and in control? and 2) Is the process normally distributed? two other key questions will be asked when interpreting the results of the capability study. These questions are

1. Is the process capable?
2. Are the process results acceptable?

The current process spread (of a steady-state process) answers the question of capability. If the customer specification limits fall 1.5 times outside the natural process variation (that is, $Cp > 1.5$), then the process is capable of producing the outcomes specified by the customer. However, a capable process's results may fall shy of actually producing the desired deliverables and hence may not be acceptable.

Producing "acceptable" results requires the process mean to be centered on or extremely close to the target value. This question focuses on the centeredness of the process—how "accurate" is the process in meeting customer requirements?

Adjustments

When using Process Capability Indices to determine what action to take to improve the process, know that it is easier to center the process (move Cpk onto the target), rather than reduce the process spread (a.k.a. dispersion, width, or variation). (Note: *Process capability* means that the process spread is capable of fitting within the customer specification limits; it does not necessarily need to be doing so. For example, if you were about to purchase a new Hummer vehicle and you live in a home built before the 1950s, you might question whether the car could fit within the width of your garage door (that is how to consider process capability).

When studying process capability, what could be impacting the variation (or standard deviation)? By understanding the cause behind the variation, one can begin to develop and focus the improvement strategies. The variance (or spread) of a steady state, in control process is explained by the common cause. Commonly used classifications of potential root causes include: Machine, Materials, Methods, Metrics, Mother Nature (or Environment), and People—also often referred to as *5Ms and P*. (**See Also** "Cause-and-Effect Diagram," in the "Hints and Tips" section, **p. 182**, for more detail on 5Ms and P.)

Manual Calculations of Process Capability (Cp and Cpk)

Example 1

The following sample data is provided:

- The process is stable, in control, and normally distributed.
- Mean (x-bar) = 40
- Standard Deviation (s) = 2.6
- Lower Specification Limit (LSL) = 25
- Upper Specification Limit (USL) = 65

The manual calculations to determine Cp and Cpk are as follows:

Step 1. Draw the bell-shaped curve and place the given values on the curve, as illustrated on Figure P-30, to serve as a double-check and minimize any inadvertent human errors in the manual calculations.

a. Presume the curve's spread is normal (that is, 3 standard deviations on either side of the mean, or 3 x 2.6 = 7.8). Therefore, the upper tail is at 40 (mean) + 7.8 (1/2-spread) = 47.8; and the lower tail is at 40 (mean)–7.8 (1/2-spread) = 32.2.

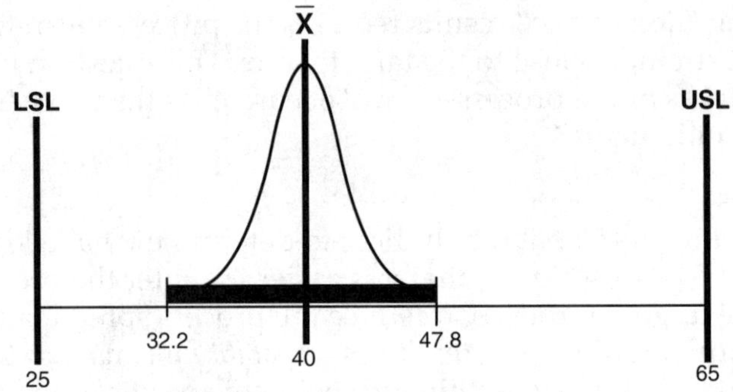

FIGURE P-30: Example 1—Process Capability Curve

Step 2. Ask the key capability questions before calculating the actual indexes and develop a "ballpark" estimate as a double-check in anticipating the results.

a. Estimate whether the process appears capable. Does the process spread appear to be more narrow than the specification spread?

ANSWER: "Yes, the process spread looks to fit inside the specification spread."

b. Estimate the Cpk. How big does it look? Which specification limit is the mean closest to, and how many ½-bell shapes are between it and the mean?

ANSWER: "Closest spec. limit is the LSL, and it looks to be about two ½-bells away (or 2 Cpk)."

Step 3. What is the Cp? Calculate **Cp** using the following formula: **[(USL-LSL)/6s]**.

a. **ANSWER**: Cp = [(65-25)/(6 x 2.6)] = (40/15.6) = 2.56; Cp @ 2.56 $\geq$ the target Cp of 1.5. Therefore, the specification spread is 256% (or about 2.5 times) wider than the process spread. "Yes, this is a very capable process."

Rule of Thumb

Cp should be $\geq$ 1.5.

Step 4. How do you calculate **Cpk**? First, recall that the process mean is closest to the Lower Specification Limit (LSL). Thus, the formula to use is **[(x-bar–LSL)/3s]**.

ANSWER: Cpk = [(40-25)/(3 x 2.6)] = (15/7.8) = 1.92. Cp @ 1.92 ≥ the target Cpk of 1.33. Therefore, the process is "acceptable." (This matches the "ballpark" estimate of about 2 determined in Step 2.)

Rule of Thumb

Cpk should be ≥1.33.

Example 2
The following sample data is provided:

- The process is stable, in control, and normally distributed.
- Mean (x-bar) = 40
- Standard Deviation (s) = 2.6
- Lower Specification Limit (LSL) = 35
- Upper Specification Limit (USL) = 65

The manual calculations to determine Cp and Cpk are as follows:

Step 1. Draw the bell-shaped curve and place the given values on the curve, as illustrated on Figure P-31, to serve as a double-check and minimize any inadvertent human errors in the manual calculations.

a. Presume the curve's spread is normal (that is, 3 standard deviations on either side of the mean, or 3 x 2.6 = 7.8). Therefore, the upper tail is at 40 (mean) + 7.8 (1/2-spread) = 47.8; and the lower tail is at 40 (mean)–7.8 (1/2-spread) = 32.2.

Step 2. Ask the key capability questions before calculating the actual indexes and develop a "ballpark" estimate as a double-check in anticipating the results.

a. **Estimate whether the process appears capable?** Does the process spread appear to be more narrow than the specification spread?

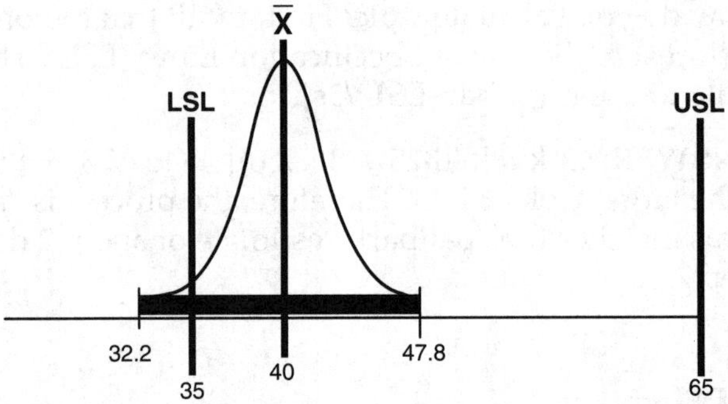

FIGURE P-31: Example 2—Process Capability Curve

ANSWER: Yes; the process spread looks to be narrow enough to fit within the specification spread.

b. **Estimate the Cpk; how big does it look?** Which specification limit is the mean closest to, and how many ½-bell shapes are between it and the mean?

ANSWER: Closest spec. limit is the LSL, and it looks to be slightly less than one ½-bells away (about one-third less, or Cpk~0.6).

Step 3. How do you calculate **Cp** using the following formula: **[(USL-LSL)/6s]**?

ANSWER: Cp = [(65-35)/(6 x 2.6)] = (30/15.6) = 1.92. Cp @ 1.92 ≥ the Cp *Rule of Thumb* of 1.5. Therefore, the specification spread is 192% wider than the process spread. "Yes, this process is capable."

Step 4. How do you calculate **Cpk**? First, recall that the process mean is closest to the Lower Specification Limit (LSL). Thus, the formula to use is **[(x-bar—LSL)/3s]**.

ANSWER: Cpk = [(40-35)/(3 x 2.6)] = (5/7.8) = 0.64. Cp @ 0.64 < the Cpk *Rule of Thumb* of at least 1.0. Therefore, the process is "unacceptable." (This matches the "ballpark" estimate of less than a 1/2-bell determined in Step 2.)

Recall that **the difference between Cp and Cpk represents the potential gain in process capability if the process were to be centered**. In this case, the difference between Cp and Cpk (1.92–0.64) is 1.28. Thus, exploring possible adjustments to improve the centeredness of the process probably would be worthwhile.

A B C D E F G H I J K L M N O P Q R S T U V W X Y Z

Automated Calculations of Process Capability (Cp and Cpk) Using MINITAB

Example 3

The following sample data is provided:

- The data set found in Table P-5, (where n = 50 and no sub-groups):

TABLE P-5: Example 3 Data Set

26.5	25.9	23.9	22.1	23.8	23.6	30.1	20.7	25.3	27.6
25.2	25.4	23.3	22.4	26.6	31.3	26.3	24.4	26.2	22.6
30.2	22.9	21.6	31.2	25.6	23.2	26.6	23.1	26.5	22.3
25.2	24.1	25.4	26.5	29.3	26.5	25.4	31.5	28.7	28.2
26.5	21.2	22.1	22.7	24.9	28.8	19.3	23.0	22.8	28.6

- Upper Specification Limit = 50
- Lower Specification Limit = 20
- The process is stable as illustrated in Figure P-32. In MINITAB, select ***Graph > Time Series Plot > Simple***. (**See Also** "Graphical Methods," **p. 323**)

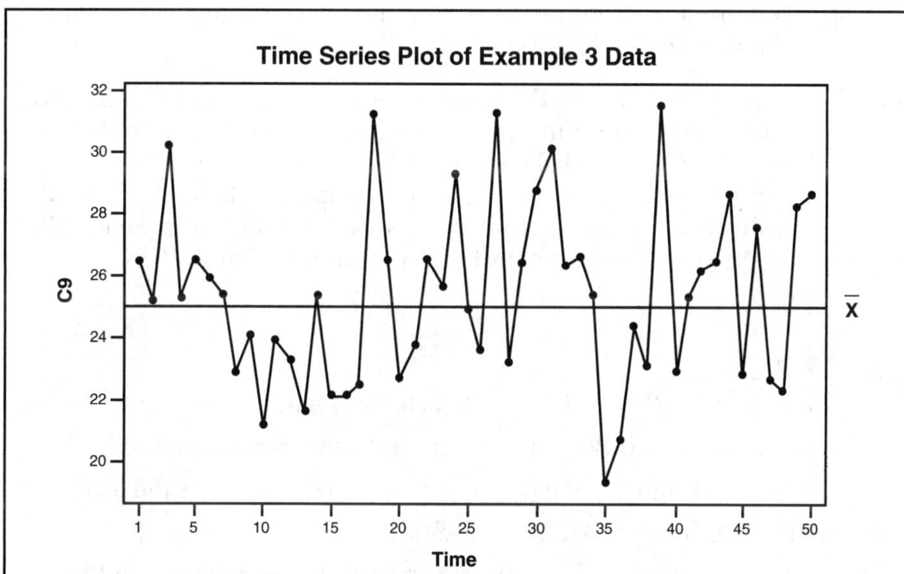

FIGURE P-32: Example 3—Time Series Plot, Using MINITAB

- The process is normally distributed, as shown in Figure P-33. [In MINITAB, for the histogram, select ***Graph > Histogram > With Fit;*** for the Normal Probability plot, select ***Graph > Probability Plot > Single.*** (**See Also** "Graphical Methods," **p. 323**)

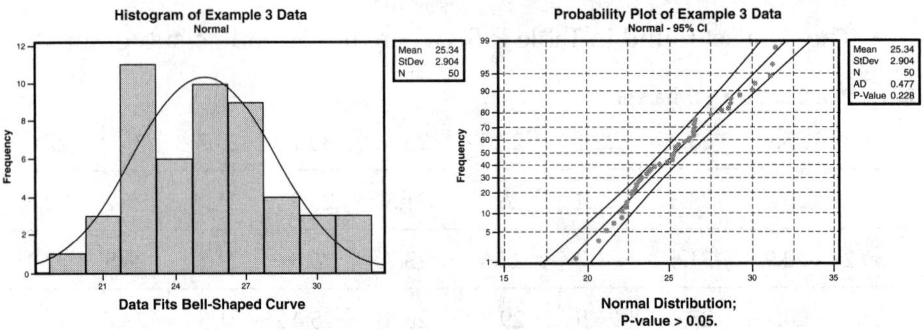

FIGURE P-33: Example 3—Histogram and Normal Probability Chart Using MINITAB

The procedure to calculate Cp and Cpk using MINITAB are as follows:

Step 1. Arrange the data in a MINITAB Worksheet, preferably in a single column. (Note that the data may be in a group of columns, with subgroups across rows; however, this book's examples follow the single column approach.)

Step 2. In MINITAB, select ***Stat > Quality Tools >Capability Analysis > Normal...*** from the toolbar at the top of the screen, and a "dialog box" will appear.

Step 3. Within the MINITAB Capability Analysis (Normal Distribution) dialog box, found in Figure P-34, select the appropriate column of data in the worksheet. Enter the subgroup size for the data set. (If no subgroups, enter 1 in the appropriate field.) Enter the values for the **upper and lower specification limits**. Select OK to generate the graph and output found in Figure P-35.

> **Note**
>
> Notice that the MINITAB dialog box contains a button in the upper-right corner, called Box-Cox, which transforms non-normal data if needed. In Example 3, the data are normal, hence by-pass the extra step of selecting the Box-Cox Transformation.
>
> Ask the key capability questions, and answer them based on the above results in the MINITAB output, found in Figure P-35.

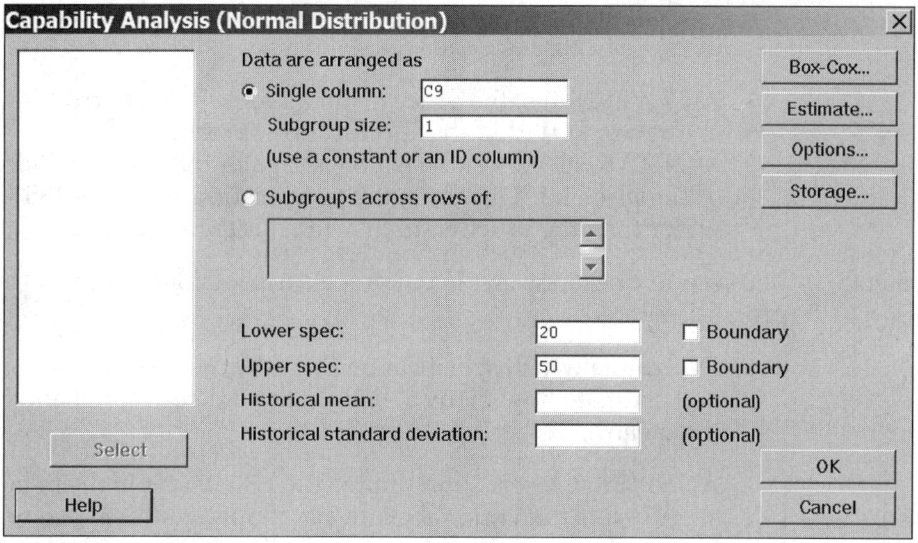

FIGURE P-34: MINITAB Dialog Box for Example 3

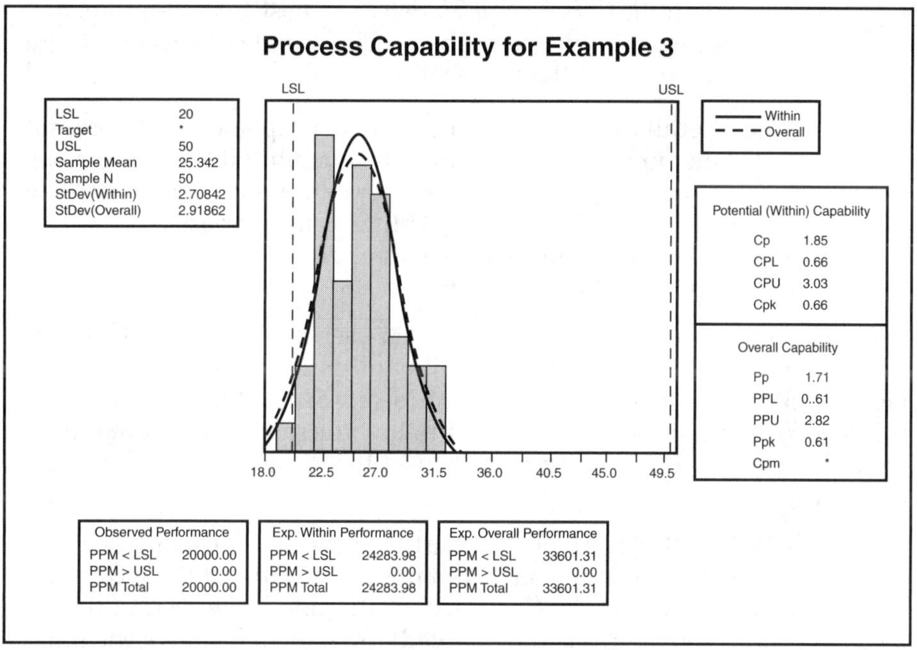

FIGURE P-35: Example 3 MINITAB Process Capability Output

Step 4. The first questions are

a. Does the process appear to be capable?

A B C D E F G H I J K L M N O P Q R S T U V W X Y Z

b. Does the process spread appear to be more narrow than the specification spread?

ANSWERS: Yes; the process spread looks to be narrow enough to fit within the specification spread. And MINITAB calculated **Cp** @ 1.85 is greater than the Cp *Rule of Thumb* of 1.5. Therefore, the specification spread is 185% wider than the process spread. *Yes, this process is capable.*

Step 5. The second questions are, Is the process acceptable? Is the process centered?

a. Determine which specification limit the mean is closest to, and estimate how many ½-bell shapes are between it and the mean?

ANSWER: Closest spec. limit is the LSL, and it looks to be slightly less than one ½-bell away (about one-third less, or Cpk~0.6).

b. **ANSWER**: MINITAB calculated **Cpk** @ 0.66, which is less than the Cpk *Rule of Thumb* of at least 1.0. Therefore, the process is "unacceptable." (This matches the ballpark estimate of less than a ½-bell.)

c. Recall that the difference between Cp and Cpk represents the potential gain in process capability if the process were to be centered. In this case, the difference between Cp and Cpk (1.85–0.66) is 1.19. Thus, exploring possible adjustments to improve the centeredness of the process probably would be worthwhile.

Another powerful display of the same data uses the MINITAB Six Pack option that provides six process capability output graphs, as illustrated in Figure P-36. From the drop-down menu, select ***Stat > Quality Tools >Capability Sixpack > Normal...*** and follow the same procedure outlined in Steps 3 through 5. Figure P-36 uses the same data as displayed in Figure P-35.

The Six Pack includes a Control chart [I Chart (Individuals Control chart), or an x-bar chart if the data contains subgroups] of the individual observations, a *Moving Range Chart* (MR chart), and a *run chart* of the last 25 observations (or subgroups) displayed down the left side of the output. The top-right corner *Capability Histogram* of the data is the same graph used in the *Capability Analysis* option (Figure P-35). Also on the right side, the output includes a *Normal Probability Plot* and a *Process Capability Plot* that provides the within and overall capability statistics Cp,

Cpk, Cpm (if a target is specified), Pp and Ppk. (**See Also** "Control Charts—7QC Tool," **p. 217** for a discussion on I charts, x-bar charts, MR charts and Normal Probability plots; "Run Chart—7QC Tool," **p. 610**; and "Histogram—7QC Tool," **p. 330**)

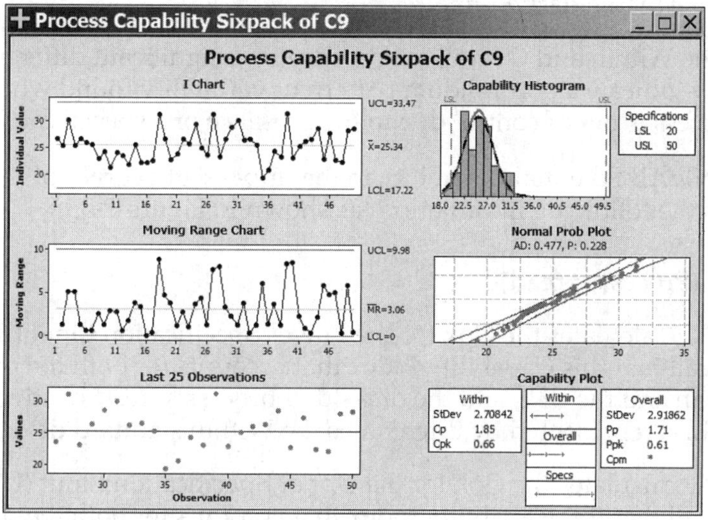

FIGURE P-36: Example 3 MINITAB Process Capability in a Six Pack

How to Analyze and Apply the Tool's Output

Additional Information from MINITAB Output

MINITAB provides output in addition to Cp and Cpk. Referring to Figure P-36, MINITAB publishes a Process Data table profiling the data set that is

- Lower Specification Limit (LSL).
- Target value (if known; if unknown MINITAB places an asterisks (*) in place of a value).
- Upper Specification Limit (USL).
- Sample mean (x-bar).
- Sample size (n).
- Standard deviation (StDev (Within))—looking at variation within the sample subgroups; also indicated on the graphical image by the

A B C D E F G H I J K L M N O P Q R S T U V W X Y Z

solid line. In case of Example 3, there were no sub-groups, but this estimates the "short-term" standard deviation (s).

- Standard deviation (StDev (Overall)) across all measures, estimating long-term or the population (σ); also indicated on the graphical image by the dashed-line.
- If the Within and Overall curves display a significant difference, it may indicate a non-random pattern of variability found when a process is out of control due to the presence of a special cause event.

The MINITAB Potential Capability in the upper-right corner of the output contains the following information (as shown in Figure P-35):

- Cp (Process spread)
- CPL (Calculates the Cpk for the Lower Specification Limit (LSL). Recall that this Capability Index that accounts for both the process mean and the LSL, and the one-sided process spread relative to the LSL, is represented by three times the Within standard deviation.
- CPU calculates the Cpk for the Upper Specification Limit (USL). Recall that this Capability Index that accounts for both the process mean and the USL, and the one-sided process spread relative to the USL, is represented by three times the Within standard deviation.
- Cpk (the smaller of the CPU and CPL).
- Pp (process spread for the overall capability).
- PPL calculates the Ppk for the Lower Specification Limit (LSL). Recall that this Capability Index accounts for both the process mean and the one-sided process spread relative to the LSL. This is represented by three times the Overall standard deviation.
- PPU calculates the Ppk for the Upper Specification Limit (USL). Recall that this Capability Index accounts for both the process mean and the one-sided process spread relative to the USL. This is represented by three times the Overall standard deviation.
- Ppk (the smaller of the PPU and PPL).
- Cpm indicates the number of standard deviations the process mean is from a target value. This number is provided only if a target is specified, otherwise an asterisk (*) will represent an empty value.

The MINITAB Process Performance results can be found in the lower bottom section of the output and provides the following information (as shown in Figure P-35):

- *Observed Performance*—Actual number of parts per million (PPM) (similar to DPMO) observed beyond the specification limits.
    - PPM < LSL—Number of PPM found less than the LSL.
    - PPM > USL—Number of PPM found above than the LSL.
    - PPM Total—The total number of PPM found outside of both specification limits.
- *Exp. Within Performance*—The *expected* values actual number of parts per million (PPM) observed beyond the specification limits and the *Within* standard deviation.
    - PPM < LSL—Number of PPM found less than the LSL.
    - PPM > USL—Number of PPM found above than the LSL.
    - PPM Total—The total number of PPM found outside of both specification limits.
- *Exp. Overall Performance*—The *expected* values actual number of parts per million (PPM) observed beyond the specification limits and the Overall standard deviation.
    - PPM < LSL—Number of PPM found less than the LSL.
    - PPM > USL—Number of PPM found above than the LSL.
    - PPM Total—The total number of PPM found outside of both specification limits.

For additional detail on the MINITAB output, explore its Help menu and its "StatGuide" found on the toolbar. MINITAB offers concise and complete explanations of terms and the statistics used. In addition, MINITAB includes a thorough Help feature within the dialog boxes, shown in Figure P-34.

Additional Process Capability Analysis and Application

The "How to Use the Tool or Technique" section of this "Process Capability Analysis" entry embeds relevant analysis and hints and tips as it introduces a concept. However, Figure P-37 summarizes the process capability concept with four diagrams illustrating capable versus incapable processes and acceptable versus unacceptable processes.

A B C D E F G H I J K L M N O P Q R S T U V W X Y Z

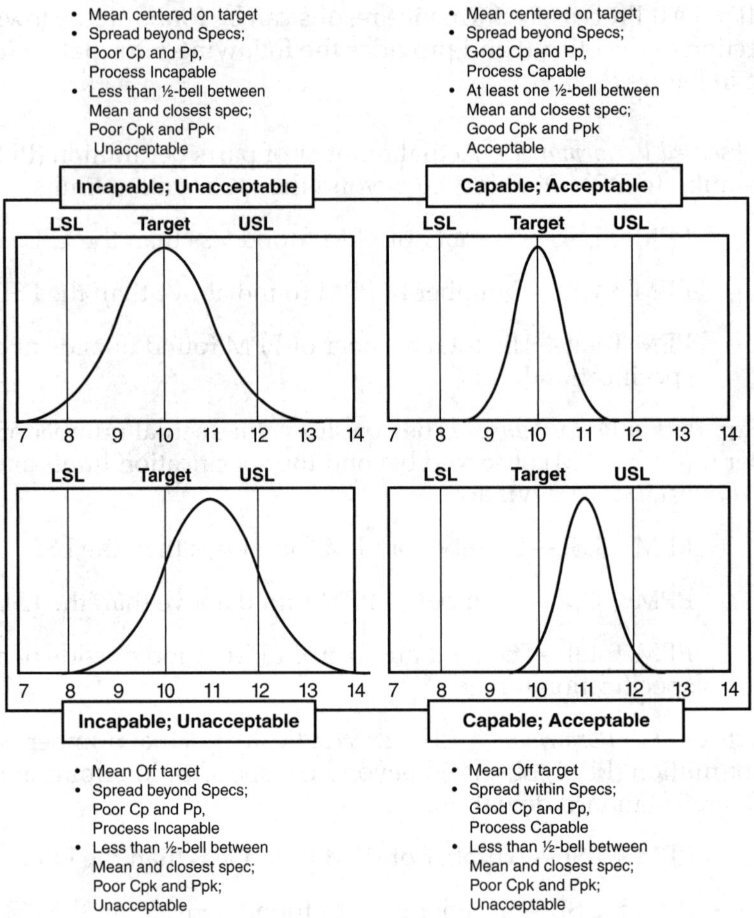

FIGURE P-37: Summary Process Capability Diagram

Hints and Tips

- Conduct Process Capability studies when the performance of a steady-state process needs to be monitored. If the focus is to reduce any *out-of-spec* products, then the following Process Capability Indices are preferred: Cpk and Cp (or Ppk and Pp).
- If the goal of monitoring the process performance is to minimize variation of the process mean from a target, then the preferred Process Capability Index is Cpm.
- Prior to beginning a Process Capability study, ensure that both the process exhibits random variation over time (using either a Control

chart or run chart) and the data are distributed normally (using either a Normal Probability chart or histogram). Otherwise, the process capability calculations will be meaningless.

- When conducting a manual Process Capability analysis, remember to:
    - Diagram the process bell-shaped curve,
    - Label the curve with the known data which should include the mean (x-bar), the standard deviation (s), and the upper and lower specification limits (USL and LSL), and
    - Calculate the process spread ($\pm$ 3-sigma, equating to 99.73% of the data) by multiplying the standard deviation by three, then both add it and subtract it from the mean, and label the curve with those values to define where the tails narrow.
- Good process performance—capable and acceptable:
    - Capability determined by variation (or spread): Cp $\geq$ 1.5; some experts accept 1.0, but with the long-term tendency to drift $\pm$1.5 sigma, a Cp $\geq$ 1.5 is preferred (or Pp $\geq$ 1.5). Then the process spread fits within the specification spread.
    - Capability and acceptability calculated by Cpk because it considers both the process variability and the proximity of the mean to the nearest specification limit. Cpk $\geq$ 1.33; some experts accept 1.0, depending on the circumstances (or Ppk $\geq$ 1.33).
    - Acceptable: Cpm: centered on the target.
    - If process perfectly centered, then Cp = Cpk.
    - If the process is perfectly centered <u>and</u> perfectly on target, then Cpm = Cp = Cpk.
    - Recall that the tails of a normal distribution extend beyond $\pm$ 3-sigma, or 99.73% of the data. Hence, the probability of "poor" product from a capable and acceptable process can happen (probably 0.27% of the time, represented by either tail of the distribution).
- Motorola's criteria for a Six Sigma process:
    - A Six Sigma capable process refers to tight process variation, wherein a 12 σ spread easily fits within the specification limits

and still has room for a 1.5 σ drift to account for long-term variation. Hence, with the process mean centered, the process spread of 6 σ on either side of the mean would fit well within the specification limits, and accommodate for a mean drift of $\pm$ 1.5.

- Process adjustments:
  - It is easier to adjust centeredness than it is to address variation (spread).
  - Recall that the difference between Cp and Cpk represents the potential gain in process capability if the process were to be centered.
  - The amount of out-of-spec product can be calculated by using DPMO and Yield calculations. It can also be calculated by using **both** Cp and Cpk, but not either index alone.
  - Two processes with identical Cpk-values may have different amounts of product out-of-spec and may need different actions to improve each of them.

Caution

As with all calculated statistics, be aware that they contain inherent tendency for error—human error. Potential error topics include sampling, sample size, measurement error, non-transformed non-normal data, and lack of confidence interval communication. Be aware of these potential "gotchas" and plan to prevent them.

Supporting or Linked Tools

Supporting tools that might provide input when developing a Process Capability Analysis include

- Histogram (**See Also** "Histogram—7QC Tool," **p. 330**)
- Normal Probability Plot, (including the Normality Test) (**See Also** "Control Charts—7QC Tool," **p. 217**)
- Control charts (including Time Series plots or run charts) (**See Also** "Control Charts—7QC Tool," and "Run Chart—7QC Tool," **p. 217** and **610**, respectively.)

A completed Process Capability Analysis provides input to tools such as

- QFD (**See Also** "Quality Function Deployment (QFD)," **p. 543**)
- Root Cause analysis tools and techniques
- Concept Generation methods (**See Also** "Failure Modes and Effects Analysis (FMEA)," in the "Hints and Tips" section, **p. 295**)
- FMEA (**See Also** "Failure Modes and Effects Analysis (FMEA)," **p. 287**)

Figure P-38 illustrates the link between the Process Capability analysis and its related tools and techniques.

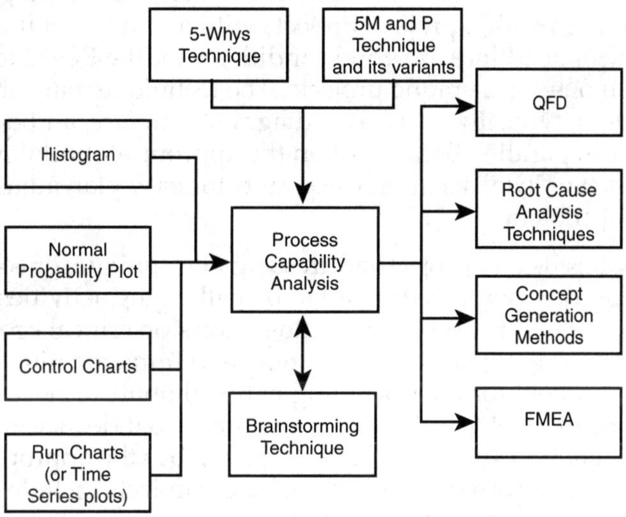

FIGURE P-38: Process Capability Analysis Tool Linkage

Process Decision Program Charts (PDPC)—7M Tool

What Question(s) Does the Tool or Technique Answer?

What might go wrong during the planning of this complex, significant project?

PDPCs help you to

- Plan and prevent problems from occurring in a high-stakes project
- Graphically plot course of action when many events or milestones are uncertain

Alternative Names and Variations

- Variations on the tool include FMEA (**See Also** "Failure Modes and Effects Analysis (FMEA)," **p. 287**)

A B C D E F G H I J K L M N O P Q R S T U V W X Y Z

When Best to Use the Tool or Technique

A Process Decision Program Chart (PDPC) is a planning and decision-making tool used when developing a large, complex project plan.

Brief Description

The PDPCs use a structured technique to identify what could go wrong in a complex, high-stakes project. Projects with a number of uncertainties and with critical deadlines are good candidates for the PDPC tool, and it works well on one-time unique projects. The technique maps the various event sequences, typically into a Tree diagram structure, and develops contingencies in parallel. Because often the appropriate countermeasures are unknown, the PDPC technique accounts for easy plan adjustments in response to a problem.

PDPC applies best to projects addressing new, unique, or complex problems, requiring perhaps difficult and challenging activities, such as the first-time planning and implementing a mission critical project—for example, the first-time planning, designing and implementing of 100% digital-voting system for the upcoming national political election for President. Also the PDPC technique is dynamic, so it deals well with the unknown, where problems may be anticipated, but the appropriate contingency remains uncertain. Examples of such projects include accident prevention and a new system (or policy) being implemented.

Given that the PDPC is a problem-prevention, risk planning management tool, it is comparable to the FMEA. Both tools explore what the potential failures could be and document countermeasures against the potential causes. Both tools are designed to be used dynamically and updated, refreshed and revised as part of an evergreen process. The PDPC evolved from a need in project planning to address risk in high-stakes, often unique projects with lots of uncertainty and critical deadlines to meet. Hence, the PDPC follows and supports the project planning process. The FMEA tends to enjoy a broader appeal because it applies simple risk scenarios in addition to the complex, and it also applies to ongoing operations, not just one-time projects. The FMEA distinguishes itself from a PDPC primarily because of its more rigorous prioritization component, but also its more detailed structure to dissect a potential root cause's impact (or effect) and to document a more complete action plan. If the scenario warrants it, **both the PDPC and FMEA techniques could apply**, and complement one another in the problem-prevention, risk management planning effort. (**See Also** "Failure Modes and Effects Analysis (FMEA)," **p. 287**)

The PDPC is a member of the 7M Tools, attributed in part to Dr. Shewhart, as seven "management" tools, or sometimes referred to as the *7MP* or *seven management and planning tools*. These 7M Tools make up the set of traditional quality tools used to analyze quantitative data. The 7M Toolset includes: 1) Activity network diagrams or Arrow diagrams; 2) Affinity diagrams; 3) Interrelationship digraphs or Relations diagrams; 4) Matrix diagrams; 5) Prioritization matrices, often replacing the more complex Matrix data analysis; 6) Process decision program charts (PDPC); and 7) Tree diagrams. *The Quality Toolbox*, by Nancy Tague, presents the 7M Tools ranked from those used for abstract analysis to detailed planning: Affinity diagram, Relations diagram, Tree diagram, Matrix diagram, Matrix data analysis (commonly replaced by a more simple Prioritization Matrix), Arrow diagram, and Process Decision Program Chart (PDPC).

Three Different PDPC Structures

A PDPC can be constructed one of three ways: 1) a forward linear (horizontal) Tree diagram, 2), a reversed linear (vertical) Tree diagram, and 3) an outline format.

The first Tree-like method depicts the sequence of prescribed steps from beginning to end. Structurally, the diagram often flows from left to right. This horizontal structure typically includes five levels of branches or detail. As the project progresses, if an event (or step) encounters difficulties, then contingency steps to resolve the issue are developed realtime and continue the progression toward the final goal. Figure P-39 illustrates a horizontally structured PDPC.

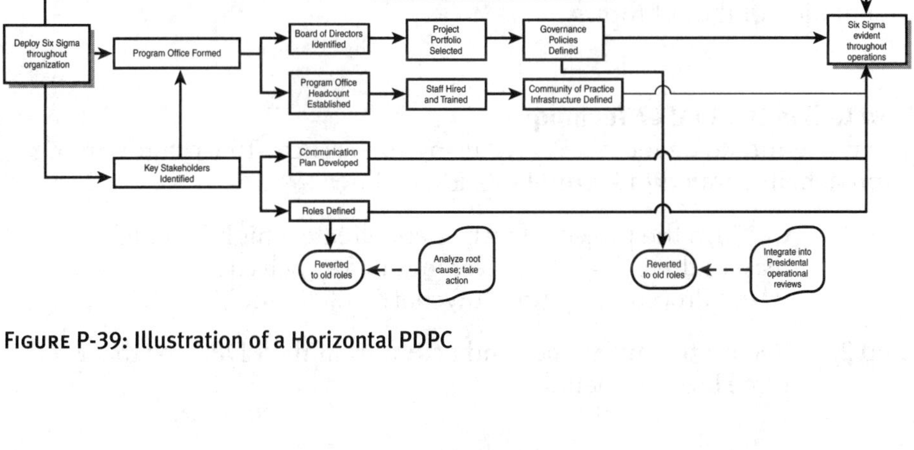

FIGURE P-39: Illustration of a Horizontal PDPC

The second Tree diagramming method starts with the end (or goal) of the project and walks backward. Hence, the mapping of events involves the various options or considerations at critical junctures. Structural layout typically flows from the top, with the higher level details, and flows downward to the bottom of the page, with the lower level details. Typically, the structure contains five levels of detail as rows and resembles a classic Work Breakdown Structure (WBS). At the lowest level, a set of conditional *what-if* questions lead to corresponding countermeasures. Figure P-40 shows a schematic of a vertical-structured PDPC.

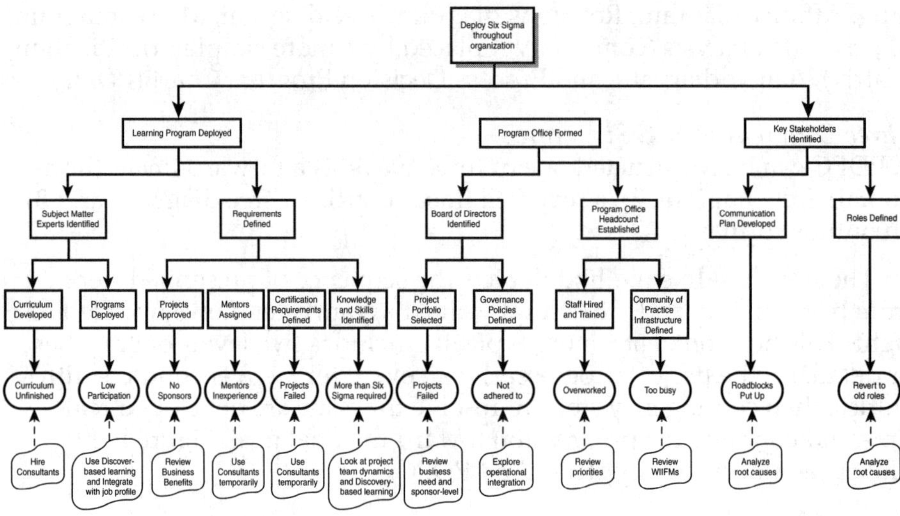

FIGURE P-40: Illustration of a Vertical PDPC

The third structure uses an outline format with the major categories representing the milestones and subheadings as the lower level of detail. This method resembles an outline of a paper, wherein the parsing of information into more granular detail is indicated by increasing the indentation of the subtopics.

How to Use the Tool or Technique

The procedure to develop a PDPC, using the vertical-Tree diagramming approach, illustrated in Figure P-40, is as follows:

Step 1. Confirm the project objective and create a high-level Process map of three to eight milestones or key activity categories. (**See Also** "Process Map (or Flowchart)—7QC Tool," **p. 522**)

Step 2. Record the milestones and activities in the PDPC for the top three levels of detail.

a. Select one of the three PDPC structures.

b. Document the high-level process contents in the PDPC—level one.

c. Brainstorm the next two levels of activities detail and document them on the appropriate hierarchical relationship of the PDPC—levels two and three. (**See Also** "Brainstorming Technique," **p. 168**)

Step 3. Develop the PDPC contingencies.

a. Referencing the lowest level of detail documented on the PDPC, from Step 2.c., brainstorm **what could go wrong**. Discuss what-if scenarios to uncover potential failures.

b. Discuss and evaluate the potential problems. Identify those that are improbable and/or have an insignificant impact and eliminate them.

c. Document the remaining problems on the PDPC—level four.

d. Brainstorm potential countermeasures for each PDPC problem. These contingencies could include modifications to the planning process or developing a respondent action plan if the problem occurs. The action plan could aim to prevent, minimize, or transfer the risk to another party. (**See Also** "Poka-Yoke," **p. 462**, for ideas on potential countermeasures.)

e. Document the countermeasures on the PDPC—level five. This level is often denoted by dashed Tree branch lines to the countermeasure, or drawing a cloud-like structure around it.

Step 4. Finalize the PDPC and communicate it to the appropriate stakeholders to incorporate the tool into ongoing planning and review sessions for the project.

a. Review, refresh, revise, and communicate the PDPC throughout the duration of the project.

Hints and Tips

- If using multiple contingency planning tools, start with the PDPC to organize and diagram potential problems. The Tree-structure serves as an easy communication vehicle to focus high-level discussions about uncertainties and risk. Use the PDPC as input to other contingency tools, such as FMEA. (**See Also** "Failure Modes and Effects Analysis (FMEA)," **p. 287**)
- Thought-starter questions used to identify potential problems in the brainstorming session include
    - What if...?
    - What inputs are required? What are their respective **operational definitions** of "good," and what could go wrong...?
    - Who is supplying the inputs? Is there a mutual understanding around supplier specifications and requirements? What are the suppliers' CTQs? Are they documented and agreed to, and what could go wrong...? (**See Also** "Measurement Systems Analysis (MSA)," for a discussion on operational definition, **p. 412**, and "Critical to Quality (CTQ)," **p. 242**)
    - What are the dependencies, minimum conditions, and assumptions for each of the inputs, processes, and outputs, and what could go wrong...?
    - What is out of scope, and what could go wrong...?
    - What are the potential impacts from the 5Ms and P (Machine (Technology), Method, Materials, Measurement, Mother Nature (environment, including cultural, political, regulatory), and People), and what could go wrong...? (**See Also** "Cause-and-Effect Diagram—7QC Tool," **p. 173**, for further discussion on triggers to uncover potential root cause.)
    - What has been the past experience, and what went wrong that could be replicated in this situation...? What is different from before...?
    - How could we sabotage this project's success if we wanted to?

A B C D E F G H I J K L M N O P Q R S T U V W X Y Z

Supporting or Linked Tools

Supporting tools that might provide input when developing a PDPC include

- Project charter (**See Also** "SMART Problem and Goal Statements for a Project Charter," **p. 665**)
- Process map (**See Also** "Process Map (or Flowchart)—7QC Tool," **p. 522**)
- VOC and VOB data gathering (**See Also** "Voice of Customer Gathering Techniques," **p. 737**)
- Brainstorming (**See Also** "Brainstorming Technique," **p. 168**)

A completed PDPC provides input to tools such as

- Brainstorming (**See Also** "Brainstorming Technique," **p. 168**)
- Control plan (**See Also** "Matrix Diagrams—7M Tool," **p. 399**)
- FMEA (**See Also** "Failure Modes and Effects Analysis (FMEA)," **p. 287**)
- Transition and/or Implementation plan (**See Also** "Matrix Diagrams—7M Tool," **p. 399**)

Figure P-41 illustrates the link between a PDPC and its related tools and techniques.

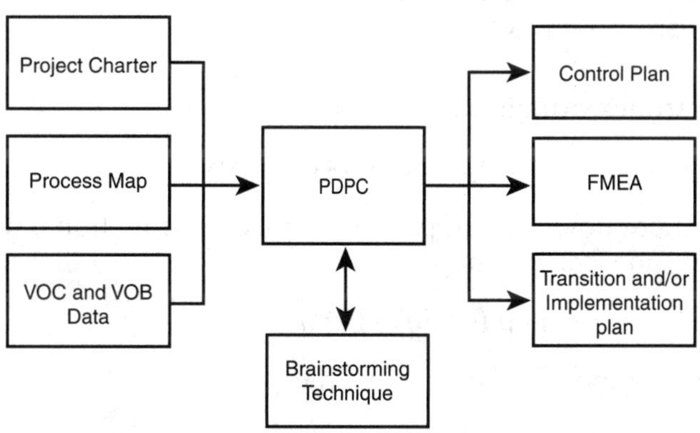

Figure P-41: PDPC Tool Linkage

Variations

FMEA (**See Also** "Failure Modes and Effects Analysis (FMEA)," **p. 287**)

Process Map (or Flowchart)—7QC Tool

What Question(s) Does the Tool or Technique Answer?

What are the components of the process; what is involved?

Process maps help you to

- Understand the process and where the opportunities for improvement exist.
- Gain a common understanding of a process and identify waste and areas where the process is poorly defined.
- Display and examine the current process and compare it with what could be.
- Communicate what is entailed in the process—its sequence of activities, inputs, outputs, and who is involved. Graphically outline a procedure.
- Plan (or simulate) the future or improved process.
- Categorize current process into its key components: activities, inputs, and outputs.

Alternative Names and Variations

This tool is also known as

- Flowchart; flow diagram
- Activity flowchart or Activity Process map
- Deployment flowchart or process map, cross-functional flowchart, or swim-lane Process map
- Detailed Process map (or flowchart)
- High-level Process map (or flowchart)
- Workflow diagram

Variations on the tool include

- Arrow diagram or Activity Network Diagram (**See Also** "Activity Network Diagram (AND)—7M Tool," **p. 127**)

A B C D E F G H I J K L M N O P Q R S T U V W X Y Z

- SIPOC (**See Also** "SIPOC (Supplier-Input-Process-Output-Customer," **p. 663**)
- Value Stream map or Process map (**See Also** "Value Stream Analysis," **p. 727**)

When Best to Use the Tool or Technique

A Process map is a communication tool that diagrams what activities occur today in a process or what could or should happen. It is a good planning and analytical tool to examine the components of a process (inputs, activities, outputs, metrics, and process players) to identify potential areas for improvement.

Brief Description

A Process map (or flowchart) **schematically** portrays a set of interrelated resources and activities to transform inputs into outputs. It identifies the steps in sequential order and its associated inputs and outputs. It may vary in degree of detail from a high-level, depicting only the critical few process categories or milestones, to the minutest detail, describing the lowest level tasks. Fundamentally, the flowchart centers on the **sequenced activities** but also may include information about its inputs, outputs, metrics and the process players. A process can be defined as a group of logically related activities and tasks involving people, procedures, and equipment needed to change information and/or materials into a specified product or service. It can include key decisions, handoffs, and conditional situations.

Some may ask why a process perspective is important, particularly if they do not believe that their work is part of a process. However, if work produces a result or outcome, that output is attributable to a process and its inputs, regardless of the size, scale, or scope of the work. Furthermore, a common business goal is to be successful. Success translates into driving some sort of result or output. Nevertheless, current or past results alone cannot predict future outcomes or success. The process must be understood to predict future outcomes. A business concerns itself with establishing a plan and targets to drive results, using the Process map to monitor results, to determine progress, and predict its ability to obtain those goals by serving as a checklist or notepad to record data. The results-orientation comprehends that *the outputs are a function of the process and its inputs* used to produce them. This translates into the mathematical equation $Y = f(X)$, where Y represents the output, and X the key process and its input variables. The Process map graphically displays this $Y = f(X)$ equation. Thus, a Process map often is the starting point for a process improvement initiative. (**See Also** "Y = f (X)," **p. 758**)

The Process map is a powerful and flexible tool because it applies to a variety of situations from manufacturing, to services; from transactional or administrative to continuous; from a program to a project; from the simple to the complex; and from the known to the planned. Some experts have modified the seven quality control tools (7QC Tools) to include the Process map. The 7QC Tools are viewed as the *core* quality improvement tools. The original list, attributed to Dr. Kaoru Ishikawa, included: Cause-and-Effect diagram; Check sheet (or checklist); Control charts; Histogram; Pareto chart; Scatter diagram; and Stratification. However, more recently, the 7QC Toolset has been revised by substituting the Stratification technique with either a flowchart (or Process map) or a run chart (or Time Series plot).

Given the flexible nature of the Process map, it can reflect four different perspectives: 1) What you *think* the process is, 2) What the process *actually* is, 3) What the process **could be**, and 4) What the process **should be**. Moreover, a complete Process map portrays two dimensions—first, the process parameters that describe the consumables required to complete the step and second, what the process players do. It also includes *product parameters*, which are the characteristics the tangible item must have to be acceptable—whether it is an actual product or documented information. In conclusion, there are three Process map applications:

- High-level Process map
- Detailed Process map representing two states:
    - *As is* state
    - *To be* state (also known as *should be, improved*, or *desired* state)

High-level Process Map

A high-level Process map consists of broad categories that describe a set of activities or milestones. If a macro view captures categories, then often only nouns are used as titles. Characteristically, the high-level map contains three to eight items that represent the end-to-end scope. In addition, a high-level Process map captures such an elevated perspective such that it generally remains unchanged, even after a process improvement has occurred.

This type of Process map serves as a good communication tool to establish project boundaries and focuses a project improvement team on the scope of interest. Thus, the start and stop points should clearly mark its boundaries to define the project scope. Figure P-42 shows a high-level Process map of completing a successful sale.

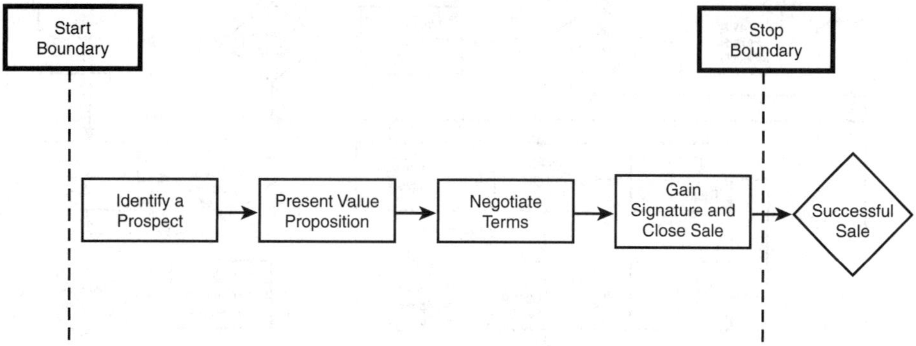

FIGURE P-42: High-level Process Map Example

Detailed Process Map

This flowchart contains the greatest amount of information. It is used to identify opportunities for improvement and therefore is as complete a picture as possible including the activities (process and sub-process steps), inputs, outputs, metrics, and often roles and responsibilities of the process players.

A detailed Process map should only contain as much information as needed. Hence, the appropriate amount of detail depends on its use. If a cross-functional, multi-disciplined management team uses the chart, it may contain fewer details than if it were being used by a function's process players. Figures P-43 and P-44 exhibit two different types of detailed Process maps for a selling process. Figure P-43 highlights one area within the high-level map and drills down into that step as the area of focus; sometimes called a *Top-down* flowchart. Figure P-44 provides a complete end-to-end picture of a selling process.

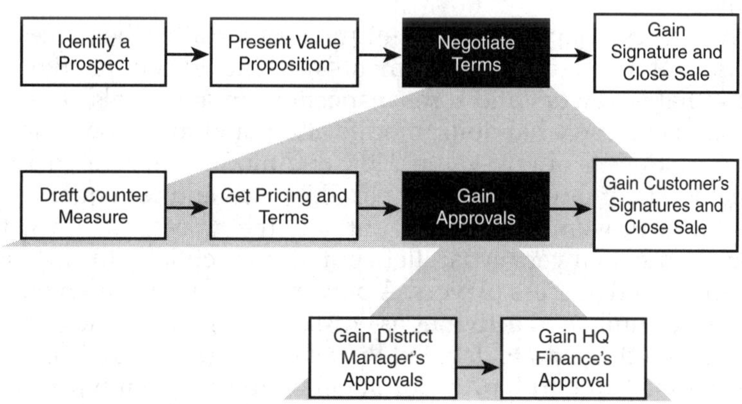

FIGURE P-43: Top-down Process Map Example

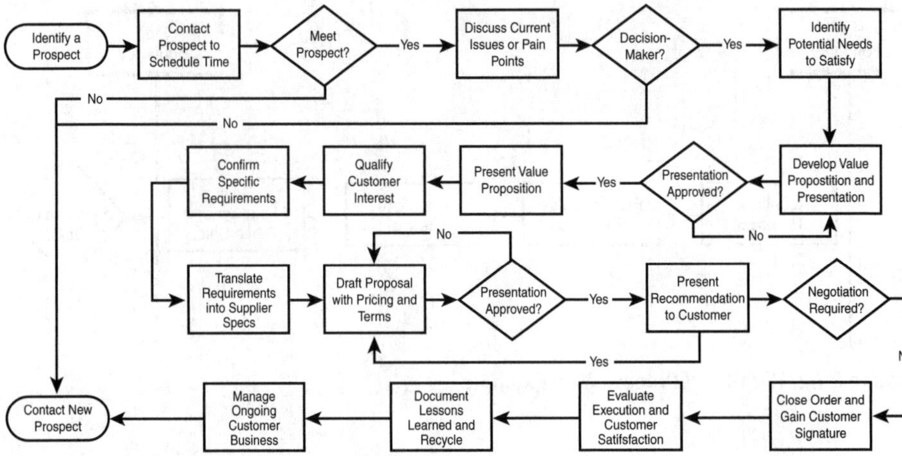

FIGURE P-44: Detailed Process Map Example

As Is State

The As Is flowchart documents actual operations (the good, the bad, and the ugly), rather than how it is supposed to operate—the as is; not the to be. Walk the process to ensure the diagram depicts what actually occurs, rather than the procedural documentation. Walking the process can catch hidden factors of rework and do-loops, workarounds, and undocumented contingencies. Thus, the content should be periodically reviewed to reflect any changes. Process changes need to be understood but may not necessarily be bad. They may be adjustments in response to a change and call for a more permanent change in the process or perhaps it could be a *kaizen* improvement, or positive deviance, that should be communicated broadly and become part of the standard operating procedure.

To Be State

The To Be Process map depicts the improved state after the process has been leaned of its wastes. Improvements may include elimination of recycle loops (that is, rework and some inspections or approvals, delays and waiting, and excessive handoffs; modification or elimination of activities that produce defects or non-value adds; and modifications to increase flow (that is, streamlining, creating parallel paths and re-sequencing). It clarifies the improvements made to not only the process steps and their sequence (sequential versus parallel), but also potentially the inputs, outputs, metrics, and process players. A powerful communication tool depicts those eliminated activities as crossed-out steps on the current As Is map, as an **intermediate** detailed Process map. (**See Also** "Lean and Lean Six Sigma," in Part I, **p. 29**, for detail on the different types of waste.)

Since the To Be Process map documents a goal or direction for the process players, it becomes a training document and part of the standard operating procedure used in the transition phase to the improved state. Careful monitoring of the process players' progress and adaptation to the change ensures that the To Be map reflects the new reality, versus merely a "should be" procedure. Upon implementation of a process change, the To Be Process map quickly becomes an As Is map. Hence, the flow should be periodically reviewed to validate adherence and accuracy.

Graphical Layout

The flowchart tool, albeit flexible, has conventions to follow with respect to the symbols and layout design. Process maps may be created by hand or by using a variety of software application packages including Microsoft Word, PowerPoint, and Visio.

Common Symbols

Process map conventions for symbols are extensive, as many as 185 shapes. The most commonly used symbols are shown in Figure P-45.

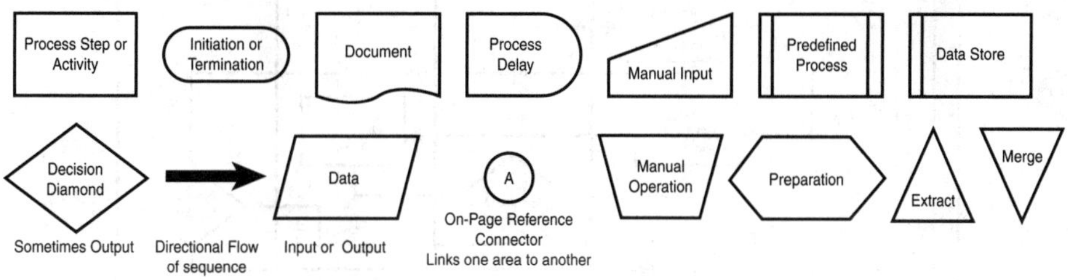

FIGURE P-45: Common Process Map Symbols

Common Layout Configurations

Its layout can flow either horizontally or vertically. The individual steps or activities always are labeled with a title or description distinguishing them from the others, typically in a verb-noun combination. Sometimes they can be numbered. The boundary conditions should be clearly identified using the Initiation or Termination symbol identified in Figure P-45. The sequence or direction of the flow must be clearly marked with directional arrows. In addition, traditionally decision diamonds ask binary questions that call for two arrows exiting it to indicate both the affirmative and negative paths (that is, yes/no or good/bad, for example).

The layout displayed in Figure P-44 is a classic activity diagram, wherein the focus is primarily activity-centric and often captures decision points, rework loops, and complexity. A *deployment diagram* or *swim-lane* Process map is another layout configuration that includes bands depicting functional groups involved in the process. Activities contained within a swim-lane are performed by that specific functional group. As the activity flow crosses these swim-lanes, it connotes a handoff from one party to another. This type of diagram is particularly useful for processes involving information flow between people or functions, as it highlights hand-off areas, which represent an opportunity for consolidation to eliminate potential waste and improve cycle time. Figure P-46 provides a vertically displayed deployment flowchart example.

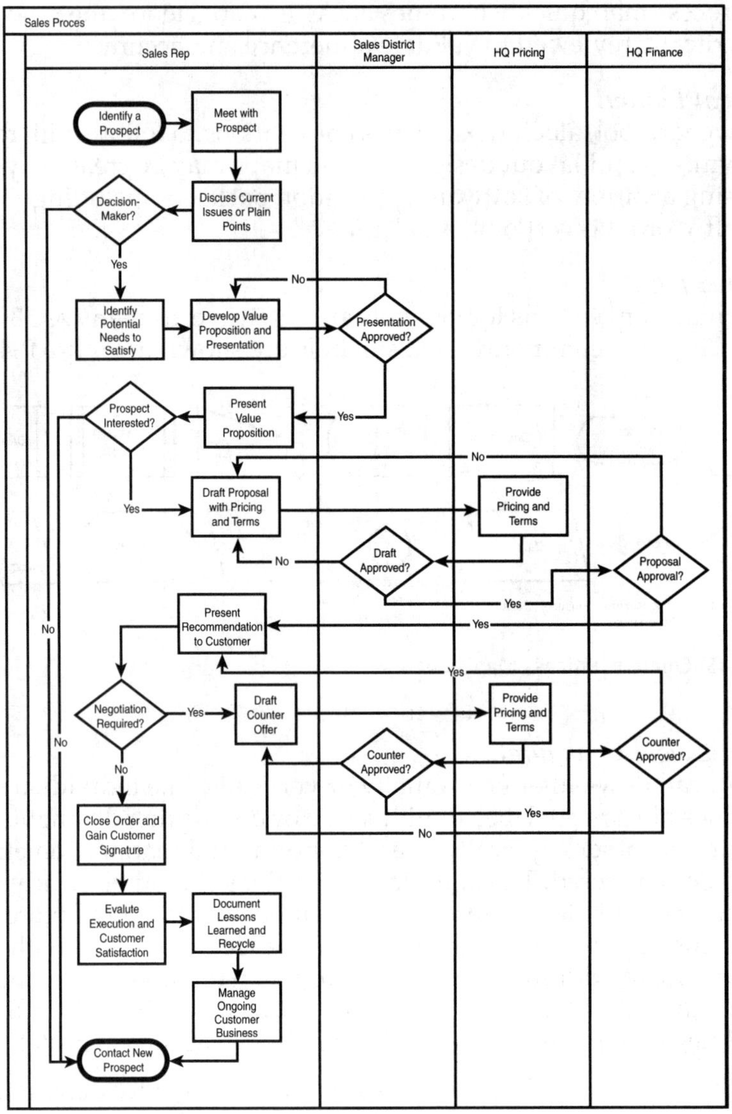

Figure P-46: Deployment Process Map (or Swim-Lane) Example

How to Use the Tool or Technique

Given the flexible nature of a Process map, the intended use should direct how the diagram is built.

The following procedures will serve as a guideline when developing a detailed Process map with a process improvement project team.

Step 1. Assemble a representative cross-section of the multiple perspectives of process players and explain the purpose of meeting.

a. Meeting preparation: Secure sticky note pads and markers to hand out to each person. The larger sticky notes often are better. You may want to consider segmenting the cross-functional team by color of sticky pad and/or marker.

b. Room preparation: Hang blank flip chart sheets around the perimeter of the room to serve as working space for the team to post the individual activities.

c. If the high-level Process map is known, label at least one flip chart sheet per each category.

Step 2. Individually, brainstorm the various activities, documenting one idea per sticky note. Ensure the person places his/her initials in the lower corner to help identify author during the clarification step. (Allow 10 to 20 minutes of this silent activity.)

Step 3. Upon completion of Step 2, invite each person to post the sticky notes on the flip charts around the perimeter of the room.

a. If high-level Process map is posted, they align the stickies with the appropriate category.

b. If flip chart sheets are blank, the team needs to group activities into affinities, and eventually identify and label the high-level categories.

Step 4. Clarify stickies by reading each one, verifying if clarification is needed, and modify sticky as appropriate.

a. If duplications exist, keep only one.

b. If gaps exist, create additional stickies to complete the representation.

c. If stickies belong in multiple places, create duplicates and put in all the appropriate places. Further clarification and process may determine differences and then document distinguishing characteristics.

Step 5. Arrange the activities in the proper sequence. Arrange each activity sticky to reflect the current flow until all are accounted for and everyone agrees with the sequence.

Step 6. Draw flow arrows to indicate direction.

Step 7. Document any supporting detail necessary and number the activities and person accountable for activity. Identify inputs, outputs metrics, and critical decision points.

a. Inputs and outputs encompass not only those at the beginning and end of the process, but also those within the process (or in-process).

b. Inputs can be further classified into what is controllable (C), part of standard operating procedure (S), or noise (N)—that which is uncontrollable.

c. Decision points represent questions that lead to multiple paths.

Step 8. Review flowchart to ensure completeness and accuracy. Dismiss the meeting.

Step 9. Document the flowchart electronically. Distribute to the meeting participants and ask them to review and edit the document over a period of time while working in (or walking) the process, and submit any edits back to the document creator for changes.

a. You may assign "observers" to walk the process and make notations on the Process map draft to look for "subconscious" activities, exceptions and "hidden factories" of rework.

b. Sometimes it is insightful to walk the process backward, from the end to the beginning.

c. Record or double-check supporting information—person accountable for activity, inputs, outputs, and metrics.

d. Upon completion of the Process map, use it as a tool to analyze improvement opportunities.

e. After improvements are made, revise and update the Process map to reflect the improved state.

Hints and Tips

Creation process

- Develop a Process map with a team of **people who work in the process**, but designate one person responsible for managing the overall process of mapping the total flow.
- Build a process from end-to-end, starting from either the beginning or end and avoiding simply pasting together individual process areas that management built.
- Allow time for reflection and edit of a Process map draft to capture potentially forgotten or infrequent activities.
- Capture the perspectives from as many different types of process players as possible—the process workers, process suppliers, process customers, process owner(s), and supervisors.
- High-level Process maps depict a macro perspective of the overall process categories that remain constant even after an improvement or leaning initiative. Conversely, a detailed Process map's activities, and sometimes flow, reflect a change after an improvement initiative.
- Add as much supporting detail to a detailed flowchart as relevant to the process analysis—cycle time (average and range) and metrics (that is, cost, quality and performance measurements).
- Sometimes a "new" set of eyes (objective observer) helps "see" the process and identify the rote activities that are performed unconsciously and often go undocumented. These often may be workarounds in response to a situation to make the process work.

Layout

- Capture the content and sequence of activities, rather than worry about what is the "right way" to draw it.
- Draw the Process map manually before capturing it electronically.
- Diagram the sequence of activities that accomplish work to depict their flow, rather than mapping the organization.
- Map the appropriate level of detail for the given business question or problem trying to be addressed, rather than every aspect of a process.

Analysis

- Identify value adding, value enabling, and non-value adding activities throughout the process.
- Use a detailed flowchart as a checklist to ensure consideration of process components end-to-end.
- Mark-up a detailed Process map to highlight eliminated steps.
- Pose **questions** (similar to the following suggestions) and document the answers on a detailed Process map:
    - **Who?** (supplier, accountable process player, decision-maker, inspector, customer)
    - **What?** (item or information, value-add versus non-value-add)
    - **When?** (what preceded it, what follows)
    - **Where?** (produced, comes from, it goes)
    - **How?** (does it flow—how is it being transported, handed off, and how it is being transformed (value add), used by downstream customer)
    - **Why?** are things done this way? It serves as a follow-on question for subsequent exploration.

Supporting or Linked Tools

Supporting tools that might provide input when developing a Process map include

- Brainstorming (**See Also** "Brainstorming Technique," **p. 168**)
- Standard operating procedure (SOP)

A completed Process map provides input to tools such as

- Activity Network Diagram (**See Also** "Activity Network Diagram (AND)—7M Tool," **p. 127**)
- Cause-and-Effect diagram (**See Also** "Cause-and-Effect Diagram—7QC Tool," **p. 173**)
- Control plan (**See Also** "Matrix Diagrams—7M Tool," **p. 399**)
- FMEA (**See Also** "Failure Modes and Effects Analysis (FMEA)," **p. 287**)

- SIPOC (**See Also** "SIPOC (Supplier-Input-Process-Output-Customer)," **p. 663**)
- Standard operating procedure (SOP)
- Training plan
- Value Stream Analysis (**See Also** "Value Stream Analysis," **p. 727**)

Figure P-47 illustrates the link between a Process map and its related tools and techniques.

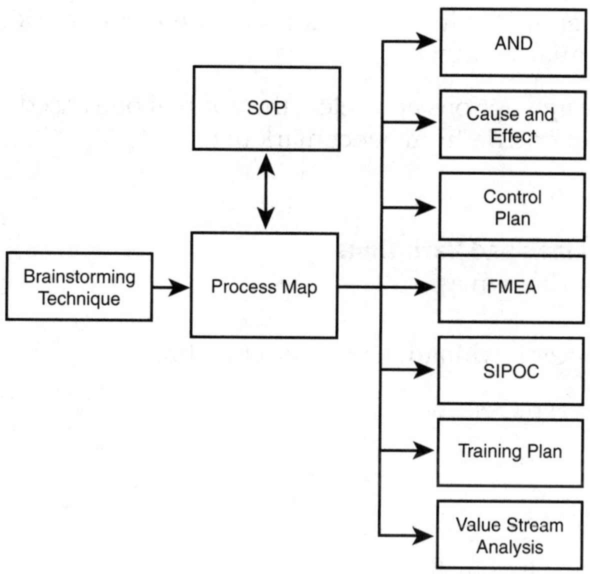

Figure P-47: Process Map Tool Linkage

Variations

- Arrow diagram or Activity Network diagram (**See Also** "Activity Network Diagram (AND)—7M Tool," **p. 127**)
- SIPOC (**See Also** "SIPOC (Supplier-Imput-Process-Output-Customer," **p. 663**)
- Value Stream map or Process map (**See Also** "Value Stream Analysis," **p. 727**)

A B C D E F G H I J K L M N O P Q R S T U V W X Y Z

Pugh Concept Evaluation

What Question(s) Does the Tool or Technique Answer?

Which design or potential solution option is best?

The Pugh Concept Evaluation helps you to

- Articulate fundamental principles important to the integrity and feasibility of a concept
- Gain better insight into the concept requirements, design problems, and potential solutions
- Refine design options and select the optimal one based on a common set of criteria from benchmark data

Alternative Names and Variations

This tool is also known as

- Pugh Concept Evaluation and Selection Process
- Pugh Concept Selection
- Pugh Concept
- Pugh process

Variations on the tool include

- Solution Selection matrix (**See Also** "Solution Selection Matrix," p. 672)

When Best to Use the Tool or Technique

When there are a set of alternative improvement designs (or potential solutions) to select from, use the Pugh Concept to improve and evaluate the options and ultimately select the winning concept.

Brief Description

The Pugh Concept Evaluation process is a structured technique to refine concept alternatives and select the best one using a benchmark. The benchmark represents a best-in-class design, and is referred to as the *datum*. The datum is defined as the evaluation criteria—the concept that the solution alternatives must beat. The datum is the best-in-class, industry showcase for a solution or solution component that is identified through benchmarking, from customers, or from industry analysts. The use of such a benchmark approach simplifies the evaluation comparison process, as opposed to comparing all concepts to all other concepts simultaneously.

The approach recommends a cross-functional team to *further develop* and *improve* the initial set of concepts. As a result, the team often generates additional new concepts. The team uses a matrix-based process in several two to three hour-long team meetings over a few weeks to converge on a superior design. The process aims to ensure that the final concept best meets the customers' new, unique, and difficult (NUD) requirements, such that often the final selection is a modification of an originally considered design. NUD requirements translate into

- **New—**A requirement the customer never asked to be fulfilled before; it is completely a fresh requirement.
- **Unique**—A requirement being fulfilled by a competitor or a substitution product or service your organization currently is not providing.
- **Difficult**—A requirement that is very difficult to fulfill (not necessarily new or unique but often is).

(**See Also** the "NUD versus ECO" section within "Kano Model: KJ Analysis," **p. 376**)

This approach is named after its developer, Stuart Pugh, a professor of Engineering Design at the University of Strathclyde in Glasgow, Scotland. Professor Pugh died in 1993 but was still viewed as a leader in product development and total design methodology. The Total Design concept envelopes seven sources of information: Market Knowledge; Requirement Knowledge; Concept Knowledge; System Knowledge; Detail Design Knowledge; Manufacturing Knowledge; and Sales/Service Knowledge. Pugh's concepts are taught internationally in many Design for Six Sigma (DFSS) programs. Don Clausing, professor at MIT and one of Pugh's biggest advocates, continues to popularize Pugh's concepts through his teachings and writings.

The Pugh process, within the Total Design methodology, is a procedural tool for controlled divergence and convergence of the best possible solution to a given set of design requirements. It involves three main phases: 1) the initial phase of *Concept Evaluation and Hybridization,* 2) the second phase of *Concept Evaluation* (often final), and 3) the additional phase of *Concept Evaluation*, as necessary. It requires several inputs that fall into five main categories:

- **Product Requirements**—defined and structured.
    - **Ranking**—from Customer Requirements Survey.
- **Quantitative Technical Requirement Prioritization.**
- **Competitive Benchmark Data**—engineering data on competitive product performance.
- **Concepts generated to equal levels of expression:**
    - **Math models**—numeric (and sometimes verbal) description of the concept.
    - **Graphical models**—pictorial (and sometimes verbal) depiction of the concept.
    - **Physical models**—visual display of the concept:
        - **Conduct initial studies in Design For Manufacturing and Assembly (DFMA) and Design Failure Mode and Effect Analysis (DFMEA).**
- **Funding and Resources**

Development teams using this process often achieve greater insight and awareness of possible solutions by building on and integrating each other's ideas. This method stimulates the generation of new and better hybrid-design concepts to better meet customers' NUD requirements. Teams other than product development teams find they benefit from this approach in enhancing and evaluating potential solutions. Regardless of the type of improvement team, its make-up should comprise diverse talent and perspectives working on a common goal to stimulate the creative process and ensure a robust design. The Pugh process draws out and highlights clear distinctions among solution alternatives.

How to Use the Tool or Technique

The Pugh Concept Evaluation technique employs the following procedure:

Step 1. Select the decision-criteria used to evaluate the various concept alternatives.

a. A good source includes a completed Quality Function Deployment (QFD) matrix, using the CTQ column (the vertical axis on the far left). (**See Also** "Quality Function Deployment (QFD)," **p. 543**)

Step 2. Discuss and clarify the different concepts. Try to discuss each concept to a similar level of detail to avoid bias and misunderstanding. Document the concepts with appropriate visuals and verbal descriptors.

Step 3. Select the benchmark or datum. Consider the technical requirements when selecting the benchmark. Select the datum concept that is among the best concepts available for the baseline. A "datum" concept is selected as the initial comparison point for all the other concepts.

a. In the absence of a strong competitive design, select the best internal design based on the consensus of the team (often this will be the current "best" design).

b. A good source includes a completed Quality Function Deployment (QFD) matrix, referencing the competitive annex.

Step 4. Construct an L-shaped matrix with the vertical axis (left side) for the criteria and the horizontal axis (top row) for the different concepts. (**See Also** "Matrix Diagrams—7M Tool," for a discussion on matrix types, **p. 399**)

a. Ensure that all the team members can view the matrix. Construct the matrix one of the following ways: electronically (using a spreadsheet, such as Excel) and project it or on a flip chart paper posted on the walls or whiteboard.

b. The first column on the far left should be entitled "Criteria."

   i. In each cell under the title of the first column, document each decision-criterion, one per cell working downward, to label each row in the matrix.

   ii. After all the criteria are listed, add three more rows and label them as follows:

      - Total Pluses
      - Total Minuses
      - Total Same

c. The second column contains the benchmark information and is labeled "Datum."

d. The remaining columns represent the various alternative concepts under evaluation and should be labeled accordingly. The matrix should reference the appropriate visuals and verbal descriptors for each concept.

A B C D E F G H I J K L M N O P Q R S T U V W X Y Z

A B C D E F G H I J K L M N O P Q R S T U V W X Y Z

Step 5. Compare each concept relative to the datum using non-numeric rating scale: "+" (a plus-sign) for better than the datum; "–" (a minus-sign) for worse than the datum; an "S" for the same as the datum.

This begins the Concept Evaluation and Hybridization phase.

a. Rate each concept using the non-numeric rating scale: +, -, and S versus the datum.

Step 6. Evaluate the ratings.

a. Total the score for each concept—the total number of plus, minus, and same scores and record in the appropriate cell.

b. Identify any pattern across the different alternatives. Certain concepts will exhibit distinct strengths; others will display clear patterns of weakness. As the concepts are evaluated, the number may get reduced. Ensure that all concepts are compared on the same basis and at the same generic level.

c. See what the positives will contribute to one's insight in the design.

Step 7. Optimize the design concepts. Attack the negatives and enhance the positives.

a. Discuss the most promising concepts.

   i. Isolate and document **weaknesses** of the strong concepts.

   ii. Identify what could be changed to **eliminate the negatives**.

   iii. Check that this does not **reverse a positive**.

   iv. Identify the positive interactions between concepts, which create new hybrid concepts. Define a new concept from this process and add it to the matrix (divergence).

   v. Leave the existing concept as is; it may be used later.

   vi. The matrix has now expanded.

b. Assess weak concepts. Attack negatives to see if they can be improved relative to the datum.

i. If they can be significantly improved, add this new concept to the matrix (leave the weak concept as it is).

ii. Synthesize the strengths of weak concepts that otherwise could not be helped. Add the new synthesized concept to the matrix.

iii. Kill or modify the negative ones.

iv. The matrix will shrink.

c. If a number of strong concepts do not emerge, assess any apparent uniformity of strengths and weaknesses.

Step 8. Select a new datum and *"rerun"* the matrix. When any one concept persists as uniquely strong, rerun the matrix with that strong concept as the datum. A new hybrid can be entered into the matrix for consideration.

a. See if the strength still persists. If it does, this is a *confirmation* of the initial conclusion.

b. If it is beaten, then repeat Step 6 until distinctly strong concepts persist.

You are DONE at this point if the decision is made to further develop the strongest concepts from phase 1.

i. The strong concepts are put through more development and testing. Ensure each concept receives equally balanced attention so as to not bias one over another.

ii. A deeper understanding and maturing may require the development requirements or CTQs to be expanded or refined.

Step 9. *Optional:* Plan for further work. At the end of the first working session, the team may need to collect more information, perform experiments, and evaluate technical capabilities.

This begins the second phase—*Concept Evaluation*.

Step 10. *Optional:* Reconvene the team to refine the concepts and rerun the matrix for further analysis based on the additional information collected. Continue this iterative process until a winning concept beats the datum.

a. With an expanded matrix (criteria and concepts), repeat the evaluation steps starting from Step 4 through 7.

b. Question the results in detail. Look at new patterns of strengths and weaknesses, especially if criteria changed significantly. The re-affirmation of the trend established in the first phase tends to give confidence in the emerging design

Figure P-48 illustrates a generic Pugh Concept Evaluation matrix.

Criteria	Product Concepts				
	Benchmark	A	B	C	D
Evaluation Criteria #1		S	-	+	S
Evaluation Criteria #2		+	-	-	S
Evaluation Criteria #3	D	+	+	+	S
Evaluation Criteria #4	A	-	-	S	-
Evaluation Criteria #5	T	-	-	S	-
Evaluation Criteria #6	U	+	-	+	+
Evaluation Criteria #7	M	+	-	+	+
Total Pluses		4	2	4	2
Total Minuses		2	6	1	2
Total Same		1	0	2	3

FIGURE P-48: Basic Pugh Concept Evaluation Matrix

Hints and Tips

- Typically, a superior concept is attained by the Convergence Phase of the process.
- The more the different points of view and perspectives explored the better.
    - The concept of safety and wisdom coming from the variety of cross-function team members provides additional confidence that vulnerability has been minimized.
- All this work is done on the large matrix in front of the entire team; all actions are clearly evident to each member.
- Everyone must scan and understand all concepts to actively participate in the evaluation. Otherwise, some opportunity will be missed.
- Try to describe and model all concepts to same level of detail; consider mathematical, graphical, and physical models if appropriate.
- Beware of poorly defined criteria, primarily due to the lack of properly understood requirements and poorly developed CTQs.

Ambiguous criteria lead to different interpretations by team members. (**See Also** "Critical to Quality (CTQ)," **p. 242**)

- Purify, redefine, or eliminate the poor criterion.
- Sometimes they are eliminated **during** the process (called a *run*).

- Persistence of uniformity of strengths across concept alternatives usually means that one or more concepts are subsets of the others (essentially the same).
- Take time to understand the feasibility of each concept alternative, as well as their associated risks and consequences.
  - Be careful that the brainstorming portion of the process is taken seriously and with responsibility to avoid perception of simply a freewheeling process.
- Introduce a facilitator to help balance the input from team members, if needed. Strong-willed team members (usually with a lot of experience whose design has fallen) may begin to become leery of the emerging dominant concept. They may suggest that early concepts were better than this new emerging dominant concept. Their defense may be based on emotion, experience, bluster, and outright denial of facts. However, if they can be won over using the process, the final design concept may be stronger.

A B C D E F G H I J K L M N O P Q R S T U V W X Y Z

Supporting or Linked Tools

Supporting tools that might provide input when developing a Pugh Concept Evaluation matrix include

- Benchmarking (**See Also** "Benchmarking," **p. 160** and "Benchmarking—Avoid Arrogance and Lethargy," in Part III, **p. PDF:789**)
- Brainstorming (**See Also** "Brainstorming Technique," **p. 168**)
- Real Win Worth (RWW) ("**See Also** "Real-Win-Worth (RWW) Analysis," **p. 560**)
- QFD (**See Also** "Quality Function Deployment (QFD)," **p. 543**)
- SWOT Analysis (**See Also** "SWOT (Strengths-Weaknesses-Opportunities-Threats)," **p. 699**)
- VOC gathering techniques (**See Also** "Voice of Customer Gathering Techniques," **p. 737**)

A completed Pugh Concept Evaluation matrix diagram provides input to tools such as

- FMEA (**See Also** "Failure Modes and Effects Analysis (FMEA)," **p. 287**)
- Pilots
- Transition and Control plans (**See Also** "Matrix Diagrams—7M Tool," **p. 399**)

Figure P-49 illustrates the link between a Pugh Concept Evaluation matrix and its related tools and techniques.

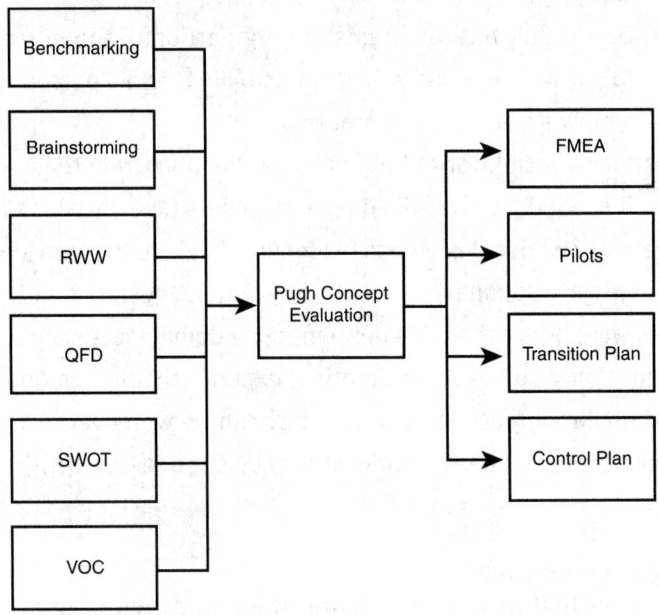

Figure P-49: Pugh Concept Evaluation Tool Linkage

Additional Resources or References

- Pugh, Stuart. *Total Design*; Massachusetts: Addison Wesley (1990).
- *Creating Innovative Products Using Total Design*; edited by Don Clausing & Ron Andrade; Massachusetts: Addison Wesley (1996) ISBN: 0-201-63485-6.

Variations

Solution Selection matrix (**See Also** "Solution Selection Matrix," **p. 672**)

Q

Quality Function Deployment (QFD)

What Question(s) Does the Tool or Technique Answer?

Which technical specifications for a product or services offering best meet a specific set of customer requirements?

QFD helps you to

- Translate customer requirements into specific offering specifications.
- Prioritize possible offering specifications and make trade-off decisions based on weighted customer requirements and ranked competitive assessment.

Alternative Names and Variations

This tool is also known as

- House of Quality (HOQ)
- QFD matrix
- Requirement matrix

When Best to Use the Tool or Technique

Use the QFD when you are designing a product or services offering and need to select among several alternative features and functionality.

Brief Description

The Quality Function Deployment is a powerful prioritization tool that combines several different types of matrices into one to form a house-like structure. Sometimes referred to as a House of Quality (HOQ), this tool captures the Voice of the Customer to identify the required quality, features, and functions needed to be deployed in a single offering. The QFD process uses a graphical format to document the information gathered and processed. In its most complete state, this information constructs the five "rooms" in the QFD house, as illustrated in Figure Q-1:

- The *What* or *Wants*—Customer requirements, needs, and priorities that form the far left-wing of the house.
- The *Competitive Assessment*—Compares customer priorities with appropriate marketplace offerings, across key competitive deployments, which forms the right-wing annex of the house.
- The *How*—The offering's technical design features, functionality, and characteristics to meet the customer requirements. This forms the attic of the house.
- The *Design Relationships*—Describes the interrelationship between the design features, which form the roof of the house.
- The *Foundation*—Uses benchmarked target values as objective measurements to evaluate each characteristic, forming the basement of the house.

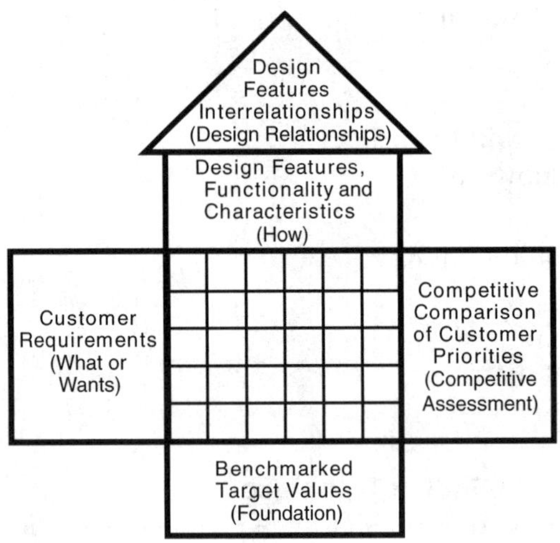

FIGURE Q-1: House Schematic of QFD

The QFD is a flexible tool, customizable for a given situation. The team using the technique can modify the guidelines that alter the scales used in the house. However, the basics of the QFD remain constant—the VOC selects the features and functionality of an offering. The tool was first used to design an oil tanker at the Kobe shipyards of Japan in 1972 by Yoji Akao and Shigeru Mizuno to design customer satisfaction into an offering before it is produced. Prior to this, quality control methods were primarily aimed at fixing a problem during or after production. In the mid-1980s, Don Clausing of MIT introduced this design tool to the United States. A classic product design application is in the automotive industry. In fact, Clausing tells of an engineer who initially wanted to place the emergency hand brake of a sports car between the seat and the door. However, VOC testing found that women drivers wearing skirts had difficulty with the new placement of the hand brake. The QFD highlighted potential dissatisfaction with the location of this feature, and the idea was scrapped.

A prerequisite to building a QFD entails collecting customer requirements. Several VOC gathering techniques include surveys, focus groups, interviews, trade shows, and hot lines. Once the VOC data has been synthesized with the customers' rankings, the QFD process is ready to begin. (**See Also** "Voice of Customer Gathering Techniques," **p. 737**)

Progress Drilldown Approach

The primary QFD component features the relationship matrix between the VOC and the potential design options. Once the design options are prioritized in a QFD, the design team can continue to drill down into more specific detail, starting with the major features to subcomponents, down to its component parts to create three separate QFD matrices. To do so, rotate the QFD 90-degrees to the left (counterclockwise), convert the prioritized attic features (the how) into the vertical axis (the what), and list the new more detailed technical specifications as the new attic items. This drill-down process continues until all the technical specification details are prioritized.

Design engineers typically start this progressive drill-down approach with planning the development and go through four phases to reach a deeper understanding of the required process control and quality. This four-phase approach produces separate QFD matrices, as shown in Figure Q-2, and includes the following sequence:

QFD 1. **Product Planning QFD**—Identifies the offering characteristics that best meets customer requirements, analyzes competitive opportunities, and establishes critical target values. Figure Q-9 illustrates this QFD matrix.

QFD 2. **Component Deployment QFD**—Identifies the critical parts and assembly components using the prioritized offering characteristics in QFD 1 and establishes critical target values.

QFD 3. **Process Planning QFD**—Determines critical process operational requirements and elements using the prioritized components in QFD 2 and establishes critical process parameters.

QFD 4. **Quality Control QFD**—Prioritizes the process control methods and parameters and establishes production and inspection methods that best support the prioritized process elements of QFD 3.

A B C D E F G H I J K L M N O P Q R S T U V W X Y Z

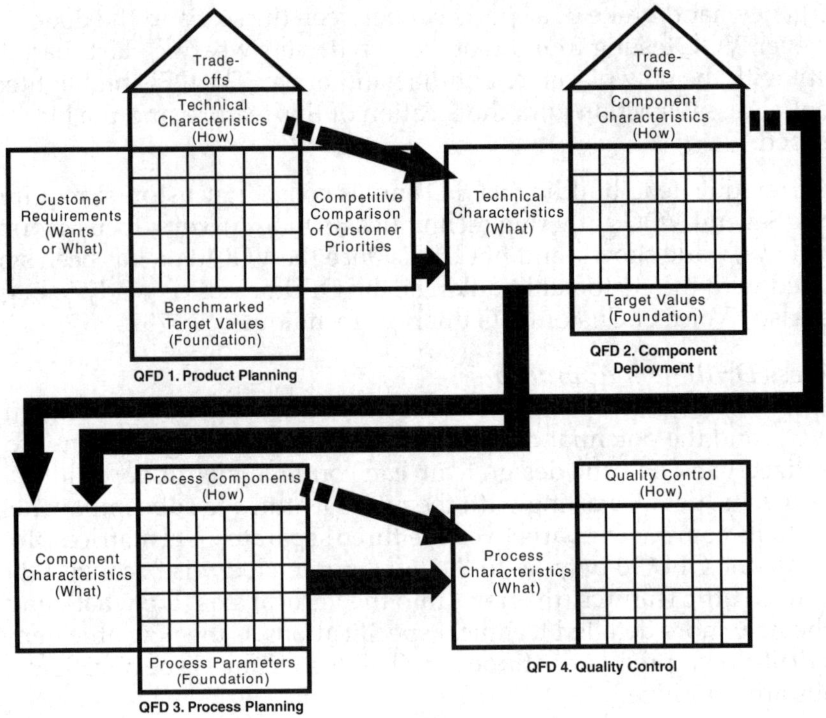

FIGURE Q-2: Four-phase Progressive Drill-down QFD Methodology

How to Use the Tool or Technique

The procedure to construct a single five-room QFD matrix is as follows. The guidelines describe how to develop each room, its purpose, and how the rooms relate to one another.

Preparation: Assemble the VOC data and a cross-functional team responsible for designing the offering. Create a blank QFD matrix template as show in Figure Q-1. (Typically use a spreadsheet software application such as Microsoft Excel.)

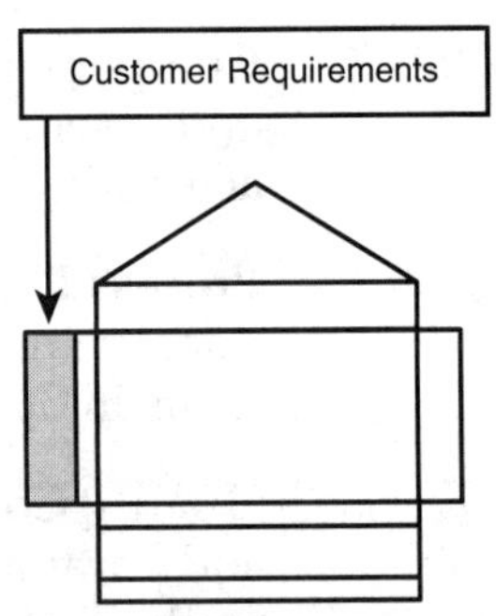

FIGURE Q-3: QFD Construction—Step 1.a.: Customer Requirements

Step 1. Customer Requirements and Priorities (What and Wants):

a. Document the customer requirements in the left-wing column of the house, which will serve as row headings as shown in Figure Q-3.

b. Adjacent to the customer requirements column, insert a column to document the customer rankings (or priorities) of each requirement relative to one another as shown in Figure Q-5.

Use an ordinal weighting scale method to represent the customer's importance ranking of their requirements. Ask the customers to stack rank the requirements during the VOC gathering initiative, avoiding "tie" rankings to better distinguish one need or want from another.

Note

The highest number reflects the most important requirement.

Ensure that the customers provide the weightings; they're not from internal judgments.

c. ***Competitive Assessment.*** In the far right-wing annex of the house, document the customer's perception of your company's offering(s) and similar competitive offerings, which serves as part of your competitive assessment. A source for this data may include customer satisfaction surveys, but also industry analysts and industry associations—and some benchmark data may provide the needed information.

Use a scale from 1 to 5, with **5 as the best performer** and **1 as the worst**. List the scale as column headings at the top of the right-wing annex.

Determine a set of icons that represent your company and the competition. Figure Q-4 provides example icons.

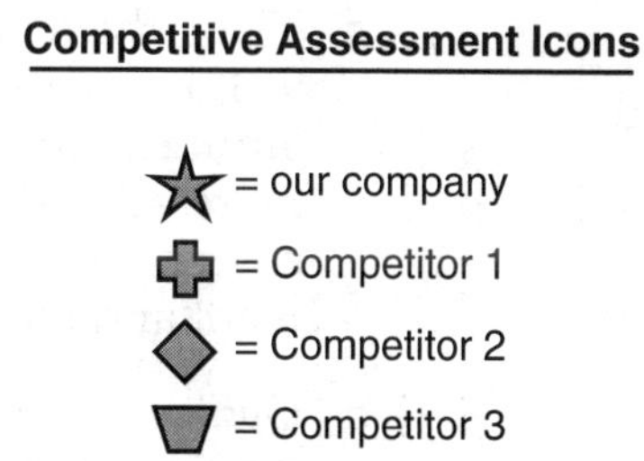

Figure Q-4: QFD Competitive Assessment Sample Icons

Place the appropriate company symbols aligned in the appropriate rating column to represent the customer's perception of the offerings from your company and the competition.

A B C D E F G H I J K L M N O P Q R S T U V W X Y Z

Figure Q-5 reflects the customer priorities for both their requirements (placed in the left-most gray column) and competitive assessment (in the right-most gray column) of the QFD. The black-shading indicates the QFD region already discussed.

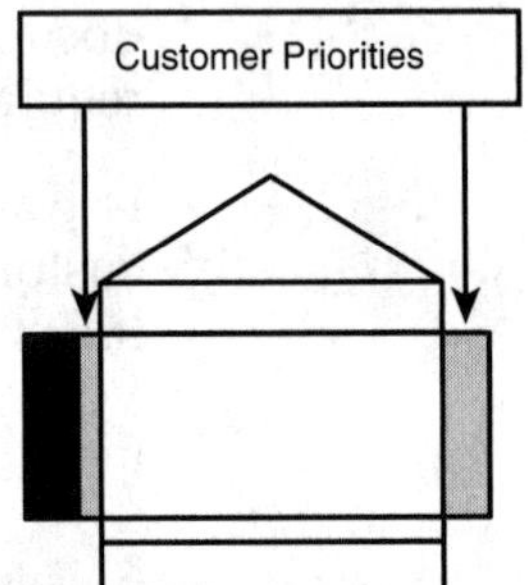

FIGURE Q-5: QFD Construction—Step 1.b.: Customer Priorities

Step 2. Technical Design Features, Functionality, and Characteristics (How):

a. Confirm the list of various design options (features, functionality, and characteristics) and refine it to ensure each directly affects a customer requirement and that each is described in terminology that is meaningful and actionable for your organization.

b. In the attic, list the various design options across the top to serve as column headings for the matrix underneath in the main part of the house, as illustrated in Figure Q-6.

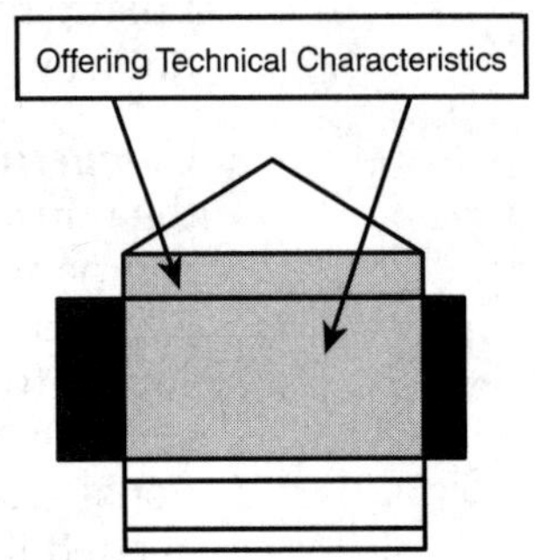

FIGURE Q-6: QFD Construction—Step 2.b. and c.: Technical Characteristics

c. Using the relationship scale of 9, 3, 1, 0 (with **9 as the strongest** relationship and **0 as no relationship**), rate the strength of each technical design option relationship with each customer requirement and record the rating in the appropriate cell.

d. *Optional*: Some QFD matrices insert a row just under the design characteristics to indicate the deployment direction of each item. Directionality defines how the requirement is best fulfilled, such that the options included hit the target requirement, or represent more or less than the targeted needs.

Hint

Typically, arrows indicate the directional relationship to the target: **up-arrow** signifies bigger is better; **down-arrow** represents smaller is better; and a zero or empty cell symbolizes that nominal is best.

Step 3. Prioritized Design Characteristics

a. Create a row just beneath the last customer requirement and label it "Total Ratings."

b. Working down one Design Feature column at a time, calculate its total prioritized score. To calculate this score, multiply each cell's relationship rating times the customer requirement weighting. Sum the arithmetic product in each cell for that entire column and record the total in appropriate cell of the Total Ratings row.

c. Repeat Step 3.b. until the Total Ratings are calculated for each of the Design Feature columns.

d. *Optional*: Sometimes the relative weighting of the Total Ratings better communicates the relative impact of one Design Characteristic over another. To calculate the Relative Rating, sum the scores of all the Total Ratings and then divide each Design Feature's Total Rating Score by the sum of all the Total Ratings. Record the appropriate proportion in each cell.

Step 4. Design Relationships (between features)

a. In the roof of the house, define the interrelationship or correlation between two different characteristics. This portion of the house sometimes is referred to as the *Correlation matrix*.

If a pair of characteristics support each other, such that as one increases so does the second to meet customer requirements, then the pair has a *positive relationship*. If one increases and the other decreases to meet customer requirements, then they have a *negative relationship*.

Negative relationships indicate requirement conflicts that are causes for concern.

b. Use a "P" to indicate a positive correlation, and an "N" for a negative correlation.

c. Rate each pair of technical features, functions, and characteristics and record the correlation rating in the appropriate cell in the roof. Take time to ensure that the proper two characteristics line up with the rating, given that the angularity of the roof may introduce alignment difficulty.

Figure Q-7 illustrates this step.

d. *Optional:* Total Correlation Ratings #1—If the design feature interrelationships have an impact on overall prioritization of the feature, the correlation rating should be factored into the overall rating.

i. Calculate the impact of the correlation. Translate these letters into a numeric score, with P equaling a +3 and N equaling a -3.

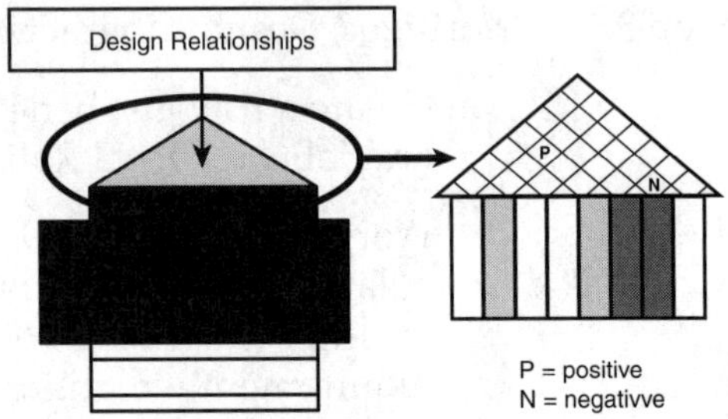

FIGURE Q-7: QFD Construction—Step 4: Design Relationships

ii. Create two rows just beneath the Total Ratings row and label the top row as the "Total Correlation Rating" and the last row as "Aggregate Rating."

iii. Sum each of the correlations scores for each Design Characteristic and record the total in the appropriate cell in the Total Correlation Rating row.

iv. Working by column, multiply the Total Correlation Rating times the Total Rating and record the arithmetic product in the appropriate cell of the Aggregate Rating row.

Repeat this sub-step until the Total Correlation Rating and Aggregate Rating scores are calculated for each of the Design Feature columns.

e. *Optional*: Total Correlation Ratings #2—If several pairs of characteristics have relationships with one another and more granularity is required, replace the P and N rating with the ±3 value with a new scale.

Use the scale of + or - 9, 3, 1, 0 to indicate the strength of the correlation, with ±9 indicating a strong positive or negative relationship and 0 signifying no relationship.

If this option is selected, follow the procedure outlined in Step 4.d. but use the ±9, 3, 1, 0 numeric scale rather than the ±3.

Step 5. The Foundation (target values)

a. The basement provides the foundation of the house of quality with objective measurements to evaluate each characteristic. Create a row at the bottom of the matrix (below the rating rows—Total Rating, Total Correlation

A B C D E F G H I J K L M N O P Q R S T U V W X Y Z

Rating, and Aggregate Rating). Label this new row "Measurement Unit."

b. Record the appropriate unit of measure for each Technical Design Characteristic. If the target value is known, record that number with the unit. Figure Q-8 illustrates Step 5.

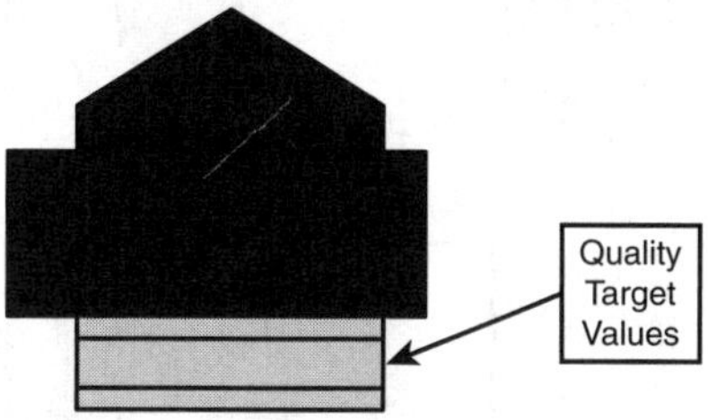

FIGURE Q-8: QFD Construction—Step 5: Target Values

Step 6. Summary

Synthesize all the information in the QFD matrix to decide the appropriate strategy and prioritization of design characteristics. Evaluate the competitive landscape and determine how the relative market positioning in the right-wing annex impacts the overall strategy and selection of features. Assess the required investment and determine whether it aligns with the business goals of the organization.

Warning

The design features yielding the highest Total Rating may not necessarily represent the best set of characteristics. Look at the QFD in total and understand the interrelationships of each room in context with the overall organizational objective.

Figure Q-9 shows a completed simple QFD matrix for a technology/professional consulting services business, comparing one company to three of its competitors.

Hints and Tips

A QFD structure can be as simple or as complex as the situation requires. When assembling a team to build a QFD, ensure that the team is cross-functional, with multiple perspectives represented.

Building a QFD rarely can be completed in one meeting. It takes several meetings, often over several weeks and even months to complete partially because of the input data required and partially due to the discussion needed to reach consensus on the relationship ratings.

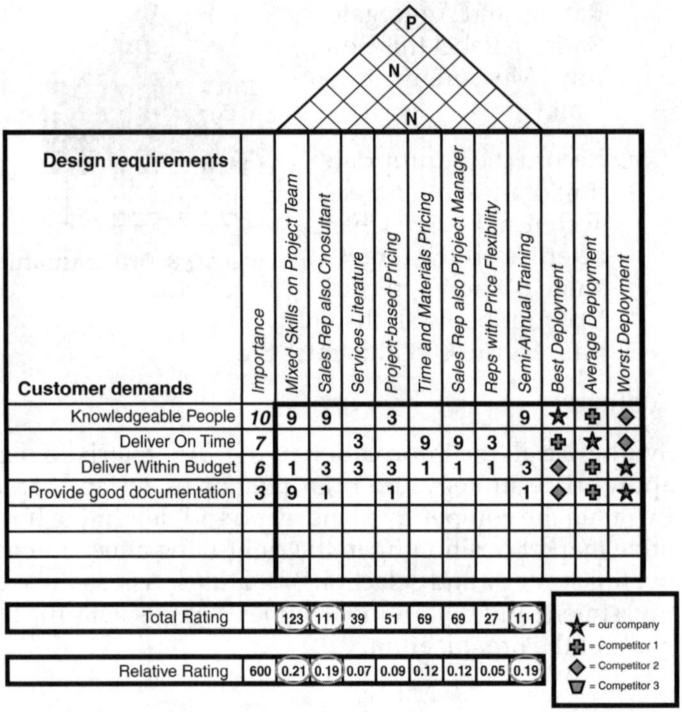

FIGURE Q-9: Simple Example of a QFD Matrix

Supporting or Linked Tools

Supporting tools that might provide input when developing a QFD matrix include

- Benchmarking (**See Also** "Benchmarking," **p. 160** and the article, "Benchmarking—Avoid Arrogance and Lethargy," in Part III, **p. PDF:789**)
- Brainstorming (**See Also** "Brainstorming Technique," **p. 168**)
- Data Collection Matrix (**See Also** "Data Collection Matrix," **p. 248**)
- Product Planning matrix (**See Also** "Matrix Diagrams—7M Tool," and "Solution Selection Matrix," **p. 399** and **p. 672**, respectively.)
- SWOT Analysis (**See Also** "SWOT (Strengths-Weaknesses-Opportunities-Threats," **p. 699**)
- VOC data (**See Also** "Voice of Customer Gathering Techniques," **p. 737**)

A completed QFD matrix provides input to tools such as

- Control Plan (**See Also** "Matrix Diagrams—7M Tool," **p. 399**)
- DOE (**See Also** "Design of Experiment (DOE)," **p. 250**)
- FMEA (**See Also** "Failure Modes and Effects Analysis (FMEA)," **p. 287**)
- Project Planning tools for development (**See Also** "Activity Network Diagram (AND)—7M Tool," **p. 127**, and "PERT (Program Evaluation and Review Technique) Chart," **p. 453**)

Figure Q-10 illustrates the link between a QFD matrix and its related tools and techniques.

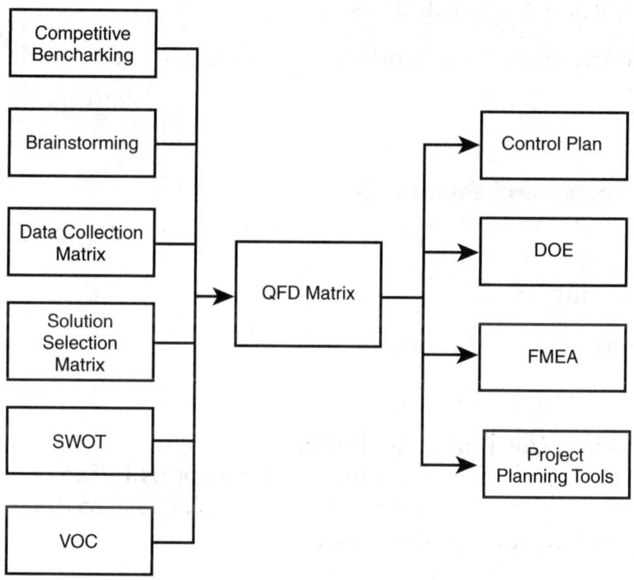

FIGURE Q-10: QFD Tool Linkage

Additional Resources or References

Cohen, Lou. *How to Make QFD Work for You*. Massachusetts: Addison Wesley Longman, Inc., 1995. ISBN: 0-201-63330-2.

R

RACI Matrix (Responsible, Accountable, Consulted, Informed)

A B C D E F G H I J K L M N O P Q R S T U V W X Y Z

What Question(s) Does the Tool or Technique Answer?

Who is responsible for what?

An RACI matrix helps you to

- Identify who is involved in producing a task and/or deliverable
- Communicate responsibilities
- Identify any gaps or redundancies associated with people's responsibilities

Alternative Names and Variations

This tool is also known as

- Responsibility chart
- Roles and Responsibilities chart

When Best to Use the Tool or Technique

To organize a project team or document the responsibilities of key process stakeholders with respect to deliverables or tasks. Use to document current state or improved state of a process.

Brief Description

Responsible, **A**ccountable, **C**onsulted, and **I**nformed (RACI) is a straightforward matrix tool that identifies who needs to be involved in a task or activity and/or who is to produce a deliverable or output. The RACI tool uses a simple L-matrix structure and may contain either all tasks, all deliverables, or a blend of the two. (**See Also** "Matrix Diagrams—7M Tool," **p. 399** for detail on different matrix structures.)

The RACI applies to two scenarios—a project team and the process players. When applied to a **project** team, the RACI designates a **person by name**. The RACI tool acknowledges the short-term nature of a project by using actual people's names to make assignments. The RACI in this context serves as a good recruiting tool to request a particular individual be assigned to the project team. By specifying a project-specific deliverable, selecting the optimal person with the appropriate expertise, knowledge, or functional background needed to produce that output becomes easier.

When applied to a **process**, the RACI specifies a **job title, role,** or **job function** associated with the process. This generic work assignment allows for individuals to be interchangeable while keeping the job description's responsibilities current. The RACI in this context defines the roles and responsibilities of the process players in each step of the workflow (task and/or deliverable). It describes either the current or the improved state of a process. The RACI serves as a good change management tool used to communicate and compare the pre- versus post-improvement conditions.

The RACI uses common language but has a specific definition of its terms:

- **Responsible (R)**—This person or role performs as the "worker bee." For any one given task or deliverable, it may have multiple people assigned to be responsible for it. These are the people required to do the work to produce the output or complete the task.
- **Accountable (A)**—This represents the **one** person to whom the responsible people are accountable. The accountable person is the single individual who must answer if the task is incomplete or fails to meet requirements. Any given task or deliverable may have only ONE accountable person assigned to it. The accountable person may also be a part of the other roles—responsible, consulted, or informed.
- **Consulted (C)**—This person is known as the subject matter expert. The consulted person possesses unique information or capabilities necessary to complete work. This person provides input—an information supplier. This role may or may not be required for any given task or deliverable.
- **Informed (I)**—This person needs to be notified of results and rarely needs to be consulted. Often this person represents the customer of a task or output either directly or downstream in the process. The informed person also may be a key stakeholder who needs to be apprised of the task completion. This role may or may not be required for any given task or deliverable.

The RACI tool is a powerful and flexible communication document. **Project**-based RACI sometimes expands the structure to include an additional column for due dates of the deliverables. **Process**-based RACI sometimes expands the structure to include an additional column for the deliverables' metrics. Figure R-1 illustrates a generic template for both a project-based and process-based RACI.

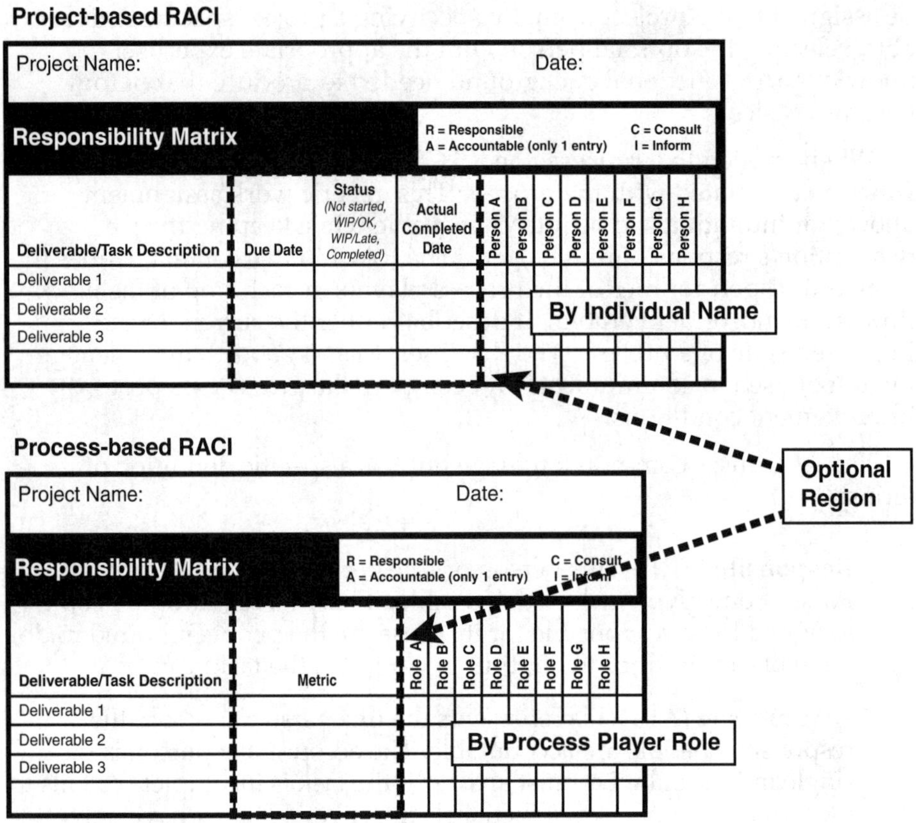

FIGURE R-1: Project Versus Process RACI Matrix Template

A B C D E F G H I J K L M N O P Q R S T U V W X Y Z

How to Use the Tool or Technique

Regardless if the RACI applies to a project or process, build it using the following procedure:

Step 1. Construct an L-shaped matrix. This can be done manually or with application software such as Microsoft Excel or the table functions in Word or PowerPoint.

Step 2. Identify the tasks or deliverables required and list them down the far-left column, the vertical axis, to set them up as row headings.

a. If constructing a task-based RACI, list the activities in the sequence that occur in the process.

b. A Process map provides good input for this step.

Step 3. Insert a blank row at the top of the matrix. Identify the job functions involved in the process and list them across the top row, the horizontal axis, as the column headings.

a. If this is a project-based RACI, substitute the roles with the project team members' names.

b. A Process map provides good input for this step.

Step 4. Working one row at a time, identify the various RACI roles for each task or deliverable and document the assignment in the corresponding cell.

a. Repeat this step until all the rows have been addressed.

b. Ensure that an accountability assignment appears in every row.

c. Review the matrix for accuracy.

Figure R-2 shows a project-based RACI, and Figure R-3 exhibits a process-based RACI.

Project: Sales Force Effectiveness

RACI Matrix

R = Responsible C = Consult
A = Accountable (only 1 entry) I = Inform

Deliverable/Task Description	Mark	Bill	Nancy	Diane	Corbin	Garret	Monica	Lauren		
Field Sales Job Descriptions	A	R	R	I	C	C	R	C		
Sales performance worldwide for 3 years		AR		I	C					
Basic Sales Training Curriculum	C			A	R	R	R	R		
Sales pipline data	AR	R	R		C					

FIGURE R-2: Project-based RACI Matrix

Project: Sales Force Effectiveness

RACI Matrix

R = Responsible
A = Accountable (only 1 entry)
C = Consult
I = Inform

Deliverable/Task Description	Sales Rep	Sales Manager	District Manager	District Controller	HQ Pricing Analyst	HQ Finance	Legal	Customer Admin		
Identify a prospect	AR	I/C	I							
Meet prospect	AR	I/C	I							
Discuss current issues / pain points	AR	I/C								
Determine if Decision-maker	AR									
Identify potential needs to satisfy	AR	I/C								
Develop value proposition and presentation	AR	I/C								
Approve presentation	R	I/C								
Present value proposition	AR	I								
Validate prospect's interest	AR	I	I							
Determine pricing	R	I	I/C	I/C	R	A	C	I		
Identify any special contract terms	AR	I	I/C	I/C	R	A	C	I		
Draft proposal	AR	I/C	I	I	R	I				
Approve proposal	I	R	A	I		C		I		
Inspect proposal	R	R	A	R	R	I/C				
Present recommendation to prospect	AR	I/C	I							
Negotiate any special conditions	R	R	A	R	R	I/C	C	I		
Gain approval signatures	R	A								
Close order	AR	I	I	I	I	I		R		
Evaluate customer satisfaction	I	I						AR		
Document lessons learned	AR	I								

Figure R-3: Process-based RACI Matrix

Hints and Tips

- One accountable person (or role) must be assigned to every task or deliverable.
- Empty cells are acceptable for a given deliverable or task, as long as the accountability is assigned.
- Multiple responsible, consulted, and informed roles may be assigned for any given task or deliverable.

- Multiple combinations may occur for a given person (or role) with an assigned task or deliverable (for example Accountable/Responsible (A/R), Consulted/Informed (C/I), Accountable/Informed (A/I)).
- A project-based RACI should identify individuals to the assignments rather than job titles, roles, or functions.
- A process-based RACI should identify job titles, roles, or functions rather than individuals' names to the assignments.
- The list of assignments may be 100% tasks, 100% deliverables, or a blend of tasks and deliverables.
- Reference a Process map to build either a task-based, deliverables-based, or hybrid RACI.
- When analyzing an RACI for improvement opportunities, look for any redundancies or gaps. Resolve the issue and revise the RACI and other supporting documentation.
- Adapt the structure with additional columns to expand its purpose, for example
    - Target completion date (for a given task and/or deliverable)
    - Progress to-date (for a given task and/or deliverable)
    - Completion date (for a given task and/or deliverable)
    - Metric (for a given output)

Supporting or Linked Tools

Supporting tools that might provide input when developing an RACI matrix include

- Brainstorming (**See Also** "Brainstorming Technique," **p. 168**)
- Process map (**See Also** "Process Map (or Flowchart)—7QC Tool," **p. 522**)

A completed RACI matrix provides input to tools such as

- Control plan (**See Also** "Matrix Diagrams—7M Tool," **p. 399**)
- FMEA (**See Also** "Failure Modes and Effects Analysis (FMEA)," **p. 287**)
- Process map (**See Also** "Process Map (or Flowchart)—7QC Tool," **p. 522**)

Figure R-4 illustrates the link between an RACI matrix and its related tools and techniques.

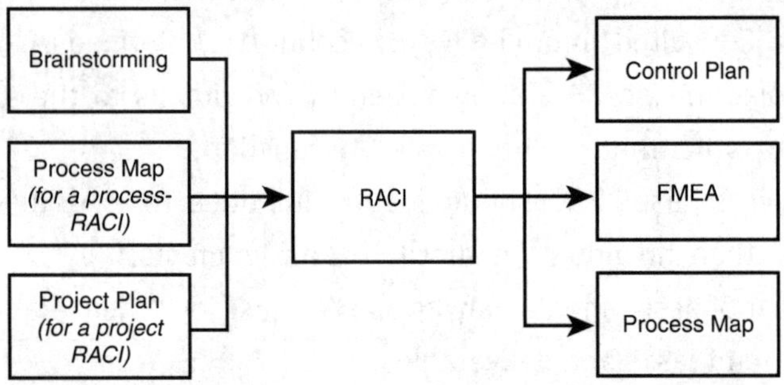

FIGURE R-4: RACI Matrix Tool Linkage

Real-Win-Worth (RWW) Analysis

What Question(s) Does the Tool or Technique Answer?

How would a potential offering be valued in the marketplace and positioned well against competition? Is it worth the investment to develop the idea? Would it be successful? Would the concept outpace competition?

An RWW analysis helps you to

- Evaluate candidate offering elements by using a numerical ranking and risk characterization technique
- Determine if the potential offering can win in the competitive environment and if it is worth the investment of time and resources to develop

When Best to Use the Tool or Technique

The Real-Win-Worth technique is best used during both the portfolio refresh process and the offering development and commercialization preparation process. During the portfolio renewal process, RWW evaluates whether an offering concept should be funded and activated as a development project. It can examine several potential portfolio concepts and prioritize them in terms of risk to help determine the best candidate(s) for investment.

The RWW tool carries over from the strategic portfolio renewal process into the tactical offering development process. The project team

A B C D E F G H I J K L M N O P Q R S T U V W X Y Z

updates the RWW analysis to continue to evaluate marketplace changes and update with any new information. The technique helps to understand whether the marketplace would value the offering, how it would fare against competition, and what related potential risks might be. This evergreen process continues throughout the time the offering is being designed, developed, and prepared for launch. (**See Also** "Six Sigma for Marketing (SSFM)," in Part I for additional detail in the sections on the IDEA and UAPL methods, **p. 73** and **p. 81**, respectively.)

Brief Description

The RWW technique is a *judgment enhancement* technique that helps teams make choices. It identifies those markets and opportunity characteristics that make investing in new elements of an offering portfolio viable. It prioritizes and identifies the RWW elements to be deployed as part of the portfolio renewal process and defines the appropriate timing. The RWW technique also aids product development teams to understand how the marketplace values a specific offering. RWW examines the offering (or offering component) relative to competition. It weights the investment needs of time and resources against the opportunity.

The technique comprises three components, each represented by its own matrix, to answer one of the following three RWW questions:

1. What are the **Real** market and opportunities?
2. Where can we **Win**?
3. Is it **Worth** our investment?

The **Real** market is defined by data that characterizes its purchasing capacity (or size), the purchasing dynamics (or growth), and the purchasing capability (or demand). The Real opportunities disclose whether or not the team gathered data that clearly defines the market need, if the concept aligns with the market need, and if the idea aligns with the firm's core competencies. Industry analysts and market segment associations can provide this data.

The ability to ***Win*** is predicated on two dimensions—the competitive environment and the firm's capabilities. RWW takes into account the timing, technologies and performance, pricing, brand identity, and market share to evaluate the competitive landscape. The firm's business capacity and capability to compete in this marketplace involve examining its core competencies in technology, design, production, and services. In addition, the firm's process and operational efficiencies are examined. Last, the firm's resources are evaluated, which include money, space, facilities, and equipment.

A B C D E F G H I J K L M N O P Q R S T U V W X Y Z

Whether the concept is **Worth** investment involves support from the firm's financial community to determine the potential Net Present Value (NPV), its potential Internal Rate of Return (IRR), and proposed pricing and costs scenarios. Opportunity costs should be explored to understand the required versus the available budget. The team assesses if the potential offering financials align with the strategic financial goals of the firm (or division). And the team evaluates the opportunity's risk relative to the other opportunities under consideration.(**See Also** "Cost/Benefit Analysis," **p. 238**, for a discussion on NPV and IRR.)

Before getting started, the project team needs to collect the following inputs:

1. Market, segment, and opportunity data
2. Competitive benchmarking; market trend; and behavioral analysis data
3. Economic business case forecasts
4. Business strategy, financial goals, and core competencies

(**See Also** "Voice of Customer Gathering Techniques," **p. 737**; "Benchmarking," **p. 160**; the article, "Benchmarking—Avoid Arrogance and Lethargy," in Part III, **p. PDF:789**; "Market Perceived Quality Profile (MPQP)," **p. 390**; and "SWOT (Strengths-Weaknesses-Opportunities-Threats," **p. 699**)

The power of the RWW technique becomes evident as the project team updates the tool. As portfolio renewal deliverables are refreshed, identifying and answering key RWW questions get updated with increasing clarity of data. As the team moves from the portfolio renewal process into the product development process, the RWW technique encourages the team to customize and add to the depth and number of questions asked by the tool. A well-implemented RWW analysis is iterative as an offering moves from concept to launch-readiness.

The RWW outputs include

1. Ranked values associated with each Real-Win-Worth criteria.
2. Ongoing development of data that builds insight on a phase-by-phase basis from portfolio renewal to offering development and commercialization.
3. Prioritized weighting of a concept risk and the worthiness of its investment needs against a real opportunity.

How to Use the Tool or Technique

Presuming that the first application of the RWW technique is in the portfolio renewal process, the project team needs to gather the appropriate input data on the marketplace, competitive environment, and internal goals and capabilities, as just mentioned.

For first-time users, a set of example RWW questions are provided in the following list to get started. However, the project team should customize the questions to better suit its unique business model and marketplace. Develop the RWW questions over time and across projects to translate the RWW questions for your own business and culture.

Step 1. Create three "Scorecard" matrices to assess each RWW dimension.

Create a set of three matrices, one for each of the three RWW characteristics: **Real** market and opportunities, likelihood to **Win**, and if the concept is **Worth** the investment.

This might best be completed in Microsoft Excel, to make for easy computation and navigation among the three matrices. Establish each Excel tab as a unique matrix—Tab 1 for Real; Tab 2 for Win; Tab 3 for Worth.

Construct each matrix with three columns. The far left column (column 1) lists the RWW questions; the middle column (column 2) contains the ranking values; and the far right column (column 3) is reserved for comments.

Step 2. Record the questions for each RWW characteristic.

Start with the baseline RWW questions and examine them by category—Real, Win, and Worth. Based on your business model, edit the baseline questions as appropriate. Add any additional questions to define the final set of RWW questions.

Record each individual question on its respective RWW matrix, placing one question per matrix cell within the far left column (column 1). Each question serves as a row heading.

a. **Matrix 1**: The **REAL** market and opportunity questions:

   i. Have we defined a legitimate market and segment arena in which to participate?

   ii. Have we quantified the size of the market?

   iii. Have we quantified the market growth rate?

   iv. Have we quantified the purchasing capacity and capability of the market players?

   v. Have we defined clear market trends and customer needs?

   vi. Do these market trends and needs align with our ideas and capabilities (for example, technologies)?

   vii. Does the market have a strategic fit with our business strategy for organic growth and core competencies?

   viii. Does the market have a strategic fit with our business strategy for external partnerships, joint ventures, and acquisitions?

b. **Matrix 2**: Can we **WIN** questions:

   i. Can we meet the launch cycle-time requirements to be first or early in the market?

   ii. Do we have technology and performance superiority relative to our competition?

   iii. Can we establish competitive pricing, and effectively win against competition?

   iv. Is our brand identity competitive?

   v. Can we gain market share?

   vi. Are our core competencies enabling us to compete?

   vii. Are our operational efficiencies enabling us to compete?

   viii. Do we have enough resources to compete in this market?

c. **Matrix 3**: Is it **WORTH** the investment questions:

   i. Is the potential NPV worth the investment?

   ii. Is the risk-adjusted NPV (ECV) worth the investment?

   iii. Does the Internal Rate of Return (IRR) potential meet our standards?

   iv. Can we balance price with our costs?

   v. Can we balance our consumption of resources in alignment with our operating budget?

vi. Is the opportunity there to support our business financial goals?

vii. Is this opportunity worth the risk in comparison to other opportunities?

Step 3. Assess each RWW question by rating the ability answer it.

The response score quantifies the level to which the project team can fulfill (or address) the needs of an RWW characteristic. The value of the score reflects either

a. The condition of the relevant marketplace, competition, or risk data that the team currently possesses

b. The team's usage level of supporting tools to gather the required data

RWW Strength-of-Fulfillment Scale—Within each matrix, using the rating scale provided, answer each question and record the number in the middle column:

a. **Score of 1** = based on internal assumptions, guesses, past experiences, opinions, feelings; or no supporting tool usage by the team

b. **Score of 3** = based on old, partial data sets and secondary sources of data (mostly internal interpretations); or some supporting tool usage by the team

c. **Score of 5** = based on new, complete data sets and primary sources of data (mostly external interpolations); or extensive supporting tool usage by the team

Step 4. Record any comments.

Identify any supporting comments, such as the tools used to collect pertinent data and the date that the data was last gathered.

a. Supporting tools

There are several supporting tools that the project team can draw upon to gather data on the RWW characteristics. The set of figures that follow show examples of some supporting tools that align with each question. Figure R-5 displays a sample **REAL** matrix with its supporting tools.

Real-Win-Worth Analysis

	REAL Criteria	Ranking Value	Comments	Supporting Tools
1	Have we defined a **legitimate market and segment arena** in which to participate?			*Market Definition, Segmentation and Opportunity Identification, SWOT, MPQP*
2	Have we qualified the **size** of the market?			*Market Definition, Segmentation and Opportunity Identification*
3	Have we quantified the **market growth rate**?			*Market Definition, Segmentation and Opportunity Identification, Mkt. Behavioral Dynamics Mapping*
4	Have we quantified the **purchasing capacity and capability** of the market players?			*Market Definition, Segmentation and Opportunity Identification, SWOT, Porter's 5 Forces*
5	Have we defined clear **market trends and customer needs**?			*Market Behavioral Dynamics Mapping, MPQP, VOC Gathering and Processing, Competitive Benchmarking*
6	Do the market trends and needs **align with our ideas and capabilities** (e.g. technologies)?			*Product Portfolio Architecting, Portfolio Ranking and Balancing, Idea Data base Documentation*
7	Does the market have a **strategic fit with our business strategy for organic growth and core** competencies?			*Innovation Strategy, Market Definition, Segmentation and Opportunity Identification, Portfolio Ranking and Balancing, Idea Data base Documentation*
8	Does the market have a **strategic fit with our business strategy for external partnerships, Joint Ventures** and **acquisitions**?			*Innovation Strategy, Market Definition, Segmentation and Opportunity Identification, Portfolio Ranking and Balancing, Idea Database Documentation*
	Total REAL Characteristic Fulfillment Score	0		

- **1** = *based upon internal assumptions, guesses, past experiences, opinions, feelings; or no supporting tool used.*
- **3** = *based upon old, partial data sets and secondary sources of data (mostly internal interpretations); or some supporting tool used*
- **5** = *based upon new, complete data sets and primary sources of data (mostly external interpolations); or extensive supporting tool used.*

FIGURE R-5: Real RWW Matrix (with Supporting Tools)

Figure R-6 displays a sample **WIN** matrix with its supporting tools.

Real-Win-Worth Analysis

	WIN Criteria	Ranking Value	Comments	Supporting Tools
1	Can we **meet the launch cycle-time requirements** to be first or early in the market?			*Portfolio Ranking and Balancing, Product Pipe-line Planning and Activation*
2	Do we have **technology and performance superiority** relative to our competition?			*Pugh Process, Competitive Benchmarking, Portfolio risk Analysis, MPQP, SWOT, idea Data base Documentation*
3	Can we **establish competitive pricing**?			*Competitive Benchmarking, Portfolio Risk Analysis, MPQP, SWOT, Business Case Analysis*
4	Is our **brand identity competitive**?			*Competitive Benchmarking, MPQP, SWOT, Porter's Forces*
5	Can we **gain market share**?			*Market Definition, Segmentation and Opportunity Identification, SWOT, MPQP, Competitive Benchmarking*
6	Can we **compete on price**?			*Market Definition, Segmentation and Opportunity Identification, SWOT, MPQP, Competitive Benchmarking*
7	Are our **core competencies enabling** us to compete?			*Market Definition, Segmentation and Opportunity Identification, SWOT, MPQP, Innovation Strategy*
8	Are our operational **efficiencies enabling** us to compete?			*Portfolio FMEA, Innovation Strategy, Product Pipe-line Planning and Activation*
9	Do we have **enough resources** to compete in this market?			*Portfolio FMEA, Innovation Strategy, Product Pipe-line Planning and Activation*
	Total WIN Characteristic Fulfillment Score	0		

- **1** = *based upon internal assumptions, guesses, past experiences, opinions, feelings; or no supporting tool used.*
- **3** = *based upon old, partial data sets and secondary sources of data (mostly internal interpretations); or some supporting tool used*
- **5** = *based upon new, complete data sets and primary sources of data (mostly external interpolations); or extensive supporting tool used.*

FIGURE R-6: Win RWW Matrix (with Supporting Tools)

Figure R-7 displays a sample **WORTH** matrix with its supporting tools.

Real-Win-Worth Analysis

	WORTH Criteria	Ranking Value	Comments	Supporting Tools
1	Is the **potential NPV** worth the investment?			*Business Case Analysis, Portfolio FMEA*
2	Is the **risk adjusted NPV (ECV)** worth the investment?			*Business Case Analysis, Portfolio FMEA*
3	Does the **Internal Rate of Return potential** meet our standards?			*Business Case Analysis, Portfolio FMEA*
4	Can we **balance price with our costs?**			*Business Case Analysis, Portfolio FMEA, MPQP, SWOT*
5	Can we **balance our consumption of resources** in alignment with our operating budget?			*Portfolio FMEA, MPQP, SWOT, Product Pipe-line Planning and Activation*
6	Is the opportunity able to support our **business financial goals?**			*Market Definition, Segmentation and Opportunity identification, Business Case Analysis, Portfolio FMEA*
7	Is this opportunity **worth the risk in comparison to other opportunities?**			*Market Definition, Segmentation and Opportunity identification, Idea Data base Documentation, Business Case Analysis, Portfolio FMEA, Pugh Process*
	Total REAL Characteristic Fulfillment Score	0		

- 1 = *based upon internal assumptions, guesses, past experiences, opinions, feelings; or no supporting tool used.*
- 3 = *based upon old, partial data sets and secondary sources of data (mostly internal interpretations); or some supporting tool used*
- 5 = *based upon new, complete data sets and primary sources of data (mostly external interpolations); or extensive supporting tool used.*

FIGURE R-7: Worth RWW Matrix (with Supporting Tools)

Step 5. Sum the scores within each RWW characteristic matrix.

Within each of the three matrices, add the scores associated with each question found in the middle column to compute the individual Real, Win, and Risk characteristic scores.

Step 6. Compute the aggregate RWW score.

Add the total scores from each individual Real, Win, and Risk matrix to determine the overall RWW strength-of-fulfillment score. The three-score tally reflects the offering's RWW opportunity, with the largest scores reflecting the least risky investment.

How to Analyze and Apply the Tool's Output

The RWW analysis can apply to two key process arenas: the strategic portfolio renewal process and the tactical offering development and commercialization preparation process.

In the portfolio renewal process, the strategy team probably examines several future offering concepts to select which idea(s) will receive investment funding and thereby activate a development project. Use the RWW analysis in conjunction with the other portfolio renewal tools to help guide the balance of your portfolio. The largest RWW score (representing the lowest risk) is a desirable score but is not necessarily the best score, but it should not necessarily eliminate offerings with smaller scores. Recall that the RWW Analysis is a *judgment enhancement technique*. Nonetheless, an RWW analysis is a judgment based on a snap-shot in time, and market conditions and competition inevitably change. The team may be blinded by what it "desires" to be funded and might answer the RWW analysis questions in a biased manner. Hence, a balanced portfolio of RWW scores may represent a less risky position. In summary, when faced with uncertainty, an investment portfolio should represent a mixture of risk and a mixture of scores.

The development/commercialization project team will continue to revise the RWW analysis from the portfolio renewal team. At this stage, the technique refines the project team's understanding of the commercialization efforts needed to win in the competitive environment.

Examples

Sample RWW Analysis

Figure R-8 shows the set of three characteristics matrices linked as one RWW analysis for one product concept completed during a portfolio renewal process.

Hints and Tips

- Use data driven facts to complete the individual RWW matrices.
- Ensure the completed RWW tool is part of an evergreen process from portfolio renewal through the end of the product development and market-launch preparation process.
- This technique enhances judgment as to the best concepts in which to invest; there is no one right answer. The largest RWW score may not necessarily eliminate a concept that received a lower score. Attempt to build a mix of RWW scores to balance the portfolio of concepts to receive funding as an activated development project.

Supporting or Linked Tools

Supporting tools, techniques, and best practices that might provide input when developing an RWW analysis include:

- Business Case Analysis
- Competitive Benchmarking (**See Also** "Benchmarking—Avoid Arrogance and Lethargy," in Part III, **p. PDF:789** for more information.)

REAL Criteria	Ranking Value	Comments
Have we defined a **legitimate market and segment arena** in which to participate?	3	*Market Definition and SWOT completed on 6/25/05*
Have we qualified the **size** of the market?	5	*Market Definition completed on 6/25/05*
Have we quantified the **market growth rate**?	5	*Segmentation and Opportunity Identification completed on 7/1/05*
Have we quantified the **purchasing capacity and capability** of the market players?	1	*SWOT and Porter's 5 Forces completed on 6/25/05*
Have we defined clear **market trends and customer needs**?	3	*Market trends tracking completed on 6/1/05*
Do the market trends and needs **align with our ideas and capabilities** (e.g. technologies)?	5	*Technical Portfolio Ranking and Balancing completed on 7/15/05*
Does the market have a **strategic fit with our business strategy for organic growth and core** competencies?	5	*Innovation Strategy documented on 5/9/05*
Does the market have a **strategic fit with our business strategy for external partnerships, Joint Ventures** and **acquisitions**?	3	*Innovation Strategy documented on 5/9/05-*
Total REAL Characteristic Fulfillment Score	**30**	

WIN Criteria	Ranking Value	Comments
Can we **meet the launch cycle-time requirements** to be first or early in the market?	5	
Do we have **technology and performance superiority** relative to our competition?	3	*Pugh Process and SWOT completed on 6/25/05*
Can we **establish competitive pricing**?	3	*Competitive Benchmarking completed 7/1/05*
Is our **brand identity competitive**?	3	*Competitive Benchmarking completed 7/1/05*
Can we **gain market share**?	3	*SWOT completed on 6/25/05*
Can we **compete on price**?	3	*SWOT completed on 6/25/05*
Are our **core competencies enabling** us to compete?	5	*SWOT completed on 6/25/05*
Are our operational **efficiencies enabling** us to compete?	3	*Portfolio FMEA completed on 7/705*
Do we have **enough resources** to compete in this market?	3	*Portfolio FMEA completed on 7/705*
Total WIN Characteristic Fulfillment Score	**31**	

WORTH Criteria	Ranking Value	Comments
Is the **potential NPV** worth the investment?	5	*Business Case Analysis, Portfolio FMEA completed on 7/7/05*
Is the **risk adjusted NPV (ECV)** worth the investment?	3	*Business Case Analysis, Portfolio FMEA completed on 7/7/05*
Does the **Internal Rate of Return potential** meet our standards?	3	*Business Case Analysis, Portfolio FMEA completed on 7/7/05*
Can we **balance price with our costs**?	5	*Business Case Analysis, Portfolio FMEA completed on 7/7/05*
Can we **balance our consumption of resources** in alignment with our operating budget?	5	*Business Case Analysis, Portfolio FMEA completed on 7/7/05*
Is the opportunity able to support our **business financial goals**?	3	*Business Case Analysis, Portfolio FMEA completed on 7/7/05*
Is this opportunity **worth the risk in comparison to other opportunities?**	5	*Business Case Analysis, Portfolio FMEA completed on 7/7/05*
Total REAL Characteristic Fulfillment Score	**29**	

- **1** = *based upon internal assumptions, guesses, past experiences, opinions, feelings; or no supporting tool used.*
- **3** = *based upon old, partial data sets and secondary sources of data (mostly internal interpretations); or some supporting tool used*
- **5** = *based upon new, complete data sets and primary sources of data (mostly external interpolations); or extensive supporting tool used.*

Total RWW Score: 90

Figure R-8: Sample RWW Analysis

- Idea Database Documentation
- Innovation Strategy
- Market Behavioral Dynamics Mapping
- Market Definition
- MPQP (estimated Market Perceived Quality Profile) (**See Also** "Market Perceived Quality Profile (MPQP)," **p. 390**)
- Offering Pipe-line Planning and Activation
- Porter's 5 Forces (**See Also** "Porter's 5 Forces," **p. 464**)
- Portfolio Architecting, Ranking and Balancing (**See Also** "Selecting Project Portfolios Using Monte Carlo Simulation and Optimization," in Part III, **p. PDF:921**)
- Portfolio Risk Analysis (FMEA) (**See Also** "Failure Modes and Effects Analysis (FMEA)," **p. 287**)
- Pugh Process (**See Also** "Pugh Concept Evaluation," **p. 534**)
- Segmentation and Opportunity Identification
- SWOT (Strengths-Weaknesses-Opportunities-Threats) (**See Also** "SWOT (Strengths-Weaknesses-Opportunities-Threats," **p. 699**)
- VOC Gathering and Processing (**See Also** "Voice of Customer Gathering Techniques," **p. 737**)

A completed RWW analysis provides input to tools such as

- Launch plan (**See Also** "Matrix Diagrams—7M Tool," **p. 399**)
- Communication plan (**See Also** "Matrix Diagrams—7M Tool," **p. 399**)
- Marketing materials and collaterals
- FMEA (**See Also** "Failure Modes and Effects Analysis (FMEA)," **p. 287**)
- Control plan (**See Also** "Matrix Diagrams—7M Tool," **p. 399**)

Figure R-9 illustrates the link between the RWW analysis and its related tools and techniques.

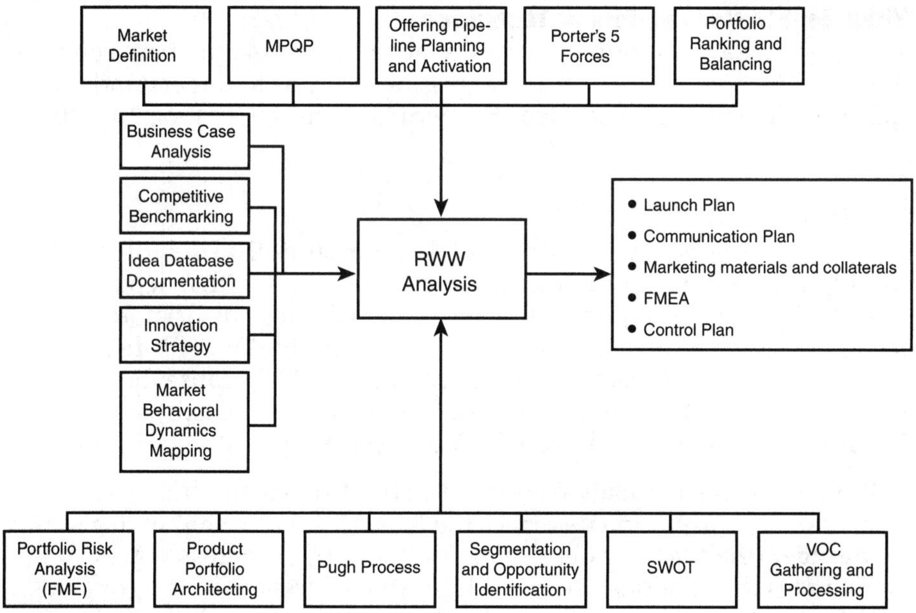

FIGURE R-9: RWW Analysis Tools Linkage

Regression Analysis

What Question(s) Does the Tool or Technique Answer?

What is the cause-and-effect model that describes the process and its critical variables?

Regression analysis helps you to

- Model and predict results (of a dependent variable) based on mathematical manipulation of one or more independent variables
- Define a statistical model to estimate the causality relationship, thus saving time, money and resources predicting, optimizing, and controlling a business outcome

Alternative Names and Variations

This technique includes different approaches, depending on the available data:

- Linear Regression
- Multiple Regression
- Logistic Regression

When Best to Use the Tool or Technique

After drawing a Scatter diagram to visualize the correlation between two numerical sets of data, conduct a regression analysis to understand any causality relationship. (**See Also** "Scatter Diagram—7QC Tool," **p. 640**)

Brief Description

Regression analysis is a statistical tool that measures the strength of relationship between one or more independent variables with a dependent variable. It develops a quantitative model that relates one or more independent variables (Xs) to a single dependent variable (Y). This relationship is often referred to as "the regression of X on Y" because the independent variable is plotted on the X-axis (horizontal axis), and the dependent variable is plotted on the Y-axis (vertical axis).

It builds on the correlation concepts to develop an empirical, data-based model. **Correlation** describes the X and Y relationship with a **single number** (the *Pearson's Correlation Coefficient (r)*), whereas **regression** summarizes the relationship with a **line**—the *regression line*. Regression fundamentally operates on the principle that "if X causes Y and X is known, then Y can be predicted." If the independent variables (X) can be adjusted and controlled, then the regression model can assist in controlling or optimizing the output (Y).

This regression analysis models the **Y = f(x)** equation (that is, Y is a function of X), where Y is the output or response variable, and the X represents the critical variables in the given process and its inputs that impact Y.

The regression model presumes that a system (either physical or behavioral) is governed by physics that can be expressed by the following equation:

$$y = \underbrace{\beta_0 + \beta_1 x}_{\text{physics}} + \underbrace{\varepsilon}_{\text{variation (noise)}}$$

Notice that this is the algebraic equation for a line (y = mx + b) plus an error-term. The terms β_0 is the slope of the line, and β_1 is the y-intercept, such that if X=0, then the β_1 would fall on the Y-axis. Both β_0 and β_1 are constants. The error-term is denoted by ε. If the equation had zero error, and X were known, then Y could be calculated.

The sample data (X) is gathered to represent a population. Thus, these observations (or sample data) resemble the physics of the population's system plus any noise. The objective is to predict the population's system. The *predicted value* (y-hat, or $\hat{y}$) represents the physics portion of the equation, denoted as the beta β variables and the observation variable (X), namely $\beta_0 + \beta_1 X$.

A B C D E F G H I J K L M N O P Q R S T U V W X Y Z

The study of baseball lends itself well to regression analysis. Baseball enthusiasts know the importance of statistics to their sport. Economics professor J.C. Bradbury, of Kennesaw State University in Georgia applies multiple regression to the volumes of baseball historical data and wrote a book (*The Baseball Economist: The Real Game Exposed*) based on his findings. Bradbury uses statistics to describe baseball facts from fiction and bust some prevalent myths as to what makes a team successful. Bradbury claims to have identified baseball's critical success factors that if adjusted and controlled can produce a winning season. (**See Also** "Science Journal: A New Study Sows How Baseball Myths Can Hurt the Game," in the *Wall Street Journal*, Friday, February 16, 2007, page B1.)

The regression analysis strives to find the best model to fit the data. The best-fit line passes through the data points where the *smallest possible* total sum of the square of all the residuals. The calculation squares the difference between the residual values and the predicted values that fall on the regression line to accommodate for the distance of the sample data on either side of the line—regardless if it is a positive or negative difference. Thus, the regression analysis (usually) uses the *Least Squares method* to determine the best fit by selecting the minimum sum of the squared deviations about the regression line. The approach uses the smallest sum of all the squared residual values calculated for each data point to determine the regression line.

This model uses a regression line (best-fit line) to estimate the physics in the presence of noise. The noise is estimated by the residual value. The sample data helps make assumptions about the error (ε), such that the observed value (Y) minus the fitted (predicted) value on the best-fit line yields the **residual value**. The noise is the variation between the observed data and the best-fit line, as shown in Figure R-10.

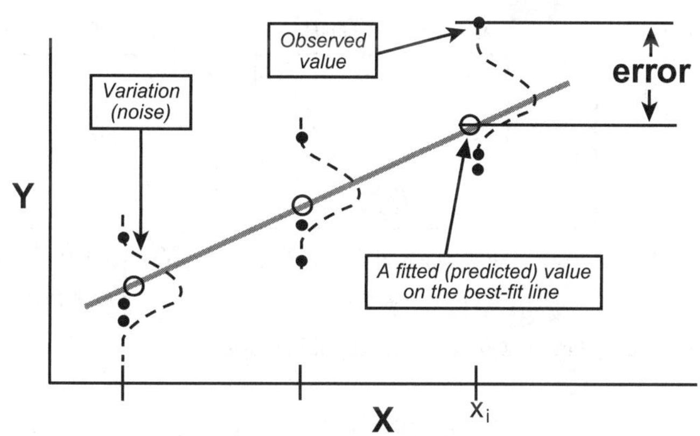

Figure R-10: Fitted Line Plot Illustrating Error

Linear Regression Assumptions

Hypothesis: The *Null hypothesis* states that the known X values are unable to predict Y. Thus any change in X has no effect on Y; the Y values remain horizontal—the slope of the regression line is zero. The regression line's slope is proportional to the correlation coefficient, such that if the one is zero, so is the other. The regression analysis can use the correlation coefficient as a surrogate for the slope of the regression line—called *Pearson's Correlation Coefficient* **(r),** and r=0. Conversely, the ***alternative hypothesis*** states that the true relationship between the variables is linear; wherein r = 1 or r = -1.

Minimum Sample Size: 20 data points.

Normally distributed data: The residuals are assumed to be normally distributed.

Residuals: The regression model assumptions about the residuals (error, ε) are three-fold: *independence, normally distributed*, and have a *constant variance*. The last assumption of constant variance is the most important of the three. The condition of independence ensures that the values of each variable are a result of **random sampling**.

Non-Linear Regression

There are alternative types of regression modeling other than linear. Multiple regression not only models multiple variables, but also may introduce interactions, quadratic, and/or cubic factors in the equation to better fit the model. An interaction comprehends the effect that two or more variables have on one another. A quadratic is the squaring of a term [X^2—multiplying by itself once (X*X)], and a cubic is the cubing of a term (X^3—multiplying by itself twice (X*X*X)). Figure R-11 compares these different model types drawn in MINITAB.

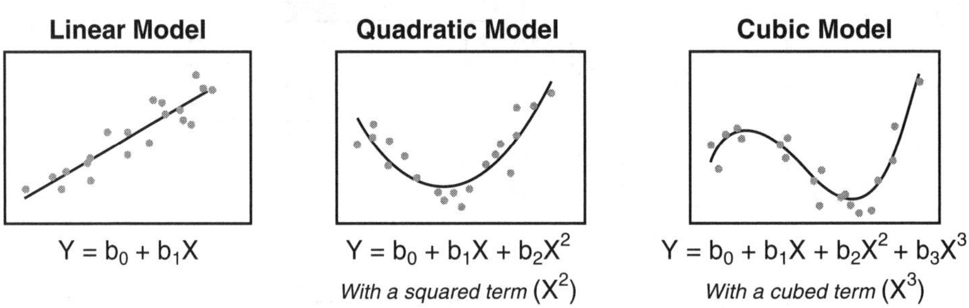

FIGURE R-11: Comparing Linear, Quadratic, and Cubic Models

Multiple regression can be dissected into multiple linear regression modeling multiple independent variables and multivariate regression modeling multiple independent and dependent variables. The formula for testing the multiple regression's Null hypothesis uses the ratio of variances as an F-statistic. (**See Also** "Analysis of Variance (ANOVA)—7M Tool," **p. 142**)

Logistic regression uses the Logistic or Logit function to model the system's scenarios involving non-normally distributed data or discrete variables such as categorical, binary, and ordinal variables.

Complex forms of regression using the *splines technique* may be useful. Splines is a statistical technique that uses a mix of different lines and curves to generate a better fitting model. Scenarios that may call for this type of modeling include product lifecycles. A product lifecycle curve may be generalized to encompass four segments. It starts with an S-curve slowly climbing as the product is introduced in the marketplace. Next, (ideally) the product experiences exponential growth. Third, the curve levels off as it matures, and sales growth flattens as the market gets saturated. Last, as the product approaches end-of-life, the curve follows a steep decline along another S-curve.

As with simple linear regression, the shape of curve (the mathematical function) is determined by the errors in predicting the dependent variables. However, some of the more complex regression models must use the *Maximum Likelihood method* to determine the best fit, rather than the Least Squares method used for linear regression.

Hints and Tips

Request the assistance of a statistician to conduct these types of complex regression analyses.

Interpreting the Residual Plots

Regression analysis can be completed manually, but it is very tedious, thus it is recommend to use a statistical software package such as MINITAB.

Graphical output is an essential component of a regression analysis. MINITAB provides a single four-in-one residual plot that assists in evaluating the three regression assumptions about the residuals (independence; normally distributed; and have a constant variance). The four-up graph, show in Figure R-12, includes a Normal Probability plot, histogram, Scatter plot around the regression line, and a Time Series plot of the residuals.

(**See Also** the "Normal versus Non-normal Data" section in the "Control Charts" entry, **p. 227**, for the Normal Probability plot; "Histogram—7QC Tool," **p. 330**; "Scatter Diagram—7QC Tool," **p. 640**; and "Run Chart—7QC Tool," **p. 610** for more detail on each graphical tool.)

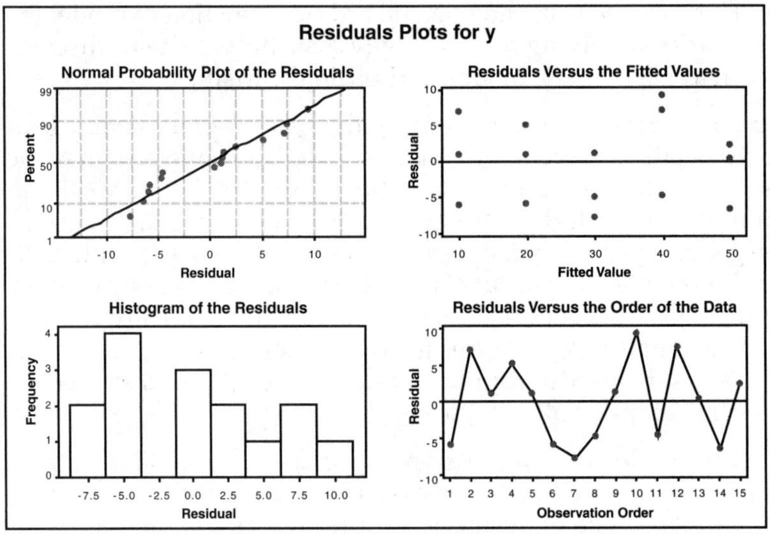

FIGURE R-12: Example of a MINITAB Residual Plot (Four-in-One)

Dissect Figure R-12 into its four components. The upper-left quadrant depicts a linear normal probability plot of residuals to support the normality assumption. The lower-left quadrant features a histogram of the residuals to provide another tool to evaluate the normality assumption. The Scatter plot of residuals versus the fitted value, displayed in the upper-right quadrant, portrays a random pattern to support the constant variance assumption. The forth quadrant in the lower-right exhibits a time-ordered plot of residuals with a random pattern to support the independence assumption. In summary, all four graphical tools displayed in Figure R-12 indicate that the regression model is adequate, based on the residual assumptions.

Common Patterns

As previously stated, the residual assumption of constant variance calls for a no pattern to be displayed in the Scatter diagram of residuals versus the fitted value. Often this random distribution is referred to as a *buckshot* pattern, shown in Figure R-13, signaling that the model is adequate.

Figure R-13 illustrates the buckshot pattern and three others—the Funnel, Smile (or Frown), and Football patterns. Each of these remaining three Scatter plots contain non-random patterns showing an inconsistent variance of the residuals. Because their patterns are non-random, the regression model is inadequate, and the graphs have an X drawn through them.

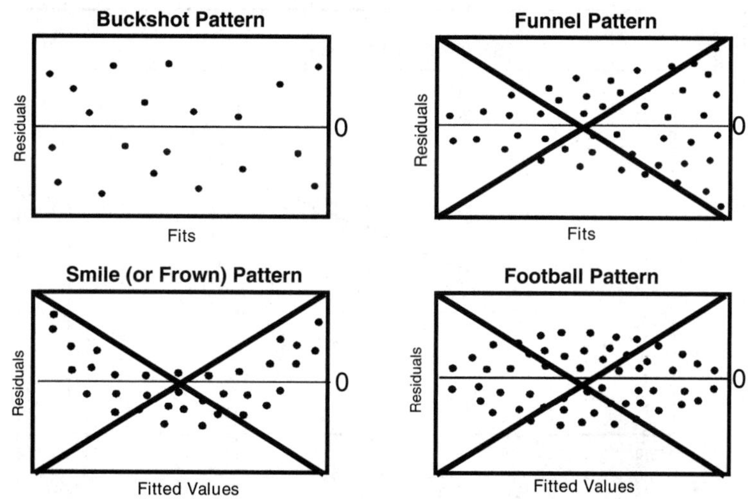

FIGURE R-13: Common Patterns of Residuals Versus Fitted Values

In the Funnel pattern, the error increases as Y increases. This model fits better at the low-Y values than at the high-Y values. A transform may be needed.

In the Smile (or Frown) pattern, the model fits better at the mid-Y values than at low- and high-Y values. A higher-order model maybe needed, such as a quadratic, as shown in Figure R-11.

In the Football pattern, the model fits better at the low- and high-Y values than at the mid-Y values. A transform may be needed.

Interpreting the Statistical Results

After creating a regression model, there are two key results that indicate whether or not the model is a good fit. The regression analysis focuses on the two statistics—the R-Square-adjusted term and the p-value.

R-Square-adjusted

Given the Scatter plot in Figure R-14, which line best fits the data? The answer is revealed in conducting a regression analysis. The objective to determine the best fitting model is to identify the largest possible R-Square (r^2 or R^2) value, also called the *coefficient of determination*. The coefficient of determination is calculated by squaring the correlation coefficient, and has a value between 0 and 1.

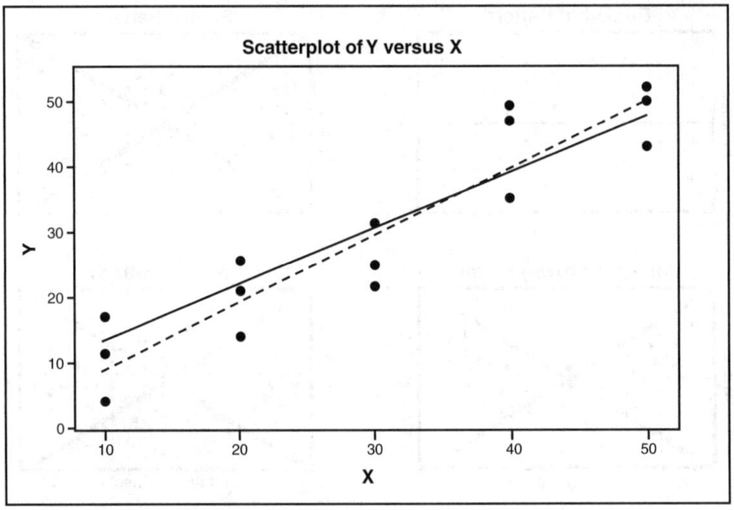

FIGURE R-14: Which Line Is the Best Fitting?

R-Square increases with every additional predictor (independent variable) added to the model even if the predictor adds little value to the model. To compensate for this effect, the statistical term called *R-Square-adjusted* ($r^2_{adjusted}$ or $R^2_{adjusted}$) is introduced. R-Square-adjusted is like R-Square except that it is allowed to decrease if non-useful predictors are added to the model. R-Square-adjusted reinforces the notion of a *parsimonious model*—the concept of striving for the most conservative, sparse, frugal, simple model. In general, the R-Square-adjusted value is the best term to assess how good a model is. The formula for R-Square-adjusted ($R^2_{adjusted}$) is as follows:

$$R^2_{\text{adjusted}} = 1 - (1 - R^2)\left[\frac{n-1}{n-p}\right],$$

where n is the sample size, and p is the number of predictors (independent variables) used in the model.

P-value

The Predictor's P-value—When building a regression model with more than one independent variable, often an iterative process of elimination is used to identify the critical few predictors needed to achieve the most parsimonious model. The p-value helps to identify which predictor in the model is useful. Those independent variables deemed useful in the regression model will yield a **p-value < 0.05**, indicating that it is statistically distinguishable from the other predictors.

Recall the Null hypothesis states that the known X values are unable to predict Y, and the Alternative hypothesis states that the true relationship between the variables is linear (r=1).

The Lack of Fit P-value for ANOVA Residual Error—The p-value also indicates the amount of fit the regression model has applied when examining the resulting Residual Error's Lack of Fit portion of the Analysis of Variance (ANOVA) test. The Lack of Fit p-value indicates if the linear predictors (Xs) alone are sufficient to explain the error in the model.

The small p-value indicates that the predictors (Xs) are not sufficient in explaining the variation in the response (Y). With a p-value < 0.05 for the Residual Error Lack of Fit, the model needs improvement, and interactions of the significant predictors should be used; for example add a quadratic or interaction term one at a time and rerun the model. The Null hypothesis is rejected.

Conversely, when the ANOVA Residual Error Lack of Fit's p-value > 0.05, it indicates that there is no lack of fit. There is insufficient evidence to reject the Null.

For the ANOVA Residual Error Lack of Fit test, the Null hypothesis states that the known X values (predictors) alone are able to predict the variance in Y (for example, variation is zero); and the Alternative hypothesis states that the known X values (predictors) are unable to predict the variance in Y. Refer to the following section on terminology for more detail. (**See Also** "Hypothesis Testing," **p. 335** for more information on the p-value and "Analysis of Variance (ANOVA)—7M Tool," **p. 142**)

Predictions

Models, such as regression, often are used to predict something, such as a cause-and-effect relationship. Regression attempts to statistically model the effect one or more (independent) variables has on another (dependent) variable.

Regression predicts (or estimates) the value of a dependent variable (Y) using independent variables (X) not found in the sample data. *Interpolation* describes the predicting of an outcome (Y) when the unknown independent variables (X) fall within the range of the known sample data values. *Extrapolation* defines predicting an outcome (Y) based on unknown independent variables (X) that fall outside the range of the sample data values. Interpolation yields far more accurate predictions, hence it is the preferred approach between it and extrapolation.

This type of prediction is not necessarily predicting the future. If the independent variable were time, then the regression analysis could only utilize historical data to extrapolate in the future. Predicting the *future* is difficult and might best be accomplished using statistical tools such as Monte Carlo simulation to forecast possible outcomes. Monte Carlo simulations are non-deterministic, providing a range of probable possibilities. (**See Also** "Monte Carlo Simulation," **p. 431**)

How to Use the Tool or Technique

Regression analysis can be conducted manually or by using computer software. The following procedure uses MINITAB to illustrate the technique. The procedure covers both simple linear regression and multiple regression.

The procedure to run a logistic regression in MINITAB follows a similar approach as that described for the multiple regression. However, logistic regression is not covered in this book. Logistic regression concepts begin to embark on some of the more complex regression models, thus proper setup and analysis may best be done with the assistance of a statistician.

Preparation: Determine the relationship that needs to be studied, collect the data, and enter the numeric data has been entered into the MINITAB Worksheet.

Simple Linear Regression

Step 1. Graph the data first.

a. Graph the relationship on a Scatter diagram and examine the fitted line. To do so, select the following commands from the MINITAB drop-down menu: ***Stat > Regression > Fitted Line Plot....***

b. On the Fitted Line Plot main screen, select the columns containing the appropriate independent and dependent variables to place in the corresponding dialog boxes. Unless the data contains a quadratic or cubic function, keep the Type of Regression Model on its default of Linear, as illustrated in Figure R-15.

c. Click OK to generate the subsequent Scatter diagram with the initial regression model and its best-fit line for the sample data, as illustrated in Figure R-13.

d. MINITAB provides both the proposed regression model and some statistics to evaluate how well the model fits the data, as depicted in the Scatter diagram found in Figure R-15. In this example, the linear regression model is: Y = 0.033 + 0.9900X. The R-Square is reported at 86.9%.

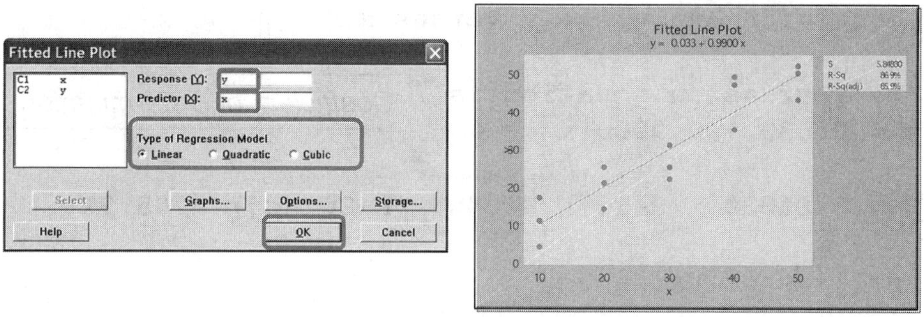

FIGURE R-15: Example of a MINITAB Regression Fitted Line Plot Main Screen and Resulting Scatter Diagram

Step 2. Analyze the regression model.

a. MINITAB's resulting session window provides the necessary statistics to evaluate how well this model fits, shown in Figure R-16.

b. The coefficient of determination (R-Square) appears reasonably strong at 86.9%, indicating an excellent relationship. (**See Also** "Rule of Thumb" in the Hints and Tips section of this entry, **p. 595**, for detail on the strength of the relationship.)

c. The ANOVA test reported a p-value of 0.000 for the regression (line), thus less than the conventional threshold of a p-value < 0.05, indicating that the regression line's slope is statistically different from zero. Based on the p-value of 0.000, the Null hypothesis is rejected.

> **Note**
>
> The Null hypothesis states that the regression line's slope is zero, showing no linear relationship between the independent and dependent variables.
>
> MINITAB's Fitted Line Plot (shown in Figure R-15) also provides the R-Square information, along with R-Square-adjusted and the standard deviation. However, the Session Window provides much more information.

```
Regression Analysis: y versus x

The regression equation is
y = 0.033 + 0.9900 x

S = 5.84830   R-Sq = 86.9%   R-Sq(adj) = 85.9%

Analysis of Variance

Source      DF       SS       MS      F      P
Regression   1  2940.30  2940.30  85.97  0.000
Error       13   444.63    34.20
Total       14  3384.93
```

Coefficient of Determination

H_0: Slope is zero, is rejected.

FIGURE R-16: Example of a MINITAB Regression Fitted Line Plot Session Window

Multiple Regression

Scenario: For purposes of illustration, the following physical system example is used to describe the procedure to conduct a multiple regression.

An engineer is developing a medical product. The technology requires heating and cooling a volume of fluid contained in a plastic pouch. The chosen system uses resistive heating and forced air cooling.

The engineer thinks that the following variables are critical to determining the time it takes to cool the product. The list of independent variables (Xs) is: Fan Speed, Fluid Volume, Ambient Temperature, Altitude, and %RH (Relative Humidity). The output (Y) is the time to cool the product.

This is a smaller is better scenario, whereby the less time it takes to cool the product, the better.

Business question: What is the most parsimonious regression model to optimize and control the cool time? (Thus, are all the independent variables critical, or should some be removed to derive the best-fit model?)

Preparation: Determine the relationship that needs to be studied, collect the data, and enter the numeric data into the MINITAB Worksheet.

Step 1. Graph the data first.

a. Graph the relationship on a Scatter diagram and examine the fitted line. To do so, select the following commands from the MINITAB drop-down menu: ***Stat > Regression > Regression....***

A B C D E F G H I J K L M N O P Q R S T U V W X Y Z

b. On the Regression main screen, select the columns containing the appropriate independent and dependent variables to place in the corresponding dialog boxes, as illustrated in "Area 1" of Figure R-17.

   i. **Select** the column of data with the dependent variable to go into the Response dialog box. In this case, the cool time.

   ii. **Select** the appropriate columns containing the different independent variables to go in the Predictor dialog box.

   In this case, initially all the variables are presumed critical until the model proves otherwise. Thus, all are selected: Fan Speed, Fluid Volume, Ambient Temperature, Altitude, and %RH (Relative Humidity).

c. **Select** the Graphs… button on the Regression main screen to determine the type of graphing output desired and the residual values of interest, as illustrated in "Area 2" of Figure R-17.

   i. Select the type of Residuals for Plots of interest in the appropriate dialog box. In this case, ***Regular*** was selected.

   ii. Select the type of layout for the Residual Plots in the appropriate dialog box. In this case, ***Four in one*** was selected.

   iii. Enter the independent variables of interest requiring residuals be calculated in the Residual Versus the Variables dialog box.

   In this case, all of the independent variables were selected, as they each are presumed significant predictors; but **not** the dependent variable (cool time).

   iv. Select OK.

d. Select the Options… button on the Regression main screen to identify any regression model options, as illustrated in "Area 3" of Figure R-17.

   i. The ***Fit Intercept*** is the default. In this case, leave the Fit Intercept selected.

   ii. In this case, select the ***Pure error*** under the Lack of Fit Tests.

A B C D E F G H I J K L M N O P Q R S T U V W X Y Z

iii. Select OK.

e. Select the OK button on the *Regression* main screen to run the regression model, as illustrated in "Area 1" of Figure R-17.

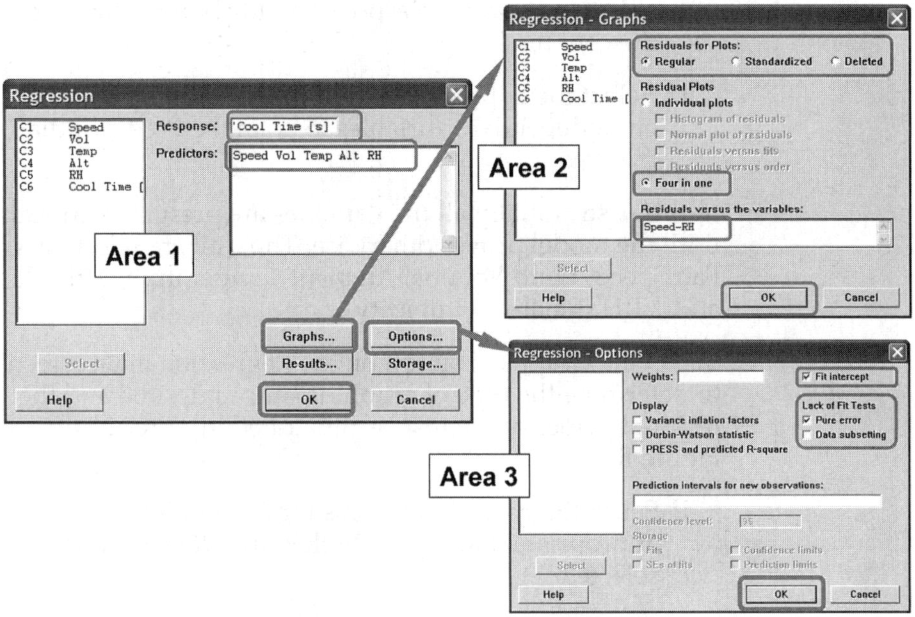

FIGURE R-17: Example of a MINITAB Regression Main Screen

Step 2. Analyze the regression model's Residual Plots Four in one graph, as illustrated in Figure R-18. Examine each of the four residual graphs one at a time to determine if the regression model assumptions for residuals are met.

a. **Normal Probability Plot of the Residuals** (in the upper-left quadrant)—Examine how the data plot appears normal such that the data is plotted close to the line. Recall the regression assumption that the residuals are to be distributed normally. (**See Also** the "Control Charts" entry in the "Normal Versus Non-normal Data" section, **p. 227**, for a more detailed discussion about the Normal Probability plot.)

In the Figure R-18 example, the residuals appear not to be evenly distributed around the normal probability line, and either extreme data points seem outside what the expected confidence interval might be. It appears the model does not fit the observations well.

b. **Histogram of the Residuals** (in the lower-left quadrant)—Examine the histogram to see if the data appears to be distributed normally, that is, a bell-shaped curve. Recall the regression assumption that the residuals are to be distributed normally. (**See Also** "Histogram—7QC Tool," **p. 330**, for more detail.)

   In the Figure R-18 example, the residuals appear to be skewed, exhibiting a long tail on the left of the plot, reinforcing that the model does not fit well.

c. **Residuals Versus the Fitted Values** (in the upper-right quadrant)—Examine how randomly the data appears to be distributed around the residual line (indicated as the "zero" line on the plot). Recall the regression assumption that the residuals are to have a constant variance around the regression line.

   In the Figure R-18 example, the residuals appear to be distributed in a pattern of an upside-down smile, or a frown, with the absence of data below the residual line in the center of the plot. Again, it appears the model does not fit well.

d. **Residuals Versus the Order of the Data** (in the lower-right quadrant)—Examine how randomly the data appears to be distributed around the residual line (again depicted as the "zero" line on the plot). Recall the regression assumptions that the residuals are to have a constant variance around the regression line and independence such that they would be distributed randomly over time around the regression line.

   In the Figure R-18 example, the residuals appear randomly distributed around the residual line for the first-half of the observed sequence but tighten and for the most part move above the line for the second-half—again reinforcing that the model does not fit well.

Step 3. Analyze each of the regression model's Residual Plots versus each of the independent variables, as illustrated in Figure R-19. Examine each of the four residual graphs one at a time to determine if the data appears to be randomly distributed or displays a pattern.

A B C D E F G H I J K L M N O P Q R S T U V W X Y Z

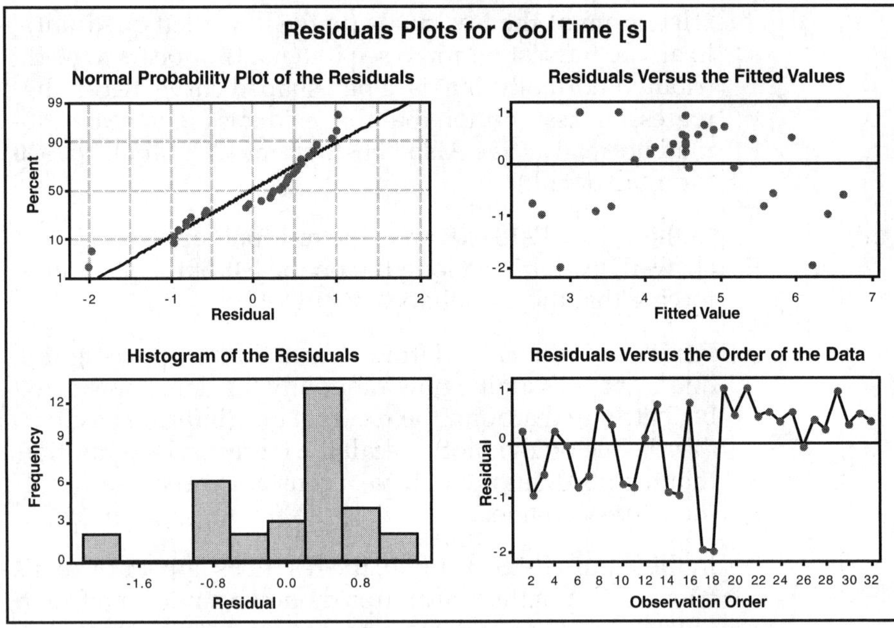

FIGURE R-18: Example of a MINITAB Residual Plot for a Multiple Regression (Four in One)

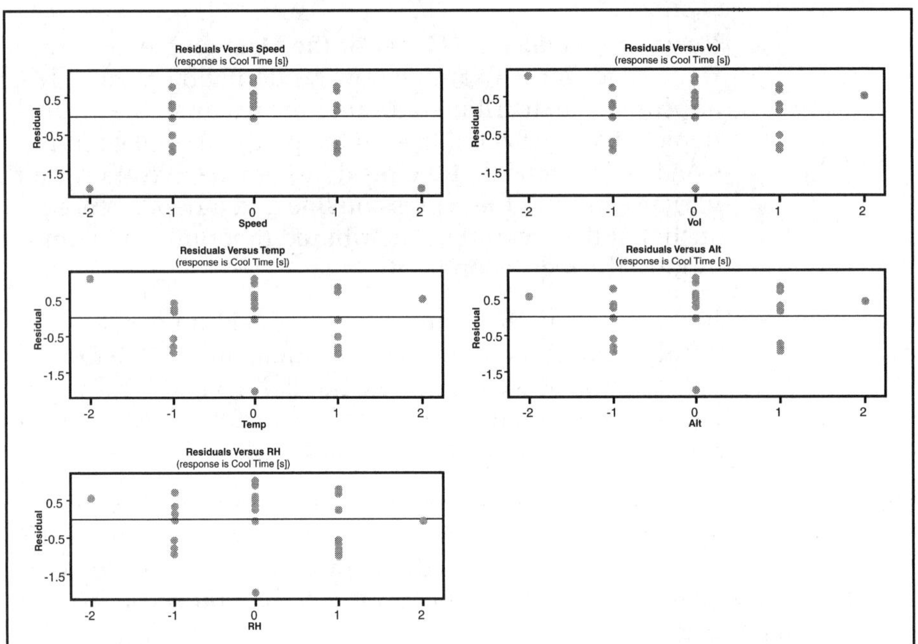

FIGURE R-19: Example of a MINITAB Residual Plot Versus Each Independent Variable

In the case of this engineering example shown in Figure R-17, one plot stands out as having strongest pattern of a frown, and that is the graph in the upper-left corner of the Residuals Versus Speed.

The pattern of a frown or smile indicates the need for a quadratic term. (Reference Figure R-11, which illustrates the different model patterns.)

Step 4. Modify the model via an iterative approach.

a. Create a quadratic for the variable with the smile/frown pattern.

   i. In the MINITAB Worksheet, identify an empty column and label it to indicate the new squared-term.

      In this example, the Speed variable exhibited a smile/frown pattern in its Scatter plot of residuals versus fitted value. Label the new column as `Speed2`.

   ii. From the Worksheet use the calculator function to square the data in the appropriate column. **Select *Calc > Calculator...*** from its drop-down menu.

   iii. In the Calculator main menu, select the column in which the new squared-term data should be stored and put it in the Store Result in Variable dialog box.

   iv. In the Calculator main menu, within the Expression dialog box, enter the column name containing the term to be squared, then select the multiplication function (indicated by an asterisks *), and then enter a second time the column with the variable to be squared.

      In this example, the dialog box should read `'Speed' * 'Speed'`.

      Click OK, and MINITAB squares the independent variable and stores the data in the selected new column. Figure R-20 illustrates this step.

      An alternative calculator dialog box entry that produces the same outcome is the `double asterisks2 (**2)`, signifying the squared-function (to the second-power). Hence the data in the Expression dialog box would be the term to be squared **2. In this example, the dialog box should read `'Speed' **2`.

b. Rerun the regression described in Step 1 (***Stat > Regression > Regression...***) and check the residuals again.

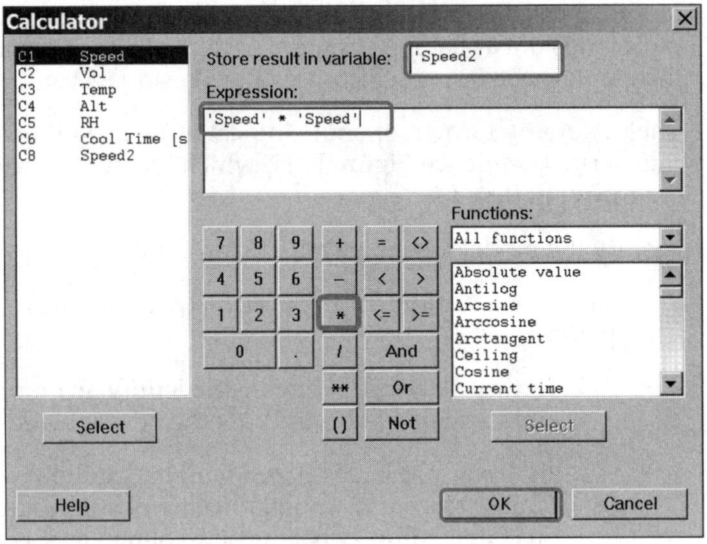

FIGURE R-20: MINITAB Calculator to Square the Speed Variable

Step 5. Analyze the new model by checking the graphical and statistical output.

a. Examine the graphical output. Figure R-21 provides the graphical output for this engineering example.

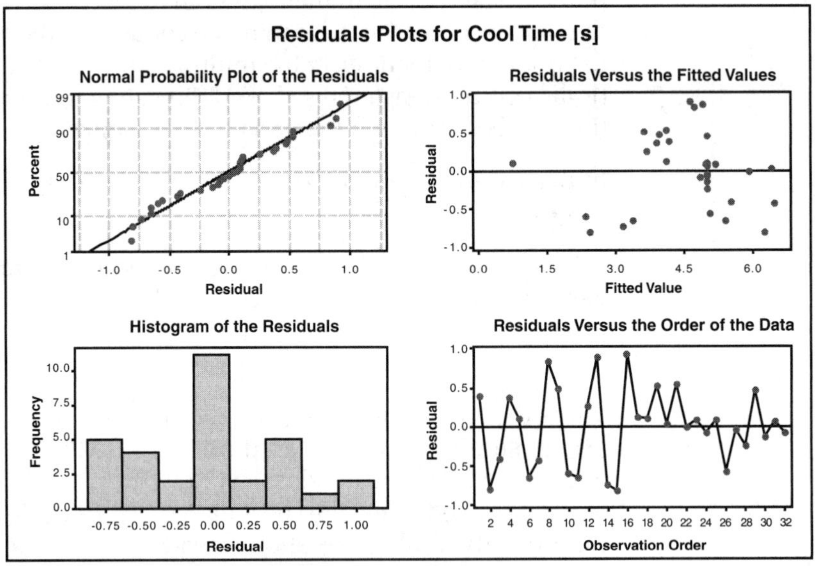

FIGURE R-21: Rerun Example of the MINITAB Residual Plot for a Multiple Regression (Four in One)

A B C D E F G H I J K L M N O P Q R S T U V W X Y Z

Referring to Figure R-21, there still appears to be a systematic pattern in the time-ordered plot such that the latter half of the graph reflects a non-random pattern.

b. Examine the statistical output in the session window. Figure R-22 shows this engineering example's session window.

```
Regression Analysis: Cool Time [s] versus Speed, Vol, ...

The regression equation is
Cool Time [s] = 5.02 - 0.841 Speed + 0.703 Vol + 0.448 Temp - 0.080 Alt
          + 0.022 RH - 0.645 Speed2

Predictor    Coef  SE Coef      T      P
Constant   5.0169   0.1234  40.65  0.000
Speed     -0.8414   0.1126  -7.47  0.000
Vol        0.7033   0.1126   6.24  0.000
Temp       0.4479   0.1126   3.98  0.001
Alt       -0.0795   0.1126  -0.71  0.487    <= Remove insignificant factors
RH         0.0223   0.1126   0.20  0.845    <=
Speed2    -0.6449   0.1008  -6.40  0.000

S = 0.551867   R-Sq = 85.9%   R-Sq(adj) = 82.5%

Analysis of Variance

Source          DF       SS      MS      F      P
Regression       6  46.3204  7.7201  25.35  0.000
Residual Error  25   7.6139  0.3046
  Lack of Fit   20   7.2995  0.3650   5.80  0.030    <= Shows lack of fit
  Pure Error     5   0.3144  0.0629
Total           31  53.9343
```

FIGURE R-22: Rerun Example of the MINITAB Session Window

Referring to Figure R-22, the engineering example has two insignificant factors included in the regression model, as indicated by their p-values being **greater than 0.05**—Alt (Altitude) and RH (Relative Humidity).

> **Note**
>
> The Null hypothesis states that there is no relationship between the independent and dependent variable. For the two factors, Alt and RH, there is insufficient evidence to reject the Null. To achieve a parsimonious model, these two factors need to be removed from the model and the model rerun.

For the ANOVA Residual Error Lack of Fit, it has a small p-value (less than 0.05). Hence, reject the Null hypothesis, which states that the predictors alone explain the variation in the output. The model still needs improvement by either adding a higher term or an interaction. In conclusion, add interaction terms from the significant predictors (Speed, Volume, Temperature).

Step 6. Modify the model. Explore if any interaction terms improve the model.

a. Using the calculator function, create interactions of each of the significant predictors by repeating Step 4 three times to create three interactions.

  i. In this example, the Expression dialog box should read ***'Speed' * 'Vol'***.

  ii. In this example, the Expression dialog box should read ***'Speed' * 'Temp'***.

  iii. In this example, the Expression dialog box should read ***'Vol' * 'Temp'***.

b. Rerun the regression model by repeating Step 1 (***Stat > Regression > Regression...***) and check the residuals again. Figure R-23 illustrates the Regression main screen to rerun the model with the three new interaction factors. Notice, that in the Regression—Graphs screen, for the in the *Residual versus the variables* dialog box, the significant predictors (X), including the three new interaction terms, are entered. (A similar procedure to Step 1.c.)

Step 7. Analyze the new model by checking the graphical and statistical output.

a. Examine the graphical output. Figure R-24 provides the graphical output for this engineering example.

Referring to Figure R-24 illustration, the residual plots look much improved over earlier models. The residuals appear to be distributed normally in both the Normal Probability plot and histogram. The residuals appear more randomly distributed around the regression line in both the Scatter plot and the Time Series plot.

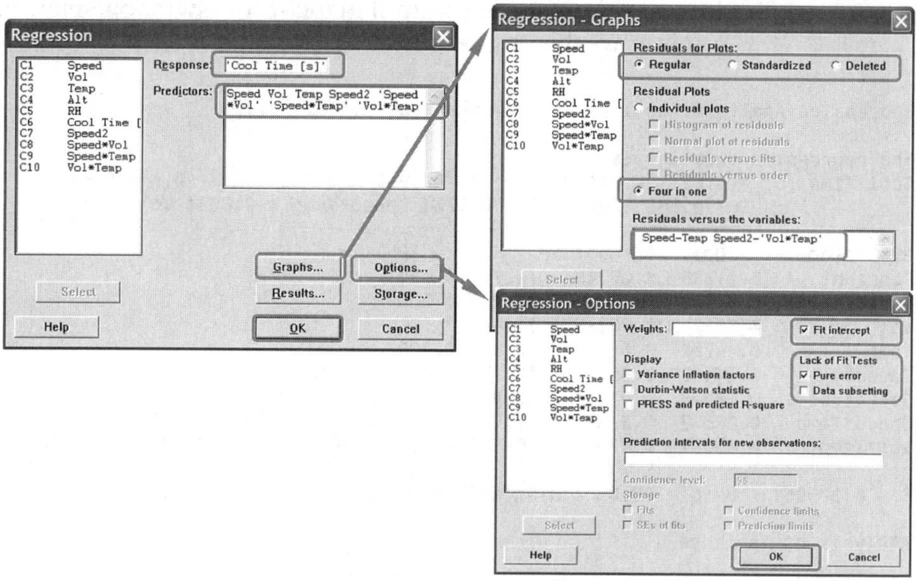

FIGURE R-23: Rerun Example with Interactions in the Regression Main Screen

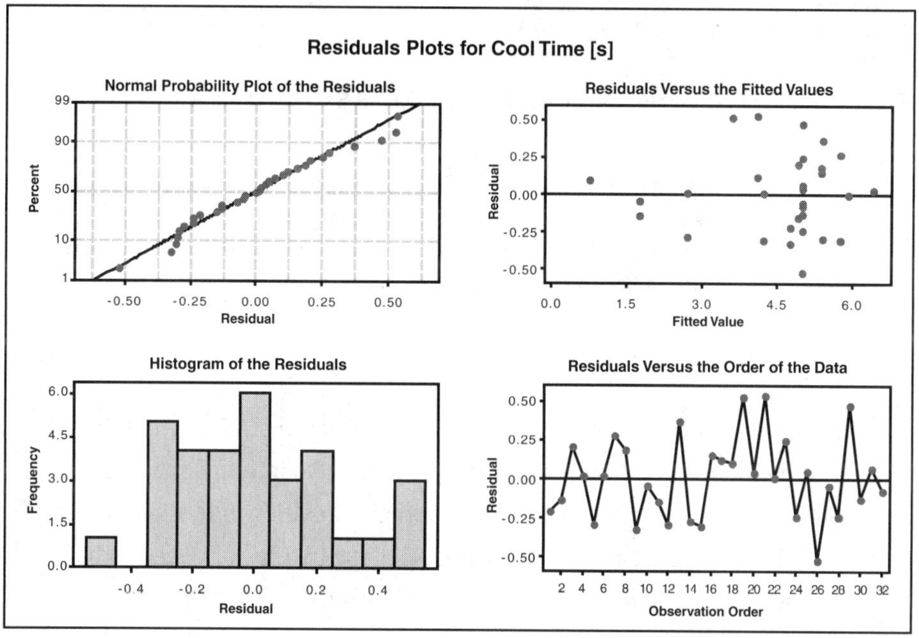

FIGURE R-24: Rerun Example with Interactions—MINITAB Residual Plot, Multiple Regression

A B C D E F G H I J K L M N O P Q R S T U V W X Y Z

b. Examine the statistical output in the session window. Figure R-25 shows this engineering example's session window.

```
Regression Analysis: Cool Time [s] versus Speed, Vol, ...

The regression equation is
Cool Time [s] = 5.02 - 0.841 Speed + 0.703 Vol + 0.448 Temp - 0.645 Speed2
                + 0.585 Speed*Vol + 0.0754 Speed*Temp + 0.0534 Vol*Temp

Predictor       Coef  SE Coef       T      P
Constant     5.01685  0.06709   74.77  0.000
Speed       -0.84142  0.06125  -13.74  0.000
Vol          0.70333  0.06125   11.48  0.000
Temp         0.44792  0.06125    7.31  0.000
Speed2      -0.64493  0.05478  -11.77  0.000
Speed*Vol    0.58525  0.07501    7.80  0.000
Speed*Temp   0.07537  0.07501    1.00  0.325 <--- Remove insignificant terms to
Vol*Temp     0.05337  0.07501    0.71  0.484 <--- produce parsimonious model.

S = 0.300051   R-Sq = 96.0%   R-Sq(adj) = 94.8%

Analysis of Variance

Source          DF       SS      MS      F      P
Regression       7  51.7736  7.3962  82.15  0.000
Residual Error  24   2.1607  0.0900
  Lack of Fit    7   0.9082  0.1297   1.76  0.161 <--- No longer any lack of fit.
  Pure Error    17   1.2526  0.0737
Total           31  53.9343
```

FIGURE R-25: Rerun Example with Interactions—MINITAB Session Window

Referring to Figure R-25, the engineering example has two insignificant interaction terms included in this regression model, as indicated by their p-values being **greater than 0.05**—Alt*Temp and Vol*Temp.

> **Note**
>
> The Null hypothesis states that there is no relationship between the independent and dependent variable. For the two interaction factors, Alt*Temp and Vol*Temp, there is insufficient evidence to reject the Null. To achieve a parsimonious model, these two factors need to be removed from the model and the model rerun.

A B C D E F G H I J K L M N O P Q R S T U V W X Y Z

The ANOVA Residual Error Lack of Fit has a p-value greater than 0.05. Hence, there is insufficient evidence to reject the Null hypothesis, which states that the predictors alone explain the variation in the output. The model is adequate, given that the significant predictors are able to explain the variation in the response.

Step 8. Modify the model. Following the procedure in Step 1 (***Stat > Regression > Regression...***), remove the insignificant predictors (that is, Alt*Temp and Vol*Temp) and rerun the regression model. Check the residuals again.

Step 9. Analyze the new model by checking the statistical output. (Because Step 7.a., Figure R-24, already provided a good set of residual plots, there is no need to look at them again.

a. Examine the statistical output in the session window.

Figure R-26 shows this engineering example's session window. Referring to Figure R-26, the engineering example the predictor variables all have p-values less than 0.05.

For the ANOVA Residual Error Lack of Fit, it has a p-value greater than 0.05, at 0.201, which is improved over the earlier run shown in Figure R-25. Hence, there is insufficient evidence to reject the Null hypothesis, which states that the predictors alone explain the variation in the output. The model is adequate, given that the significant predictors are able to explain the variation in the response.

The statistics indicate that this is a parsimonious model. The final model's equation is found at the top of the session window in Figure R-26:

Cool Time (Y) = 5.02 - 0.841 Speed + 0.703 Vol + 0.448 Temp - 0.645 Speed2 + 0.585 Speed*Vol.

Regression Analysis: Cool Time [s] versus Speed, Vol, ...

The regression equation is
Cool Time [s] = 5.02 - 0.841 Speed + 0.703 Vol + 0.448 Temp - 0.645 Speed2
+ 0.585 Speed*Vol

Predictor	Coef	SE Coef	T	P
Constant	5.01685	0.06647	75.48	0.000
Speed	-0.84142	0.06067	-13.87	0.000
Vol	0.70333	0.06067	11.59	0.000
Temp	0.44792	0.06067	7.38	0.000
Speed2	-0.64493	0.05427	-11.88	0.000
Speed*Vol	0.58525	0.07431	7.88	0.000

S = 0.297245 R-Sq = 95.7% R-Sq(adj) = 94.9%

Analysis of Variance

Source	DF	SS	MS	F	P
Regression	5	51.637	10.327	116.89	0.000
Residual Error	26	2.297	0.088		
Lack of Fit	9	1.045	0.116	1.58	0.201
Pure Error	17	1.253	0.074		
Total	31	53.934			

Figure R-26: Rerun Example with Fewer Interactions—MINITAB Session Window

Hints and Tips

The X values represent the independent variables that impact the Y values. (In a Scatter plot, the independent variable is plotted on the X-axis and the dependent variable on the Y-axis.) Be certain to properly assign the data using this convention when conducting a regression analysis.

Examine the Scatter plot **prior** to conducting a regression analysis to visualize the relationship. If the curve reflects a non-linear curve, a linear regression will not be applicable. The importance of plotting the data first is discussed further in the Scatter diagram entry. (**See Also**, "Scatter Diagram—7QC Tool," **p. 640**)

The Scatter diagrams of the residuals versus the fitted values should display a random (or buckshot-like) pattern. If it reveals any potential patterns in the data that might distort the statistical calculations of a regression analysis

for example, apply a quadratic to a smile (or frown) pattern and apply a cubic to an S-shape pattern. (Refer to Figures R-11 and R-13.)

Rule of Thumb—The strength of the relationship determined by the coefficient of determination (R^2 or r^2—where it falls between 0% and 100%, or 0 to 1) is as follows:

- **Excellent** relationship:
    - Mechanical/Physical System yielding an $\mathbf{r^2}$ of 91% to 100%.
    - Behavioral/Human System yielding an $\mathbf{r^2}$ of 71% to 100%.
- **Good** relationship:
    - Mechanical/Physical System yielding an $\mathbf{r^2}$ of 76% to 90%.
    - Behavioral/Human System yielding an $\mathbf{r^2}$ of 51% to 70%.
- **Weak** relationship:
    - Mechanical/Physical System yielding an $\mathbf{r^2}$ of 60% to 75%.
    - Behavioral/Human System yielding an $\mathbf{r^2}$ of 30% to 50%.

Statistical Terminology Associated with Regression

- **Best-fit line**—The regression line (derived from the regression model) drawn in a Scatter diagram, as shown in Figure R-12.

  Note that the slope of the regression line in Figure R-27 indicates a positive relationship as it goes from the lower-left to the upper-right corner of the graph, trending upward. Thus as X increases, so does Y.

- **Binary logistic regression**—Describes the relationship of independent variables with a dependent variable that is a binomial (e.g. good/bad, and yes/no).

- **Causation**—Means that if the value of one (independent) variable is changed, then it induces a change in the dependent variable—a *cause-and-effect relationship.*

- **Coefficient of determination (r2)**—Determines how well the Scatter plot's best fit line fits the data. This number falls between 0 and 1, with 1 representing a perfect fit and 0 representing no linear fit. No linear relationship is depicted as a horizontal best-fit line on a Scatter diagram.

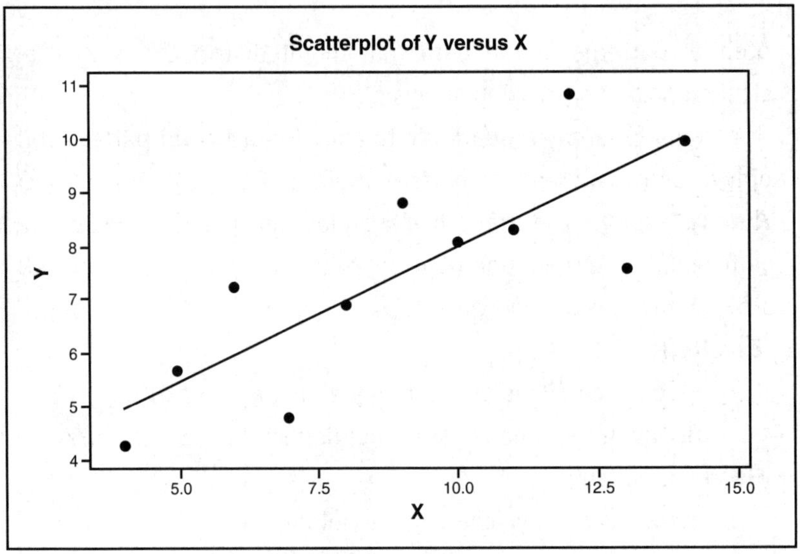

FIGURE R-27: Scatter Plot with a Regression Line

The coefficient of determination represents the proportion of Y's variation that is explained by the regression line. Because most data points fall around, rather than on the regression line, the remainder of the proportion (1-$\mathbf{r}^2$) is error. Hence, $\mathbf{r}^2$ measures the percent of variability in the response explained the by linear relationship with the predictor (best fit line or model) —ranging between 0% and 100%, or 0 and 1.

- **Confidence interval**—The range of values on either side of the residual line that the represents the confidence within which the true regression line falls. The default usually set at 95%.
- **Correlation**—A metric that measures the linear relationship between two process variables.
    - A strong correlation does **not** imply causation (a cause-and-effect relationship). For example, the number of car accidents on a rainy day may be strongly correlated with the sale of umbrellas, but the buying of umbrellas did not cause the accidents.
    - Lack of correlation does **not** necessarily mean that no causal relationship exists. The range of data may be too narrow to detect relationship.

A B C D E F G H I J K L M N O P Q R S T U V W X Y Z

- **Pearson's Correlation Coefficient (r)**—Defines the correlation metric and falls between (-1) and (1), where 0 indicates no linear relationship, (-1) describes a perfect negative correlation, and (1) a perfect positive correlation. Thus, the larger the absolute value of r (either + or -), the stronger the linear relationship. Graphically, this strong relationship is characterized by a tight distribution of data around a best-fit line plotted in a Scatter diagram.

  Figure R-28 displays different Scatter diagrams with their respective Pearson's Correlation Coefficient identified with each.

  The Pearson's Correlation Coefficient **(r)** is proportional to the slope of the regression line adjusted for the differences in both X and Y's standard deviations. Its formula is as follows:

$$r_{xy} = \frac{\sum_{i=1}^{n} (x_i - \bar{x})(y_i - \bar{y})}{\sqrt{\sum_{i=1}^{n} (x_i - \bar{x})^2 \sum_{i=1}^{n} (y_i - \bar{y})^2}}$$

  Square the Pearson's Correlation Coefficient to derive the coefficient of determination (**r^2** or **R^2**) used for the regression analysis. (**See Also** "Scatter Diagram—7QC Tool," **p. 640**)

- **Error**—The difference between the *observed variation (noise)* between the *expected value,* derived from the average of the entire population (which typically is *unobservable*). **Error** *technically* is ***unobservable***. Errors are independent random variables of each other. Also called residual.

- **Extrapolation**—In a stable process, the model *extends beyond* the range of two known input variable data values to predict a third unknown independent variable and its response variable. It presumes the unknown independent variable *falls outside* the range of the known modeled sample data—projecting.

A B C D E F G H I J K L M N O P Q R S T U V W X Y Z

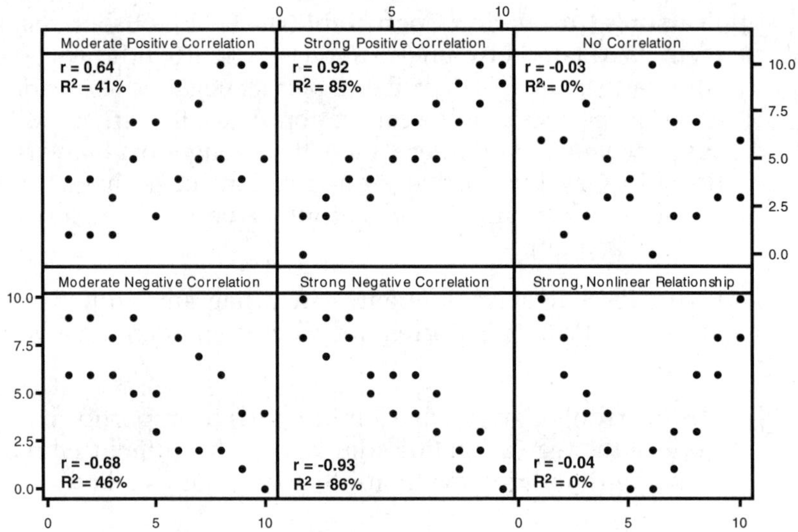

FIGURE R-28: Scatter Plots with Corresponding Pearson's Correlation Coefficient (r) and Coefficient of Determination (R^2) Values

- **Interaction**—Defined as a when the response of one variable is dependent on another, and depicted mathematically as the product combination of those variables (multiplied). Two-way interaction combines two variables, three-way is for three, and so on. (**See Also** "Analysis of Variance (ANOVA)—7M Tool," and "Design of Experiment (DOE)," **p. 142** and **p. 250**, respectively.)

- **Intercept (of a line)**—Where the line crosses (or intercepts) the y-axis, and is represented by the constant **b** in the linear equations **Y = mX + b**.

- **Interpolation**—In a stable process, the model uses the range of two known input variable data values to predict a third unknown independent variable and its response variable. It presumes the unknown independent variable *falls within* the range of the known modeled sample data.

- **Linear regression**—A quantitative model building tool that relates one or more independent variables (xs) to a single dependent variable (y).

- **Logistic ordinal regression**—Describes the relationship among ordinal variables (e.g. high-medium-low, or excellent-good-bad).

- **Multiple linear regression**—Models the relationship between multiple independent variables (Xs) and a dependent variable (Y).

  The simplest model (with one independent variable) is written as:

  $$\mathbf{Y = b_0 + b_1X_1 + b_2X_2 + error}$$

  where **b** is a coefficient; $\mathbf{b_1}$ and $\mathbf{b_2}$ are the respective slopes, and $\mathbf{b_0}$ is the intercept.

- **Regression line**—The line of best fit drawn through a Scatter plot, as determined by the regression model.

- **Regression model**—A statistical tool that measures the strength of relationship between one or more independent variables with a dependent variable.

- **Residual**—The observable estimate of the unobservable error. Calculated as the difference between the observed value and an estimated value (observable sample average), derived from the model's line of best fit and the actual data points. **Residuals are observable** and are dependent of each other.

- **Scatter diagram**—Graphical tool displaying the qualitative strength of the relationship (linear or curvilinear) between two continuous or discrete variables; but **not** indicating cause-and-effect. The stronger the relationship (positive or negative), the greater the likelihood that the manipulation of the independent variable(s) may affect the dependent variable. (**See Also** "Scatter Diagram—7QC Tool," **p. 640**)

- **Simple linear regression**—Models the relationship between a single independent variable (X) and a dependent variable (Y).

  The simplest model (with one independent variable) is written as:

  $$\mathbf{Y = a_0 + a_1X + error}$$

  where **a** is a coefficient; $\mathbf{a_1}$ is the slope and $\mathbf{a_0}$ is the intercept.

- **Slope**—Calculated as rise over run; or y divided by x, and is represented by the constant **m** in the linear equations **Y = mX + b**.

  - **Positive slope**—Indicates that the line of the curve goes from the lower-left to the upper-right corner of the graph, trending upward.

   - **Negative slope**—Indicates that the line of the curve goes from the upper-left to the lower-right corner of the graph, trending downward.

- **Variables**—Two main types:
   - Dependent or response variable (Y).
   - Independent variable (X).

A B C D E F G H I J K L M N O P Q R S T U V W X Y Z

Supporting or Linked Tools

Supporting tools that might provide input when developing a regression analysis include

- Cause-and-Effect Diagram (**See Also** "Cause-and-Effect Diagram—7QC Tool," **p. 173**)
- Data Collection Sheets (**See Also** "Data Collection Matrix," **p. 248**)
- Sampling (**See Also** "Sampling," **p. 618**)
- Scatter diagram (**See Also** "Scatter Diagram—7QC Tool," **p. 640**)

A completed regression analysis provides input to tools such as

- DOE (**See Also** "Design of Experiment (DOE)," **p. 250**)
- FMEA (**See Also** "Failure Modes and Effects Analysis (FMEA)," **p. 287**)
- Control Plan (**See Also** "Matrix Diagrams—7M Tool," **p. 399**)

Figure R-29 illustrates the link between a regression analysis and its related tools and techniques.

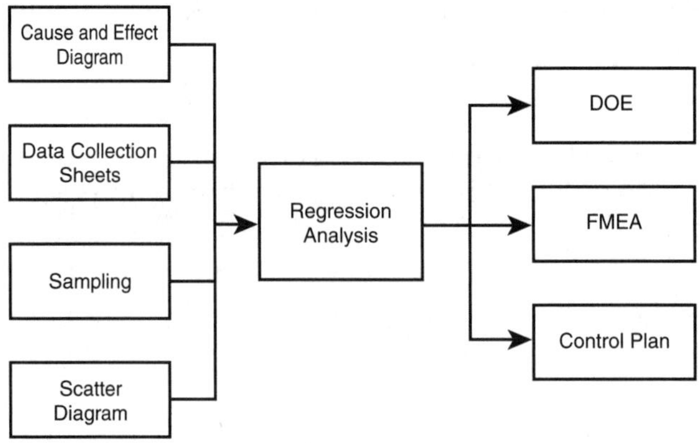

Figure R-29: Regression Analysis Tool Linkage

Risk Mitigation Plan

What Question(s) Does the Tool or Technique Answer?

How can you best plan for, manage, and mitigate unforeseen risk?

A risk mitigation plan helps you to

- Prepare for risk events—through risk identification, quantification, response planning, and response control
- Stimulate thinking when developing the list of the responses to a potential problem
- Organize the planning process and use of various risk tools

Alternative Names and Variations

Risk management tools include

- Cause-and-Effect diagram; Ishikawa diagram; Fishbone diagram (**See Also** "Cause-and-Effect Diagram—7QC Tool," **p. 173**)
    - Brainstorming (**See Also** "Brainstorming Technique," **p. 168**)
    - Cause categorization, checklists, or techniques such as 5MandP, 6Ms, and 4Ps (**See Also** "Cause-and-Effect Diagram—7QC Tool," **p. 173**)
    - 5-Whys Technique (**See Also** "5-Whys," **p. 305**)
    - Cause enumeration diagram (**See Also** "Cause-and-Effect Diagram—7QC Tool," **p. 173**)
    - CEDAC® (Cause-and-Effect diagram and Cards) (**See Also** "Cause-and-Effect Diagram—7QC Tool," **p. 173**)
    - Desired-results Fishbone (**See Also** "Cause-and-Effect Diagram—7QC Tool," **p. 173**)
    - Process Fishbone (**See Also** "Cause-and-Effect Diagram—7QC Tool," **p. 173**)
    - Reverse Fishbone diagram (**See Also** "Cause-and-Effect Diagram—7QC Tool," **p. 173**)
    - Time-delay Fishbone (**See Also** "Cause-and-Effect Diagram—7QC Tool," **p. 173**)
- Cause and Prevention Matrix or Cause-Prevention diagram (**See Also** "Cause and Prevention Diagram," **p. 198**)

- FMEA (**See Also** "Failure Modes and Effects Analysis (FMEA)," **p. 287**)
- Fault Tree Analysis (**See Also** "Fault Tree Analysis (FTA)," **p. 309**)
- Process Decision Program Charts (PDPC) (**See Also** "Process Decision Program Charts (PDPC)—7M Tool," **p. 515**)

When Best to Use the Tool or Technique

Risk planning, mitigation and planning is a process that identifies, analyzes and responds to risk, and should be done throughout all phases of a project, program and/or operational processes.

Brief Description

Risk management is a planning process rather than a tool or technique in and of itself. Risk management tools are covered elsewhere in this book.

Risk mitigation planning, in its broadest sense, attempts to prepare for the unknown and respond when it occurs with forethought to either maximize the results of positive events minimize the consequences of adverse events. Yes, risk equals both opportunities and threats. It involves the art and science concerned with identifying, analyzing, and responding to risk. It is part of doing business, given that uncertainty and change exist. Risk management helps to overcome situations wherein uncertainty hinders decision-making.

Risk planning should focus on *gathering information* to proactively combat uncertainty. This involves talking to both internal and external experts and reviewing historical data about past similar scenarios, with the intent of heightening awareness as to the possible sources of risk. Figure R-30 illustrates the inverse relationship of information and uncertainty. Information comes from two primary sources. The first is objective sources of recorded experiences and data on current and past events. The second is subjective sources based on knowledgeable experts. This latter source is particularly helpful with early planning. Risk planning should be a continuous, formal activity, conducted on a regular basis, throughout a project, program, or operation.

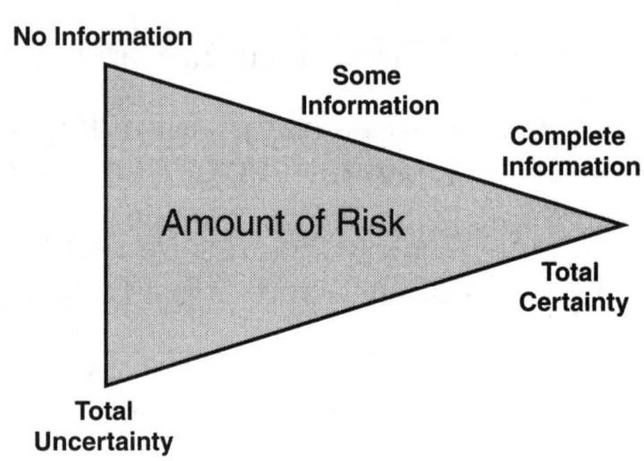

Figure R-30: The Inverse Relationship of Information

Warning

Less successful risk planning tends to focus on past experiences or past incidences. People tend to relate risk to things that are highly chancy or hazardous, but many risks are so commonplace we scarcely notice them. Rarely does risk planning address uncertainty explicitly or formally; however, it should so as to avoid too many surprises.

Establishing Common Terminology

Risk is defined as an element or factor that involves uncertain hazards. It connotes probability of an event and its consequences—the confidence level with which an outcome will occur. *Probability* ranges from the objective (informed by data or experimentation) to the subjective (relying on judgment or opinion).

Certainty might be defined as having availability of all the information necessary to make the right decision, such that the outcome can be predicted with confidence.

Uncertainty is the complete absence of information, where nothing is known about the possible outcomes. With the lack of information, probability distributions based on experience cannot be developed; they would be done only in the abstract.

Risk management is the formal process in which risk factors are systematically identified, assessed, and controlled. It is a systematic method to identify and control risk events that potentially cause unwanted change or improve any positive results. Risk management quantifies the risk level and its impact of an action so it can be related to the organization's tolerance level. The tolerance level may vary depending on the type of impact, for example financial, legal, regulatory, quality, safety.

Risk Management Process

The risk management process consists of four phases: identification, quantification, response, and control. Sometimes risk identification and quantification are treated as a single process called risk analysis or risk assessment. Risk response is sometimes called risk mitigation. And risk response development and risk response control sometimes are treated as a single process called risk management. Regardless of how the phasing, they are iterative and may happen in parallel. To unleash the power of this process and its structure, it should become ubiquitous with the ongoing operations or work and not treated as an independent, one-time event or function.

Risk Identification

This phase determines which risks are likely to have the most significant impact and ascertains their characteristics. Along with identification,

documenting the essential risk characteristics and **communicating** them are essential elements of risk planning. The process owner typically is the person accountable for this phase and drives the refresh and renewal process to keep the information current.

Risk identification is a systematic process involving the following approach.

Step 1. Risk classification.

Classify potential risks according to their cause or source, **not** the effect or impact. Understanding the potential root cause dictates the appropriate countermeasures. Sources of risk vary by business type or business model. Identify and categorize appropriate sources for your business. Considerations might include

a. External unpredictable causes: government regulations, natural hazards.

b. External predictable causes: raw materials availability, business and financial risk.

c. Internal (non-technical): labor stoppages, cash flow issues, safety issues, poor estimates, requirement changes, poor planning, poorly defined roles or processes.

d. Technical (systems, software): changes in technology, design issues, operations, and maintenance issues.

e. Legal: licenses, patent rights, lawsuits, contractual issues.

Risk classification takes on additional dimensions other than its cause. The appropriate classifications needed for your risk planning depends on factors such as your business size, scale, and scope. However, at minimum, utilize risk source as a primary classification.

Another risk classification is business versus insurable risk. *Business risk* encompasses an opportunity of **profit or loss**, whereas *insurable risk* represents a chance for **loss only**. Given the one-sided nature of insurable risk, it is also known as *pure risk*. Other risk classifications include level of uncertainty, magnitude of impact, and nature of risk (defined by its characteristics). Risk characteristics consist of the probability of occurrence, range of possible outcomes, expected timing, and anticipated frequency.

Step 2. Rank risks.

Define an appropriate rating system for the situation that weights or prioritizes the identified categories.

The rating system might represent an integration of the three classic FMEA scales of severity, frequency of occurrence, and detect-ability and integrate them into one risk priority number (RPN)—or the scale could rank the ability to manage effective responses. For example, consider the impact that a cumulative effect might have. If several risk events occur in conjunction, the risk impact of the combination may be more severe than the individual events alone; thus, managing this cumulative risk response could present more challenges.

Step 3. Identify tools and techniques.

Determine what additional information is needed and the best way to organize and present the content for purposes of response planning and decision-making. Outputs from risk identification are four-fold:

a. list of potential risk events
b. risk symptoms
c. sources of risk
d. inputs to other processes

Risk Quantification

This phase evaluates risk to assess the range of possible outcomes. It determines the probability or likelihood of occurrence of a risk event or several risk events in combination, as well as the magnitude of impact if it occurred. Its primary purpose is to determine which risk warrants a response.

The person or role accountable for this phase depends on the organization and type of risk involved. More than likely, this phase's activities would be parsed out to several people or functions to quantify and qualify the entire portfolio of risk opportunities. If that were the case, the process owner serves as the overall customer for this phase.

Risk quantification collects inputs from key stakeholders as to their risk tolerances, various sources of risk, potential risk events, cost estimates, and event duration estimates. This phase uses several tools and techniques such as statistical data, expected monetary value, decision trees, simulation, and expert judgment.

Given the emphasis on probability in this phase, statistics plays a critical role. *Descriptive statistics* (mean, median, mode, range, variance, standard deviation) help to describe critical parameters. Moreover, statistical concepts such as normal distribution and Monte Carlo simulations add a depth of perspective to the probability of an event or set of events. A Monte Carlo simulation models the behavior or performance of a system and produces a statistical distribution of expected results.

Financial Quantification: The *expected monetary value* (EMV) of a risk event is calculated by multiplying the probability of occurrence times the estimated financial gain or loss if that event occurs. Rather than simply relying on a deterministic estimate (one number), a Monte Carlo simulation provides more information with a range of probable outcomes. (**See Also** "Monte Carlo Simulation," **p. 431** and "Selecting Project Portfolios Using Monte Carlo Simulation and Optimization," in Part III, **p. PDF:921** for more detail on the technique and its various applications.)

A
B
C
D
E
F
G
H
I
J
K
L
M
N
O
P
Q
R
S
T
U
V
W
X
Y
Z

Hints and Tips

A way to explore the impacts of a decision and its multiple paths of possibilities is to use a *decision tree*. It is not as robust as a Monte Carlo simulation, but it does diagram key interactions among decisions and associated chance events as understood by the decision-maker and calculates the expected monetary value (EMV) of the different paths. Each branch of the tree represents either a decision or chance event, wherein the tree, in aggregate, represents the sum total of 100% of the possibilities, as shown in Figure R-31.

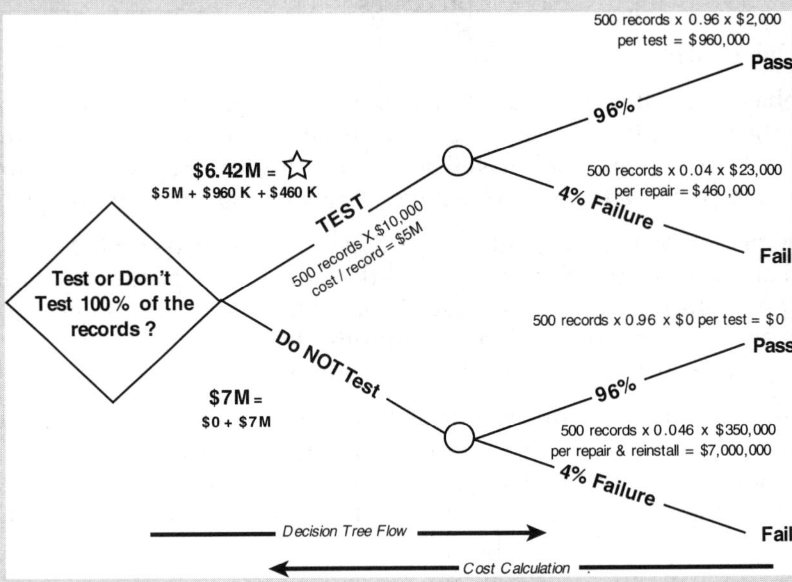

FIGURE R-31: Decision Tree Example

Schedule Quantification: A schedule may represent either business risk or insurable risk. Two simple tools to assist in schedule risk calculation are the Activity Network Diagram (AND) and PERT (with its most likely, pessimistic and optimistic values to triangulate a quantification estimate).

A more sophisticated tool available to understand the probability of a range of schedule outcomes is the Monte Carlo simulation. (**See Also** "Activity Network Diagram (AND)—7M Tool," **p. 127**; "PERT (Program Evaluation and Review Technique) Chart," **p. 453**; and "Monte Carlo Simulation," **p. 431**)

Risk Qualification: If risk quantification is difficult, a qualification rating system should be developed using an ordinal scale. Develop a qualification risk rating system appropriate for your business. An example of a qualification system is as follows:

- **High**—Risk represents significant probability of cost, schedule, quality, and/or performance disruption.
- **Medium**—Risk represents moderate probability of disruption, and the difficulties could possibly be overcome.
- **Low**—Risk represents a small disruption potential, and if it occurred, the resulting difficulties probably could be overcome.

Outputs from risk quantification are two-fold: 1) opportunities to pursue (or those threats to respond to) and 2) opportunities to ignore (or those threats to accept).

Risk Response

This phase determines the appropriate course of action for each individual risk opportunity of interest. Risk planning often calls for a prioritization of risk opportunities, based on the quantification and qualification of risk. Those prioritized risks at minimum receive the benefit of subsequent response planning.

Typically the *process owner* is the person accountable for this phase, to ensure ongoing management and monitoring of appropriate response activities.

Risk response falls into three main categories:

- **Risk Avoidance**—The elimination of a specific threat, usually by eliminating the cause.
- **Risk Mitigation**—The reduction of the expected monetary value of a risk event by minimizing the probability of occurrence and/or the risk event value. Another mitigation approach is the transference of all or a portion of the risk to another party (outsourcing, partnering, insurance are examples). Insurance transfer examples include:
  - Direct property damage (for property such as auto collision, equipment coverage, fire insurance)
  - Indirect consequential loss (contractor protection for debris removal, equipment replacement)

- Legal liability (such as design errors, workplace injuries, performance failures)
- Personnel (worker's compensation, employee replacement costs)

- **Risk Acceptance (or Retention)**—The acceptance of the consequences of a risk event by either actively or passively allowing it to occur without intervention. Active acceptance may result from developing a contingency plan. The consequence likely would either be positive or moderately negative.

Note

Idea generation tools and techniques help to keep the response planning creative and fresh. Brainstorming techniques and interviewing of experts can yield appropriate potential risks and corresponding response approaches. A powerful interviewing method utilizes a panel of experts to develop convergent solutions, called the Delphi technique, and often convenes periodically over a period of years. Other techniques to explore are procurement strategies, contingency planning, alternative strategy planning, and insurance strategies.

Key questions to ask about the possible risk events that warrant response planning include

- How could this risk be avoided?
- Can the risk be reduced?
- Can the risk be shared or transferred (to insurance for example)?
- Should the risk be accepted, and if so, does it warrant a financial reserve or schedule allowance or contingency?
- Can the risk be contained?

Outputs from risk response development may include: 1) risk management plan, 2) inputs to other processes, 3) contingency plans, 4) reserves, and 5) contractual agreements.

A risk management plan documents the procedures to manage risk. It covers the roles and responsibilities associated with managing risk and managing its evergreen process. It defines how contingency plans will be implemented and how reserves are to be allocated, if necessary.

Risk Control

This phase responds to changes in risk and keeps the risk planning process alive. This phase involves executing the risk management plan to respond to a given risk event. The approach used to manage risk should be documented and communicated widely. Risk control monitors the environment for change and information both internally and externally and responds to changes in risk as it occurs.

Typically the process owner is the person accountable for this phase, to ensure ongoing management and monitoring of appropriate response activities. Even if the monitoring, reporting, and controlling tasks are delegated, the process owner provides the direction, validates its necessity, and energizes the process.

Workarounds are a type of risk response control. They are unplanned responses to negative risk events. This type of response is not defined in advance of the risk event occurring, but it may trigger a best practice that would be important to document and communicate. Additional response development is another important technique that reinforces the importance of actively managing and monitoring for information and changes. It is triggered when unanticipated risk events occur or when the effect of the event exceeds the expected magnitude. A supporting tool to this technique is the Process Decision Program Charts (PDPC). (**See Also** "Process Decision Program Charts (PDPC)—7M Tool," **p. 515**)

There are a host of tools available to aid in the various stages of risk mitigation planning. Some of the soft tools are featured as entries in this book and also are found on the preceding list of variations. In addition to those soft tools, mostly matrix-like tools, there are several software application tools to assist with risk management.

Outputs from risk response control are two-fold: 1) corrective action (implementing contingency plans or workarounds) and 2) updates to the risk management plan.

Hints and Tips

How do you know when to take a risk? Consider the following dimensions to help answer the question:

- Quantify the risk level.
- Determine its impact (magnitude)—positive or negative.
- Define the risk tolerance level for the organization.

Additional Resources or References

- *A Guide to the Project Management Body of Knowledge*. Pennsylvania: Project Management Institute, Inc., 2000 (for more on risk management).
- U.S. Defense Acquisition University, Acquisition Community Connection—for a list of Risk Management tools (Risk Management Systems; Standalone Tools; Analysis Tools; Quick Links to other Risk Topics); https://acc.dau.mil/CommunityBrowser.aspx

Rolled Throughput Yield

See Also "Process Capability Analysis," **p. 486**

Run Chart—7QC Tool

What Question(s) Does the Tool or Technique Answer?

How does the data look over time? Are the data randomly distributed over time? Does the process look stable and random over time?

A run chart helps you to

- Graphically display a data set to detect trends, shifts, cycles, or extreme values in the data
- Determine if the process is stable over time and randomly distributed

Alternative Names and Variations

This tool is also known as

- Time Plot; Time Series plot

Variations on the tool include

- Line graph

When Best to Use the Tool or Technique

A Time Series plot should be one of the first graphical tools used to display the key output variable(s) in the sequence it was produced to examine if the process appears to be stable over time and if the data are randomly distributed over time.

Use a run chart versus a control chart when the data set contains less than the 20 data point minimum for a control chart.

Brief Description

A run chart displays data in a time sequence by placing the units of measure on the Y-axis (vertical axis) and time on the X-axis (horizontal axis). It is a simple, easy to create, easy to read line chart of continuous or discrete data representing a product or service characteristic. The graph plots the process output data as it is produced in the same sequence. Thus, it reveals any trends, shifts, cycles, or extreme values in the data over time—the absence of which indicates process stability and randomly distributed data.

The run chart's trend line represents the process data. Sometimes it features a second line to represent the average of the data for that specific period of time. This average line serves as a reference point to identify any data patterns. Essentially a run chart is a limitless control chart (no control limits). A trend or shift may indicate a process influx, either improving or degrading, and need to be brought under control. Extreme values may indicate a special cause event needing to be eliminated. The cyclical nature may indicate different work procedures from shift-to-shift, or operator-to-operator, or location-to-location. Regardless, eliminate data patterns to steady the process. Utilization of other statistical tools such as process capability and Hypothesis testing assume process stabilization as a prerequisite.

The run chart is a member of the revised 7QC Tools, (or seven "Quality Control" tools), attributed to Dr. Kaoru Ishikawa. The 7QC Tools sometimes are called the *seven basic tools* given that they were the first set of tools identified as the **core** quality improvement tools. Ishikawa's original 7QC Toolset includes: Cause-Effect diagram; Check sheet (or checklist); Control charts; Histogram; Pareto chart; Scatter diagram; and Stratification. Recently, the 7QC Toolset was modified by substituting the Stratification technique with either a flowchart (or Process map) or a run chart (or Time Series plot).

How to Use the Tool or Technique

Use the following procedure to develop a run chart using MINITAB. Note that MINITAB refers to a run chart as a *Time Series plot.*

Step 1. Given that numeric data has been entered into the MINITAB Worksheet, select the following commands from its drop-down menu: ***Graph > Time Series plot....***

Figure R-32 displays sample MINITAB screens to create a run chart, which is displayed in Figure R-33.

A B C D E F G H I J K L M N O P Q R S T U V W X Y Z

Step 2. In the Time Series plot main screen, select the appropriate *graph type* ("Area 1") and click the OK button. In this example, the *simple* plot was selected.

Step 3. The Time Series plot—Simple screen opens. Select the appropriate process variable data of interest (Yield, for example), shown in "Area 2" of Figure R-32.

Step 4. Click the Time/Scale button to select the appropriate *time scale* to be displayed along the X-axis, select the column containing the X-axis time data and click the OK button, as shown in "Area 3" of Figure R-32.

Step 5. Click the OK button on the Time Series plot—Simple screen, shown in "Area 2" of Figure R-32, and MINITAB generates a run chart. Figure R-33 illustrates a sample run chart output.

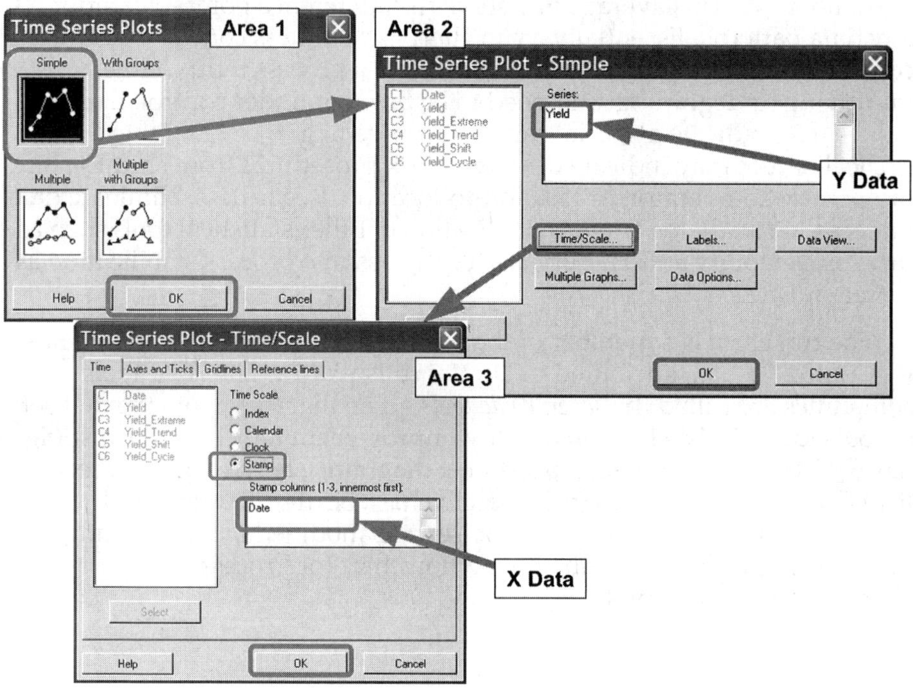

FIGURE R-32: Example MINITAB Run Chart Main Screen

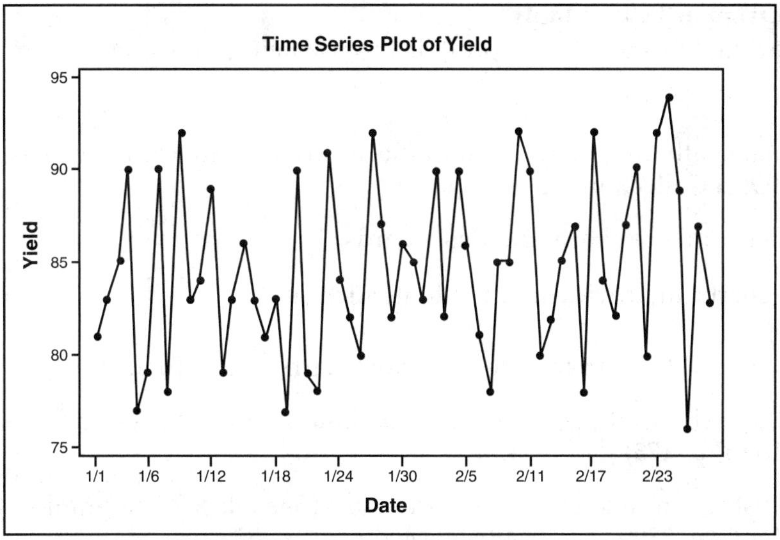

FIGURE R-33: Example MINITAB Run Chart

Based on the Figure R-33 graph, the data appears to be randomly distributed over time, thereby indicating stability.

Hints and Tips

Run charts are simple to create and read. Use them as a preliminary quick view of process stability and randomness, particularly when the data set contains less than the required 20 data point minimum for control charts.

Run chart data must maintain the same time sequence as to when it was produced and observed.

Run chart data provides less information than a control chart. A control chart's upper and lower control limits enable the calculation of different out-of-control statistical tests and the standard deviation. The confirming tests of stability and randomness require statistical analysis, such as that used in control charts. If possible, use a control chart when analyzing the performance of a process. (**See Also** "Control Charts—7QC Tool," **p. 217**)

Supporting or Linked Tools

Supporting tools that might provide input when creating a run chart include

- Data gathering plan to collect the appropriate metrics. (**See Also** "Data Collection Matrix," **p. 248**)
- Performance charts and dashboards.

A run chart can provide input to tools such as

- Control charts (**See Also** "Control Charts—7QC Tool," **p. 217**)
- Cause-Effect diagrams (**See Also** "Cause-and-Effect Diagram—7QC Tool," **p. 173**)
- Histograms and other graphical tools (**See Also** "Histogram—7QC Tool," **p. 330** and "Graphical Methods," **p. 323**)
- FMEA with follow-on action planning (**See Also** "Failure Modes and Effects Analysis (FMEA)," **p. 287**)

Figure R-34 illustrates the link between a run chart and its related tools and techniques.

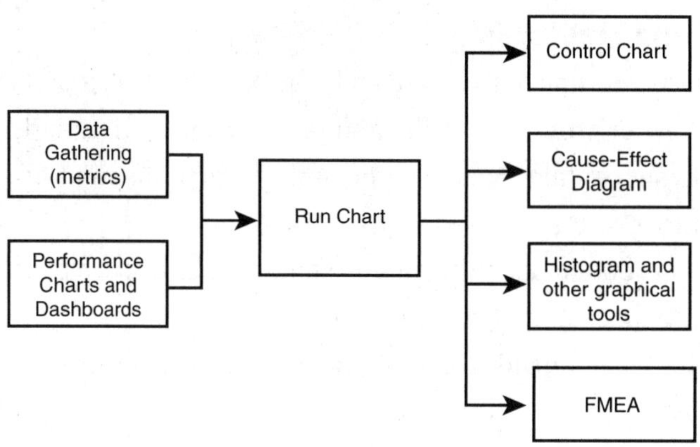

FIGURE R-34: Run Chart Tool Linkage

A B C D E F G H I J K L M N O P Q R S T U V W X Y Z

S

7M—Seven Management Tools

What Question(s) Does the Tool or Technique Answer?

What tools can you use to analyze qualitative data?

This categorization (or acronym) helps you to double-check which tools apply to qualitative data.

Alternative Names and Variations

This category is also known as

- 7MP or seven management and planning tools

Variations on the category include

- 7QC Tools, or seven quality control tools

When Best to Use the Tool or Technique

To recall what fundamental quality tools are used to analyze, organize, manage, and plan using qualitative data.

Brief Description

In the mid-1970s, the JUSE (Union of Japanese Scientists and Engineers) assembled a suite of seven management tools, also known as the *7M Tools*, to ensure that projects comprehended qualitative information. The toolset focused on creativity, communication, and planning with qualitative data. Most of the 7M Tools are attributed in part to Dr. Shewhart who introduced and popularized many of the "traditional quality tools" to analyze quantitative data. The 7M Toolset includes

- **Activity Network diagrams or Arrow diagrams**—A technique that depicts the process flow to evaluate the time needed to complete a project. (**See Also** "Activity Network Diagram (AND)—7M Tool," **p. 127**)
- **Affinity diagrams**—A technique that dissects and organizes verbal data into its natural groupings to summarize themes and help to make the input actionable. (**See Also** "Affinity Diagram—7M Tool," **p. 136**)
- **Interrelationship digraphs or Relations diagrams**—A diagramming tool used to plot a complex situation and to parse it into its natural cause-and-effect relationships. (**See Also** "Interrelationship Diagram—7M Tool," **p. 369**)

- **Matrix diagrams**—A set of different shaped grids that display relationships between and among different factors to align, dissect, and rank. (**See Also** "Matrix Diagrams—7M Tool," **p. 399**)
- **Prioritization Matrices**—Often replacing the more complex Matrix data analysis. A set of matrices that rank relationship strengths to aid in the selection process. (**See Also** "Prioritization Matrices—7M Tool," **p. 470**)
- **Process Decision Program Charts (PDPC)**—A risk mitigation technique that charts what might go wrong during a development project and the corresponding countermeasures to prevent or mitigate the impact. (**See Also** "Process Decision Program Charts (PDPC—7M Tool," **p. 515**)
- **Tree diagrams**—A technique that displays the hierarchical relationship among topics. (**See Also** "Tree Diagram—7M Tool," **p. 712**)

The Quality Toolbox, authored by Nancy Tague, presents the 7M Tools ranked from those used for abstract analysis to detailed planning: Affinity diagram, Relations diagram, Tree diagram, Matrix diagram, Matrix Data analysis (commonly replaced by a more simple Prioritization Matrix), Arrow diagram, and Process Decision Program Chart (PDPC).

7QC—Seven Quality Control Tools

What Question(s) Does the Tool or Technique Answer?

What basic tools best assure quality control?

Alternative Names and Variations

This category is also known as

- 7QC or seven quality control tools
- Seven basic tools

Variations on the category include

- 7M or seven management tools
- 7MP or seven management and planning tools

When Best to Use the Tool or Technique

To recall the minimal set of tools used to analyze, improve, and maintain quality control.

Brief Description

Kaoru Ishikawa purported that "...as much as 95% of quality related problems...can be solved with seven fundamental quantitative tools." Ishikawa, the creator of the Fishbone diagram (sometimes referred to by his last name), promoted a set of seven basic tools to ensure quality control. While the list of tools are rather simple, in aggregate they are powerful in analyzing, improving, and maintaining quality control. The *original* 7QC Toolset includes

- **Cause-and-Effect diagram (or Fishbone, or Ishikawa diagram)**—It is used to analyze, organize, and illustrate the cause-and-effect relationship to a problem. (**See Also** "Cause-and-Effect Diagram—7QC Tool," **p. 173**)
- **Checklists or Check sheets**—They serve as simple vehicles to record data that provides a snapshot of the frequency patterns and to remind people of standard operating procedures. (**See Also** "Checklists—7QC Tool," **p. 204**)
- **Control charts**—They are used to monitor how a process operates, if it behaves consistently and predictably, or if any special cause variation events occur. (**See Also** "Control Charts—7QC Tool," **p. 217**)
- **Histogram**—A graphical tool displaying frequency to investigate centeredness (location), variation, and shape of the data. (**See Also** "Histogram—7QC Tool," **p. 330**)
- **Pareto chart**—A graphical tool that ranks causes from greatest to least magnitude, to analyze the percent contribution of one item over another. (**See Also** "Pareto Chart—7QC Tool," **p. 445**)
- **Scatter diagram**—A graphical tool that displays the relationship between two variables (X and Y). (**See Also** "Scatter Diagram—7QC Tool," **p. 640**)
- **Stratification***—A technique that culls out the different sources of data to better evaluate patterns. (**See Also** "Stratification—7QC Tool," **p. 697**)

See Also "Process Map (or Flowchart)—7QC Tool," **p. 522** and "Run Chart," **p. 611**

* The 7QC tools often now replace the Stratification technique with either Flowchart (for example, Process map) or run chart, depending to the type of data or topic that is of interest.

Sampling

What Question(s) Does the Tool or Technique Answer?

What is the best method of collecting a representative sample of the population?

Sampling helps you to

- Collect a portion of all the data
- Use that portion of the data to make conclusions
- Save on time, resources, and money

When Best to Use the Tool or Technique

Use sampling when the population size is too large to study each member, but some decisions or conclusions are needed. Sampling saves on time, money, and resources versus studying each member of an entire set of data from a population.

Brief Description

Sampling is an efficient and effective alternative to looking at all the data. The term *population* represents all items or observations of a particular group of interest. A *sample* is a representative subset used to estimate a larger group—either a population or process.

Studying the entire population can be expensive, take time, and require resources. If the group is particularly large, just because of its sheer size, studying it in total may be difficult or near impossible (that is, the entire population of the United States or China). Depending on the type of study being conducted, the test may need to destroy the item (destructive testing), such as taste tests and some quality tests such as a car crash-test to check the safety mechanisms. Destructive testing is expensive and consumes much of the tangible item to be tested, and so is thereby prohibitive to conduct on the entire population. In summary, general reasons behind sampling involve

- Economic factors
- Time factors
- Population size
- Partially inaccessible populations and/or processes
- Destructive nature of the observation
- Accuracy

Another advantage in sampling is that the averages from several samples will tend to distribute normally. This describes the Central Limit Theorem, which when plotting the sample averages produces a predictable normal distribution that serves as a base assumption for a large majority of the popular statistical tools. (**See Also** "Statistical Tools," for a discussion on the Central Limit Theorem, **p. 684**)

Sample characteristics can be studied and analyzed and used to infer or approximate a population characteristic, called a *parameter*. Sound conclusions can often be drawn from a relatively small amount of data. Given appropriate sampling techniques and sample size, the sample information can accurately portray a population. Use of both descriptive and inferential statistics describe and relate a sample and its population. Inferences are made with a certain degree of confidence, referred to statistically as a *confidence interval*. Confidence intervals support the inferred conclusions about a population.

(**See Also** "Hypothesis Testing," **p. 335**, for more discussion on confidence intervals; and "Statistical Tools," **p. 684**, for a discussion on descriptive and inferential statistics.)

Representative Sampling

Constructing a sample that is representative is the first priority. For conclusions to be valid, samples must be representative of the population or process. *Representative* means that the sample's characteristics reflect that of the population's. If the focus is a subset of characteristics within a population, the representative sample reflects all of those characteristics. There should be no systematic differences between the data in the sample and the data not in the sample (that is, inaccuracy of exit polling data in recent elections).

A representative sample requires planning. A *data collection plan* helps organize sampling activities, ensuring that all of the needed data is gathered and ensuring data integrity. Consider the following questions when planning to collect samples:

- Are there natural groupings in the data? If so, what proportion of each group should be sampled?
- When is it best to sample?
- How frequent should you sample?
- Where is the appropriate place to sample?

Sampling involves an understanding of the types of data needing to be collected to ensure data integrity. Data integrity requires attention on collecting random, unbiased data, adhering to the time sequence of the data production or observation to capture and record the data in that order. If the item's characteristics change over time, record the measurement or classification as quickly as possible after production and again after the stabilization period. Inspect for data recording, entry, or transfer errors. Validate the operational definition. Calibrate and analyze the measurement system and select the appropriate sampling technique to ensure accuracy, precision, and unbiased data. (**See Also** "Data Collection Matrix," **p. 248** and "Measurement System Analysis (MSA)," **p. 412** for discussion on operational definition, accuracy, precision, and bias.)

Appropriate sample size addresses reliability. The larger the sample size, the more reliable the sample. Sample size will be discussed later in this entry.

A sampling plan involves consideration about the technique of extracting samples from a population. Those techniques vary by circumstance—the population to be studied and what is known about it. A sampling plan attempts to avoid *sampling error*. Sampling error results from an issue of validity (issues of bias), not reliability. Sampling strategies include the following techniques.

Population Versus Process Sampling

Valid conclusions require a representative sample of the population or the process. However, population sampling and process sampling have different objectives and approaches. The sampling strategy depends partially on whether the population is static or dynamic. (The term *static* refers to a relatively fixed number of members in the group, such as the number of annual flights per airline in and out of Chicago O'Hare airport; whereas, *dynamic* refers to a frequently changing number, such as the human population.) Dynamic refers to a process; wherein the population is an outcome of a process.

Static populations require considerations of appropriate sample size. Sample size formulas use assumptions based on a static population and what is known about it. See the "sample size"later in this section. Most sampling applications involve sampling from dynamic processes.

Population Sampling

When clear distinct boundaries can be defined to describe a population in its entirety, population sampling techniques can be used. Population sampling requires that each item contained in that population can be identified and numbered.

Population sampling technique describes the characteristics of the items within a known population group. For example, if sampling were required of an organization's employees or of a complete batch of cookies, population sampling could be used because the population boundaries are known, and each member of the population can be identified and counted. Figure S-1 illustrates the population sampling technique.

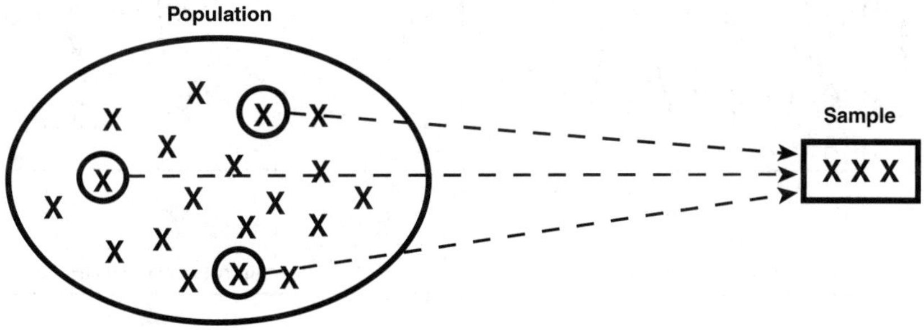

FIGURE S-1: Population Sampling

Process Sampling

Process sampling involves a dynamic scenario wherein all units cannot be identified as some may not yet exist (that is, items that will be produced from the process tomorrow). In this case, the process sampling technique should be used. It characterizes a process so as to predict future outcomes and/or behavior and to make future improvements, if necessary. For example, a customer call center process might want to examine the total resolution time of customer calls. Figure S-2 illustrates the process sampling technique.

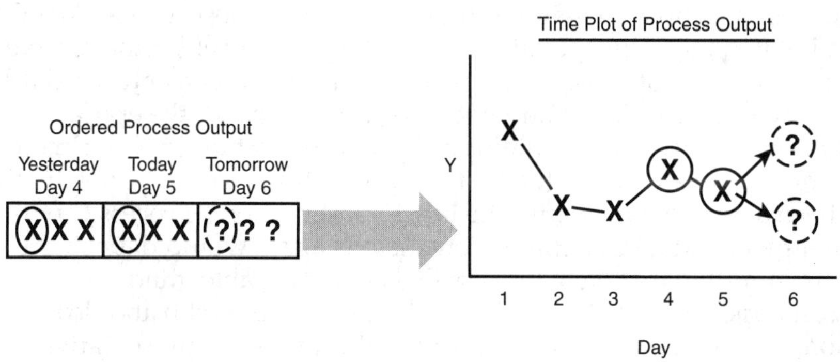

FIGURE S-2: Process Sampling

Process *stability* is an important consideration whether to sample or not. Different conclusions can be drawn based on when in the process a sample is extracted. If the process were unstable, then timing determinations become difficult. Consider the different run charts shown in Figure S-3 and determine when best to sample in the four process scenarios illustrated.

A B C D E F G H I J K L M N O P Q R S T U V W X Y Z

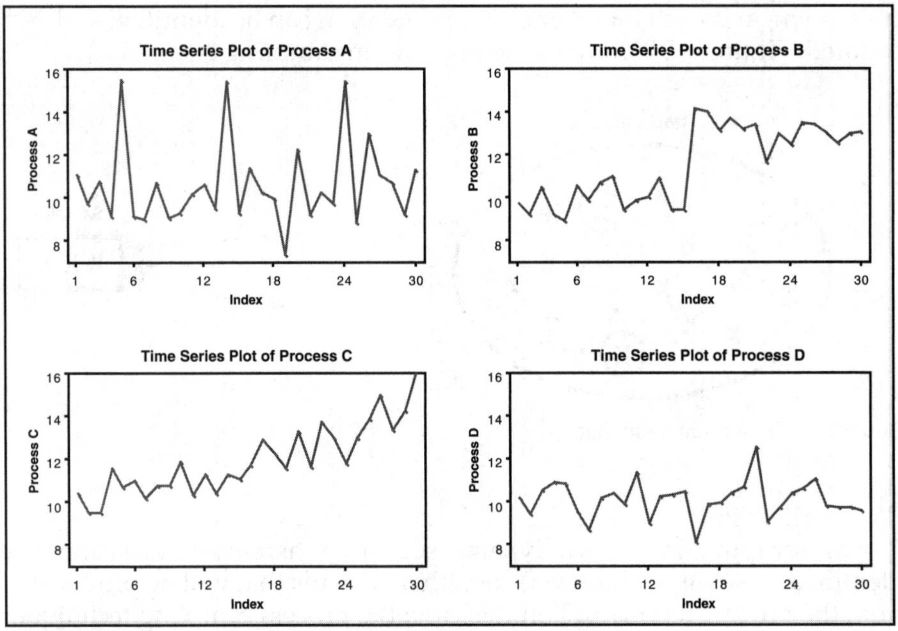

FIGURE S-3: Run Charts of Different Processes

Depending on at which point a sample is taken, the conclusions may vary. The run charts in Figure S-3 provide an unstable picture for Processes A, B, and C. Process A shows signs of *special causes*, given its three spikes. A sample containing these three spiked observations would give an inflated mean and standard deviation. Conversely, if the sample did not include any of those three spiked points, the mean and standard deviation would be deflated and not be representative of the entire process scenario. Process B contains a *shift* at observation 16. Conclusions drawn for a sample taken in the first 15 observations would be vastly different from a sample taken after the last 15 observations. Process C portrays an upward trend over time. Conclusions again would depend on where the samples are taken. Process D displays a stable, random process; it appears predictable within a band of natural ebb and flow variation. Various sampling methods would provide a representative sample from Process D.

Sampling Methods

Sampling approaches fall into four main categories—random sampling, stratified random sampling, systematic sampling, and subgroup sampling. Both random and stratified random sampling techniques apply to population sampling. Process sampling methods use either systematic or subgroup sampling. Furthermore, market research also uses quota sampling and self-selected sampling techniques. The different scenarios determine which method best ensures optimal representation. The sampling technique defines the different kind of samples produced.

Random Population Sampling
Random sampling assumes a homogeneous population. Regardless of which item is selected, it poses similar characteristics of interest as the remaining items in the population.

Random sampling requires that each population item has an equal likelihood of being selected. Sampling must be representative and not just the result of what was easy to obtain. Often emphasis is placed on the mechanics of using a sampling plan and not on the identification and selection itself. However, both aspects of sampling are important. Sampling without randomness compromises a good sampling plan.

Statistically, *random* sampling has a different definition than *haphazard*. Haphazard means at the whim of the sampler. If the sampler determines the sampling process, it contains inherent bias. Examples of inherent sampling include closing your eyes and throwing a dart or pointing, picking a number out of your head, sampling what is convenient. Random sampling avoids sampling bias. It employs the use of a statistical random number table or random number generation software that either produces a table or selects data randomly. An example of how to create a random numbers table in MINITAB can be found in the subsequent "How To Use the Tool or Technique" section of this entry.

Stratified Random Population Sampling
Stratified random sampling takes into account a stratified population, with a distinctive mix of segregated characteristics. The characteristics of interest determine subgroups within the population. This is a sophisticated technique that requires careful consideration of sample size.

Use this technique when the sampling needs to preserve the relative proportion of each distinct subgroup to the population. This technique deliberately samples a proportionate number of items from each sub-population randomly. The random sampling technique alone *tends* to arrive at the same result; however, the stratified random sampling technique *ensures* it. Figure S-4 compares the difference between random sampling and stratified random sampling concepts.

Systematic Process Sampling
Systematic sampling selects items according to a *pattern rule.* The sample selection follows a non-random procedure, such as pick every tenth item produced in a process. Notice only *a* single item is selected at a time. The key is that the selection maintain the relative sequence as the process' occurrences or production. This relates to manufacturing processes as well as service processes. For example, every fourth customer caller could receive a satisfaction survey after contacting a customer call center.

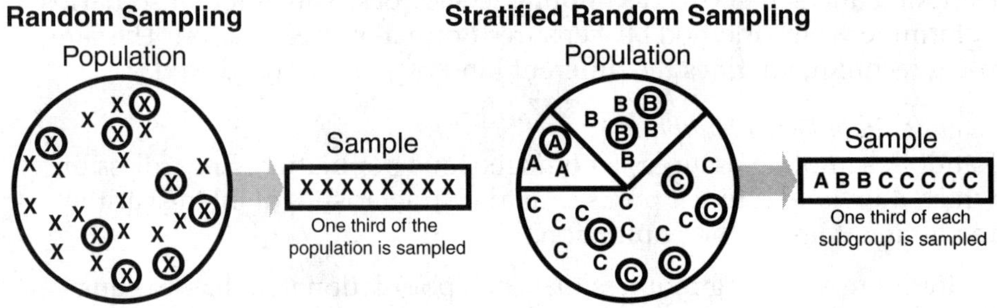

FIGURE S-4: Comparison of Random and Stratified Sampling

Process knowledge and sampling needs (size of sample and throughput of process) define the pattern rule. Systematic sampling ensures a sample is representative in the presence of process drift (frequency of sampling needs to be often enough to capture drift). Thus, systematic process sampling can be as good as, or even better than, a random sample. Figure S-5 illustrates the systematic process sampling method.

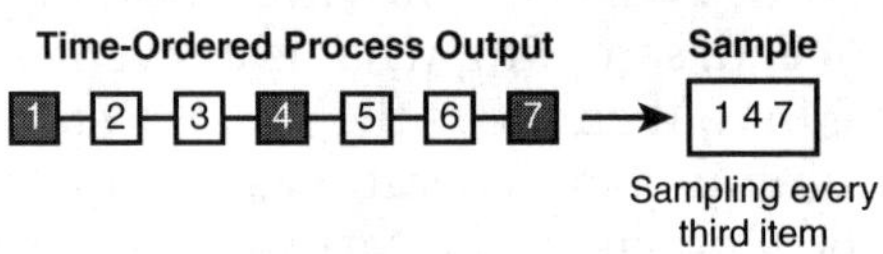

FIGURE S-5: Systematic Process Sampling Method

Subgroup Process Sampling
is very similar to the systematic process sampling method, but instead of subgroup sampling one item at a time several samples are taken from a batch or subgroup. This approach often is used in high-volume processes, such as the manufacturing of pharmaceutical tablets. Subgroup sampling involves the same size sample from each batch, and selects the samples at predetermined regular time intervals.

Subgroup sampling relies on the assumption that a *rational subgroup* exists. A rational subgroup requires that the items occur close in time proximity to one another. The items **within** a subgroup tend to be more similar to one another than items **between** subgroups. For example, a high volume commercial bakery bakes several cookies in a week. Sampling rational subgroups of cookies could be defined by when a batch's ingredients were mixed to test for thoroughness of mixing, or when the cookies were baked to check the conveyor belt timing through the ovens and test for doneness. In a pharmaceutical plant, sample a set number of tablets every half hour to measure the mass and compare to the product requirements.

Quota Sampling
This is a form of convenience sampling used by some survey techniques such as door-to-door, mall surveying, or election polling after voting. Quota sampling empowers the person responsible for collecting the survey data to select the potential survey subjects. Each surveyor is assigned a quota of surveys to complete. The potential subject may decline; however, the selection as to when to survey and who to survey within a set of guidelines or restrictions, are left up to the convenience of the surveyor. Different data collectors may employ different collection techniques. To prevent bias, the more data collectors the better. Unfortunately, what the surveyors perceive as convenient tends to be more similar than not, leading to issues of bias. This approach should be used with caution, and the guidelines should be carefully designed and communicated to the survey collectors.

Self-selected Sampling
This is a form of convenience sampling wherein the subjects decide whether or not to participate in the sampling process. This technique is used frequently with either inbound or outbound call centers. For example, if a customer places a phone order or asks for help, at the end of the call, the caller may choose to participate in a satisfaction survey (inbound); or if the recipient of an unsolicited call agrees to participate in a survey (outbound).

In general, the more self-selection inherent in the sampling process, the more potential for bias. In the case of Internet and television polling, they employ self-selection sampling and have the inherent bias of the person first selecting to watch the program or go to that web site in the first place, before deciding whether to participate in the survey. A large sample size might help to minimize some bias; however, proceed with caution when drawing conclusions and inferring the findings onto a population. Some say these polls are "not scientific."

Sample Size
One of the most frequently asked question of statisticians is, *How many measurements are required to have a statistically representative sample of the population?* Surprisingly to many, there is no correct answer because often the answer is a matter of tradeoffs. The size of the population and what is known about the population (What kind of data can be collected? Are there subgroups?) factor into the calculation of sample size. The tradeoffs factor in risk and economics. Many of these considerations must be determined in advance of calculating the sample size.

Sample size does matter. Because sample data is used to make inferences about populations, how is the validity of that projection checked? Sample size, denoted by **n**, is a key factor. Large sample sizes minimize inference error. However, new students of statistics typically are surprised to learn that there is no one correct answer when it comes to sample size.

Sample size calculations are structured to minimize risks and maximize the ability to find useful differences. In addition, more is always better when it comes to sample size. However, the economics of time and money play a part in tempering the size of a sample. How many samples are needed depends on the required confidence level (related to Type I and Type II Errors) and the type of data available (variable data is measured, attribute data is counted). In general, various tradeoffs are considered among four factors that affect sample size, which are

- **Standard Deviation** (denoted by **s**)—A direct relationship, such that as standard deviation increases, the sample size also increases. This tradeoff relates to the value of the population standard deviation.
- **Detectable Difference** (known as **delta** or Δ)—An inverse relationship, such that as the detectable difference decreases, the sample size increases. This tradeoff relates to the ability to detect a difference between values represented by the Null and Alternative hypotheses.
- **Type I Error** (**alpha** or α)—An inverse relationship, such that as alpha decreases, the sample size increases. The *default* value is set at an ***alpha of 0.05.*** This tradeoff relates to risk of rejecting a true hypothesis.
- **Type II Error** (known as **beta** or β)—An inverse relationship, such that as beta decreases, the sample size increases. The *default* value is set at a ***beta of 0.2.***

  This tradeoff represents the risk of accepting a false Null hypothesis when the alternative hypothesis is true.
- Alternately, this can be stated as a direct relationship by using *Power*, which is defined as 1–beta, such that as Power increases, the sample size increases. (**See Also** "Hypothesis Testing," **p. 335** for additional information on Type I and Type II Errors.)

The sampling plan examines these four factors: standard deviation, detectable difference (or delta), alpha, and beta to determine the sample size **prior to beginning the sampling process**. This is called *prospective* planning.

Using historical data requires a retrospective sample size calculation. Much of Hypothesis Testing involves a retrospective scenario, wherein assessing the Power and detectable difference occurs after a sample has already been collected and analyzed. Often a Hypothesis test shows no significance, and the follow-on question is asked, "What was the Power of the test?" If the Power is low, then having additional sample data would increase Power, which is the ability to detect a difference if one truly exists.

Sample size is not linearly related to precision, as shown in Figure S-6. For example, a statistically significant survey of likely voters found that 45% (±4%) would vote for the incumbent. The sample size for this survey was 601.

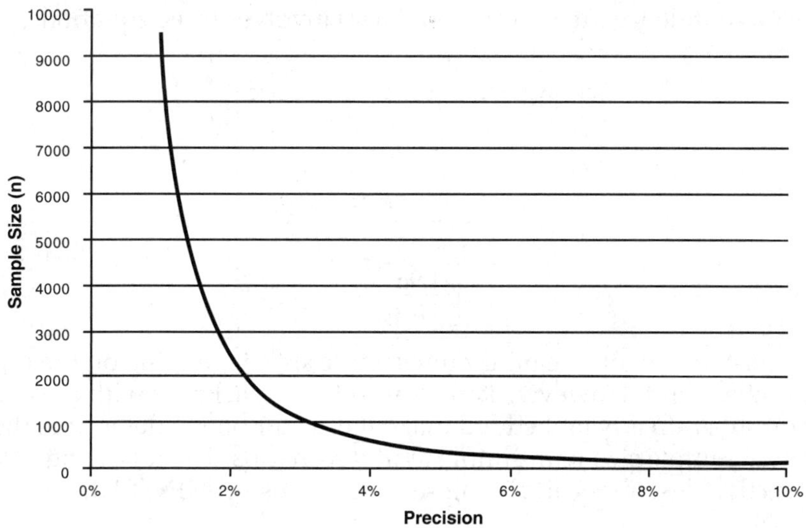

FIGURE S-6: Sample Size for Proportional Data

The formulas for calculating the minimum sample size (**n**) vary by the data type and thus the type of Hypothesis test. The various formulas include the following:

- Normal distribution, where the standard deviation is known:
- Two-tail test: $n = (Z_{\alpha/2} + Z_{\beta})^2 \, (\sigma/\delta)^2$
- One-tail test: $n = (Z_{\alpha/2} + Z_{\beta})^2 \, (\sigma/\delta)^2$

These formulas use critical values of the standard normal distribution.

- Student t-distribution, where the population standard deviation is unknown:
    - Two-tail test: $n = (t_{\alpha/2} + t_{\beta})^2 \, (s/\delta)^2$
    - One-tail test: $n = (t_{\alpha/2} + t_{\beta})^2 \, (s/\delta)^2$

Recall that the critical values for a t-distribution ($t_{critical}$) rely on knowing the *degrees of freedom*. And degrees of freedom depend on the sample size, which it is yet to be determined.

- Estimate average daily yield of a chemical or mixture process with a 95% confidence interval (two-tail test) involves three equations.

$$\text{Yield avg.} - \delta \leq \mu \leq \text{Yield avg.} + \delta$$

$$\delta = \frac{\sigma}{\sqrt{n}} Z_{0.025}$$

$$n \geq \left[\frac{1.96}{\delta}\right]^2 \sigma^2$$

Tables exist to look up the minimum sample size depending on the type of Hypothesis test. However, statistical software packages with these various formulas already embedded make it easy for one to determine the minimum sample size with simple *what if* scenarios. The following *"How to…"* section describes calculating sample size using MINITAB.

How to Use the Tool or Technique

The next set of procedures use MINITAB's *Power and Sample Size* capabilities and different Hypothesis tests to answer questions about sample size or any of the other four factors that affect sample size. (**See Also** "Hypothesis Testing," **p. 335**)

The process to calculate sample size in MINITAB basically follows a similar "what if" scenario for the various Hypothesis tests. Enter the known summary information into the appropriate dialog boxes for the four sample size factors (standard deviation, detectable difference, alpha, and beta), leave the sample size dialog box empty, and MINITAB provides the minimum sample size. If sample size is known along with any three out of the four dependent factors, the fourth factor can be calculated.

Determine Sample Size: 1-Sample t-Test

Example scenario background: You are responsible for a soft drink bottling process. Your current equipment operates at a throughput of 600 bottles per minute on average. There is a new model coming on the market that touts a throughput of 700 bottles per minute. A business case shows that at that speed it would be cost effective to make the purchase. You plan to go to the supplier to verify the throughput on your product. Based on your current equipment, the machine-to-machine variability (standard deviation) is 50 bottles per minute. How many machines would you request the supplier to demo for you to be confident of this new capital purchase?

Assumptions:

- Question: What is the sample size of new machines to demo?
- *Differences* = 700–600 bottles per minute = **100**.
- *Power Values* = (1-Beta); in this example the default is used; (1–0.2) = **0.8**.
- *Standard Deviation* (s) = **50** bottles per minute.
- *Significance Level* (Alpha) = **0.05**.
- *Alternative hypothesis* = **greater than**; assume that the new machine is better, so conduct one-sided test.

MINITAB procedure:

Step 1. From MINITAB's main drop-down menu, select the following sequence of commands, ***Stat > Power and Sample Size > 1-Sample t...***

Step 2. Enter the assumption data into the appropriate dialog boxes, as illustrated in Figure S-7.

a. In the main screen of the Power and Sample Size for 1-Sample t, enter the following three items—`Differences`, `Power values`, and `Standard deviation`. **Keep empty** the Sample Sizes dialog box.

b. Click the Options button to open the Power and Sample Size for 1-Sample t Options window. Select the appropriate Alternative hypothesis. Enter the appropriate Significance Level in its dialog box. Click the OK button.

c. Click the OK button on the main screen.

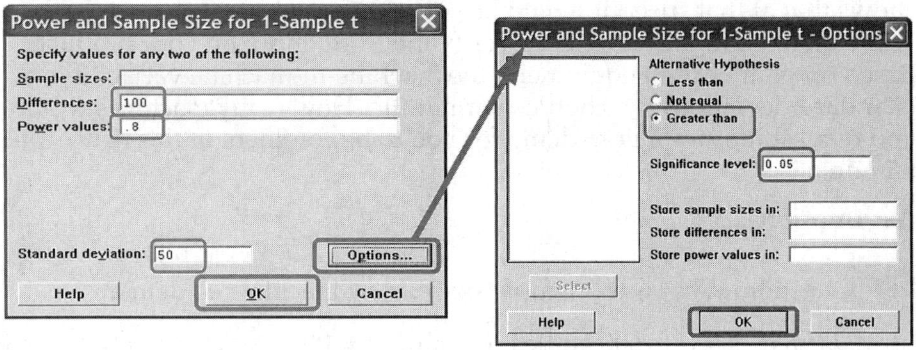

FIGURE S-7: MINITAB Screens for 1-Sample t-Test Sample Size

Figure S-8 provides the MINITAB output in its session window, using the soda bottle example data.

Power and Sample Size

1-Sample t Test

Testing mean = null (versus > null)

Calculating power for mean = null + difference

Alpha = 0.05 Assumed standard deviation = 50

Difference	Sample Size	Target Power	Actual Power
100	4	0.8	0.908447

FIGURE S-8: Bottle Example in MINITAB Session Window for 1-Sample t-Test Sample Size

MINITAB determined that **four machines** should be demonstrated. If they had a throughput greater than or equal to 700, with a standard deviation (s) no larger than 50, there is a 90% chance of concluding that the new model can achieve 700 bottles per minute.

Determine Power: 2-Sample t-Test

Example scenario background: You are considering a second supplier for pizza boxes. You have a standard test to ensure the box can hold 5lbs. of weight because the moisture, heat, and oil weaken the cardboard material. It is determined that if this critical deflection force for the new supplier differs by more than 50mm from the current supplier, there would be excessive variability in the box design. If the new supplier has 10 boxes available for testing, what would be the power for standard deviation (*s*) = 30mm and alpha = 0.05?

Assumptions:

- Question: What would be the power for standard deviation (s) = 30mm, and alpha = 0.05?
- *Sample Sizes* = **10** boxes.
- *Differences* = **50**mm.
- *Standard Deviation* (s) = **30** mm.
- *Significance Level (Alpha)* = **0.05**.
- *Alternative hypothesis* = **not equal test.**

MINITAB procedure:

Step 1. From MINITAB's main drop-down menu, select the following sequence of commands, ***Stat > Power and Sample Size > 2-Sample t...***

Step 2. Enter the assumption data into the appropriate dialog boxes, as illustrated in Figure S-9.

a. In the main screen of the Power and Sample Size for 2-Sample t, enter the following three items—`Sample size`, `Differences`, and `Standard deviation`. **Keep empty** the Power values dialog box.

b. Click the Options button to open the Power and Sample Size for 1-Sample t Options window. Select the appropriate Alternative hypothesis. Enter the appropriate Significance Level in its dialog box. Click the OK button.

c. Click the OK button on the main screen.

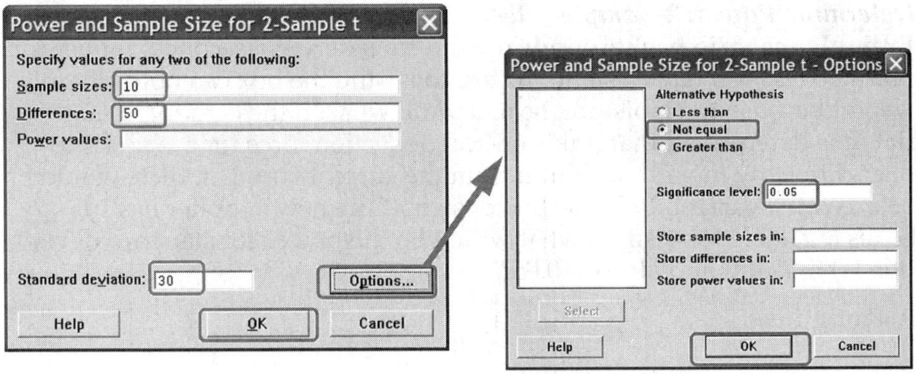

FIGURE S-9: MINITAB Screens for 2-Sample t-Test for Power

Figure S-10 provides the MINITAB output in its session window, using the pizza box example data.

```
Power and Sample Size

2-Sample t Test

Testing mean 1 = mean 2 (versus not =)
Calculating power for mean 1 = mean 2 + difference
Alpha = 0.05  Assumed standard deviation = 30

                Sample
Difference        Size      Power
        50          10   0.940808

The sample size is for each group.
```

FIGURE S-10: Pizza Box Example in MINITAB Session Window for 2-Sample t-Test for Power

MINITAB determined that for a sample size of 10 boxes, we have very good power (0.94). If the boxes differ by at least 50mm, we will have a good chance of detecting it with the standard test.

Determine Sample Size: 2-Proportion

Example scenario background: A candy company makes a brand of candy that contains five different colors in one bag. It publishes the proportion of different colors of its candies on its web site. According to the company information, the proportion of blue candies is 0.24. The proportion of pink candies is 0.2. If you wanted to test the company's assertion, how many samples would be required to detect this difference for alpha = 0.05?

Assumptions:

- Question: How many samples are required to detect an alpha level of **0.05**?
- *Proportion 1 values* = **0.2** pink candies.
- *Power values* = **0.8**.
- *Proportion 2* = **0.24** blue candies.
- *Significance Level (Alpha)* = **0.05**.
- *Alternative hypothesis* = Less than test; less pink candies than blue candies.

> **Note**
>
> If blue candies (0.24) were placed in the Proportion 1 value dialog box, and the pink candies (0.20) in the Portion 2, then the *Alternative hypothesis* test would need to be greater than.

MINITAB procedure:

Step 1. From MINITAB's main drop-down menu, select the following sequence of commands, ***Stat > Power and Sample Size > 2-Proportions…***

Step 2. Enter the assumption data into the appropriate dialog boxes, as illustrated in Figure S-11.

a. In the main screen of the Power and Sample Size for 2 Proportions, enter the following three items—`Portion 1 values`, `Power values`, and `Portion 2`. **Keep empty** the Sample Sizes dialog box.

b. Click the Options button to open the Power and Sample Size for 2 Proportions—O… window. Select the appropriate Alternative hypothesis. Enter the appropriate Significance Level in its dialog box. Click the OK button.

c. Click the OK button on the main screen.

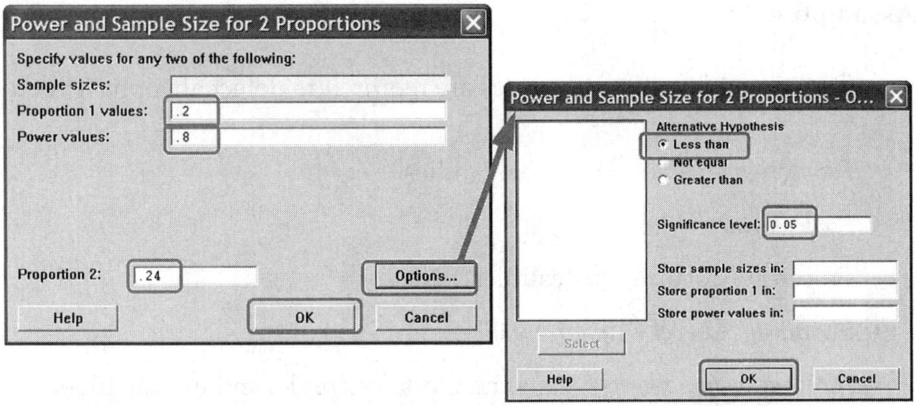

FIGURE S-11: MINITAB Screens for 2-Portions Sample Size

Figure S-12 provides the MINITAB output in its session window, using the soda bottle example data.

```
Power and Sample Size

Test for Two Proportions

Testing proportion 1 = proportion 2 (versus <)
Calculating power for proportion 2 = 0.24
Alpha = 0.05

                 Sample  Target
Proportion 1       Size   Power  Actual Power
         0.2       1326     0.8      0.800233

The sample size is for each group.
```

FIGURE S-12: Candy Example in MINITAB session window for 2-Portions Sample Size

MINITAB determined that a sample containing at least 1326 of each color of candy is needed to detect a difference of just 4% between the number of blue and pink candies. Blue candies represent 20% of the total in the bag. So the total number of candies needed to sample is approximately the five different colors times the required sample size, which is 6630 individual candies (5 x 1326 = 6630); now that is a lot of candy.

Determine Sample Size—Retrospective Scenario

Example scenario background: A sales rep was considering two different routes for a new target list. She tested each route five times throughout the day to better understand the different traffic patterns. The mean for route 1 was 22.4 minutes. The mean for route 2 was 24.8 minutes, and the standard deviation for both was about three minutes

She performed a 2-Sample t-test in MINITAB using the commands of ***Stat > Basic Statistics > 2 Sample t (Test and Confidence Interval)…*** loaded this five-run data from her two different routes, selected ***Summarized data,*** and got the following results in MINITAB's session window, as shown in Figure S-13.

Stat > Basic Statistics > …
2 Sample t (Test and Confidence Interval)…
Select ***Summarized data***

```
Two-Sample T-Test and CI
Sample  N  Mean   StDev  SE Mean
1       5  22.40  3.00   1.3
2       5  24.80  3.00   1.3
Difference = mu (1) - mu (2)
Estimate for difference:  -2.40000
95% CI for difference:  (-6.77534, 1.97534)
T-Test of difference = 0 (vs not =): T-Value = -1.26  P-Value = 0.242  DF = 8
```

Figure S-13: MINITAB Session Window for 2-Sample t-Test for the Sales Rep Example

The sales rep was disappointed with the results. She was hoping that one route would prove to be faster, but the *T-Test of difference* is zero. Moreover, the p-value is 0.242, indicating that there is insufficient evidence to reject the Null hypothesis, meaning that the two route times were indistinguishably different. (**See Also** "Hypothesis Testing," **p. 335** for a discussion about p-value.)

Calculate the power of the test. If the times for the routes remain stable, how many additional trips would need to be made in order to show a difference with a reasonable power?

Assumptions:

- Question: How many additional trips are needed to indicate a difference with reasonable power?
- *Differences* = 24.8–22.4 = **2.4** minutes.

- *Power values*—Try multiple alternatives by entering all three numbers with a space in between each = **0.7 0.8 0.9.**
- *Standard deviation* = **3.0** because times remain stable, same as before.
- *Alternative Hypothesis* = Not equal test, hoping route 1 is faster.

MINITAB procedure:

Step 1. From MINITAB's main drop-down menu, select the following sequence of commands, ***Stat > Power and Sample Size > 2-Sample t...***

Step 2. Enter the assumption data into the appropriate dialog boxes, as illustrated in Figure S-14.

a. In the main screen of the Power and Sample Size for 2-Sample t, enter the following three items—`Differences`, `Power values`, and `Standard deviation`. **Keep empty** the Sample Sizes dialog box.

b. Click the Options button to open the Power and Sample Size for 2-Sample t - Options window. Select the appropriate Alternative hypothesis. Enter the appropriate Significance Level in its dialog box. Click the OK button.

c. Click the OK button on the main screen.

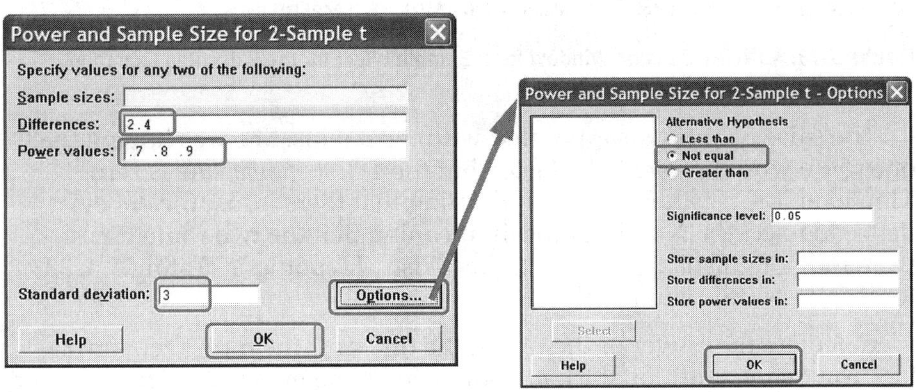

FIGURE S-14: MINITAB Session Window for 2-Sample t-Test for the Sales Rep Example

Figure S-15 provides the MINITAB output in its session window, using the sales rep example data.

Power and Sample Size

2-Sample t Test

Testing mean 1 = mean 2 (versus not =)
Calculating power for mean 1 = mean 2 + difference
Alpha = 0.05 Assumed standard deviation = 3

Difference	Sample Size	Target Power	Actual Power
2.4	21	0.7	0.715555
2.4	26	0.8	0.807487
2.4	4	.9	0.901502

The sample size is for each group.

FIGURE S-15: Sales Rep Example in MINITAB Session Window for 2-Sample t-Test and Power

MINITAB determined that a total of 26 trips were necessary to have adequate Power of 80%. The sales rep has already taken five, so there are 21 remaining trips yet to take.

ANOVA Sample Size

Calculating the sample size for a variety of Hypothesis tests is similar to the preceding examples. (**See Also** "Analysis of Variance (ANOVA)—7M Tool," '**p. 142**)

If a three-level ANOVA test needed to be conducted, and the business required the defaults of an 80% Power level and 5% alpha level, the sample size can be calculated for different values of detectable difference.

From MINITAB's drop-down menu, select the sequence ***Stat > Power and Sample Size > 2-Sample t....***

For example purposes, enter in the following data into the different dialog boxes on the Power and Sample Size for One-Way ANOVA main screen, as illustrated in Figure S-16:

- *Number of levels* = **3**
- *Sample sizes* = (Leave this box empty.)
- *Values of the maximum difference between means* = **20 30 40 50 60**
- *Power values* = **0.8**

- *Standard deviation* = **20**
- Options button: Leave the default of 0.05 in the Significance level dialog box.

Select the OK button, and the following session window appears, as shown in Figure S-17. Notice that there is an inverse relationship between sample size and the detectable difference; such that as sample size increases, the detectable difference decreases.

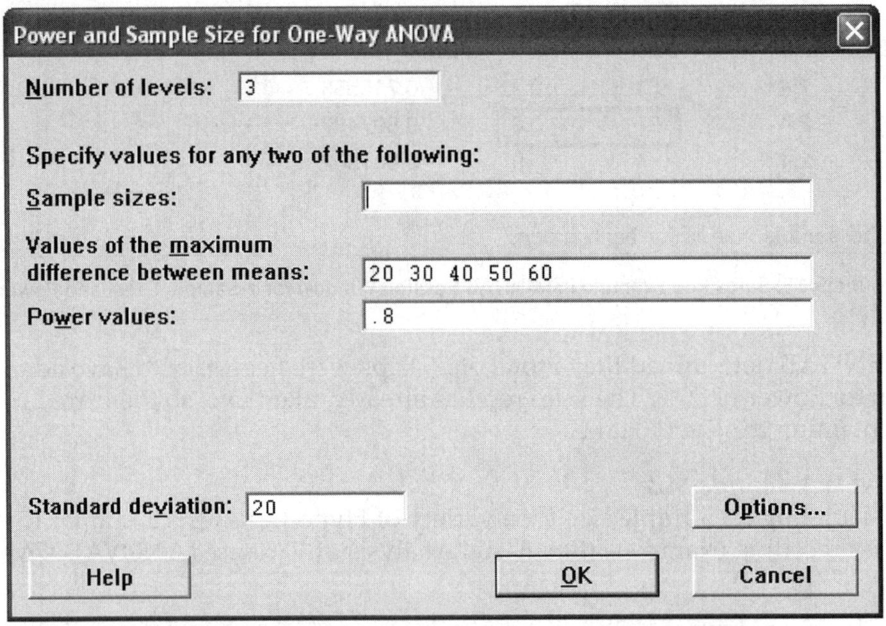

FIGURE S-16: MINITAB One-Way ANOVA for Sample Size

One-way ANOVA
Alpha = 0.05 Assumed standard deviation = 20 Number of Levels = 3

SS Means	Sample Size	Target Power	Actual Power	Maximum Difference
200	21	0.8	0.814770	20
450	10	0.8	0.817278	30
800	6	0.8	0.805317	40
1250	5	0.8	0.883287	50

FIGURE S-17: MINITAB Session Window for One-Way ANOVA—Sample Size

Random Numbers Generation

MINITAB can create a random numbers table. Select the following sequence from its drop-down menu: ***Calc > Random Data > Normal…***.

A B C D E F G H I J K L M N O P Q R S T U V W X Y Z

Decide how many columns and rows of data you wish to be generated. Determine the mean and standard deviation needed. Select the OK button. MINITAB will populate its worksheet with the amount of random numbers requested, which yield the specified mean and standard deviation. Figure S-18 shows a sample MINITAB main screen for random number generation of normally distributed data.

> **Note**
>
> Non-normal random numbers may also be generated by specifying the appropriate parameters, not the mean and standard deviation.

Select *Calc > Random Data* and then make the appropriate selection.

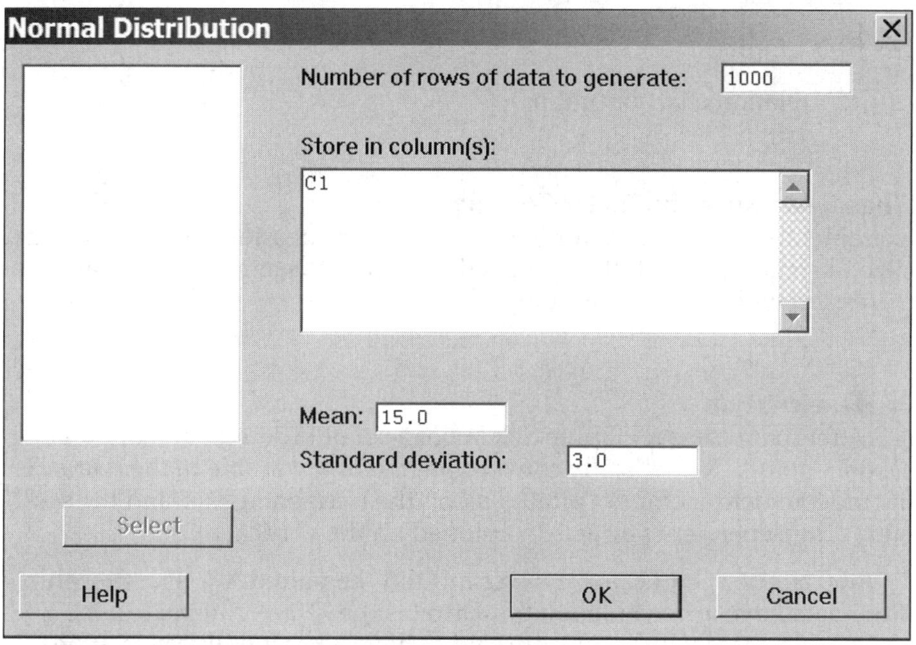

FIGURE S-18: MINITAB Random Number Generator for Normal Distribution

Because every random number generation table would be different, an illustration of the final MINITAB Worksheet would not be useful.

Scatter Diagram—7QC Tool

What Question(s) Does the Tool or Technique Answer?

Are these two factors correlated? What is the relationship between these two factors?

A Scatter diagram helps you to

- Visualize the relationship between two variables—either linear or a curve

Alternative Names and Variations

This tool is also known as

- Scatterplot
- X-Y graph
- Correlation chart or graph

When Best to Use the Tool or Technique

Examine the Scatterplot **prior** to conducting a regression analysis to visualize the relationship. If the curve reflects a non-linear curve, a linear regression will not be applicable.

Brief Description

A Scatter diagram is a graphic display of a set of ordered pairs of numerical data points (X and Y). By convention, the first variable in the pair is the independent factor (X) plotted along the horizontal axis, and the second is the dependent variable (Y) plotted on the vertical axis.

Analysis of a Scatterplot evolves around the paired X-Y's relationship. If the data have an **inverse relationship** (as the X factor increases, the Y-variable decreases), then they are said to have a *negative* linear relationship. A negative linear relationship displays its points in a downward trending slope from upper-left to lower-right corner of an X-Y graph.

If they have a *direct relationship*, such that both X and Y increase simultaneously, it is a *positive* linear relationship. A positive linear relationship displays its points in an upward trending slope from lower-left to upper-right corner of an X-Y graph.

If X increases and Y holds fairly constant, then it is said they have no relationship. No linear relationship displays its points in a flat, horizontal, or zero slope, parallel with the X-axis of an X-Y graph.

The Scatter diagram can graph both linear or non-linear relationships. It supports two statistical tools—the *correlation analysis* for linear relationship and the *regression analysis* for non-linear relationships.

Linear Relationships—Use with Correlation Analysis

A Scatter diagram describes the correlation between two factors. *Correlation* is a metric that measures the linear relationship between two process variables. The tighter the display of data points, the stronger the relationship. A strong relationship, or correlation does **not** imply causation (a cause-and-effect relationship). For example, the number of car accidents on a rainy day may be strongly correlated with the sale of umbrellas, but the buying of umbrellas did not cause the accidents. Lack of correlation does **not** necessarily mean that no causal relationship exists. The range of data may be too narrow to detect relationship. Consider the graph in Figure S-19 if it simply revealed the center band, it would appear to display no relationship. However, if the perspective expands to encompass the entire graph, it represents a moderately negative relationship.

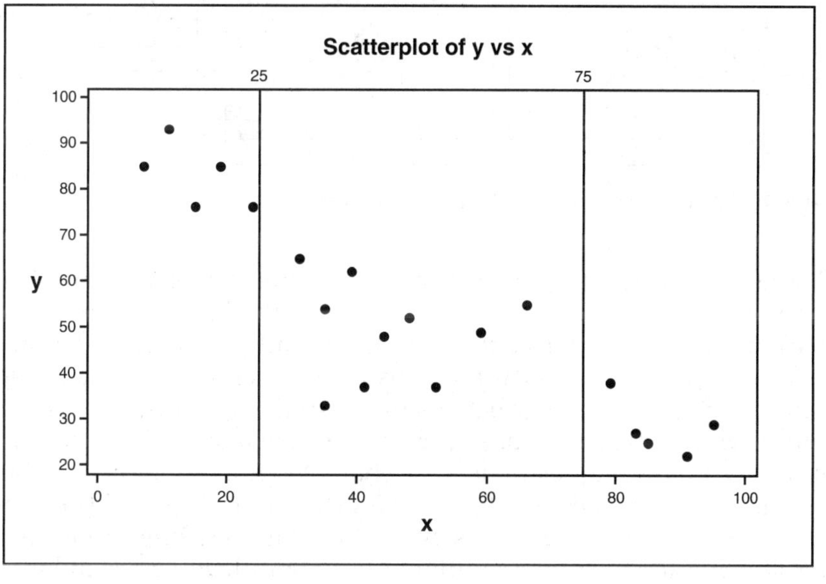

FIGURE S-19: Scatterplot with Bands

A poor-man's test for a strong linear relationship is called the pencil test. Once the data is individually plotted on an X-Y graph, sit in a chair at a table or desk and place the graph on top of the flat surface. Hold up a pencil at arms-length, if the pencil covers up the dots, it is said to be a strong relationship. If dots can be seen on either side of the pencil, the relationship is moderate to weak. Other than the pencil test, another visual technique is to look in the opposite corners to see how much white space exists. The more white space in the opposite corners from the data, the more likely a strong linear relationship exists.

Figure S-20 illustrates the different types of linear relationships and their strength.

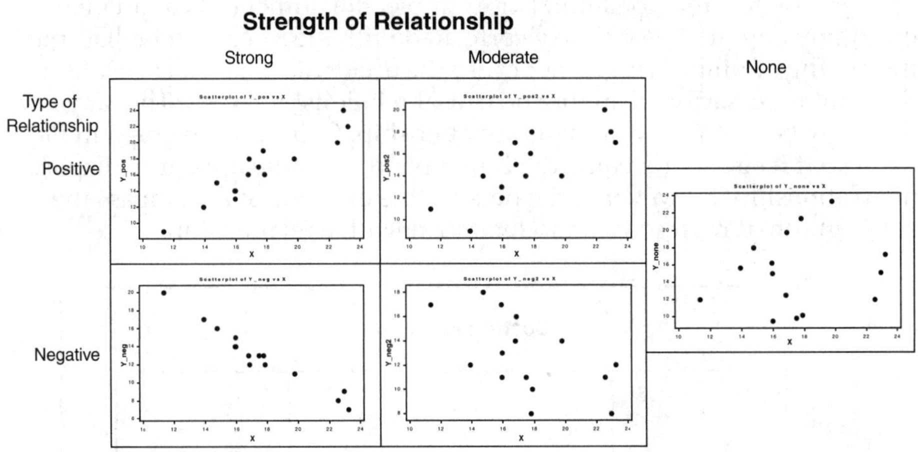

FIGURE S-20: Scatter Diagrams with Different Relationships

There is a statistic that also determines the strength of a linear relationship between two quantitative variables, called the *Pearson's Correlation Coefficient (r)*. It defines the correlation metric and falls between (-1) and (1), where 0 indicates no linear relationship, (-1) describes a perfect negative correlation, and (1) a perfect positive correlation. Thus, the larger the absolute value of **r** (either + or -), the stronger the linear relationship. Graphically, this strong relationship is characterized by a tight distribution of data around a best-fit line plotted in a Scatter diagram. The Pearson's Correlation Coefficient (r) is proportional to the slope of the regression line adjusted for the differences in both X and Y's standard deviations. Its formula is as follows:

$$r_{xy} = \frac{\sum_{i=1}^{n}(x_i - \bar{x})(y_i - \bar{y})}{\sqrt{\sum_{i=1}^{n}(x_i - \bar{x})^2 \sum_{i=1}^{n}(y_i - \bar{y})^2}}$$

Correlation analysis and its Pearson's Correlation Coefficient (r) is useful in describing the strength of a linear relationship. It is a relatively simple, straight-forward, quick analysis compared with its cousin, the regression analysis. Recall that Scatter diagrams support both types of statistical analyses and graph both linear and non-linear relationships.

Non-Linear Relationships—Use with Regression Analysis

The regression analysis applies to both linear and non-linear relationships. Use the correlation analysis for a simple test of magnitude for a linear relationship. However, if modeling is required to predict (extrapolate or interpolate) results, then regression analysis should be used. And for non-linear relationships, use regression and its R-squared value (R^2) [for example, the square of the Pearson's Correlation Coefficient (r)], to describe the relationships. (**See Also** "Regression Analysis," **p. 571**, for more detail on Pearson's Correlation Coefficient, correlation, and linear and non-linear relationships.)

Figure S-21 illustrates several Scatter diagrams with their corresponding Pearson's Correlation Coefficient (r).

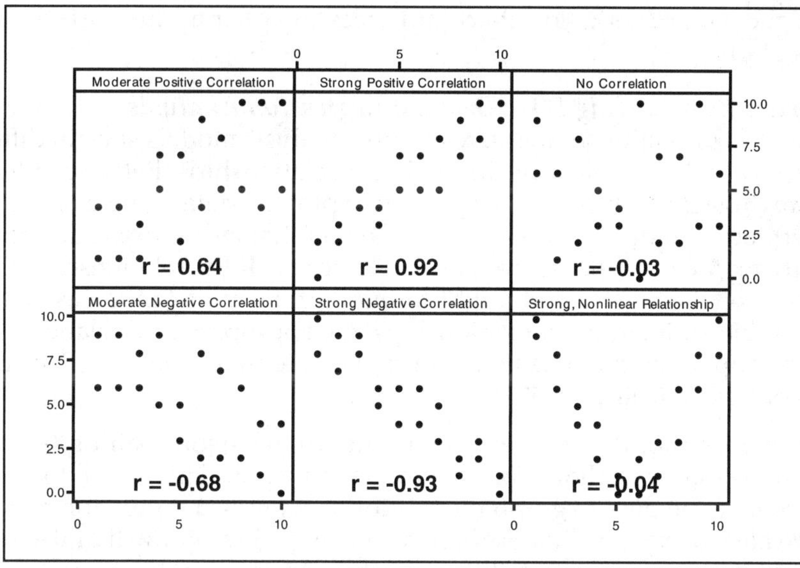

FIGURE S-21: Scatter Diagrams with Corresponding Pearson's Correlation Coefficient (r)

The Scatter diagram is a useful tool because it plots either a linear or non-linear relationship. It also can depict a quadratic or cubic relationship. As a result, subsequent regression analyses yield non-linear models with additional terms in the equation. Figure S-22 compares these different relationship Scatter diagrams drawn in MINITAB.

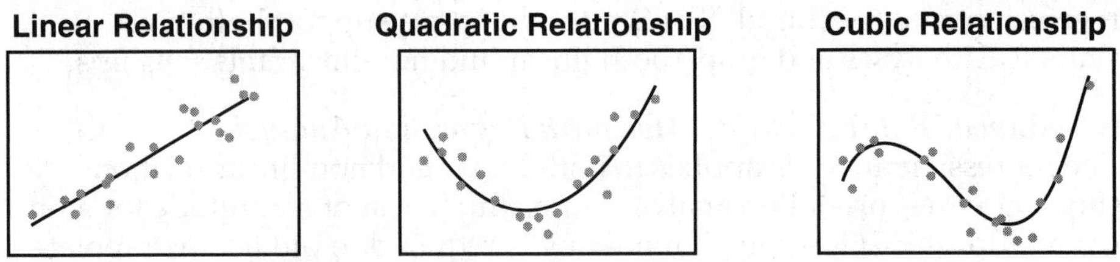

FIGURE S-22: Comparing Linear, Quadratic, and Cubic models

The Scatter chart is a member of the *7QC Tools*, (or seven *Quality Control* tools), attributed to Dr. Kaoru Ishikawa. The 7QC tools sometimes are called the *seven basic tools* because they were the first set of tools identified as the *core* quality improvement tools. Ishikawa's original 7QC toolset includes: Cause-and-Effect diagram; Check sheet (or checklist); Control charts; Histogram; Pareto chart; Scatter diagram; and Stratification. More recently, the 7QC toolset is modified by substituting the Stratification technique with either a flowchart (or Process map) or a run chart (or Time Series plot).

Importance of Plotting Data Before a Regression Analysis

Unlike the correlation analysis, regression analysis models several different types of relationship, not simply linear relationships. Before conducting a regression analysis, it is important to plot the data using a Scatterplot, to understand what type of data relationship exists before identifying the potential regression model terms. If the relationship is linear, the model is relatively straight-forward. If the model involves quadratics, cubic, or interaction relationships, the appropriate calculations must be applied to the data before testing a potential equation. (**See Also** "Regression Analysis," **p. 571**)

Brief Scenario: We are pressed for time. We have four pair of X-Y data to examine that we believe come from the same population and don't want to take the time to graph them prior to running a regression analysis. We have been to a Lean Six Sigma course, and we know that if samples have the same means that they most likely are from the same population.

We run a quick analysis in MINITAB to check that the four sets have the same mean and the same standard deviation; and they do. (MINITAB's Display Descriptive Statistics can be found from its drop-down menu by following this sequence: ***Stat > Basic Statistics > Display Descriptive Statistics****....*)

We run another quick analysis to verify that the correlation statistics are the same; and they are. (MINITAB's Display Descriptive Statistics can be found from its drop-down menu by following this sequence: ***Stat > Basic Statistics > Correlation….***)

Figure S-23 displays the MINITAB session window containing both the Descriptive Statistics and Correlation information for this example.

Descriptive Statistics: X1, Y1, X2, Y2, X3, Y3, X4, Y4

```
Variable    N    Mean  StDev
X1         11    9.00   3.32
Y1         11   7.501  2.032
X2         11    9.00   3.32
Y2         11   7.501  2.032
X3         11    9.00   3.32
Y3         11   7.500  2.030
X4         11    9.00   3.32
Y4         11   7.501  2.031
```

All Xs have the same mean and standard deviation.

All Ys have the same mean and standard deviation.

Correlations: X1, Y1, X2, Y2, X3, Y3, X4, Y4

```
        X1        Y1        X2        Y2
Y1   0.816
X2   1.000     0.816
Y2   0.816     0.750     0.816
X3   1.000     0.816     1.000     0.816
Y3   0.816     0.469     0.816     0.588     0.816
X4  -0.500    -0.529    -0.500    -0.718    -0.500    -0.345
Y4  -0.314    -0.489    -0.314    -0.478    -0.314    -0.155     0.817
```

The correlations among the four datasets are the same.

FIGURE S-23: Example MINITAB session window of Descriptive Statistics and Correlation

We feel very confident about our data and launch right into running regression analyses. We would expect that the regression equations for the four X-Y pairs also would be the same; the intercepts are the same; the slopes are the same; and the Coefficients of Determination (R2) are the same. But much to our amazement, the R-square and R-square-adjusted tell us that our model is inadequate. We are getting frustrated and running out of time.

In general, use the value of R-square-adjusted to assess how good a model fits. R-square increases with every additional predictor (or variable) added to the model even if the predictor adds little value to the model. (**See Also** "Regression Analysis," **p. 571** for more information.)

We decide to plot the data to see if graphs display any unusual patterns in the four different data sets. Figure S-24 shows the Scatter diagrams for each set, and we find that in fact they are very different.

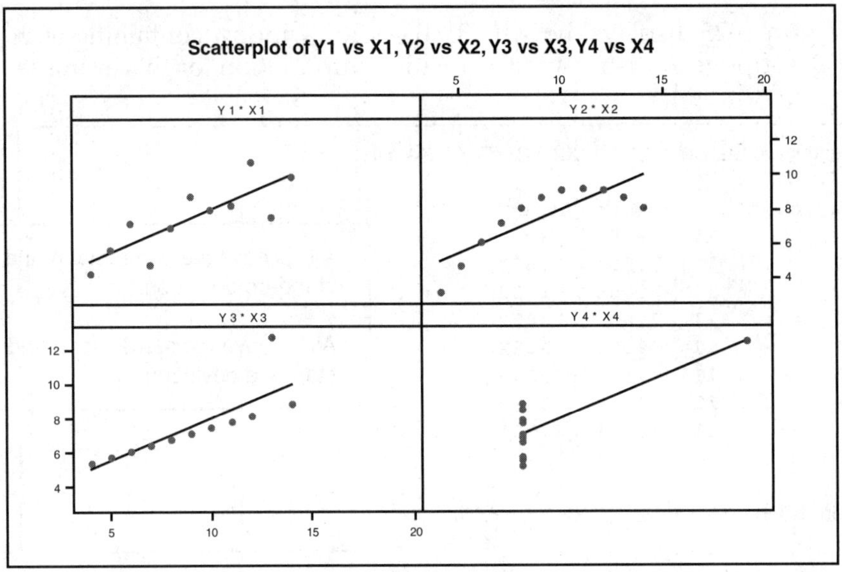

FIGURE S-24: Example Scatterplot of Four Data Sets with the Same Means and Correlations

How to Use the Tool or Technique

Use the following procedure to develop a Scatter diagram using MINITAB.

Given that numeric data has been entered into the MINITAB Worksheet, select the following commands from its drop-down menu: ***Graph > Scatterplot....***

In the Scatterplot main screen, select ***Simple,*** for a simple X-Y plot. Click the OK button.

In the Scatterplot—Simple main screen, enter the appropriate columns of X and Y data in the corresponding cells of the dialog box. Click the OK button. Figure S-25 illustrates both the MINITAB main screen and the corresponding output graph.

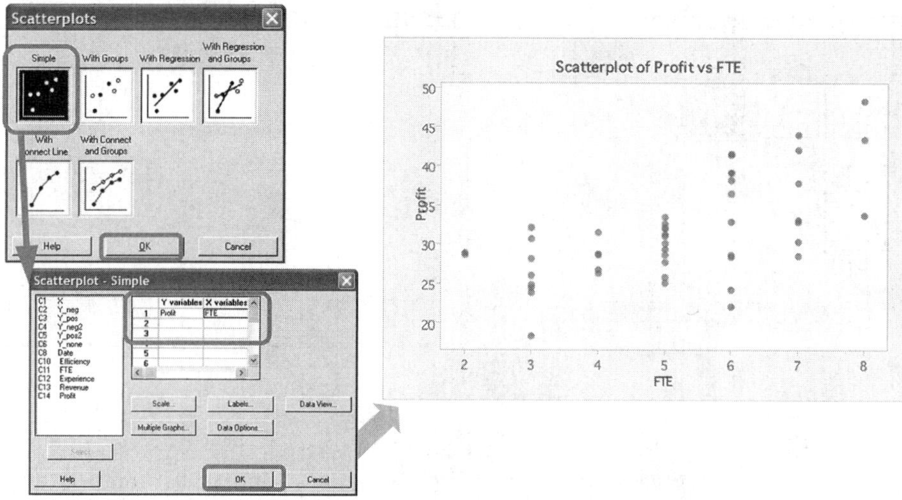

FIGURE S-25: Scatter Diagram—MINITAB Main Screen and Graph

Figure S-25 shows a moderate positive relationship between Profit and FTE (Full-time Equivalents).

The same plot can be drawn with a regression (or fitted) line. From MINITAB's drop-down menu, select the following sequence: ***Graph > Scatterplot...***.

In the Scatterplot main screen, select ***With Regression*** for a simple X-Y plot and fitted line. Click on the OK button.

In the Scatterplot—With Regression main screen, enter the appropriate columns of X and Y data in the corresponding cells of the dialog box. Click the OK button. Figure S-26 illustrates both the MINITAB main screen and the corresponding output graph.

Correlation Analysis—for Linear Relationships

Other than calculating the Pearson's Correlation Coefficient (r) manually, MINITAB also calculates it. Using MINITAB to determine the correlation statistic for the data plotted on a Scatter diagram, select the following sequence from MINITAB's main drop-down menu ***Stat > Basic Statistics > Correlation...***.

Enter the appropriate X and Y variables in the Correlation dialog box and click the OK button to calculate the Pearson's Correlation Coefficient (r), which MINITAB provides in its session window.

Figure S-26 shows the MINITAB Correlation main screen and the session window for the same example used in Figure S-26. The Profit and FTE correlation results indicate that the correlation is positive and moderately strong with an **r** of 0.621 (with 1.0 being very strong) and that the two variables are statistically correlated with a p-value of 0.000.

A B C D E F G H I J K L M N O P Q R S T U V W X Y Z

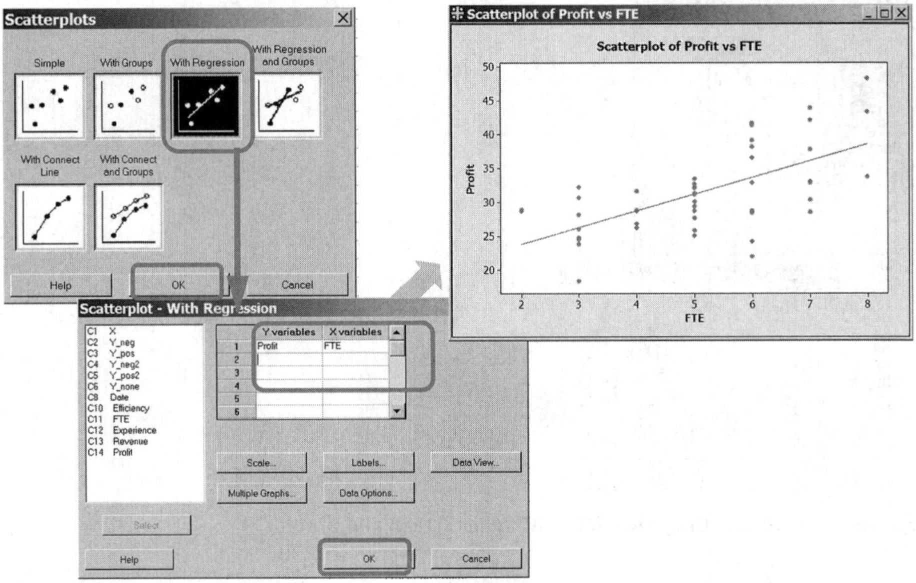

FIGURE S-26: Scatter Diagram with Fitted Line—MINITAB Main Screen and Graph

> **Note**
>
> The Null hypothesis for correlation: H_0 states that the correlation equals zero (r = 0); but in this case because the p-value is low, there is sufficient evidence to reject the H_0 and accept the Alternative Hypothesis that the pair of variables are correlated. Recall that when p is low, H_0 can go. (**See Also** "Hypothesis Testing," **p. 335**, for details on p-value and Null hypothesis.)

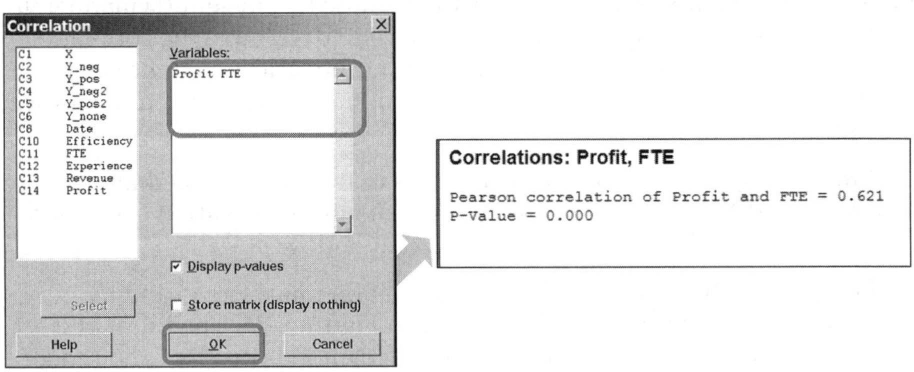

FIGURE S-27: Fitted Line Plot—MINITAB Main Screen and Graph

Fitted Line Plot or Regression Line

The same Scatter diagram exhibited in Figure S-26 can also be drawn using MINITAB's *Fitted Line Plot*. The main difference is that using the Fitted Line Plot feature, MINITAB also provides additional information with the graph, namely the regression equation [found under the graph's title), the standard deviation (s), R-square (R-Sq), and R-square-adjusted (R-Sq(adj.)]. (**See Also** "Regression Analysis," **p. 571**)

From its main drop-down menu, select the following sequence ***Stat > Regression > Fitted Line Plot....*** In the main screen, select the appropriate columns of data that correspond with the X and Y data and enter each in the corresponding dialog box. Unless the relationship between the paired data is known, select ***linear relationship*** under *the* **Type of Regression Model** option. Click the OK button. Figure S-28 illustrates both the MINITAB main screen and the corresponding output graph.

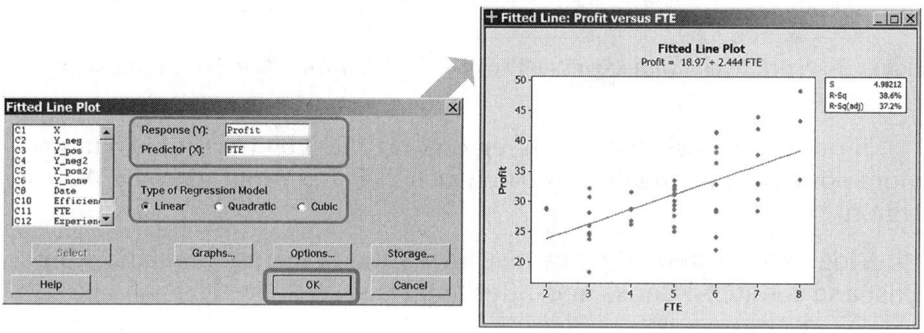

FIGURE S-28: Fitted Line Plot—MINITAB Main Screen and Graph

Subgroup Stratification

MINITAB's Scatterplot can distinguish among different subgroups to display stratification. Within the FTE category, subgroups of Region office location exist. From MINITAB's drop-down menu, select the following sequence: ***Graph > Scatterplot....***

In the Scatterplot main screen, select ***With Regression and Groups*** for a stratified X-Y plot and a fitted line. Click the OK button.

In the Scatterplot—With Regression and Groups main screen, enter the appropriate columns of X and Y data in the corresponding cells of the dialog box. Click the OK button. Figure S-29 illustrates both the MINITAB main screen and the corresponding output graph.

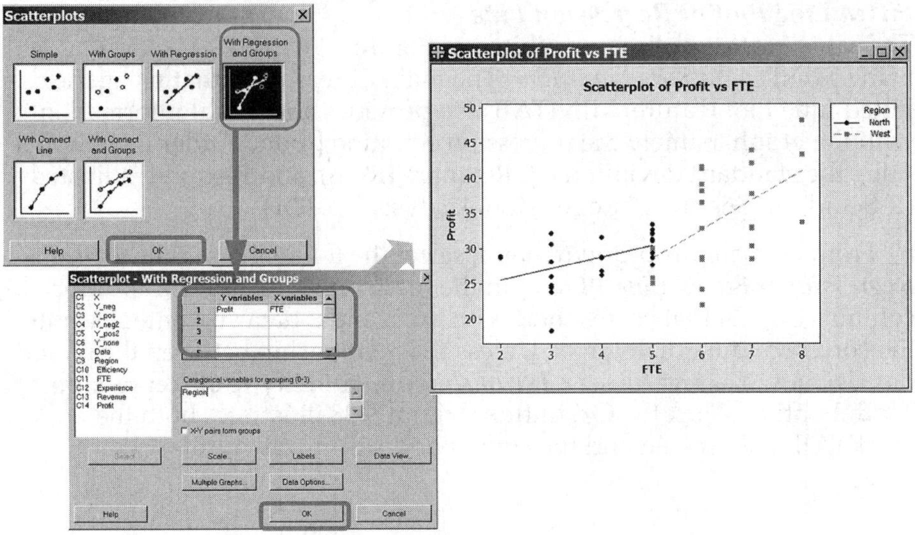

FIGURE S-29: Scatterplot with Regression and Groups—MINITAB Main Screen and Graph

Figure S-29 shows that North region (indicated by the squares) has a more positive linear relationship between FTE and profit than the West region.

A logical next step might be to ask why the difference in relationships exist and conduct a Cause-and-Effect analysis.

Matrix Plots

MINITAB offers a useful tool for when several different data exist, called a matrix plot. A matrix plot is several individual Scatter diagrams displayed via one graph. This tool is meant for analysis purposes only and is not a good presentation tool because it can become an eye-chart in that is packed with a lot of detail, making it difficult to read quickly; it requires careful scrutiny to glean any messages. To illustrate this point, see Figure S-30. However, it serves as a quick peek at a lot of data to identify if any relationships of interest exist. A matrix plot narrows the field of possible paired relationships of interest before spending time on creating individual graphs or conducting regression analysis.

To create a matrix plot, select the following sequence from MINITAB's drop-down menu: ***Graph > Matrix Plot***....

In the Matrix Plot main screen, select ***With Groups***. Click the OK button.

In the Matrix Plot—Matrix of Plots, With Groups main screen, enter all the columns of data that may be of interest in the dialog box.

A B C D E F G H I J K L M N O P Q R S T U V W X Y Z

Notice that MINITAB allows a range of data columns to be selected as indicated by a **hyphen** between two data column names as a shortcut, rather than listing each individual column name. To select the range, click in the first data column name in the far left window and hold the shift key down while clicking the last data column name of interest. Click the Select button, and the range of names will display in the dialog box.

Enter the subgroup data name to stratify. Click the OK button.

Figure S-30 illustrates both the MINITAB main screen and the corresponding output graph.

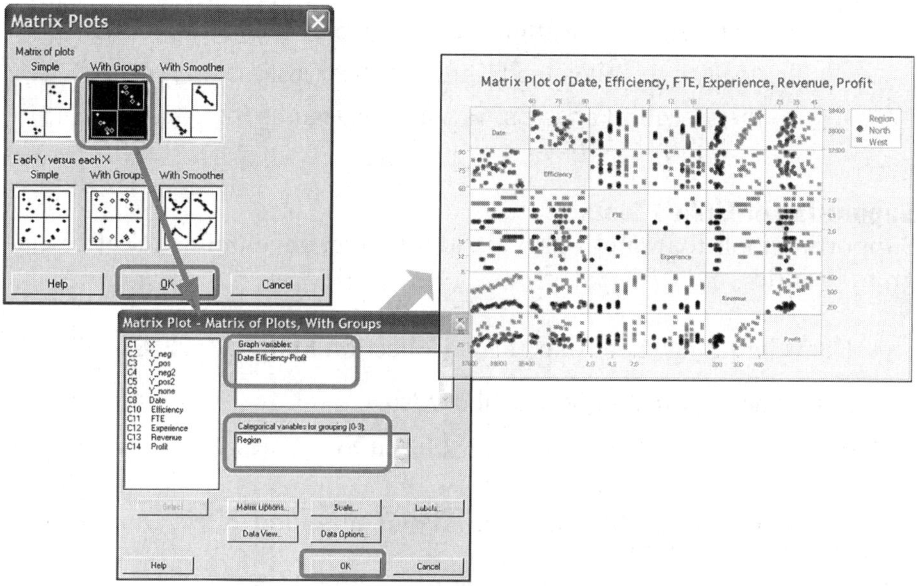

FIGURE S-30: Matrix Plot with Groups—MINITAB Main Screen and Graph

Figure S-30 illustrates an eye chart; however, it is intended only as a quick and easy tool to narrow the array of potential relationships and identify some of interest for further exploration and analysis. A matrix plot is an analysis tool, not a presentation tool.

Hints and Tips

- The X values represent the independent variables that impact the Y values. (In a Scatter plot, the independent variable is plotted on the X-axis, and the dependent variable on the

Y-axis.) Be certain to properly assign the data using this convention when conducting a regression analysis.

- Correlation does **not** mean cause-and-effect relationship.
- If the data shows no relationship, determine if subgroups exist and examine if stratification changes the findings. (**See Also** "Stratification—7QC Tool," **p. 697**)
- Correlation equates to the strength or magnitude of a relationship between the X and Y variables.
- Plot data in a Scatter diagram before running regression analysis to understand if any patterns exist that may distort the statistical analysis, as shown in Figures S-23 and S-24.

Supporting or Linked Tools

Supporting tools that might provide input when developing a Scatter diagram include

- Data Gathering (**See Also** "Data Collection Matrix," **p. 248**)
- Performance Charts and Dashboards

A completed Scatter diagram provides input to tools such as

- Cause-and-Effect diagram (**See Also** "Cause-and-Effect Diagram—7QC Tool," **p. 173**)
- Correlation analysis (**See Also** "Cause-and-Effect Prioritization," **p. 188**)
- Regression analysis (**See Also** "Regression Analysis," **p. 571**)
- FMEA (**See Also** "Failure Modes and Effects Analysis (FMEA)," **p. 287**)

Figure S-31 illustrates the link between a Scatter diagram and its related tools and techniques.

A B C D E F G H I J K L M N O P Q R S T U V W X Y Z

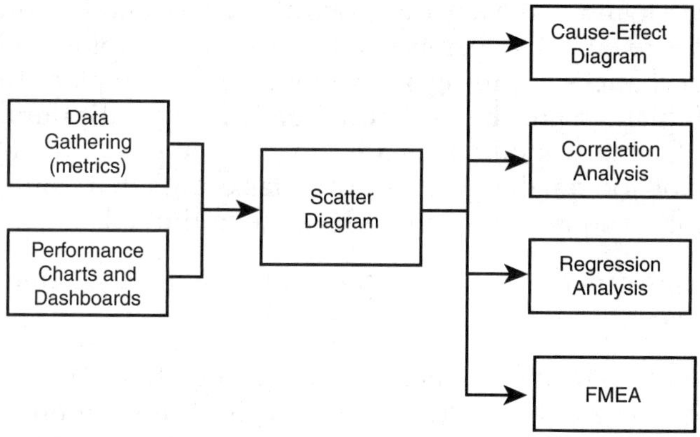

FIGURE S-31: Scatter Diagram Tool Linkage

Scorecards

What Question(s) Does the Tool or Technique Answer?

What is the progress of the project and project team or the process and process players?

Scorecards help you to

- Monitor the healthiness of a task, project, process, or entire business
- Plan and guide the team's work
- Understand if the project improvement team members, or process players, are using the right tools at the right time

When Best to Use the Tool or Technique

Use a system of scorecards throughout the lifecycle of the improvement project or process to monitor and manage the flow of work.

Brief Description

Similar to an adventurer's reliance on a compass, a business person's most powerful guiding tool to ensure the business stays on course, or adheres to its plan, is a scorecard. A scorecard is the primary predictive tool for both in-process measures and performance results. Regardless of the type of work, people should measure progress against their goals. There are two distinct reasons to establish formal methods for measuring team performance:

- Manage risk and make key decisions at reviews and key project milestones to ensure adherence to the plan or to explore indications of critical changes requiring adjustments to stay on plan. To avoid going off plan, **make changes based on leading indicators to stay on plan.** Changes should be made to fulfill the plan, rather than being made too quickly. The tools and tasks are what change, **not** the deliverables and requirements, which are immutable.

- Pay for performance in accordance with specific requirements and deliverables.

A tool is as good as the information that is supplied to it. The old adage, "garbage in, garbage out," could not be more applicable. Scorecards must track the right information to be useful to a data-driven leader. Taking the time to determine the critical risk accrual requirements defines the appropriate information to design into and track in a scorecard. The requirements may vary based on the tool, task, or gate deliverable. Hence, a **system of scorecards** works best to address: 1) the right tool, 2) applying the right tool to the right task at the right time, and 3) delivering the right summary data for risk management and decision-making. Requirements are the business questions asked before a phase of work is conducted. The goal is to design measurable work in light of the requirements **before** you start measuring.

In general, there are two types of tracking tools: checklists, the simpler version, and the scorecard. More complex scorecards also can be referred to as dashboards, depicting that all the essential key indicators—the critical parameters—needed to "drive" a business are together in one spot.

Checklists assess two states of task completion—done or not done. Checklists serve well as a reminder tool rather than as a predictive tool. Checklists prompt people to recall what needs to get done and track if it is done. The checklist catalogs the expected requirements, tasks, or deliverables. Checklists monitor if an item has been completed or not (binary *yes* or *no* response) to help avoid duplication of effort. Checklists suffer from a lack of discrimination. They fail to provide information about quality or quantity. Checklists lack predictive specifics about how well a tool was used or the percent completion of a task against its original target, and they fail to discriminate on the details of risk accrual. (**See Also** "Checklists—7QC Tool," **p. 204**)

Scorecards help to scrutinize the healthiness of the business in terms of its performance and risk accrual. The scorecard is a tool that drills down into the quality of processes or deliverables against respective targets and requirements. Scorecards can predict trends and alert business people to potential changes that require a response to maintain the planned course. A system of scorecards presents those vital few indicators needed to understand how well a business is performing against its

A B C D E F G H I J K L M N O P Q R S T U V W X Y Z

goals; whereas a subset of scorecards probes deeper into an individual process or functional task/tool application area. **The concept of scorecards applies to any discipline to monitor the healthiness of a task, project, process, or entire business.**

Use both a system of scorecards and a dashboard, as they complement one another but are unique. The dashboard contains the critical success metrics that require monitoring on a regular basis to maintain the healthiness of the ongoing operations. The dashboard takes a snapshot of the business appropriate leading and lagging indicators and organizes things typically in one place (a document and/or web site). The system of scorecards examines the work approach used by the process players, monitoring the interconnected tool-task-deliverables link to the requirements. The system of scorecards informs the review processes not only for ongoing operations (as with a dashboard), but also the phase-gate governance process of a project. Dashboards refer to the data critical to driving the business. The literature on dashboards is voluminous; however, little is written on the scorecards system. An organization should utilize both systems to aid in monitoring its business' state of health. (**See Also** "A Process for Product Development," **p. PDF:887** in Part III, for more detail on the governance and phase-gate review process.)

A four-level, hierarchical flow starts with the criteria or requirements that dictate the deliverables and then steps down to the tasks needed to complete the right outputs at the right time. Then it flows down to the appropriate tools, techniques, and best practices that support the tasks. Figure S-32 illustrates these four criteria in sequence.

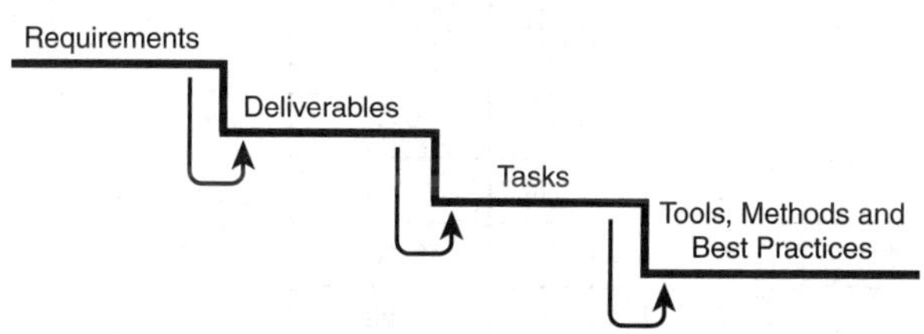

Figure S-32: Hierarchical Flow of Work

This four-level flow is well-suited to measurement by using an integrated system of scorecards. A system of scorecards can be designed and linked so that each subordinate level adds summary content up to the next level. Use several types of scorecards to measure performance and managing risk along the work hierarchy: Tool Scorecards, Task Scorecards, and two types of Gate Review Scorecards.

Tool Scorecards

The most basic level of scorecard is filled out by team members or process players who are actually using specific sets of tools, methods, and best practices to help complete their tasks. This is called a *Tool Scorecard*. They are easy to fill out and should not take more that 20 minutes or so to complete at the end of a tool application. Typically they are filled out in collaboration with the team leader for supervisory concurrence. At the conclusion of using a tool, the person or team members responsible for applying the tool to a task should account for the following measurable items: quality of tool usage, data integrity, and results versus requirements. Table S-1 provides a sample Tool Scorecard to track these items.

Tool	Quality of Tool Use	Data Integrity	Results versus Requirements	Average Score	Data Summary (incl. Type & Units)	Task Requirements
1.						
2.						
3.						
4.						

TABLE S-1: Sample Tool Scorecard

The first column of the Tool Scorecard simply records the name of the tool used.

The Quality of Tool Usage column can be scored on a scale of 1 to 10 based on the following suggested criteria and levels, as shown in Table S-2. Adjust these rankings as appropriate for your applications.

Rank	Right Tool	Fullness of Use	Measurement System Capability
10	X	High	High
9	X	Medium	High
8	X	High	Medium
7	X	Low	High
6	X	Medium	Medium
5	X	Low	Medium
4	X	High	Low
3	X	Medium	Low
2	X	Low	
1	Wrong Tool		

TABLE S-2: Suggested Quality of Tool Usage Ranking

The Integrity of the Data produced by the tool usage can be scored using the following suggestions shown in Table S-3.

Rank	Right Type of Data	Proper Units	Measurement System Capability	% Data Gathered
10	Excellent	Direct	High	High%
9	Excellent	Direct	High	Medium%
8	Excellent	Direct	Medium	High%
7	Good	Close	High	High%
6	Good	Close	Medium	Medium%
5	Good	Close	Medium	Low%
4	Weak	Indirect	Medium	High%
3	Weak	Indirect	Low	Medium%
2	Weak	Indirect	Low	Low%
1	Wrong	Wrong	None	N/A

TABLE S-3: Suggested Integrity of Data Ranking

Adjust the nature of the scoring criteria as appropriate for your applications. The key is to make clear delineation between various levels of measurable fulfillment of the criteria. This scoring stratification transforms business requirements into actionable performance criteria, thereby clarifying expectations.

The ability of the Tool Results to Fulfill the Task Requirements is scored with the help of the criteria found in Table S-4.

Scale	Criteria
10	The task is complete in all required dimensions, a well-balanced set of tools have been fully applied to 100% completion.
9 to 8	The task is approximately 80-90% complete, some minor elements of the task are not fully done,a well-balanced set of tools have been used but some minor steps have been omitted.
7 to 4	The task is not complete somewhere across the range of 40-70%,
3 to 1	The task is approximately 80-90% complete, some minor elements of the task are not fully done,a well-balanced set of tools have been used but some minor steps have been omitted.

TABLE S-4: Suggested Tool Results To Fulfill the Task Requirements Scale

The rating scale in Table S-4 accounts for how well the data fulfills the original requirements. It is acceptable to find that a full set of data cannot meet the requirement a task was designed to fulfill. If this is the case and the results are bad news, congratulations for communicating the truth. **The intent is to avoid false positives and false negatives** when making decisions about a project's viability. This metric helps quantify the under-development of data and facts that can lead to poor decisions.

Task Scorecards

Task scorecards have the capability of discriminating performance relative to its requirements at both the aggregate level of tool completion and the summary level for each major task. A sample task scorecard is shown in Table S-5.

TABLE S-5: Sample Task Scorecard

Task	Average Tool Score	% Task Fulfillment	Task Results versus Deliverable Requirements	Red	Yellow	Green	Deliverable Requirements
1.				X			
2.					X		
3.						X	
4.							

The Average Tool score averages the tool scores together that align with each major task. A very insightful metric for each major task within a phase is the *Percent Complete* or *Percent Task Fulfillment;* wherein the real mechanics of cycle-time are governed. If a team is overloaded with projects or not given enough time to use tools, it is almost certain that they will not be able to fully complete their critical tasks. Under-completed tasks usually are a leading indicator that the schedule likely will slip. Especially if too few tools are being used, and the ones that are being used are not being fully applied. So there is a double effect: 1) poor tool use leading to incomplete data sets and 2) tasks that simply are unfinished. The Average Tool Score will also tend to be low. This means risky decisions are being made on the wrong basis. It is fine to face high risk situations in projects but not because the team is too busy to do things right. Task incompletion is a major contributor to why teams make mistakes and fail to grow on a sustainable basis. Table S-6 suggests a ranking scale from 1 to 10 to quantify the level of risk inherent in the percent of uncompleted tasks.

The column comparing the Task Results versus Deliverable Requirement identifies how well the work satisfies project requirements. If 100% of the critical tasks is completed, and a balanced set of enabling tools are used to underwrite the integrity of the deliverables, the team is doing its best to control risk. Outstanding deliverables can be produced with full integrity and clarity, but simply fail to meet the requirements for the project and its business case. The team has great data saying that the goals cannot be met. This is a good reason to kill a project with confidence. Not many companies kill projects very often and even fewer do it with tool-task-deliverable confidence.

Scale	Criteria
10	The task is complete in all required dimensions, a well-balanced set of tools have been fully applied to 100% completion.
9 to 8	The task is approximately **80-90% complete**, some **minor** elements of the task are not fully done,a well-balanced set of tools have been used but some **minor** steps have been omitted.
7 to 4	The task is **not complete** somewhere across the range of 40-70%, **moderate to major** elements of the task are not done; tool selection and use has been moderate to minimal; selected tools are not being fully used, **significant** steps are being skipped.
3 to 1	The task is **not complete** across the range of 10-30%, a very few tools have been selected and used, their steps have been **heavily truncated**, and **missing major** steps altogether.

Table S-6: Suggested Score on Risk Level Inherent in Uncompleted Tasks

Indicate fulfilled requirements with a positive *green light* to continue investing in the project. Depict unfulfilled requirements using a caution *yellow* or a negative *red* light that signals a re-directing of the project or an outright discontinuance.

A color-coded scheme of classifying risk can be defined as follows:

- **GREEN**—100% of the **major** deliverables are properly documented and satisfy the Gate requirements. A few minor deliverables may be lagging in performance but present no substantive risk to the success of the project on three accounts: time, cost, and quality.

- **YELLOW**—A very few **major** deliverables are incomplete or falling short of fulfilling their requirements. A corrective action plan is documented, and there is a very high probability the problems can be overcome in a reasonable and predictable amount of time.

- **RED**—One or more **major** deliverables is unfinished or fails to meet requirements, and **no corrective action plan exists** to close this gap. The project is to be killed, redirected, or postponed until a corrective set of specific actions is defined and a predictable path to project timing is in hand.

Table S-7 suggests a set of ranking values to quantify the risk associated with varying levels of mismatch between what a major task delivers and the requirement.

Scale	Criteria
10	Results deliver all data necessary to completely support the fulfillment or lack of fulfillment of the requirements.
9 to 8	Results deliver a major portion of the data necessary to support the fulfillment of lack of fulfillment of the requirements.
7 to 4	Results deliver a moderate portion of the data necessary to support the fulfillment or lack of fulfillment of the requirements.
3 to 1	Results deliver a very limited amount of the data necessary to support the fulfillment or lack of fulfillment of the requirements

TABLE S-7: Suggested Score on Mismatch of Task to Requirements

In summary, the Tool Scorecard documents the quality of tool use, integrity of the data, and the fulfillment of a task requirement. The next level of scoring risk is a *Task Scorecard*, where the project leaders quantify how well one or more enabling tools have contributed to completing a major task; what percentage of the task has been completed; and how well the deliverables from the task fulfills the gate requirements. Scorecards help to answer the questions, *"How well did the team do in meeting the requirements?"* and *"Is the team done and ready to prepare for a Review, or **not**?"* A positive affirmation to these questions indicates the following conditions: "Tools fulfilled the Task Requirements" and the "Tasks fulfilled the Gate Requirements." We are now ready to look at the final summary scorecards that a Gate Keeping team will use to quantify risk accrual at the end of a phase of work.

Review Scorecards

The project sponsor or process owner uses a set of *Review Scorecards* to quantify accrued risk and make decisions at each Review or major milestone. Figure S-33 illustrates the two kinds of reviews: 1) Functional-Level Reviews and 2) Executive Level Reviews.

Functional Reviews are detailed and tactical in nature and prepare the team for an Executive Review. An Executive Review is strategic in nature and looks at macro-level risk. The risk management at the executive level delves into how a particular project contributes to the overall portfolio of commercialization projects or how it manages risk in the post-launch environment. Thus, functional reviewers worry about micro-details within their particular project or process, while the executive reviewers worry about accrued risk across all projects that represent the future growth potential for the business **as a portfolio**. The former looks at alignment of risk within the specific project's tactics, while the latter looks at alignment of project risk across the business strategy. Functional Reviews can be done for technical teams and marketing teams as independent events. Executive Reviews are summary presentations that should integrate both technical and marketing perspectives, as well as any other applicable macro-gate deliverables.

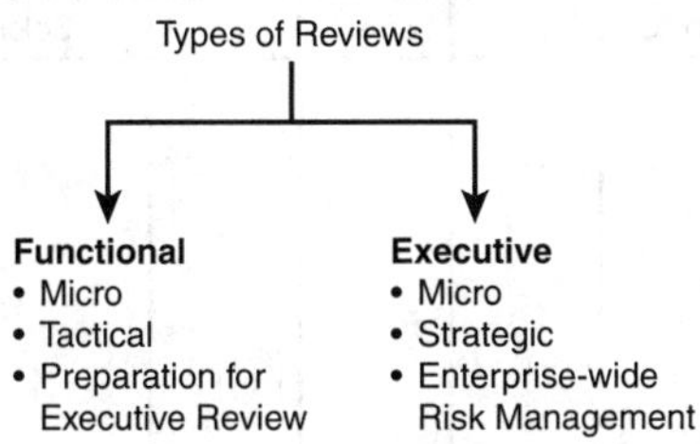

FIGURE S-33: Types of Reviews

Table S-8 shows an example of a generic template for a Functional Gate Review.

Deliverables	Grand Average Tool Score	Summary of Tasks Completion	Summary of Task Results versus Deliverable Requirements	Red	Yellow	Green	Accountable	Target Completion Date	Corrective Action & Risk Assessment
1.				X					
2.					X				
3.						X			
4.									

TABLE S-8: Sample of a Functional Gate Review Scorecard

The integrated system of Tool, Task, and Deliverable Scorecards provide a control plan for quantifying accrued risk in a traceable format that goes well beyond simple checklists. Control plans are a key element in Six Sigma. The Task Scorecard feeds summary data to this functional form of a Deliverable Scorecard.

The next review format is used for *Executive Reviews* where deliverables are summarized for rapid consumption by the executive and sponsor team. Figure S-34 provides a sample of a scorecard used at the executive level.

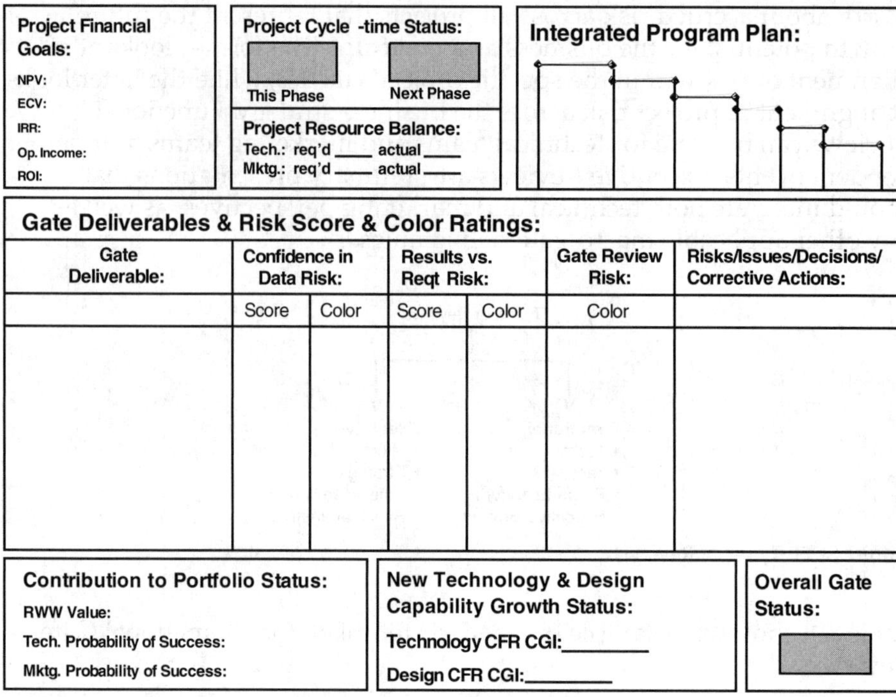

FIGURE S-34: Sample Executive Review Scorecard

The template in Figure S-34 is a common element employed by numerous companies that have a commitment to strategic portfolio management. The executive team looks at a number of these scorecards to balance risk across its portfolio of projects while driving growth to the top-line in accordance with the business strategy.

Integrated System of Scorecards

Modify the proposed integrated system of scorecards to suit your organization's process. Figure S-35 illustrates this interlocking scheme. Each hierarchical level should feature a scorecard designed to mirror the respective business model from one of the following four areas: 1) your tool-task groups (clusters), 2) your deliverable-requirement groups as they are aligned with your phases of your portfolio renewal process, 3) product commercialization process, and 4) your post-launch line management process. (**See Also** "Six Sigma for Marketing (SSFM)," in Part I, **p. 67**)

A system of scorecards should be used to manage the portfolio renewal process, with the first three being sufficient to manage risk within a project. As each product commercialization project is activated from the portfolio renewal team, an Executive Summary Scorecard is initiated and updated during commercialization.

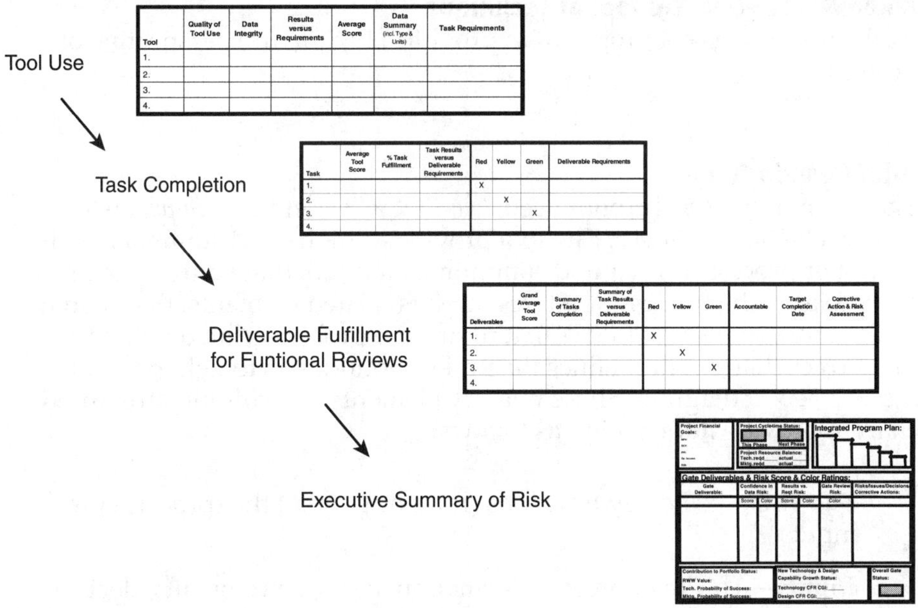

FIGURE S-35: Summary of Scorecards

As the old saying goes, "That which gets measured gets done." Monitor performance with scorecards. Use checklists as reminders of what to use or complete, **not** how well that expectation is being met. A custom designed scorecard set spanning the four-level hierarchy *of* tool-task-deliverables-requirements serves as both a quantitative and a qualitative barometer. Scorecards identify not only **what** to measure (the critical performance elements), and **how well** these critical parameters satisfy a set of criteria (or requirements), but also **when** to measure so that you have a much higher probability of preventing down-stream problems as you seek to sustain growth across your enterprise.

SIPOC (Supplier-Input-Process-Output-Customer)

What Question(s) Does the Tool or Technique Answer?

What is the scope of the project or process?

The SIPOC technique helps you to

- Communicate a high-level view of the project scope or process
- Prevent scope creep in a project

When Best to Use the Tool or Technique

At the beginning of a project or to provide a big picture perspective of a process.

Brief Description

The acronym SIPOC [pronounced *"sie-POCK"*] stands for *Supplier-Inputs-Process-Outputs-Customer.* The tool provides a high-level summary of the project or process. It is a good communication tool that sums up the focus of a project—the process of interest and its related elements. Based on the *input-process-output model* (IPO), it recaps the process-related detail from the project charter and defines the project boundaries to help prevent scope creep. It diagrams these five key elements in a column-structured chart to provide information as follows:

- **Suppliers**—The key functions (roles or people) that produce the inputs
- **Inputs**—The key process information, parts, components, decisions contributions required **prior** to beginning or completing an activity or task
- **Process**—The high-level process activities (typically three to eight steps) that transform the inputs to produce the outputs
- **Outputs**—The key process deliverables or tangible outcomes
- **Customers**—The key customers (both external and internal) requesting the outputs (or deliverables)

Often this tool is built backward, starting with the customer (the "C" of SIPOC, as if it were named COPIS) and working back through to the suppliers, given that it is the customers that determine or define the deliverables of interest. Customers provide the Voice of the Customer (VOC). The process outputs represent the key deliverables (both final and in-process) of interest that may require improvement. Hence, the outputs are measurable (either variable or attribute data). The high-level process elements bound the scope of the project or process and contain elements considered as potential root cause sources. The process inputs also represent potential sources of root causes. The key suppliers should be engaged in the improvement initiative.

Figure S-36 illustrates a SIPOC example for dining at a restaurant. Notice that the process activities are high-level and are sometimes referred to as a *high-level Process map*. The suppliers, inputs, and outputs are directly related to the process and the customer(s). (**See Also** "Process Map (or Flowchart)—7QC Tool," **p. 522**)

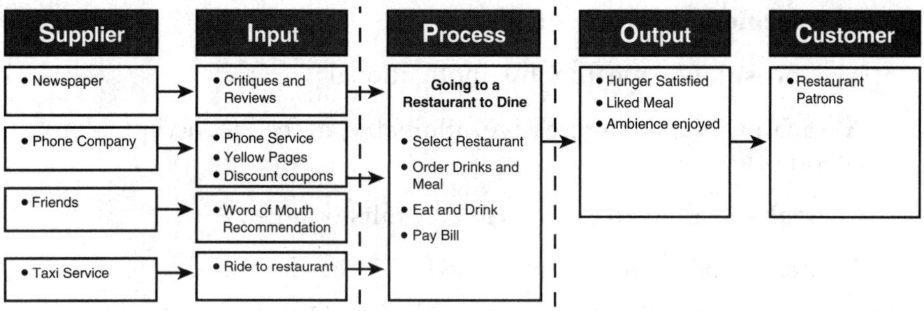

FIGURE S-36: Dining at a Restaurant SIPOC

A SIPOC serves as a high-level snapshot and serves as an effective communication tool to explain what could be a complex process in simple terms.

SMART Problem and Goal Statements for a Project Charter

What Question(s) Does the Tool or Technique Answer?

What is the most succinct description of the project's goal and problem statements?

SMART helps you to

- Communicate a clear, concise, complete description of a project's goal
- Identify the problem it addresses

When Best to Use the Tool or Technique

During the planning stage that establishes and defines a project, use this technique to better communicate its main points and goals.

Brief Description

The SMART acronym stands for *Specific-Measurable-Achievable (but Aggressive)-Realistic-Time-bounded.* Some literature defines the SMART acronym with the "A" representing *Agreed to* or *Achievable, but Aggressive* or *Attainable* and the *R* as *Relevant* to the business, therefore it is agreed to by management/sponsor. In summary, the acronym translation is fluid, but the key elements remain constant—*specific, measurable, time*-related and some combination of *aggressive, but realistic*. The following options include bolded terms to indicate the more common translation:

- **S—specific**, significant, stretching
- **M—measurable**, meaningful, motivational
- **A—achievable**, but aggressive, attainable, agreed to, acceptable, action-oriented
- **R—realistic**, **relevant**, reasonable, results-oriented
- **T—time**-bound, timely, time-based

This technique is applied to both a project's goal statement and its problem statement to ensure that both contain the essential elements to succinctly communicate their vital elements. *Specific* describes precise, concise language. *Measurable* requires a metric, target, or unit of measure. *Achievable* but *Aggressive* describes the objectives, which are *Realistic*. The last element, *Time-bounded* or *Timely* takes on different dimension for the goal and problem.

With respect to the *goal statement*, time-bounded seeks a **specific date** (day, month, year) that the project targets for its completion. Vague language providing the month, quarter, or year that the project intends to achieve its objective leaves room for ambiguity and misunderstanding between the project team and its sponsor. Ambiguity leads to improper expectation setting, wherein the project team could intend the end of a month, or worse yet a quarter, while the sponsor anticipates completion at the beginning of the month, or quarter. Misalignment leads to frustration, disappointment, and at times more serious consequences. If the project completion date slips, the team is better off communicating that to the sponsor and why, rather than hoping for a miracle.

The time element for the *problem statement* describes the duration of the problem. It defines how long this pain has persisted—is it a one-time event or a systemic problem. The time dimension bolsters the team's *case for change* when trying to secure project funding and resources and implementing and sustaining the improvement. It adds a sense of magnitude and is best accompanied with a control chart or run chart illustrating the key metric plotted over time. (**See Also** "Control Charts—7QC Tool," **p. 217** and "Run Chart—7QC Tool," **p. 611**)

The problem statement describes what is wrong over time. Alternatively, at times the problem statement is replaced by an opportunity statement that describes the possibilities. An example of a problem statement for a restaurant might be

The number of restaurant patrons have declined over the last six months, the number of restaurants in the area have increased, and food costs have increased 7% over last year.

The *goal statement* is part of a larger purpose, as a key element of a project charter. It articulates the team's improvement objective—project success criteria and duration (how long it should take to achieve). It determines when the project team knows it has achieved its mission and gauges what "good" looks like. However, it is **not the solution** or the answer on how to achieve the desired state. An example of a goal statement for a restaurant might include

> *Within the next two months, increase the number of services offered that can yield last year's 15% target profit margin.*

The example is specific (= increased offerings), measurable and achievable (= prior year's 15% profit margin), time-bounded (= within in next two months), and relevant to the business (= restaurant services). However, the time element ideally should be more specific and give a specific target date the new services would either be identified or implemented. The same example rewritten more tightly is:

> *By May 20, 2008, recommend new services offerings (with a business plan) showing how they can help us achieve last year's 15% target profit margin within two months of launching them. Upon approval, new services to be operational by July 1, 2008.*

Project Charter

A *project charter* is a sponsor document. It essentially is a contract for work between the project sponsor and the project team. This document authorizes the project, names its key stakeholders, and defines its scope to address a business need on a temporary basis. (A project has a finite duration with a definite beginning and end.) It drafts the people to work on the project and defines the project budget and timeframe. A project charter is incomplete until the project sponsor approves it and communicates it to the appropriate parties. Developing and publishing a project charter is the first step in commissioning a project—before any work begins. Often the charter or a subsequent memo issued to the organization formally recognizes the initiation of a project.

The project charter describes the high-level requirements for project success and provides the project manager the authority to apply organizational resources to project activities. A charter, its layout, and format are unique to an organization, but ideally are standardized throughout the organization. In general, charters contain any combination or all of the following components (the more commonly included categories are **bolded**):

- Header Information—Project Title, sometimes a tracking number, issuance date, and Project Sponsor establishing the project.
- Overview Statement—Aligning the project with overall company strategy/vision.

A B C D E F G H I J K L M N O P Q R S T U V W X Y Z

- **Problem Statement** (or Opportunity for Improvement).
- **Goal Statement** [with the critical parameters, improvement target, timeframe (target completion date), and sometimes the critical deliverables]—It may describe the approach to achieve the opportunity statement or solve the problem statement defined in the project scope, but **not** provide the answer or the solution.
- **Project Scope**—Describes the boundary conditions; identifies key parameters covered or not covered by the project—what's inbounds/out-of-bounds.
- Target Audience—Defines who the project scope addresses (that is, target populations, functions, disciplines, departments) or who the customer is.
- Project Deliverables—Defining the key project outputs, sometimes listed separately from the goal statement.
- **Project schedule**—The proposed high-level timeline with project milestones (including Milestone Review Meetings or phase-gates) and target completion date for each (key milestones). (**See Also** "Activity Network Diagram (AND)—7M Tool," **p. 127**, and "PERT (Program Evaluation and Review Technique) Chart," **p. 453**)
- Business Case—Definition of the quantifiable benefit of the project and how it aligns with business strategy or goals.
- Project Budget—The proposed project funding required to complete the deliverables.
- Project Resources:
    - **Project Manager**
    - Project Team Members (usually only the core team members, even if sub-teams also are involved) (**See Also** "RACI (Responsible, Accountable, Consulted, Informed) Matrix," **p. 554**)
- Approval Signatures—Project Sponsor and any critical Key Stakeholders, if applicable

Who are Stakeholders?

The *stakeholder* is defined as anyone (or any function) impacted by the project, its outcome, or the process. Examples include management; upstream and downstream process players (including suppliers and third party partners); regulatory; process players not participating in the project, but who have representatives on the project team.

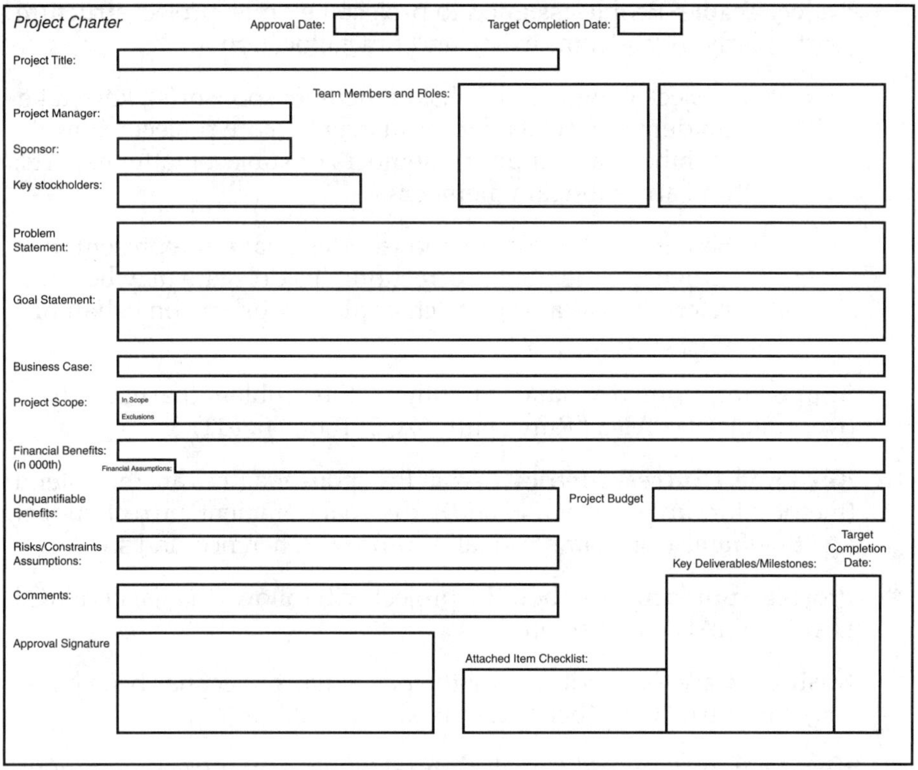

FIGURE S-37: Sample Project Charter Template

There are five basic requirements for conducting a successful project. In order of importance, they are as follows: choosing the right people; choosing the right people; choosing the right people; setting up the right organization; and using the appropriate methodology, tools, and systems.

Additional information that is beneficial to add to the project charter or provide as addendums include

- **SIPOC**—Serves as a summary document that captures a higher level snapshot in presentation format. (**See Also** "SIPOC (Supplier-Input-Process-Output-Customer)," **p. 663**)
- **High-level Process map**. (**See Also** "Process Map (or Flowchart)—7QC Tool," **p. 522**)
- **Project RACI**—Identifies who is accountable for the key deliverables. (**See Also** "RACI (Responsible, Accountable, Consulted, Informed) Matrix," **p. 554**)
- **Project Communication Plan**—(**See Also** "Matrix Diagrams—7M Tool," **p. 399** for a brief discussion on a communication plan.)

- **Project Team**—People assigned to project, key role, project structure (particularly if sub-teams exist), and time allocation.
    - Project Team member list to include: sponsor(s), key stakeholders, project leader (or manager), core project team members, any support members or subject matter experts, and any sub-team members.
    - Sometimes the project sponsor designates a representative to oversee the project direction; this delegate may be referred to as a project "champion" working on behalf of the sponsor.
- **Supporting summary data**—Highlights the problem (particularly over time). (**See Also** "Run Chart—7QC Tool," **p. 611**)
- **Key Project Success Metrics**—What the sponsor agrees are the criteria (metrics) for success over and above the goal statement, project budget, and timeframe (for example, quality, process adherence, and so on).
- **Project Approach**—Method the project will follow (that is, DMAIC, UAPL, DMADV, and so on).
- **Business Case**—Describes why the particular project has been chosen. (**See Also** "Cost/Benefit Analysis," **p. 238**)
- **Key Assumptions and Constraints**—Defines any project limitations or conditions.

Summary

A project charter and its problem and goal statements are inseparable. Together they summarize and document the fundamental business thinking about the problem and the overall project roadmap to address the problem. The charter defines project boundaries—in scope, out-of-scope, constraints (that is, budget). It identifies who is accountable to resolve the problem and what roles (or departments) are affected. The charter defines when the project will be completed; based on what criteria, and in what timeframe.

How to Use the Tool or Technique

Develop a project charter and its problem and goal statements using the following procedure. A common project charter template should be deployed consistently across an enterprise or at minimum, a division.

A common approach to completing a project charter includes

Step 1. Draft charter content based on current understanding of project scope and on lessons learned from prior projects of similar scope, scale, and complexity.

a. Start with the problem statement, drafting a concise statement that articulates the pain that the issues have been causing the business, the magnitude (or effect), and for how long, using the SMART technique.

b. Define the goal statement, applying the SMART technique.

Step 2. Review charter with sponsor, make modifications, and gain approval.

Step 3. Sponsor should publish and distribute the project charter, naming the project lead and key project members, to the organization(s) affected (particularly the stakeholders).

Step 4. Project charter is reviewed and revised on an as-needed basis.

Step 5. Project charter is revisited in Stage Gate Reviews or distributed as pre-read for a Stage Gate Review.

Step 6. At close of project, the project charter gets its final review to ensure it reflects the actual project events and includes any lessons learned. The final document is filed with the other administrative documents to close the project.

Hints and Tips

- Examine the problem statement to ensure that the SMART technique has been applied. Double-check that the described pain and its magnitude create a rallying point for the project team to focus on to resolve. The magnitude, in part, should include a time component describing how long this problem has persisted.
- Ensure that the goal statement fits the SMART by specifying a specific day of the month and year for the project completion date and that it describes what "good" looks like or how to approach solving the problem—but avoids providing a solution.
- Verify that the sponsor approved the charter.

- Verify that the project charter has been published and distributed to the key stakeholders (including project team members and respective managers).
- Use in Gate Reviews to ensure project remains within scoped boundaries, within budget, and producing on-time deliverables.
- Modify project charter if "new" circumstances change or better define the original conditions.
- At project close, ensure that an updated, accurate project charter is filed with the project papers.
- Reference charters from previous projects to determine if they had similar problems, scope, scale, budget, and so on and whether the historical knowledge in their project files suggest applicable and invaluable lessons learned.
- A project charter template should be approved by the head of the organization and deployed as a standard throughout the enterprise.
- The project charter template may be developed as part of an intellectual capital management system (for example, a project database) or by using application software such as Microsoft Word or Excel.

Solution Selection Matrix

What Question(s) Does the Tool or Technique Answer?

Which solution option best meets requirements?

A Solution Selection Matrix helps you to

- Prioritize the solution options and select the most optimal one
- Communicate the decision made in the selection process and why

Alternative Names and Variations

Variations on the tool include

- Prioritization Matrix (Also See "Prioritization Matrices—7M Tool," p. 470)
- Pugh Concept Evaluation (**See Also** "Pugh Concept Evaluation," p. 534)

A
B
C
D
E
F
G
H
I
J
K
L
M
N
O
P
Q
R
S
T
U
V
W
X
Y
Z

When Best to Use the Tool or Technique

Selecting from a set of solution options, apply this tool to rank and select the best option, **after** requirements are thoroughly understood, the problem, current state and root causes are analyzed, and potential solution concepts have been generated.

Brief Description

The Solution Selection Matrix structure and purpose is very similar to a Prioritization Matrix; however, its purpose is specifically to rank solution candidates and select the most appropriate. The criteria used to evaluate the candidates should tie to the critical-to-quality (CTQ) characteristics of the customer and business.

Selection Criteria

The selection criteria used to evaluate the candidates could be any appropriate combination of customer and business requirements. Typically the voice of the customer (VOC) prioritizes or serves as the first filter in the selection process. The voice of the business (VOB) typically dominates the final selection to ensure that the end result aligns with the business goals. By convention VOB criteria include three main considerations:

- **Feasibility**, which encompasses
  - Designing, developing, implementing, and maintaining the solution
  - Alignment with the organization's mission, goals capabilities and capacity [of funds, resources, technology, and knowledge (know how)].
- **Effectiveness** in addressing either a key area of opportunity (potential benefits), root cause of a problem, or a specific set of requirements
- **Cost** to design, develop, implement, and maintain the solution

The criteria also may include the ease of implementation and potential risk (pitfalls or adverse side effects).

Defining the criteria is a critical step in using this tool. Developing a common understanding among those who will evaluate the candidates eliminates any inconsistency in the scoring. Moreover, document this criteria definition to communicate to the stakeholders why a certain candidate was selected over the others.

Evaluation and Selection

The simple matrix approach helps to organize and focus the process of evaluating each option against a common set of criteria. This is best done with a cross-functional, multi-disciplined team to provide different perspectives. Either the consensus or equal voting approach can produce a meaningful outcome.

Use the *consensus approach* when the team make-up is comprised of similar talkativeness, experience, and reporting relationship. Each team member discusses and recommends a proposed *criteria weight* and *relationship score,* and then the team collectively arrives at a single number per appropriate cell, as illustrated in Figure S-38.

Solution	Criteria and Weights					Total
	Easy	Quick	Tech	Hi Impact	Customers	
	8	*7*	*5*	*9*	*5*	
A	1	9	3	9	3	182
B	0	9	3	9	3	174
C	1	9	1	3	3	118
D	0	3	1	9	3	122

FIGURE S-38: Sample Solution Selection Matrix Using the Consensus Approach

Use the equal voting technique when the team includes members with dissimilar experience, mixed reporting levels, and when some members tend to dominate the conversation more than others. Each member votes for a *criteria weight* and *relationship score,* either silently or aloud. The respective votes for each type are tallied (added) and divided by the number of voters to calculate the average score. That single average number is entered per appropriate cell. (An alternative to the average score would be to use the *total score;* however, sometimes the large numbers it can create seem unwieldy.) Figure S-39 shows a Solution Selection Matrix created with the *total scoring* approach to the equal voting technique.

A B C D E F G H I J K L M N O P Q R S T U V W X Y Z

Criteria	Bill	Amir	Julie	Total
1. Ease of Implementation	4	1	2	7
2. Time to Implement	1	5	2	8
3. Customer Notices Change	4	3	5	12
4. Defect Reduction	2	1	2	5

	Criteria and Weights				
	Easy	Time	Customer	Defect	
Solution	**7**	**8**	**12**	**5**	**Total**
A	1,3,3 **49**	9,3,3 **120**	1,0,3 **48**	3,1,3 **35**	**252**
B	3,9,9 **147**	3,3,3 **72**	1,1,3 **60**	9,3,9 **105**	**384**
C	1,0,1 **14**	3,1,3 **56**	9,9,3 **252**	3,1,1 **25**	**347**

Individual Scores Bill, Amir, Julie

14 = 7 x (1 + 0 + 1)

347 = 14 + 56 + 252 + 25

FIGURE S-39: Sample Solution Selection Matrix Using Equal Voting Technique with Total Scores

A third type of evaluation technique matches a solution against its original root cause to ensure the solution candidate's effectiveness. This root cause technique focuses the team better than the other two approaches. It encourages brainstorming during the idea generation step, but the ideas stay tied to the root cause, rather than symptoms. Free-wheeling, far-fetched ideas appear less frequently. After the initial creation of ideas, this approach encourages a synthesis step to hone in on more practical candidates. Moreover, this technique uses a unique three-way criteria scoring that multiplies feasibility, effectiveness, and cost to calculate an overall score. The team still needs to define and rank the specific criteria within the three categories. Figure S-40 shows a Solution Selection template for the root cause technique.

Selection

Consider the highest scoring solutions first and then weigh other intangible factors that may lead to a secondary solution candidate being considered and perhaps further tested. As with all soft tools, such as matrices, avoid blindly selecting the top score getter as the final answer without further scrutiny. Sometimes the final scores are close and require evaluation and refining of the concepts.

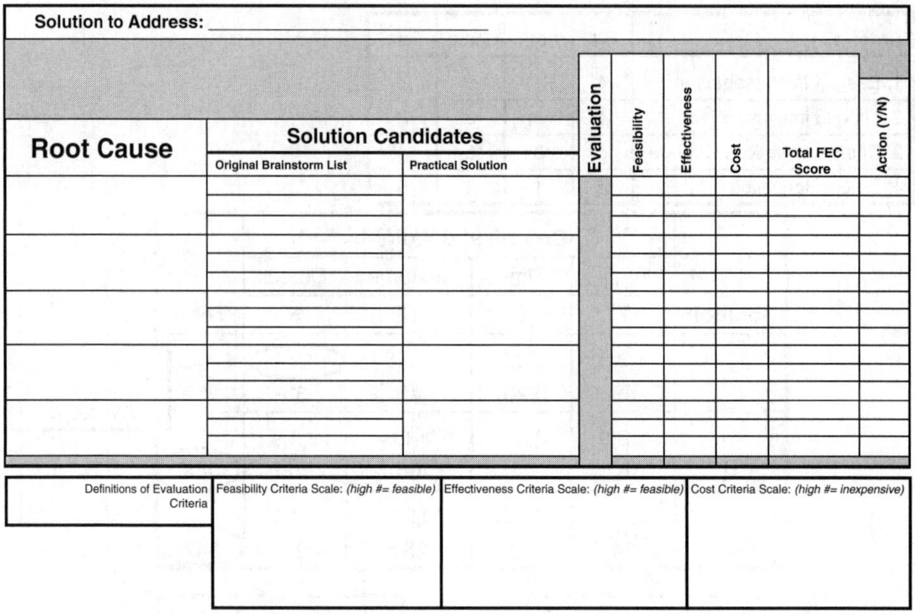

FIGURE S-40: Solution Selection Matrix template using the Root Cause Technique

Other Solution Selection Methods

If the Solution Selection Matrix fails to provide a clear choice (or perhaps even if it does), consider getting more information about the final selected solution or set of valid solution candidates. For example, model or simulate the top or some of the top solution candidates to identify additional features or dimensions to modify and help confirm the final selection. Trials (small scale tests) also provide additional insight about the proposed option(s). If an organization currently utilizes a similar solution, observe their situation and gather any best practices and lessons learned.

How to Use the Tool or Technique

Develop a Solution Selection Matrix using the following procedure:

Preparation: Review what you know about the process and the confirmed root causes.

Step 1. Idea Creation. Create potential solutions, if not available already.

a. List ideas sparked by the discoveries made while analyzing the process.

b. Generate other possible solutions that address the root causes via

   i. Various brainstorming and creativity techniques, such as challenge the rules, get rid of excuses, challenge paradigms. (**See Also** "Brainstorming Technique," **p. 168**)

   ii. Utilizing Lean concepts. (**See Also** "Lean and Lean Six Sigma," in Part I, **p. 29**)

   iii. Other potential sources of ideas including benchmarking, best practices, literature search, trade journals, prior experiences of people, industry experts, databases, and surveys.

   iv. Affinity diagrams are useful for grouping possible solutions into categories.

Step 2. Create a matrix using an L-shaped structure and list the ideas down the first column, to serve as row headings.

Construct the matrix template in Excel, Word, or on a flip chart, or any other appropriate media as long as the team members can collectively and simultaneously view the document.

Step 3. Define the selection criteria and criteria weight and document the output.

a. Agree on the selection criteria and their respective definitions.

b. Determine the weight for each criteria, using a numeric scale where the highest number is the most preferable or most important, using either the consensus or equal voting technique.

   Common criteria weight scales are 1 to 3, 1 to 5, or 1 to 10. However, we recommend stack ranking the criteria, with the largest number representing the most important to avoid tie weightings to ensure better discrimination among the various solutions.

c. Document the criteria and respective weights as a reference key for the matrix.

d. List the criteria and their respective weights across the very top row of the matrix to serve as the column headings.

Step 4. Prioritize the solution candidates.

a. Use the scale of 9-3-1 to define the relationship score for each solution candidate and the criteria and record the rating in each appropriate cell, using either the consensus or equal voting technique.

The scale translates such that nine equals strong relationship, three represents moderate, one connotes weak, and zero or a blank cell indicates no relationship. Another way to define this scale might include cost savings such as 1 for < \$100K, 3 for \$100-500K, 9 for > \$500K.

The 9-3-1 scale helps to better discriminate among the options versus a 3-2-1 scale.

b. Multiply the cell relationship score times the corresponding criteria weight to calculate the weighted criteria relationship.

c. Sum the weighted criteria relationship scores for a given solution candidate to calculate the total candidate score.

Step 5. Select the optimal solution(s).

a. Identify the highest total candidate score, which will probably be the preferred solution.

b. Avoid turning your brain off and search the matrix for the next closest candidates to see if any of them are possible candidates as well. Consider combining the concepts to create a stronger proposal.

c. Collect more information about the solution concept before doing a full-fledged implementation.

Hints and Tips

The Solution Selection Matrix contains three distinct areas: idea creation, selection criteria, and the actual prioritization and selection of a final solution or set of viable solutions to be further tested. Find the Hints and Tips for the latter two topics embedded within the "Brief Description" section of this entry.

Idea Creation:

- Involve the right people in the idea generation activity; be inclusive.
- Although this is a team exercise, individuals generate ideas, not teams. Ensure each team member contributes.
- Focus on one root cause at a time.
- Prioritize the root causes and start with the one determined to have the biggest impact on the desired response (big Y).
- Generate as many potential solutions as possible for each root cause because the more potential solutions available to consider, the more opportunities for uncovering ideas for improvement. Exhaust all the ideas that the team can produce.
- Consider combining ideas or the best characteristics from multiple solution ideas to generate a stronger solution.
- Work with far-fetched ideas by modifying and alternating perspectives to uncover a potentially hidden, more practical solution candidate.

Supporting or Linked Tools

Supporting tools that might provide input when developing a Solution Selection Matrix include

- Benchmarking (**See Also** "Benchmarking," **p. 160**)
- Brainstorming (**See Also** "Brainstorming Technique," **p. 168**)
- Fishbone diagram (**See Also** "Cause-and-Effect Diagram—7QC Tool," **p. 173**)
- RWW (**See Also** "Real-Win-Worth (RWW) Analysis," **p. 560**)
- SWOT (**See Also** "SWOT (Strengths-Weaknesses-Opportunities-Threats)," **p. 699**)
- VOC/VOB (**See Also** "Voice of Customer Gathering Techniques," **p. 737**)

A completed Solution Selection Matrix provides input to tools such as

- DOE (**See Also** "Design of Experiment (DOE)," **p. 250**)
- Pilots
- Simulations (**See Also** "Monte Carlo Simulations," **p. 431**)
- Transition and control plans (**See Also** "Matrix Diagrams—7M Tool," **p. 399**)

Figure S-41 illustrates the link between a Solution Selection Matrix and its related tools and techniques.

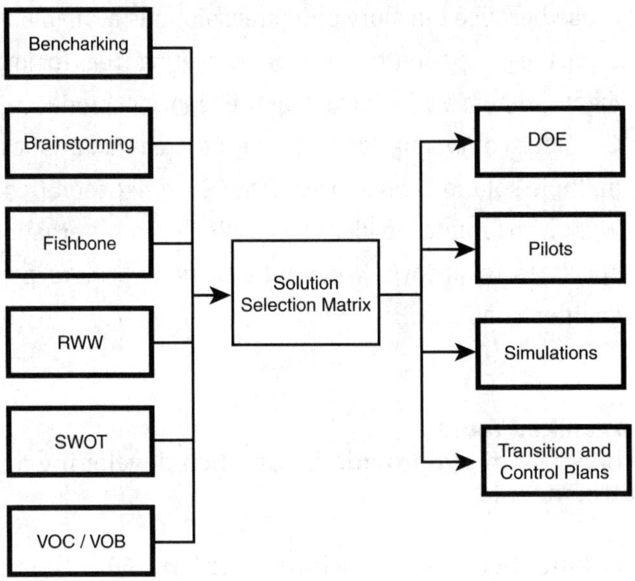

FIGURE S-41: Solution Selection Matrix Tool Linkage

Variations

Prioritization matrix (**See Also** "Prioritization Matrices—7M Tool," **p. 470**)

Pugh Concept Evaluation (**See Also** "Pugh Concept Evaluation," **p. 534**)

Stakeholder Analysis

What Question(s) Does the Tool or Technique Answer?

Who supports the project, who doesn't, and why?

Stakeholder analysis helps you to

- Assess the root cause behind why someone opposes or is indifferent to the project, its mission, its recommendation, and/or its ongoing implementation, utilization, management, and control—plus develop a specific action plan as a countermeasure
- Enhance the ownership of the project's success (including sustaining the improvement) among the stakeholders
- Improve communications
- Avoid political landmines

When Best to Use the Tool or Technique

Develop a stakeholder analysis at the beginning of a project and implement, revise, and monitor it throughout the life of the project.

Brief Description

The stakeholder analysis tool examines details around the Voice of the Business (VOB), to evaluate the **key** stakeholders' acceptance of the project scope, boundaries, deliverables, and timing. The technique is meant for the project sponsor and project manager to assist in strategizing the optimal change management and communication approaches. (This tool assumes the project sponsor holds a senior rank relative to the project manager/Six Sigma black or green belt.) The technique helps to focus attention on an action plan grounded in business issues, rather than non-actionable discussions about individual personalities.

A stakeholder is anyone impacted by the project; however, the project sponsor and project manager need to identify the key stakeholders needed to support, promote and sustain the project and its improvement. The executive sponsors often fund the project but are not the only stakeholders. Stakeholders also include the process owner, who may be a different person or role, and the suppliers to the process and customers of its outputs. **Select only the vital few stakeholders who have key implications on (or influence on) the project.** Because most key stakeholders' organizational reporting relationship will be on par with or higher than the sponsor and the project manager/Six Sigma black or green belt, it is imperative to involve the sponsor in this activity.

Personalize this process by identifying actual people, not roles within an organization. Conduct the analysis by person so that the diagnosis and countermeasure is tailored for that one individual. For this reason, keep the stakeholder analysis tool **confidential**. Given the importance of personalization, try to understand the given stakeholder's position by direct input, if possible. Determine current stakeholder positions by actually collecting data; talk to them at a conceptual level about the project and how it could impact them. If a direct, personal conversation is not possible, ask a close colleague to have the discussion. Be careful not to assume—do not guess.

The analysis technique has two parts—the *acceptance chart* and *influence planning* chart:

> **Acceptance chart**—Identifies **current level** of stakeholder support and defines the **desired level** of stakeholder support for the project to be successful. The chart is structured as an L-shaped matrix, with the key stakeholders as row headings and the different support positions as column headings. Typically, the chart graphically depicts the current level of stakeholder support by an icon and indicates the desired support level by a second icon. The convention for the support positions comprise the following four categories: strongly against, lets it happen (or neutral), helps it happen, and makes it happen (strongly supports). Figure S-42 shows a sample stakeholder acceptance chart. (**See Also** "Matrix Diagrams—7M Tool," **p. 399**, for a discussion on different shaped matrices.)

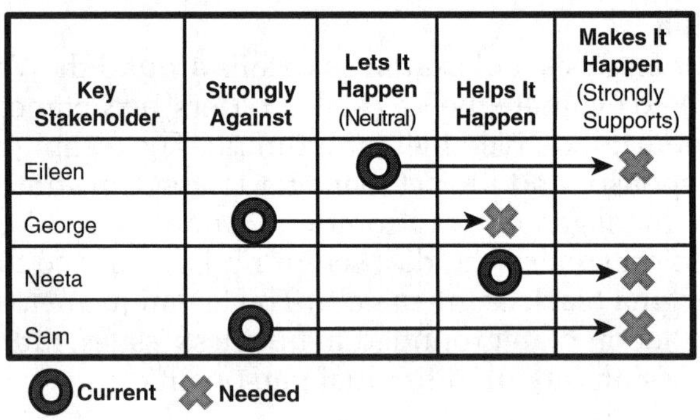

FIGURE S-42: Sample Stakeholder Acceptance Chart

Key Stakeholders

Recognize that all stakeholders **do not** have to be in the Makes It Happen category. Some stakeholders may be just as effective in the Let it Happen category, based often on the fact that their functional area is not directly related to the area. For example, if a product development project were about to launch, perhaps the functional areas of finance or human resources might fall into the Let it Happen category.

Influence planning chart—Tracks the current and desired positions. This chart indicates the source of resistance and strategy to move the stakeholder to desired level of support. Take the time to properly diagnose the type of resistance so as to employ the appropriate response strategy. The resulting action plan will either change the resistor's way of thinking or modify the project or solution. Some stakeholders may exhibit more than one type of resistance; if so, identify the most dominant type and address that one first. Figure S-43 illustrates a sample influence planning chart.

Key Stakeholder	Type of Resistance	Underlying Issue	Strategy
Eileen	Technical	Lack of Knowledge	Educate; Involve in meetings.
George	Political	Loss of Control	Emphasize project benefits.
Neeta	Organizational	Control	Stakeholders get credit for solution.
Sam	Individual	Overworked	Educate over time via colleagues.

FIGURE S-43: Sample Stakeholder Influence Planning Chart

Notice some of the types of resistance identified in Figure S-43 and the corresponding root cause (or underlying issue) and its countermeasure. For example, some strategies may include

- **Lack of knowledge**—Educate on project and personally involve in review meetings.
- **Threat of loss of control**—Focus on the project benefits to the organization.
- **Desire to control**—Credit the stakeholders for designing and implementing the solution.
- **Overworked, seen as additional work**—Educate over time and examine possible process redundancies or gaps that may cause workload issues.

Customize the strategy and action plan for the specific individual and corresponding reason for resistance. As a follow-on, a more general communication plan may result from a stakeholder analysis. This more general outcome would complement the individual strategies. (**See Also** "Matrix Diagrams—7M Tool," **p. 399**, for a brief discussion on a communication plan matrix.)

Statistical Tools

What Question(s) Does this Tool or Technique Answer?

What statistical tools are available to use, and when should you use them?

Brief Description

The various different Six Sigma methodologies view work through a consistent set of lenses that examine three core elements: **process**, **variation**, and **data** to help make business decisions. *Statistical thinking* views all activities as part of process and that all processes have inherent variability. It uses data to understand variation and to drive decisions to improve these processes. (**See Also** Part I for a discussion on the four main categories of Six Sigma methodologies.)

Statistics help examine these three core elements of work with less economic burden (time, cost and resources). Statistics enable quicker and cheaper analysis of a system. There are several statistical concepts that involve a combination of numeric and graphical techniques. A general overview of the most common ones used by Six Sigma include the following.

Descriptive Statistics

Descriptive statistics are used to describe a population, process, and sample and their characteristics, such as location (central tendency), spread (variation), and shape (skewedness and peaks). Common terms include the following:

Central Tendency

This describes the location of data. It is measured three different ways to describe the values around which the data tends to congregate.

- **Mean**—The sum (Σ) of the measurements (X) divided by the number of measurements (N or n), sometimes referred to as the *arithmetic average*, denoted by μ or $\overline{X}$. The median is affected by extreme values, such that one extreme value pulls the mean toward it. The formula for a population and sample mean where x is the number of observations (data), N is the population size, and n is the sample size):

$$\mu = \frac{\sum_{i=1}^{n} Xi}{N} \qquad \overline{X} = \frac{\sum_{i=1}^{n} Xi}{n}$$

Population Sample

A B C D E F G H I J K L M N O P Q R S T U V W X Y Z

- **Median**—The value representing an equal number of observations (data) above and below it when the data are arranged in sequence or rank order (increasing or decreasing). Half of the data exist above and half below the median. If there are an odd number of observations, the median is the middle value. If there are an even number of observations, the median is the average of the two middle-most values (for example, two middle-most values divided by two). The median is denoted by **M** or **Md**.

The median is unaffected by extreme values, so it is a good measure of central tendency for highly skewed data.

- **Mode**—The value measuring the most frequently occurring observations (count) in the data set, denoted by **Mo**.
- **Sample Calculations**—For the following set of data (2, 2, 2, 3, 3, 4, 5, 7, 10, 12), the central tendency values include
  - Mean = 50/10 = 5
  - Median = 3.5
  - Mode = 2

Figure S-44 illustrates the mean, median, and mode on a right-skewed distribution. The central tendency characteristics of a normal distribution possess equal mean, median, and mode.

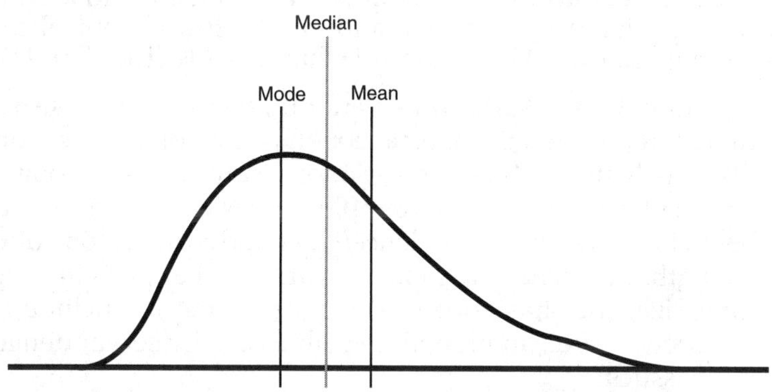

FIGURE S-44: Mean, Median, Mode of a Right-Skewed Distribution

Variation

These measurements indicate how the data is dispersed or spread out. When evaluating a model, variability measures quantify the response's robustness to sources of variation (both controlled and uncontrolled). It can be measured in three ways:

- **Range**—The difference between the data set's extreme values (the largest and smallest measurements), denoted either by an **R** or **r**. The unit of measure is the same as that of the data set. The formula is maximum value minus the minimum value of process data, or **R** = $X_{max} - X_{min}$.

The range, minimum and maximum provide a simple way of measuring the amount of consistency in a data set—the spread.

- **Variance**—Measures the sum of the squared deviations from the mean, divided by the number. It is expressed as the data's squared units of the standard deviation. Variances can be algebraically added and subtracted.

$$\sigma^2 = \frac{\sum_{i=1}^{n} (X_i - \mu)^2}{N} \qquad s^2 = \frac{\sum_{i=1}^{n} (X_i - \overline{X})^2}{n-1}$$

Population — Sample

Variation is tolerated if the process or system remains within or on target and stable over time. Thus, it is acceptable if within the process specifications or control limits. Minimizing, controlling, or eliminating variation requires an understanding of its sources. Classifying variation helps to understand its sources. Control charts help to isolate between the two types of variation—common cause and special cause.

In Six Sigma, variation is the enemy. Variation aims to identify and eliminate special cause variation and minimize and control common cause variation. (**See Also** "Control Charts—7QC Tool," **p. 217**)

- **Common Cause Variation**—Random, stable, and consistent over time. It represents the natural ebb-and-flow of a process or system. There will always be some variation in all processes—natural, inherent oscillation. Everyday examples of variation occur in human or behavioral systems as evidenced by handwriting, tone of voice, and strength. In mechanical systems, variation happens in weight, volume, size, and shape of an item. Other examples include a person's experience level, an individual's illness, daily server demand, and daily issues.

  Within a work environment, common cause variation is viewed as the responsibility of management. Management owns and creates the process, and only management can intervene to change or improve the system. An estimated 85% of the reasons for failure to meet customer expectations are related to deficiencies in systems and processes rather than the employee.

- **Special Cause Variation**—The opposite of random, and it appears erratic or unpredictable over time. On average, 5 to 15% of variation is special cause. Examples include worker errors, computer problems, phone system issues, procedural changes, workers out on holiday, and uncommon, unusual, unplanned, often uncontrollable occurrences or circumstances.

Business views special cause variation as a local workforce issue that should be handled by those involved in the process if they have proper tools and operation environment.

- **Standard Deviation**—Measures the positive square root of the variance. It is expressed in the data's units of measure. Standard deviation **cannot** be algebraically added or subtracted.

$$\sigma = \sqrt{\sigma^2} = \sqrt{\frac{\sum_{i=1}^{N} (Xi - \mu)^2}{N}} \qquad s = \sqrt{s^2} = \sqrt{\frac{\sum_{i=1}^{n} (Xi - \overline{X})^2}{n-1}}$$

Population Sample

- **Sample Calculations**—for the following set of data (2, 2, 2, 3, 3, 4, 5, 7, 10, 12); the variation calculations include
  - Range = 12–2 = 10
  - Variance = 114/(10–1) = 12.67
  - Standard deviation = 12.67 = 3.56

Shape

This describes what the curve of the data looks like, whether it's normal (and standard normal), t-Distribution (for less than 20 values), bimodal, exponential, or quadratic. Measures of skewedness and kurtosis (peakedness) help to describe the shape of the curve. (**See Also** "Histogram—7QC Tool," **p. 330** and "Regression Analysis," **p. 571**, for more discussion on shapes and corresponding illustrations.)

- **Normal Distribution**—Represents a symmetric shape and is unimodal, often referred to as a bell-shaped curve. Two parameters define it—the mean and standard deviation. Figure S-45 illustrates a normal distribution's symmetry by using the number of standard deviations on either side of its mean. Note that the standard deviation also refers to the *sigma level* and the percent area under the curve.

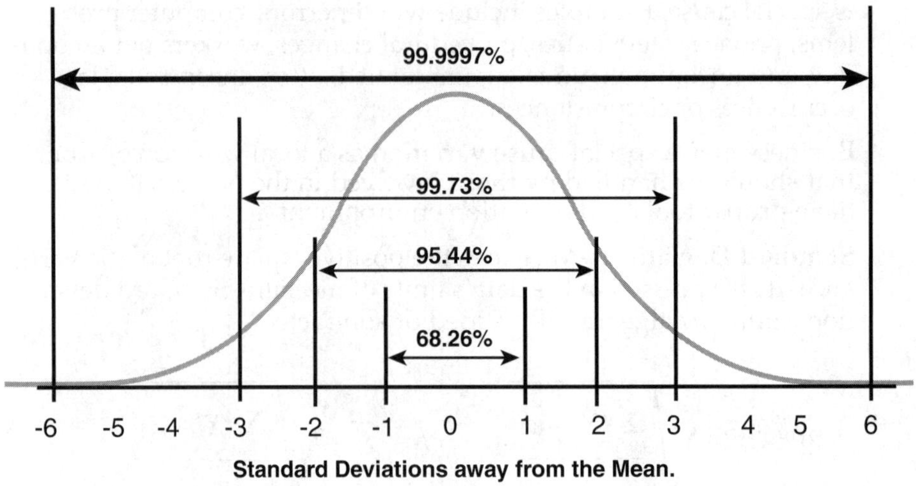

FIGURE S-45: Normal Distribution

Normal distribution is important because of the relationship between the shape of the curve and the standard deviation. Over time many processes follow a normal distribution; thus, it assists in predicting performance over time.

The applications of a normal distribution are many. Not only is it helpful in sampling techniques, it also applies in quality management when constructing the center line, the upper and lower control limits of a control chart, and the PERT chart when examining the probability of completing the project.

- **Standard Normal Distribution**—A special case of a normal distribution where its mean equals zero, and its standard deviation equals one. It serves as a base distribution to compare other distributions.

The Standard Normal curve represents a conventional scale for a normally distributed bell-shaped curve that has a central tendency of zero, and a standard deviation of one unit, wherein the units are called *sigma* (σ). The standard deviation (sigma) is the unit of measure, and the mean, mode, and median all equal zero sigma.

Figure S-45 illustrates a standard normal curve, with the Z-scale plotted along the horizontal axis. For every increment in the Z-scale, for every standard deviation, it represents one sigma (σ) unit. Six sigma (6σ) spread translates into 99.9997% of the area under the curve. Hence, the standard normal curve with its Z-scale represents the Six Sigma philosophy.

This standard normal scale is used as a common scale to which all other normal distributions with different scales can be translated. An analogy might be that the standard normal scale serves as a common currency. As an individual travels the globe, the currency of their home country needs to be converted into that of the other countries to conduct commerce. Upon return home, the traveler converts the other nations' currency back to that of the home country. The standard normal scale serves as that common currency for all normal distributions. Regardless of the unit of measure, normal distributions can be re-scaled, or converted, into a standard normal scale to compare distributions, areas under the curve, and proportions between and across other distributions, using a common sigma scale, as shown in Figure S-46.

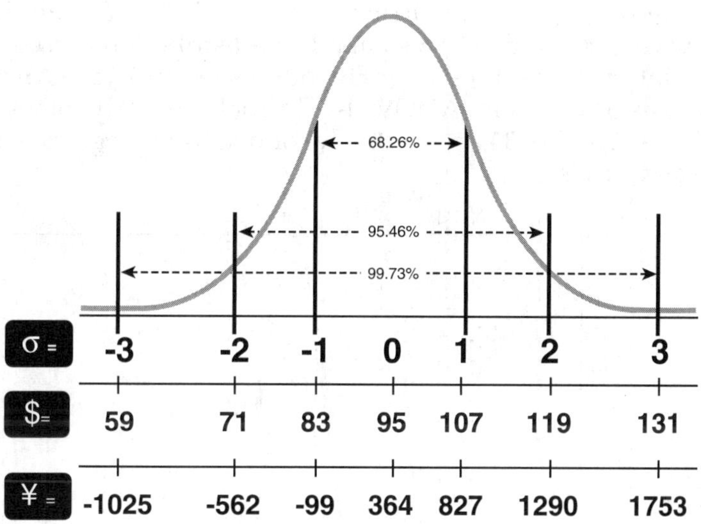

FIGURE S-46: Standard Normal Curve

The specific bell-shaped distribution (for example, the *shape* of the normal data) remains constant, hence the proportion of the area under the curve remains constant. The area in a Standard Normal curve is delineated in sigma units (recall that standard deviation is one). If two points on its X-axis are known, they demarcate the corresponding probability that a given variable will fall between them, as shown in Figure S-46. Hence, regardless of the scale (or units of measure), the shape of the Standard Normal distribution can be used to calculate the probability of an event and predict the range of its value—statistical inference.

The various values of a standard normal curve are called Z-scores. A Z-score can be calculated or looked up in a Standard Normal table. The formula to convert a distribution into a standard normal distribution is as follows, regardless of units of measure. (It is a similar procedure to converting currency from one denomination to another.)

$$Z_i = \frac{(X_i - \mu_N)}{\sigma_N}$$

Wherein, Z is the variable of interest minus population mean, divided by the population standard deviation. (**See Also** "Standard Normal Distribution," in Appendix B, **p. 975**)

- **Student t-Distribution**—Used for small sample sizes of 30 or less and when the standard deviation (sigma) is known. Its shape is affected by the sample size, such that as sample size increases, the degrees of freedom (dof) increase, and the shape approaches a normal distribution, as shown in Figure S-47. The values cluster near a single most frequently occurring value, and 95% of the area under the curve is covered in 1.96 sigma. The t-distribution is used in confidence intervals and t-tests (for Hypothesis testing). (**See Also** "Analysis of Variation (ANOVA)—7M Tool" and "Hypothesis Testing," **p. 142** and **p. 335**, respectively for details on degrees of freedom and t-tests.)

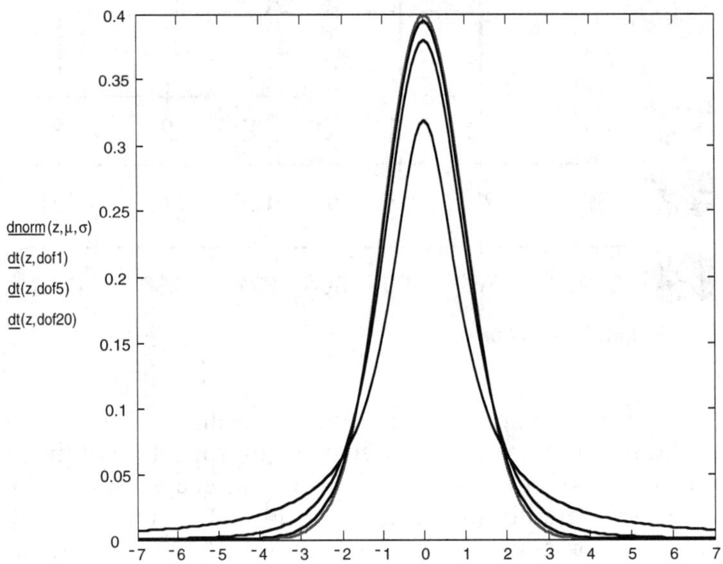

FIGURE S-47: Student t-Distribution Compared to a Standard Normal Distribution

- **Binomial Distribution**—Models two outcomes of discrete data as part of the same distribution. The equation involves a constant probability of success, denoted by **p** and a constant probability of failure, denoted by **q**, which equals 1–p. The equation for the probability of X number of successes in **n** trials is:

$$P(x) = \frac{n!}{x!(n-x)!} p^x(q^{n-x})$$

This concept applies in quality management, to create control charts for attribute data, namely the np-chart and the p-chart. (**See Also** "Control Charts—7QC Tool," **p. 217**)

- **Poisson Distribution**—Models random occurrences over time of discrete data with essentially no upper bound, where the number of occurrences of an event during an interval is independent among intervals. Examples include time, distances, volumes, area, rates, and flows. This applies to systems where it is difficult to count all non-occurrences (that is, defect counts), to approximate the binomial distribution.

  The probability of occurrence is constant among intervals, and in a small interval is small and proportional to the length of the interval. The equation is:

$$P(x) = \frac{\lambda^x e^{-\lambda}}{x!}$$

  , where e = 2.71828.

  This concept applies in quality management in constructing control charts for attribute data based on the Poisson distribution, namely the c-chart and the u-chart. (**See Also** "Control Charts—7QC Tool," **p. 217**)

- **Down-Side Variance (DSV)**—Examines the sample values smaller than the mean (X_{is}). Its formula is:

$$DSV = \frac{\sum_{i=1}^{n} (X_{is} - \overline{X})^2}{n-1}$$

  , where **n** is the number of data values in the entire set.

- **Skewness (SK)**—Examines the shape of a sample distribution. Its formula is the following ratio, where $\mathbf{s^2}$ is the sample's variance:

$$SK = \frac{S^2}{2(DSV)}$$

  If SK > 1; then the distribution is skewed to the right (positively skewed).

  If SK < 1; then the distribution is skewed to the left (negatively skewed).

  If SK = 1; then the distribution is symmetrical.

When using MINITAB and Crystal Ball software packages, note that they utilize a different formula to calculate skewedness; hence, the skewedness determination becomes zero:

$$SK = \frac{\sum_{i=1}^{n} (X_i - \overline{X})^3}{(s^3 \times N)}$$

If SK > 0; then the distribution is skewed to the right (positively skewed).

If SK < 0; then the distribution is skewed to the left (negatively skewed).

If SK = 0; then the distribution is symmetrical.

- **Pearson's Coefficient of Skewness (γ–gamma)**—another formula to calculate skewness using population data, where the population mean (μ), the median (M), and population's standard deviation (σ) are used:

$$\gamma = \frac{3(\mu - M)}{\sigma}$$

Central Limit Theorem
This theorem states that regardless of the shape of the population distribution, the distribution of the means calculated from several samples tends to become normal as the sample size increases. This concept can be expressed as an equation:

$$\sigma_{\overline{X}}^{2} = \frac{\sigma_{x2}}{n}$$

, where **n** is the number of observations.

Thus, with increasing sample size, the mean of the distribution of sample means eventually will equal the mean of the population, as illustrated in Figure S-48. The standard deviation of the distribution of sample means is equal to the population standard deviation divided by the square root of sample size.

Figure S-48 illustrates how the graph of the means from several samples is a normal distribution and tighter around the mean than the plot of individual data points.

The significance of the Central Limit Theorem on control charts is that the distribution of sample means approaches a normal distribution, regardless of the shape of the parent population. If a control chart of sample averages is plotted, on average only about 0.27% of the time the sample mean falls outside the control limits if the process remains unchanged. This is calculated as 1–99.73% for three sigma on either side of the mean.

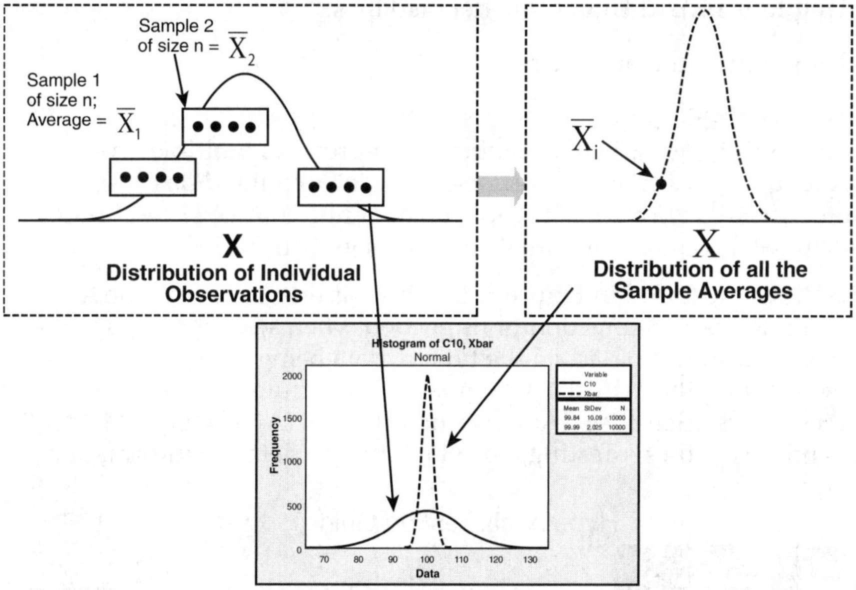

FIGURE S-48: Central Limit Theorem

Inferential (or Comparative) Statistics

Inferential statistics are used with sampling to infer knowledge about the sample and its characteristics onto that of the population. (**See Also** "Sampling," **p. 618**)

Parameters

This refers to the summary measures for the population. Parameters are usually represented by lower case Greek letters, for example

- **Population or Process Mean,** denoted by μ (*pronounced mu*).
- **Population or Process Variance,** denoted by σ^2 (*pronounced sigma-squared*).
- **Population or Process Standard Deviation,** denoted by σ (*pronounced sigma*).
- **Population Size,** denoted by **N**.

Sample Statistics

This set of measures provides summary information about the sample. They are usually represented by lower case English letters, for example:

- **Sample Mean,** denoted by $\overline{X}$ (*pronounced x-bar*).
- **Sample Variance,** denoted by s^2 (*pronounced s-squared*).

A B C D E F G H I J K L M N O P Q R S T U V W X Y Z

- **Sample Standard Deviation,** denoted by **s**.
- **Sample Size,** denoted by **n**.

Hypothesis Testing
This is a robust topic, as statistical analysis represents multiple types of tests. A unique set of tests cover *enumerative* data (counted data), such as attributed, classification, or categorical data. Enumerative Hypothesis tests include Chi Square, binomial, and Poisson distribution.

The schematic found in Figure S-49 serves as a summary guide to the various questions and considerations needed when selecting the appropriate Hypothesis test based on the type of data being analyzed. Figure S-49 also includes the MINITAB commands to execute the various Hypothesis tests. (**See Also** "Analysis of Variation (ANOVA)—7M Tool," **p. 142**, and "Hypothesis Testing," **p. 335**, for more detailed information.)

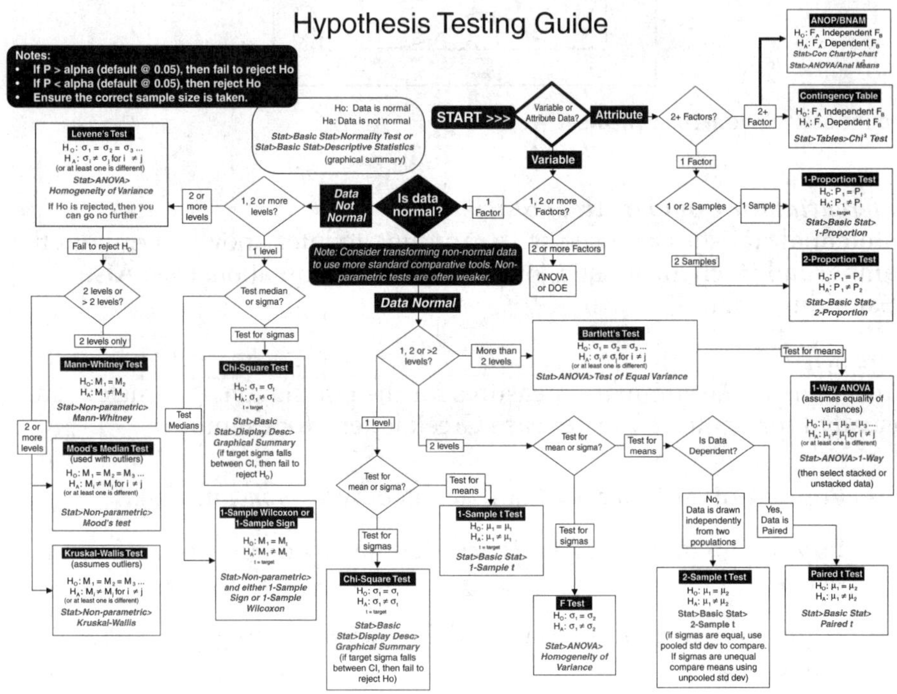

FIGURE S-49: A Guide for Hypothesis Testing

Correlation and Regression Analysis
Regression analysis quantifies the type and shape of the relationship between two or more variables to model a population or system (either mechanical or behavioral). (**See Also** "Scatter Diagrams—7QC Tool" for correlation details, **p. 640**, and "Regression Analysis," **p. 571**, for more detail.)

Design of Experiment (DOE)

DOE is another important statistical tool set to demonstrate cause-and-effect relationship. It encompasses different test designs such as screening, full-factorial, fractional factorial, optimizing, response surface, robustness (Taguchi), confirming, mixture and conjoint analysis. (**See Also** "Design of Experiment (DOE)," and "Conjoint Analysis," **p. 250** and **207**, respectively.)

Probability

The probability of an event is the value that ranges between 0 and 100% (or 0 to 1). If the event were called A, then the probability of that event occurring is denoted as P (A). The probability of A occurring would equate to: **P (A) = (# of occurrences)/(# of total possibilities)**.

Hence, using the preceding definition, an event that cannot occur (an *impossible event*) could be written P (A) = 0. Conversely, an event that were certain to occur (a *certain event*) could be written P (A) = 1. Typically, the probability of an event falls between impossible and certain and is expressed as $0 \leq P(A) \leq 1$.

The sum of the probabilities of a fixed number events (n), would be mutually exclusive and collectively exhaustive events, where the probability of n occurring is 100% or 1. When two events represent all the possible events, they are referred to as *complementary events*. If the total number of possibilities were binary (good or bad), then the probability of any one of those two events occurring would be complementary, in that P (good) = 1–P (bad). When writing the Null and Alternative hypothesis statements, they are expressed as mutually exclusive and collectively exhaustive of all the possible outcomes. For example, if the Null hypothesis were "mean 1 equals mean 2," then the Alternative hypothesis must be "mean 1 does not equal mean 2." Thus, the Null hypothesis is statistically complementary to the Alternative hypothesis, which can be expressed as P (Null) = 1–P (Alternative). (**See Also** "Hypothesis Testing," **p. 335**)

Additive Law

Risk management often applies the additive law of probability. If a risk event were independent from other events, such that the probability of its occurrence remained constant, historical data could be used to determine the likelihood of future occurrences.

In probability terms, this describes a *union*, denoted by a $\cup$ symbol and the portion that covers the overlap or *intersection*, denoted by a $\cap$ symbol.

If they two events are mutually exclusive, then the probability can be expressed as: $P(A \cup B) = P(A) + P(B)$.

A B C D E F G H I J K L M N O P Q R S T U V W X Y Z

If the two events (A and B) are **not** mutually exclusive, they can be added together. The equation expressing the additive law is: $P(A \cup B) = P(A) + P(B) - P(A \cap B)$. (**See Also** "Risk Mitigation Plan," **p. 601**)

Multiplicative Law
Quality management uses the multiplicative law of probability to examine reliability of two components. If the components are assembled in series, the reliability of both is less than the reliability of any one component by itself, such as throughput yield. If any two events (A and B) are independent, the probability of A and B occurring can be written as: $P(A \cap B) = P(A) \times P(B)$. (**See Also** "Process Capability Analysis," **p. 486**, for a discussion on throughput yield.)

The components also operate in parallel (A_p and B_p), where one is operational and the other is redundant, the reliability of these two components is higher than the reliability of any one component. The equation is: $P(A_p) = 1 - P(A)$; $P(B_p) = 1 - P(B)$; and $P(A_p \cap B_p) = P(A_p) \times P(B_p)$. Sometimes A_p and B_p are denoted as A-bar and B-bar, respectively.

Graphical Tools
The primary graphical tools include

- **Boxplot**—A type of frequency plot used to summarize data (continuous Y and discrete X). (**See Also** "Boxplots—Graphical Tool," **p. 165**)
- **Control Chart (Super-Charged Run Chart)**—Used to monitor a process and requires continuous data in sequence of production or observation. (**See Also** "Control Charts—7QC Tool," **p. 217**)
- **Dotplot**—A type of frequency plot used to stratify data (continuous Y and discrete X). (**See Also** "Dotplot," **p. 280** and "Stratification Tool—7QC Tool," **p. 445**)
- **Frequency Plot (Histogram)**—Used to determine the shape, center, and range of continuous or numeric data. (**See Also** "Histogram—7QC Tool," **p. 445**)
- **Pareto Chart**—Used to determine the relative contribution of different items (such as defects or problem areas) for discrete or categorical data. (**See Also** "Pareto Chart—7QC Tool," **p. 445**)
- **Run Chart (or Time Series Plot)**—Used to identify data patterns such as special causes, shifts, and trends. Continuous data in sequence of production or observation are required. (**See Also** "Run Chart—7QC Tool," **p. 611**)

A B C D E F G H I J K L M N O P Q R S T U V W X Y Z

- **Scatterplot**—Used to determine if a relationship exists between two discrete or continuous variables; *correlation*. Also used with Stratification. (**See Also** "Scatter Diagram," **p. 640** and "Stratification," **p. 697**)

See "Graphical Methods," on **p. 323** for a summary of the different tools.

Process
A statistical tool helpful to evaluate a process is the Process Capability techniques, including Cp, Cpk, Pp, Ppk and throughput yield analysis. As previously mentioned, process tools examine the inputs and outputs of a process for variation and defects. Other statistics applied to process cycle time span the relatively simple tools such as *graphical tools* and *variation analysis*, to robust tools such as *Monte Carlo simulation*. (**See Also** "Process Capability Analysis," **p. 486**, and "Monte Carlo Simulation," **p. 431**)

Measurement System Analysis (MSA)
Because statistical thinking focuses on three core elements—process, variation, and data, it is critical that the data collected, graphed, analyzed represent what really happens in the system or process. The MSA technique ensures that the measurement system is accurate and reliable for either continuous or discrete data, and not introducing any noise into the data. It validates that whatever variation is observed is due to variation in the item of interest, and not due to a sloppy measurement system. The measurement system is comprised of the measuring device, the operator (or measurer), and the item of interest. (**See Also** "Measurement System Analysis (MSA)," **p. 412**)

Stratification—7QC Tool

What Question(s) Does the Tool or Technique Answer?
Are there any patterns in the data? Is the data set homogeneous?

Stratification helps you to

- Reveal patterns in the data
- Understand why the data set in aggregate does not make sense

When Best to Use the Tool or Technique
The stratification tool should be combined with other data analysis tools; however, forethought is required to ensure that the proper stratification information is collected in case this technique proves helpful during analysis in seeing patterns otherwise hidden.

A B C D E F G H I J K L M N O P Q R S T U V W X Y Z

A B C D E F G H I J K L M N O P Q R S T U V W X Y Z

Brief Description

Stratification is a data analysis technique that parses data into its natural subgroups—or *strata*. Combine this technique with other data analysis tools to segregate the data into its different sources. Examples of different data sources include: different suppliers, shifts, departments, locations, equipment types, products, time of day, day of week, month of year. For process data, stratification cuts the data into its different sources like layers of a stratified rock. Interestingly, these different stratifying sources can be further segregated into subgroups, often linked to a special cause. Stratifying the data allows for *patterns to be seen*, that otherwise would be masked by the larger population.

Examples of different applications might involve the following situations:

- Stratifying survey data into the demographics of the respondents may reveal patterns that otherwise would be unseen when analyzed in aggregate.
- Examining profit per employee may be inconclusive until the data is stratified to geography to uncover economic differences and buying behaviors by region.
- Evaluation of throughput yield may appear homogeneous until the data is parsed by shift.
- Understanding an organization's revenue growth potential or weaknesses may seem unclear until the offerings portfolio was analyzed by offering with its respective competitive threats.

In the world of quality, there is a fine distinction between stratification and segmentation. Stratification is related to segmentation but is different. Stratification defines the different **sources of data**, while segmentation often represents a sampling technique. Segmentation has a broader meaning describing categorization, such as classification or types of an item. Examples of segmentation might be kinds of fruit, kinds of defects, market types, customer types.

If stratification appears to be a potential influence, plan accordingly *prior* to data collection so as to properly label the source during the data gathering activities. Moreover, the data collection tool will need to track and record the different sources to enable strata analysis later.

Presuming that the proper information has been collected, software tools such as MINITAB make it easy to explore stratification. After looking at a graph, such as a Dotplot, Multi-vari plot or Scatterplot, if the output appears inconclusive, MINITAB allows the pointer to be dragged over an area of data for quick viewing of common strata. If it appears to play a factor, simply rerun the graph specifying the stratification be identified by other different data icons and sometimes unique regression lines. (**See Also** "Dotplot," **p. 280**; "Multi-vari Chart" **p. 439**; and "Scatter Diagram—7QC Tool," **p. 640**)

Figure S-50 compares a pair of Scatter diagrams with the same profit against FTE (full-time equivalents) data—one with versus one without stratification. The one without stratification appears inconclusive. The stratified graph exhibits a different FTE to profit relationship in the North (represented by a circle) versus the West (represented by a square) regions, as indicated by their respective best-fit lines.

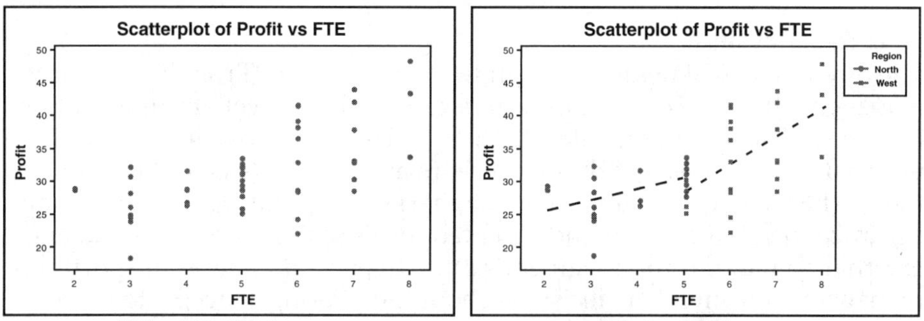

Without Stratification With Stratification

FIGURE S-50: Comparing Scatter Diagrams With and Without Stratification

Stratification is an original member of the *7QC Tools*, (or seven *Quality Control* tools), attributed to Dr. Kaoru Ishikawa. The 7QC Tools sometimes are called the *seven basic tools* given that they were the first set of tools identified as the **core** quality improvement tools. Ishikawa's original 7QC Toolset includes: Cause-and-Effect diagram; check sheet (or checklist); control charts; histogram; Pareto chart; Scatter diagram; and stratification. More recently, the 7QC Toolset has been modified by substituting the stratification technique with either a flowchart (or Process map) or a run chart (or Time Series plot).

SWOT (Strengths-Weaknesses-Opportunities-Threats)

What Question(s) Does the Tool or Technique Answer?

How does the organization's strengths and weaknesses compare with the competitive opportunities and threats?

SWOT analysis helps you to

- Identify the critical marketplace characteristics and organize them as a balanced scorecard to make strategic decisions

- Evaluate the balance between internal and external factors for a particular organization
- Identify and prioritize which market segments fit the organization

When Best to Use the Tool or Technique

During the strategic planning process, use a SWOT analysis to aid in formulating strategic direction and management goals and objectives.

Brief Description

The Strengths-Weakness-Opportunities-Threats (SWOT) analysis focuses management on the competitive landscape and how well its organization competes. The SWOT tool identifies, organizes, and ranks internal and external factors into a quadrant matrix to analyze the balance between them. The technique examines which market segments optimally fit the organization's capabilities and accepted levels of risk. It sets internal and external boundaries for a segment's attractiveness and helps the portfolio management team make decisions about the offering concepts to fund and to initiate as development projects.

The SWOT analysis presumes that an organization tries to fit within its external environment. This technique produces a summary plot of segments on a grid of ranked opportunity space in relation to the competitive position. It lists the organization's strengths, weaknesses, opportunities and threats related to a particular market segment being evaluated. Conduct a unique SWOT analysis for each given region of interest, as the external environment differs from place to place.

The SWOT analysis embodies two challenges. First, the organization must objectively evaluate its internal strengths and weaknesses and relative position to competition. Management tends to amplify strengths and underplay weaknesses, thereby negating the SWOT process. Collecting the perspective of objective observers may serve the organization well. Second, the analysis represents only the initial phase in a strategic plan; the next phase requires the execution and delivery of an action plan.

External Environment

SWOT analysis requires an understanding of the marketplace. Benchmark data is a prerequisite to determine the competitive environment, the emergence or exit of competitive players, and how they affect your organization. Include marketplace data about the technological environment to understand the available or emerging technologies. A social environment snapshot may be important, particularly if the organization considers entering a new market. Social elements include the value system, social and demographic patterns that might impact the organization's ability to operate in that region. For example, evaluate whether the region possesses the skills and knowledge workers required. Closely related to the social landscape is the socio-political environment. Evaluate the state (stability and influence) of and potential impact of the different governing bodies (local, regional, national), special interest groups, and/or legal and regulatory agencies. Understand the state of the economic environment for the region and marketplace, the trends, and any key events that might impact the organization.

The acronym **PEST** acts as a reminder of the main external forces: Political, Economic, Social, and Technological. Apply this tool to each country or region in which the organization operates (or plans to operate), as these PEST forces differ by location. Variations on the PEST analysis tool include **SLEPT** (Social, Legal, Economic, Political, and Technological) and **STEEPLE** (Social/demographic, Technological, Economic, Environmental (natural), Political, Legal, and Ethical). Again, these multiple factors each may alter as the geographic location changes, so repeat the SWOT analysis for each regional situation.

Opportunities are characterized by three attributes: financial, market behavior, and disruptive technology. Fulfilling market behavior needs results from a proper balance between marketing and technical innovation. An organization can benefit from taking advantage of disruptive technology opportunities by having a strong focus on technical innovation. Opportunities provide an environment for ideas to be tested, ranked, and prioritized for portfolio development, renewal, and balancing for the desired level of financial growth.

Threats are usually traceable to a gap between your organization and some form of competitive force in the served markets and segments, such as actions from direct competitors or indirect competitors using a disruptive technology. Threats can come from uncontrollable sources of variation in the market environment. Currency fluctuation, price elasticity, social change, political upheaval, international affairs, and man-made or natural disasters all represent sources of external threats. Threats add risk to your ability to exploit on opportunity.

Opportunities and threats represent the external factors that may create or destroy value in the marketplace. These represent uncontrollable factors to a large extent. Comprehend these external dynamics to determine how best to operate within this environment. An organization's strategy must align with the external opportunities to best leverage its capabilities and defend against the external threats. The organization's strategy should consider possible changes (trends) in the external environment to determine its appropriate direction and positioning.

Internal Environment

The organization's strengths and weaknesses evolve from its internal environment made up of organizational assets, resources, and skills. These internal factors create (or destroy) marketplace value relative to competition. They are measured and evaluated not only by the organization itself, but also the marketplace.

An organization's **strengths** usually come from its core competencies. A strength is something an organization is doing well. It is characterized by the capabilities and capacities to exploit an opportunity within or across a market. Strengths underwrite the organization's ability to successfully design, develop, distribute, and deliver value (product and/or services) to the market. Strengths exploit the external opportunities and fend off the external threats.

Internal strengths have many dimensions—anything that provides a competitive advantage. They can be depicted as new ideas, current products and services to leverage from, proprietary technologies, patents, intellectual property, skills, key resources, and expertise. Marketing capability and capacity to create brand awareness and position and a reputation for quality are all strengths. Portfolio development and balancing, commercialization and launch, channel management, distribution, sales, support, and product-line management also describe internal capabilities and capacities. Strengths include technical capability and capacity, research and development, design, manufacturing, and the service and supply chain. Moreover, an organization's financial resources, partnerships, alliances, and joint ventures are invaluable assets.

Weaknesses are areas wherein the organization lacks something or a condition that puts it at a disadvantage relative to its competitors. A weakness usually is traceable to a gap in core competencies or to poor execution within product development or post-launch processes. In addition to poor quality, weakness also may include poor cash flow, outdated technology, high overhead expenses, or unavailable skilled labor. Weaknesses represent barriers to the organization's ability to exploit an opportunity or ward off a competitive threat.

Competitive Advantage

In summary, an organization's competitive advantage embraces the aggregate internal strengths and weaknesses to counter the competition's opportunities and threats. Michael Porter proposes that there are two types of competitive advantage—*cost leadership* (low cost) and *differentiation*—which in combination create a third dimension—*focus*.

Economies of scale play an important role in cost leadership. Wal-Mart is an example of a firm identified as a cost leader. If more than one firm tries to achieve cost leadership in a market segment, eventually it leads to disastrous conditions.

Differentiation often serves as a counter-balance to a cost leadership position, as an organization attempts to deliver a product and/or service of **unique** value. To do so, it requires addressing new, unique, and difficult (NUD) requirements. At minimum, price the differentiated offering at a premium to its production costs. Apple's iPod is a good example of a differentiated MP3 player that delivers on NUD requirements with its design and iTunes music offerings. (**See Also** the "Kano Model: NUD Versus ECO" section under "KJ Analysis," **p. 376**)

Achieving focus describes the degree with which an organization drives for results—best-in-class. The competitive strategy may call for a narrowing or widening approach to best address a particular market segment. Focus may alter a strategy to go after a different market niche. Again, Apple's current market success may be the impetus behind its focus in launching the new products such as the cost-competitive Shuffle, the price premium video-iPod, or the iPhone that takes it into a new market segment.

Ultimately, an organization's strategy must match its internal capabilities with the external environment. The target market opportunities complement its capabilities, its defenses are suited to foil the external threats, and it's suited for any changes.

How to Use the Tool or Technique

The process to conduct a SWOT analysis, assemble a cross-functional, multi-discipline team, is as follows:

Preparation: Work on flip chart paper so that everyone on the team can see the work simultaneously. You may choose to write directly on the flip charts or use sticky notes to post on the wall or a white board. Use a round-robin approach by arming each person with a pen or marker or giving each person a pad of sticky notes, which aids in capturing everybody's input.

Ensure that the proper data sources are collected and brought to the meeting.

SWOT Matrix

Step 1. Construct the SWOT matrix. Create a 2x2 matrix by dividing a page into quadrants.

a. Label each quadrant. Convention places strengths in the upper-left corner, weaknesses in upper-right corner, opportunities in lower-left corner, and threats in the lower-right corner.

A B C D E F G H I J K L M N O P Q R S T U V W X Y Z

Step 2. Create a list of Opportunities and Threats (O/T) and document them on a flip chart or in the appropriate cell of the SWOT matrix. Utilize the PEST, SLEPT, or STEEPLE techniques. Sources include

a. Market and segmentation analysis

b. Primary and secondary market research data

c. Long range, over-the horizon market trend and behavioral analysis

d. External benchmarking studies

e. Market environmental force analysis

Step 3. Create a list of Strengths and Weaknesses (S/W) and document them in the appropriate cell of the SWOT matrix. Sources include

a. Internal assessments, audits and analysis, capability analysis of design, development, manufacturing, launch, sales, and service and support capabilities

b. Benchmark studies of internal strengths and weaknesses

c. Business and innovation strategy development, documentation, and execution of the strategic plan

d. Product portfolio performance analysis/audit

e. Training and use of tools and best practices

f. Project management skills and cycle-time performance

g. Financial reports

h. Product management effectiveness over life-cycles

Step 4. Post the completed SWOT matrix in the room for reference. This chart helps to communicate the high-level scenario, as illustrated in Figure S-51.

Candidate Market Segment Evaluation Matrix

Given a couple of candidate segments, apply the SWOT matrix to begin developing a unique strategy for each. Similar factors may exist in multiple segments, but deal with each segment independently and duplicate the item of interest.

Strengths (internal)	Weaknesses (internal)
• Brand recognition • Patents • Intellectual property • Good quality • Feature innovation	• Lack of design innovation • Too many products poorly differentiated • Only "industrial strength" version • Limited sales channels
Opportunities (external)	**Threats** (external)
• Developing markets • Mergers and acquisitions • Association market segment • Internet sales	• Innovative marketing and selling • Fashion designs

FIGURE S-51: SWOT Matrix

Step 5. Create the Market Segment Evaluation Matrices to further dissect the SWOT matrix, as illustrated in Figure S-52. (**See Also** "Matrix Diagrams—7M Tool," **p. 399**, for a discussion on different shapes.)

a. Using an L-shaped structure, create two five-columned matrices, presuming three market segments to evaluate. If there are more, increase the number of columns accordingly; or if less, decrease the number accordingly.

b. Label one "Opportunity Market Segment Evaluation" matrix.

   i. Label the column headings: Column 1 as "Opportunities/Threats (O/T)"; Column 2 as "Weight"; Columns 3, 4, 5 as "Market Segment 1, 2, 3," respectively.

   ii. Subdivide Columns 3, 4, 5 into two more columns each. Label each of the sub-column pairs from left to right, as "Rating" and "Score."

c. Label the second "Competitive Position Market Segment Evaluation" matrix.

   i. Label the column headings: Column 1 as "Opportunities/Threats (O/T)"; Column 2 as "Weight"; Columns 3, 4, 5 as "Market Segment 1, 2, 3," respectively.

   ii. Subdivide Columns 3, 4, 5 into two more columns each. Label the each of the sub-column pairs from left to right, as "Rating" and "Score."

Opportunities/ Threats	Weight	Segment 1		Segment 2		Segment n	
		Rating	Score	Rating	Score	Rating	Score

This Table will represent the Opportunities axis on the SWOT Summary Chart...

Strengths/ Weaknesses	Weight	Segment 1		Segment 2		Segment n	
		Rating	Score	Rating	Score	Rating	Score

This Table will represent the Competitive Position axis on the SWOT Summary Chart...

FIGURE S-52: Template for Market Segment Evaluation Matrices

Step 6. Determine the Weight of the O/T and S/W factors, as illustrated in Figure S-53.

a. Create two lists, one by combining the SWOT opportunities and threats into a consolidated O/T list and the other by combining the SWOT strengths and weaknesses into a consolidated S/W list.

b. Working with one list at a time, stack rank the items within the list in order of importance. Document the items in the first column of the appropriate Market Segment Evaluation matrix by placing the most important item at the top and then flow down the remaining items in descending order.

c. Repeat Step 6.b. for the other list and fill the appropriate Market Segment Evaluation matrix.

d. Working with one list at a time, allocate 100 points across each item within the list and record the appropriate pairing in the Weight column.

   i. Assigning numeric weight represents the amount of attention or investment the issue must receive for the business to succeed in the identified market.

   ii. Some factors may be irrelevant to a segment and would then get a weight of 0.

e. Add a row at the bottom of the matrix to hold the totals and total the Weight column to double-check that all 100 points have been dispensed.

f. Repeat Step 6.d. for the other list, to weight the items.

Step 7. Rate each candidate market segment, as illustrated in Figure S-53.

a. Market Opportunity rating. Use the following rating scale to measure the impact of each line item:

   i. 3 = Most Impact on the segment; very good

   ii. 2 = Better than most; above average impact

   iii. 1 = Normal impact

   iv. 0 = No impact at all; not good enough

   v. -1 = Detrimental impact; threat is high

b. Rate each item in the Opportunity Market Segment Evaluation matrix, using the preceding Market Opportunity rating scale. Record the score in the appropriate cell and total the Score column.

   i. Opportunities can exist now or appear during the planning phases.

   ii. Focus on present position, trends, and future expectations.

   iii. A realistic projection is the goal.

c. Repeat Step 7.b. for each of the market segments until all are rated and totaled for their market opportunity.

d. Competitive Position rating. Use the following rating scale to measure the impact of each line item:

   i. 3 = Our position provides a maximum effect and impact.

   ii. 2 = Our position has an above average effect on the market.

   iii. 1 = Our position has an expected, normal impact.

   iv. 0 = Our position would have no competitive effect or impact.

   v. -1 = Our position signifies poor effect—we are not competitive.

e. Rate each item in the Competitive Position Market Segment Evaluation matrix using the preceding Competitive Position rating scale. Record the score in the appropriate cell and total the Score column.

i. Strengths and weaknesses exist now.

f. Repeat Step 7.e. for each of the market segments until all are rated and totaled for their competitive position.

Step 8. Calculate the SWOT Value Scores.

a. Working with one Market Segment Evaluation matrix at a time, by row, multiply the weight by the rating score for each factor to calculate each factor's T/O score and a S/W score, respectively.

b. Total each Score column, by segment, for each Market Segment Evaluation matrix.

Opportunities/ Threats	Weight	Segment 1		Segment 2		Segment n	
		Rating	Score	Rating	Score	Rating	Score
	45	3	135				
	25	2	50				
	20	1	20				
	10	-1	-10				
Totals	100		**195**				

Step 6 Step 7 Step 8

Strengths/ Weaknesses	Weight	Segment 1		Segment 2		Segment n	
		Rating	Score	Rating	Score	Rating	Score
	50	1	50				
	35	1	35				
	15	3	45				
	5	-1	-5				
Totals	100		**125**				

FIGURE S-53: Market Segment Evaluation Matrices

Strategic Decision Grid

Step 9. **Assess** the results and document the findings, as illustrated in Figure S-54.

a. Create a 3x3 matrix, label it "Strategic Decision Grid."

b. Label the vertical axis "Opportunity" to represent the market opportunities found in the respective Market Segment Evaluation matrix (the strengths/weaknesses).

c. Label the horizontal axis "Competitive Position" to represent the competitive positions found in the respective Market Segment Evaluation matrix (the opportunities/threats).

d. Using the following alphabetical scale, code each cell accordingly.

    i. A = Good Opportunity and position.

    ii. B = Moderate to reasonably good opportunity/position.

    iii. C = Less than optimal opportunity/position.

    iv. Top Row, from left to right, label each cell B, A, and A, respectively.

    v. Middle Row, from left to right, label each cell C, B, and A, respectively.

    vi. Bottom Row, from left to right, label each cell C, C, and B, respectively.

e. Plot the two summary scores for each segment on the graph.

f. Use the market size data (that is, potential sales volume) to scale an icon (such as a circle) that represents each market segment. Plot the appropriately sized market segment icons in the appropriate grid on the graph.

g. Determine which segment is providing the strongest opportunities and least resistance to entry or competition.

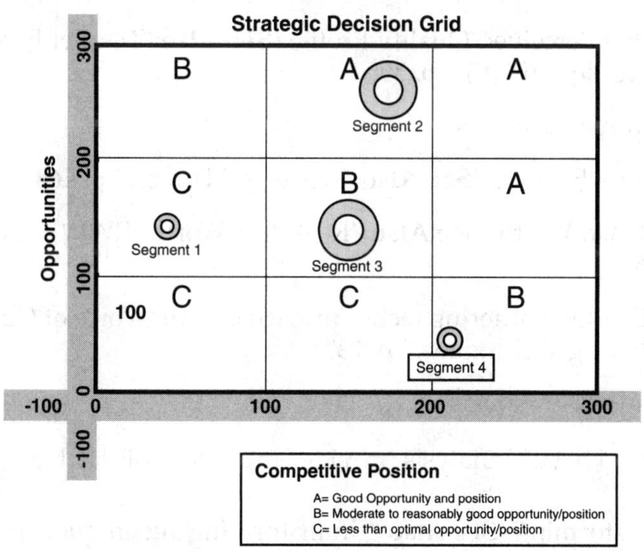

FIGURE S-54: Market Segment Evaluation Matrices

Step 10. Identify the Best Segment-Opportunity matches.

a. Document the best opportunities for each segment under analysis. Identify the issues that enhance or limit the opportunity

   i. The plot of SWOT data provides a map of the best opportunities as they relate to market segments and the size of the segment. Segment-specific weighting enables this process and helps to focus the voice of the technology (VOT) and voice of the business (VOB) requirements and constraints.

b. Identify the key issues for strategic improvement actions for improving the strengths and eliminating weaknesses for future enabling of growth.

c. Create tactics and identify local improvements that will help define how the opportunity can be exploited with the least risk relative to the size of the financial opportunity.

Supporting or Linked Tools

Supporting tools that might provide input when developing a SWOT analysis include

- Brainstorming (**See Also** "Brainstorming Technique," **p. 168**)
- Market Perceived Quality Profile (**See Also** "Market Perceived Quality Profile (MPQP)," **p. 390**)
- Performance charts
- Porter's 5 Forces (**See Also** "Porter's 5 Forces," **p. 464**)
- Real Win Worth (**See Also** "Real-Win-Worth (RWW) Analysis," **p. 560**)
- VOC/VOB gathering techniques (**See Also** "Voice of Customer Gathering Techniques," **p. 737**)
- Won/Loss Analysis

A completed SWOT analysis provides input to tools such as

- Brainstorming (**See Also** "Brainstorming Technique," **p. 168**)
- GOSPA (**See Also** "GOSPA (Goals, Objectives, Strategies Plans and Actions)," **p. 320**)

- Portfolio Strategic Plan
- FMEA (**See Also** "Failure Modes and Effects Analysis (FMEA)," **p. 287**)

Figure S-55 illustrates the link between the SWOT analysis and its related tools and techniques.

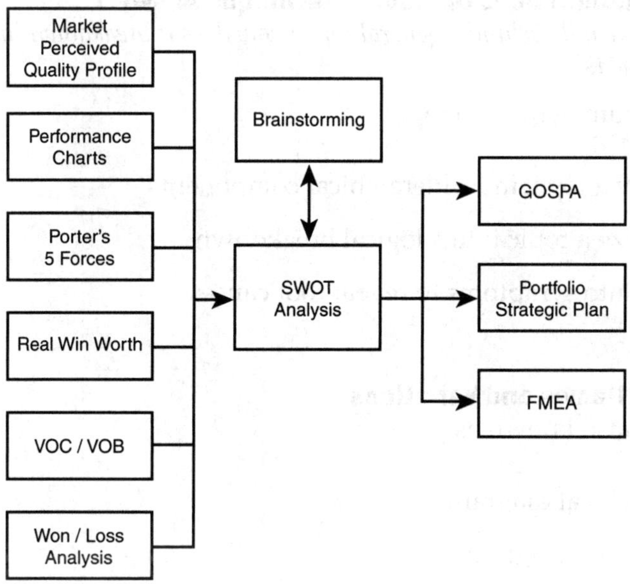

Figure S-55: SWOT Tool Linkage

T

Tree Diagram—7M Tool

A B C D E F G H I J K L M N O P Q R S T U V W X Y Z

What Question(s) Does the Tool or Technique Answer?

What are the details behind a general topic; how does it breakdown into its component-parts?

A Tree diagram helps you to

- Display a topic in its hierarchical components
- Organize a topic into a logical breakdown
- Probe into symptoms to reveal root causes

Alternative Names and Variations

This tool is also known as

- Hierarchical diagram
- Systematic diagram
- Tree analysis, analytical Tree

Variations on the tool include

- Work Breakdown Structure (WBS) (**See Also** "Work Breakdown Structure (WBS)," **p. 753**)

When Best to Use the Tool or Technique

When there's a need to parse a topic into supporting details or component parts, use a Tree diagram to systematically analyze and illustrate the hierarchical relationship.

Brief Description

The Tree diagram is a flexible tool that displays the hierarchical structure of a high-level topic broken down into its subcomponent details or related topics. It is a generic tool that illustrates the logical, sequential flow from a general topic down its multiple, related branches or paths. A completed diagram resembles a multi-branched Tree, stemming either from the top and branching down or from the side and branching horizontally (conventionally from left-to-right.) Using the Tree-structure systematically dissects a topic into greater detail and communicates an often complex and multi-faceted hierarchy with a single picture. Its flexibility allows it to apply to several different scenarios.

For example, a Tree diagram helps to analyze a process of flow into its logical paths and map out the next steps from a specific starting point. The tool also drills down into symptoms to reveal the root cause(s), as in a Cause-and-Effect diagram or a 5-Whys Tree—or analyzes a general customer requirement into specific critical-to-quality (CTQ) metrics, as in a CTQ Tree. It can display a systematic flow of an action plan, starting with the requirements and logically stepping through its related deliverables-tasks-tool linkage, similar to a Work Breakdown Structure (WBS). It can organize a response plan to a potential failure, as in a Cause and Prevention Diagram or similar to a Fault Tree analysis. (**See Also** "Cause-and-Effect Diagram—7QC Tool," **p. 173**; "Cause and Prevention Diagram," **p. 198**; "Critical-To-Quality (CTQ) Matrix," **p. 242**; "Fault Tree Analysis (FTA)," **p. 309**; and "Work Breakdown Structure (WBS)," **p. 753**)

Tree diagrams are a member of the 7M Tools, attributed in part to Dr. Shewhart, as seven "management" tools, or sometimes referred to as the *7MP* or *seven management and planning tools*. These 7M Tools make up the set of traditional quality tools used to analyze quantitative data. The 7M Toolset includes: Activity network diagrams (AND) or Arrow diagrams; Affinity diagrams; Interrelationship digraphs or Relations diagrams; Matrix diagrams; Prioritization Matrices, often replacing the more complex Matrix data analysis; Process decision program charts (PDPC); and Tree diagrams. *The Quality Toolbox*, by Nancy Tague, presents the 7M Tools ranked from those used for abstract analysis to detailed planning: Affinity diagram, Relations diagram, Tree diagram, Matrix diagram, Matrix data analysis (commonly replaced by a more simple Prioritization Matrix), Arrow diagram, and Process Decision Program Chart (PDPC).

How to Use the Tool or Technique

A Tree diagram can be built by an individual or with a team. The procedure to build a Tree diagram, as illustrated in Figure T-1, is as follows:

Step 1. Document the topic of interest at the top of the page (presuming a vertically displayed Tree). If diagramming the Tree horizontally, start to the left and work toward the right.

Step 2. Identify the subcomponents of that topic and align them by similar hierarchical levels and within natural affinity groupings.

a. Work either down one path at a time or across one hierarchical level at a time. Exhaust ideas and then proceed to either the next branch of the Tree or the next level.

b. Use brainstorming techniques, documentation, illustrations, and Process maps to assist with identifying the different detailed topics. (**See Also** "Brainstorming Technique," **p. 168**)

c. Ask probing questions to help identify subcomponents, such as the 5-Whys technique if exploring potential root causes. (**See Also** "5-Whys," **p. 305**)

d. Continue until the ideas are exhausted.

Figure T-1 illustrates a simple horizontally-displayed Tree diagram of what is critical-to-quality for car fuel mileage.

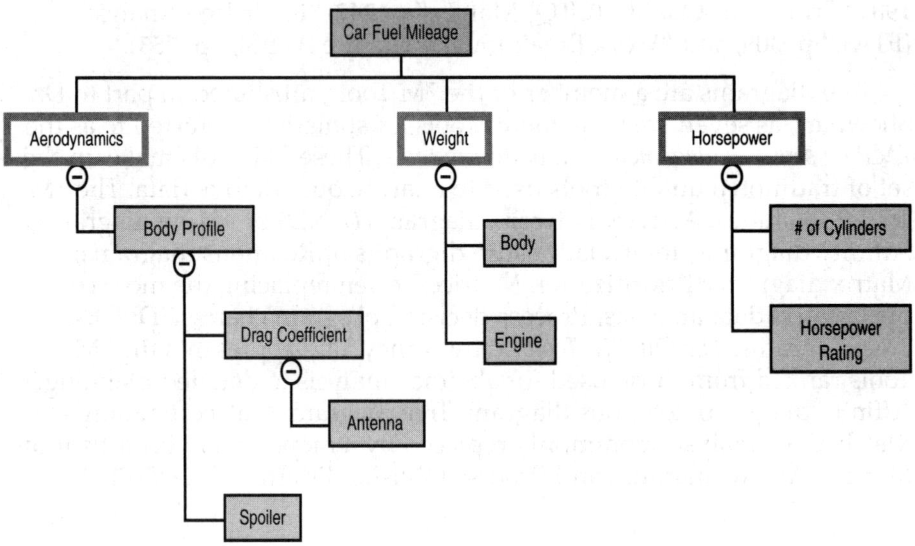

Figure T-1: Car Fuel Mileage CTQ Tree Diagram Example

Hints and Tips

Complex topics mapped into a Tree diagram often include a number system (similar to that used in an outline format) to identify the different levels and cells within the structure.

Language should reflect the purpose of the Tree diagram such that if it is a general topic worked as a noun, describe the subcomponents as nouns or noun-verb (in past-tense). If the Tree diagram represents an action plan, define the detail using verb-noun.

Diagram the Tree down to as much detail as necessary, depending on the topic or purpose of the analysis; however, as a general rule when considering the diagram as a communication vehicle, the simpler the better.

Variations

Work Breakdown Structure (WBS) (**See Also** "Work Breakdown Structure," **p.** 753)

TRIZ

What Question(s) Does the Tool or Technique Answer?

What is the best solution to address this problem and create a competitive advantage?

TRIZ helps you to

- Invent ideas that solve technical or business issues
- Analyze concepts to bolster their design, such as patents, from competitive alternatives

When Best to Use the Tool or Technique

Use the TRIZ method to rapidly generate solution concepts to problems, either technical or business in nature.

Brief Description

The TRIZ method (pronounced "*TREEZ*") is a Russian acronym representing a theory of inventive problem-solving. It was developed by Genrich S. Altshuller, a Russian scientist and engineer, in the 1940s to develop new "inventive" ideas by considering possible solutions from other disciplines and minimizing trial-and-error. TRIZ is a structured approach (algorithm) used to rapidly generate plausible concepts that solve technical problems. Altshuller believed that inventiveness can be taught and that creativity can be learned.

Altshuller claimed that traditional inventing was by trial and error, which produced much wasted time, effort, and resources. He developed a theory that one solves problems through a set of approaches that have certain patterns. He found that people apply common sense, logic, and some physics to problem-solving. After studying and analyzing over 1.5 million patents from around the world, he discovered three groups of methods to solve technical problems: 1) using various tricks (a reference to a technique), 2) methods based on utilizing physical effects and phenomena to change the state of the physical properties, and 3) complex methods (combining the tricks technique and physics).

Informed by this pattern discovery, Altshuller developed this rigorous inventive problem-solving process. While serving in the Soviet Navy as a patent expert, he found that 95% of the knowledge to solve most technical problems already existed elsewhere. He discovered that two problems from different technologies often had the same model and similar solutions. Altshuller was able to condense technical problems into a generic set of standards.

He categorized the patents and respective solutions into five levels that depict the patterns. These patterns were governed by objective laws to develop a system along its path of technical evolution, which are known as the *Laws of Technical Systems Evolution*. These five levels of determining and implementing innovation include descriptors and the percent contribution of the given category to the total solutions:

- **Level one**—Routine designs of well known methods, with few variants, wherein no invention is needed (about 32% of the total).
- **Level two**—Minor improvements or conflict partially eliminated by known industry methods usually involving some compromise (about 45% of the total). The means of solution are found from within the same field of technology as the system.
- **Level three**—Fundamental improvements, or complete conflict elimination, drawing primarily upon knowledge from another industry, wherein contradictions resolved the problem (about 18% of the total).
- **Level four**—A new principle introduced as a primary function, usually from science rather than technology (about 4% of the total).
- **Level five**—A rare scientific discovery or invention fundamentally creating a new system (about 1% of the total).

The TRIZ philosophy builds on five core elements:

- **Ideality**—A system evolving to increasing good and decreasing bad.
- **Resources**—Maximizing the effectiveness of things inside and around system.
- **Space and Time**—Viewing a system in the context of space and time.
- **Functionality**—The importance of function when thinking about systems; wherein solutions change, but functionality remains constant.
- **Contradictions**—All systems contain contradictions; look to eliminate them as a primary evolution driver.

40 Inventive Principles

This inventive problem-solving process initially involved 27 TRIZ tools, which was expanded to 40 innovated tools in the late 1990s. The TRIZ 40 inventive principles solve contradictions (problems) without compromising quality.

These principles, originally intended for technological innovation, have been generalized for practical application in business, education, and everyday life. While the core principle language (the principle name) remains constant, depending on the application, the principles are adapted to a problem's context and intent of the solution. For example, the 40 Inventive Principles for marketing, sales and advertising are adapted as follows:

- **Segmentation**—Stratify or divide into groups or independent parts or sections such as market segmentation by customer profile, demographics, geography, and buying habits.
- **Extraction** or **taking out**—Separate two things, extract or remove something such as differentiate competitive products, outsourcing, and even separating data (that is, qualitative from quantitative).
- **Local quality**—To alter an object or system from uniform (homogeneous) to non-uniform (heterogeneous), to make something better suited for its operation, or to fulfill a different and useful function. For example, personalized advertising or customized cuisine offerings to include regional recipes.
- **Asymmetry**—Change the shape or function of an object or system from symmetrical to asymmetrical or increase its degree of asymmetry. For example, offer right- and left-oriented products or move to offering male- and female-oriented services.

A B C D E F G H I J K L M N O P Q R S T U V W X Y Z

- **Merging** or **combining**—Bring closer together (or merge) identical or similar objects or make them contiguous or parallel (in operations), such as business partnership/alliance, cross-selling, and joint-coupons (that is, milk and cereal)
- **Universality**—Make an object or system perform multiple functions and standardize on them to eliminate the need for other parts. For example, new standardized forms to collect customer profile data, or cell-phones with built-in camera or email capabilities.
- **Nesting** (that is, nested doll)—Putting a product (or system) inside of another or allow it to pass through another, such as a grocery stores selling greeting cards or liquor or a coffee shop selling music CDs.
- **Counterweight** or **anti- weight**—To compensate for weight (downward tendency) to provide uplift, such as advertising or endorsements for a movie about a difficult topic.
- **Prior counteraction** or **preliminary anti-action**—To perform a counter-action in advance; to control harmful effects in the future. Take action to create beforehand stressors that will counter or oppose them. For example, gathering competitive information in anticipation of them releasing a new product, settling a lawsuit out of court, or using patents or licenses to protect intellectual capital, Poka-yoke (mistake-proofing). (**See Also** "Poka-Yoke," **p. 462**)
- **Preliminary action** or **prior action**—Prearranged action or change of an object or system before it is needed, such as preliminary market research of an offering concept (before it is designed), pre-paying for something, or an introduction to a document in anticipation of questions.

  Or *conversely*, **Afterward action**—The opposite (or inverted) of preliminary action, such as post-paying (payment in arrears), after-sales services, Q&As after reading a document, or a coupon for purchasing an item for the next purchase.
- **Beforehand cushioning** or **cushion in advance**—Prepare emergency means prior to an event or compensate for the relatively low reliability of an object by countermeasures taken in advance. For example, utilize Poka-yoke (mistake proofing) strategy, excess inventory, contingency planning or clauses in a contract and split commissions paid upon order-taking and customer acceptance of an offering. (**See Also** "Poka-Yoke," **p. 462**)
- **Equi-potentiality**—Change the working conditions so that an object need not be raised or lowered, or in a potential field, limit position changes to make something balanced or equal. For example, leveling

of a relationship by translating language or terminology, seamless transition from a trial period to permanent ownership, and toll-free customer service phone numbers.

Or conversely, **Potentiality gap**—Such as building up market entrance barriers against competition.

- **Inversion** or "**the other way around**'"—Invert the action used to solve an issue or do the opposite or turn something upside down or view from the opposite perspective. For example, instead of speeding up, slow down; push versus pull; and a rebate for the return of a competitive product.
- **Spheroidality** or the **use of curves**—Use spherically-shaped objects, curvature, or using a rotary motion or centrifugal forces. For example, use rollers or balls, round prices to the nearest dollar, implement revolving credit, or make rolling forecasts of sales volumes.
- **Dynamics**—Design or allow an object or system to be flexible and adaptive and change to achieve optimal operations. For example, implement seasonal pricing, flexible pricing to account for quantity purchases, moving billboard, and a library van to bring books to the reader/neighborhood, or ice cream truck.
- **Partial** or **excessive actions**—If something is difficult to achieve, use slightly more or less of that same method to try to make it easier or simplified. For example, pricing at $1.99, rather than $2, compromising in a conflict resolution, and under-promising but over-delivering on a project completion date or customer service.

  Or conversely, **All or nothing**—"We pay for your meal if we forget to give you a receipt," or "We pay for your personal-pan pizza if it is delivered to your table late."
- **Another dimension** or **moving to a new dimension**—Take something to another plane, re-orient something, use another side of something. For example, use multiple market research sources (that is, passive, primary, and secondary), or customer and non-customer feedback. Use multiple sales distribution channels, use a multidimensional tool or analysis (that is, QFD or multiple regression analysis). (**See Also** "Regression Analysis," **p. 571** and "Quality Function Deployment (QFD)," **p. 543**)
- **Mechanical vibration**—Cause an object to oscillate or vibrate; increase its frequency. For example, video gaming controllers that shake when a shooting sequence happens. Also frequently communicating with customers using multiple modes such as phone, surface mail, and email.

- **Periodic action**—Use infrequent or pulsating rather than continuous actions. For example, instead of calling on a customer on a regular basis, stagger or call infrequently, or air a television commercial with just music or silence versus one with talking.

- **Continuity of useful action**—Carry out work continuously with all portions of the system fully operating or eliminate any idleness or intermittent activity. For example, leverage long-term business relationships or alliances, focus on customer retention or loyalty, and continue successful traditions.

- **Rushing through** or **skipping**—Perform harmful or hazardous operations at a very high speed, such as getting through a liquidation sale as quickly as possible.

  Or *conversely*, **Lagging**—Delay in action.

- **Convert harm into a benefit**, or "**blessing in disguise**"—Utilize harmful factors or environmental effects to obtain a positive effect or until it ceases to be harmful. For example, using customer complaints as opportunities for improvement or minimizing the fear of switching (IT) systems by introducing fear of competition.

  Or *conversely*, **Cursing in disguise**—Lack of customer complaints indicating lack of caring or candor.

- **Feedback**—Introduce cross-checking or refer back to improve a process, or if it already exits, reverse it (or change its magnitude or influence). For example, use Voice of the Customer input and customer complaints to help design a solution or including engineers with marketing to gather customer input.

  Or *conversely*, **Feed-forward**—Marketing forecasting to anticipate future needs.

- **Mediator** or **Intermediary**—Use an intermediary object or transfer or carry out an action; a temporary connection. For example, utilize a two-tiered sales distribution model using wholesalers or retailers or sales agents or using a third-party logistics company (that is, UPS or FedEx) or a consultant.

  Or *conversely*, **Intermediary removal**—Conduct an interview without a professional interviewer or conduct online Internet sales without a sales force.

- **Self-service**—Make an object service itself or make use of wasted materials or energy. For example, conduct self-benchmarking or encouraging customers to use word-of-mouth about your services/products.

- **Copying**—Use a simple and inexpensive replica instead of the original, which could be complex, expensive, fragile, unavailable, or inconvenient to operate. For example, utilize video-conferencing to replace a face-to-face meeting, franchise to replace full-fledged ownership.

  Or *conversely*, **Anti-copying**—Avoid negative associations in advertising or political campaigning.

- **Inexpensive short-lived objects**—Replacing expensive, durable objects by collecting inexpensive ones. For example, using inferential statistics or sampling techniques, using temporary staff or student interns rather than full-time employees, and selling second-hand goods cheaper than new items.

- **Mechanics substitution**—Replace a system by an optical, acoustical, or olfactory (odor) system; use an electrical, magnetic or electromagnetic field for interaction with an object, such as electronic communication, electronic banking, e-cards, faxing, or scanning.

- **Pneumatics and hydraulics**—Replace solid parts of an object with a liquid or gas, utilizing air or water for inflation, or use air or hydrostatic cushions. For marketing or business, it is a bit of a stretch, but an example could be introducing "breathing spaces" into contracts or sampling expansion during a survey.

- **Flexible shells and thin films**—Replace traditional constructions or isolate an object, using flexible membranes or thin film. For marketing or business, it is a bit of a stretch, but an example could be walking a fine line during a tough negotiation.

- **Porous materials**—Make an object porous or add porous elements (that is, inserts, covers); or if it is already porous, fill the pores in advance with something. For marketing or business, it is a bit of a stretch, but an example could be empowering the sales people to negotiate pricing or empowering customer-facing support people to resolve customer issues.

- **Color changes**—Change the color of something or its surroundings to improve visibility (or mask flaws). For example, change the company colors and logo to indicate a new brand image or implement different colors (as color coding) to assist with mistake-proofing. (**See Also** "Poka-Yoke," **p. 462**)

- **Homogeneity**—Make objects that interact with a primary object out of the same material or characteristics, such as a family of products.

- **Discarding and recovering**—After something has completed its function or become useless, reject, discard, dissolve, or evaporate it, such as a manufacturer's warranty. Or immediately restore, modify, or regenerate something if exhausted or depleted.

- **Parameter changes**—Transform the physical and chemical states of an object—its density, flexibility, or temperature (that is, liquid, gas, solid), such as virtual shopping or virtual banking.
- **Phase transitions**—Implement an effect developed during the phase transition, for instance, during the change of volume or lifecycle stages of an offering.
- **Thermal expansion**—Use a material that expands or contracts with heat. For marketing or business, it is a bit of a stretch, but expanding or contracting marketing efforts depending on the success of an offering's revenues or profitability.
- **Use of strong (boosted) interactions**—Replace common air with oxygen-enriched air or atmosphere. For marketing or business, it is a bit of a stretch, but hiring very creative marketing professionals or using an industry award to bolster a marketing (advertising) campaign.
- **Inert atmosphere**—Replace the Norman environment with an inert one; carry out the process in a vacuum by adding neutral parts or inert additives to a system. For example, utilizing an anonymous survey or interview, neutral, indifferent line of questioning, or using a neutral third party during a difficult negotiation.
- **Composite materials**—Replace or change a homogeneous material with a composite or multiple structures. For example, utilize a cross-functional representation on a process improvement team or use a "good cop/bad cop" combination on a negotiating team.

TRIZ Toolset

The TRIZ toolset shares many tools with the Six Sigma discipline. Some of the common tools and techniques include

- Brainstorming (**See Also** "Brainstorming Technique," **p. 168**)
- 5-Whys technique (**See Also** "5-Whys Technique," **p. 305**)
- FMEA (**See Also** "Failure Modes and Effects Analysis (FMEA)," **p. 287**)
- Pugh Concept Selection (**See Also** "Pugh Concept Evaluation," **p. 534**)
- QFD (**See Also** "Quality Function Deployment (QFD)," **p. 543**)

TRIZ also introduces a whole host of unique tools. In addition to the 40 inventive principles, some of the flagship TRIZ tools and techniques include: Contradiction matrix, Conflicts (both technical and physical),

Ideal Final Result (IFR), Trends of Evolution, Size Time Cost Operators, Anticipatory Failure Determination, Knowledge/Effects, Resources, Trimming/Functional Analysis, Measurement Problems, Subversion Analysis, Omega Life Views, Optimization methods, Re-focus/Re-frame, Paradigm Paralysis, ARIZ and Psychological Inertia (PI) tools. Note that this is a **partial list** of tools and techniques.

The TRIZ *Ideal Final Result* (IFR) is a powerful tool. It leads to a high-level solution that establishes the boundaries of a system at its extreme result of **ideality**. Typically it is the place one starts identifying the solution concepts. It focuses on both customer needs and functional requirements. The benefits have been fully delivered, costs have been reduced to zero, and harmful effects have been eliminated. It is independent of the currently used equipment or processes. The use of the IFR technique encourages out-of-the-box thinking to help the innovation team reach breakthrough solutions.

For example, consider cutting the grass. The **system** would be a lawn mower, the **product** is grass, and the **purpose** is to control the height of the grass so that the lawn looks nice. The IFR for this system might be that the grass grows to the desired height and stops growing; no lawn mower is needed!

The *Principle of Ideality* states that systems evolve in the direction of increased ideality. It is the sum of all the benefits divided by the total sum of its costs and harmful effects. Evolution increases ideality by increasing benefits and decreasing both costs and harmful effects or eliminating harmful effects all together. Real solution concepts move the system in the direction of ideality.

The TRIZ *Functional Analysis and Trimming* is a technique that first identifies the important tangible components of a system and how they interact with each other and the environment. Second, the trimming identifies candidates and removes a component from the system that presents problems. The overall technique assists in determining the "right" problem to work on, which typically is one that also moves the system in the direction of ideality.

The *Zone of Conflict* helps to distinguish what kind of conflict and identifies where and when a conflict arises. Several questions help to identify the conflict: Who has the problem? What is the problem? When does the problem occur? Where does the problem occur? Why does the problem occur? How does the problem occur? The 5-Whys technique works as well. Note that there are two types of conflicts—technical and physical. Technical contradictions use Inventive Principles, Standard Features, and Contradiction Matrices; while physical contradictions use Separation Principles. (**See Also** "5-Whys Technique," **p. 305**)

The *ARIZ technique* stands for an Algorithm for Inventive Problem Solving. It is a structured technique that evolves a complex problem into a simple problem to solve. It is a complicated multi-step process of questions that integrates different elements of TRIZ and reformulates the problem to get a fresh look at it.

Typically an improvement involves at least one of three approaches:

- Elimination of a harmful effect
- Modifying a useful effect
- Increasing ideality

The TRIZ standard solutions are used to eliminate or modify an effect. Altshuller identified 76 standard solutions. *Trends of Evolution* can bring a system to the next evolution step and help accelerate the move to ideality. The 76 standard solutions include

- Improving the system with no or little change (13)
- Improving the system by changing the system (23)
- System transitions (6)
- Detection and measurement (17)
- Strategies for simplification and improvement (17)

Conclusion

It takes time to appreciate and utilize all of the TRIZ concepts. The TRIZ toolset is plentiful, with many variations available for different circumstances. TRIZ is a flexible technique that can used by everyone—the technical community, business, marketing, sales, education, and in everyday living. Common tools applied to a large variety of scenarios include the Contradiction Matrix or Trends of Evolution.

Altshuller developed tools that could be used to break down technical systems and describe their underlying characteristics. For each generic problem set, Altshuller identified *solution techniques* that were used to solve these problems. This information forms the TRIZ toolset that initially was applied to technical innovation but has been expanded to apply to business and education.

TRIZ formulates the problem in an abstract (non-technology-specific) manner. It helps to identify the key technical and physical contradictions. It provides a detailed list of inventive principles to ensure that the inventor is able to find the solutions needed to solve a problem even if it is out of his or her field of expertise. TRIZ uses known general trends in technology development to predict system evolution.

The technique involves typically a small team of two to six people. After clearly defining the problem, 10-30 concepts are generated in a few hours. The concepts include strong candidates from previously run brainstorming sessions plus additional, more novel concepts. It leverages other best practice tools such as Quality Function Deployment (QFD), robust design, design for manufacturing, and the Pugh process. (**See Also** "Quality Function Deployment," (QFD), **p. 543**; "Design for Six Sigma," in Part I, **p. 45**, "The Anatomy of Quality Loss in a Product," in Part III, **p. PDF:763**, and "Lean for Six Sigma for Fast Track Commercialization," in Part III, **p. PDF:835** for a brief discussion on Taguchi's robust design; and "Pugh Concept Evaluation," **p. 534**)

Solution concepts can be identified at any point in the TRIZ process. Be prepared to record any concept solutions that are generated as soon as they get generated, to capture the idea and not loose it.

Everyone can use the TRIZ technique and receive some benefit from it. Continue to explore the TRIZ literature or consult with an expert to identify the best translation for your specific application or purpose.

How to Use the Tool or Technique

At a high-level, the TRIZ process involves the following four steps:

Step 1. Identify and analyze the problem. Identify the system being studied, its environment, resource requirements, primary function, problem effect, and ideal result.

a. Identify the important components of a system at the fundamental level.

b. Define the Ideal Final Result (IFR) using the correct TRIZ tool for solution concept generation.

i. Write a statement that describes the function the system performs on the product.

ii. Consider that this result must preserve the original function; eliminate the system deficiencies; does not make the system more complicated (using free or available resources); and does not introduce new disadvantages.

c. Determine how they interact with each other and the environment—functional analysis and trimming.

i. A product or process can be described by identifying the components and defining the functions that exist. (Components are tangible objects.)

A B C D E F G H I J K L M N O P Q R S T U V W X Y Z

ii. Functions describe the interaction between two components, such as a tool-action-object. In a function, an object is acted on using a tool, and the action typically involves a parameter change for the object.

Example: An auto transports a passenger; a chair supports a person; and an oven heats food.

d. Find the zones of conflict.

Step 2. Select the right problem to solve. Restate the problem in terms of contradictions to formulate it. Identify problems that could occur. Understand the interrelationships between fixing one characteristic and its downstream (or upstream) impact. Identify any conflicts or trade-offs, similar to the roof of a QFD. (**See Also** "Quality Function Deployment (QFD)," **p. 534**)

a. Propose a solution.

b. Identify the new problem created by the solution.

Step 3. Select the appropriate TRIZ principles and tool(s).

a. Identify the standard solutions; research if the problem or similar problem has been solved before.

b. State the standard conflict or contradiction to eliminate the conflicts; find the contradicting principle that might need to be changed.

c. Identify the principle representing any undesirable secondary effects.

d. Identify improving and worsening standard features and determine the best tools needed and apply.

e. Identify inventive principles to consider using in the contradiction matrix.

Step 4. Generate solution concepts.

a. Look for the analogous solutions and adapt to the solution.

Additional Resources or References

- www.triz-journal.com
- www.aitriz.org
- www.aia-consulting.com

V

Value Stream Analysis

What Question(s) Does the Tool or Technique Answer?

Where is value being added in the process? Conversely, where does waste exist in the process?

A Value Stream analysis helps you to

- Identify the value-add and non-value-add activities in a process and discover improvement and/or redesign opportunities
- Analyze where waste occurs in a process and develop a strategy to improve the process flow by eliminating the waste
- Make the processes work for you, instead of you working for your processes

Alternative Names and Variations

Variations on the tool include

- Process map (**See Also** "Process Map (or Flowchart)—7QC Tool," **p. 522**)
- Value Stream map

When Best to Use the Tool or Technique

When analyzing a current process, use the Value Stream Analysis to identify waste to eliminate any pockets of excellence (where the process flows smoothly and work is done correctly the first time, every time).

Brief Description

Value stream analysis is a critical Lean tool that examines a process and identifies improvement opportunities. This tool analyzes which activities contain waste and which truly create value, with the aim being to enhance those value-adds and eliminate the waste. The end-user, paying-customer defines what is or is not of value. This is one of the few times in Six Sigma that internal customers cannot define pure value-add activities, but they can determine what items fall into the non-value-add business category.

A *value stream* is defined as a set of activities required to deliver a product to the customer from raw material input. The total value stream may encompass multiple companies and organizations. It is a **big-picture** perspective required to avoid selective implementation resulting in isolated islands of Lean within a overall non-lean process. The Value Stream analysis output typically entails all or some of the following calculations:

- Total number of steps
- Total process cycle time
- Percent and number of value-add steps
- Percent value-add time
    - Percent and number non-value-add activities (often split into non-value-add business and non-value-add)
- Percent non-value-add time

As a rule of thumb, a typical process includes only 0–17% value add steps and 0–5% value-add time because it is defined by **what the end-user customer is wiling to buy**. Value stream analysis embraces two perspectives when examining a process—the people working in the process and the item that travels through the process as it is transformed to its final commercial-ready state. Figure V-1 illustrates the two-pronged perspective and displays a value stream of 35 total steps from the perspective of the tangible item and only five value-added activities (denoted by the white, bolded squares).

The tool examines each process step and groups the activity in one of three categories—Value Add, Non-Value Add Business, or Non-Value Add. *Value added* work is defined as the activity that physically changes the a product or adds important information. The activity must **not** be reworked and the customer must be willing to pay for it. *Non-Value Add Business (NVAB)* represents an activity required by the business but not the customer. These NVAB activities help the business operate efficiently, such as legal requirements, regulatory requirements, recording financials, and maintaining an intellectual capital management system. Examples of NVAB include completing an expense report, filing for a patent, and entering status and lessons learned information into the project database. *Non-Value Add* is pure waste, also called "muda" in Japanese.

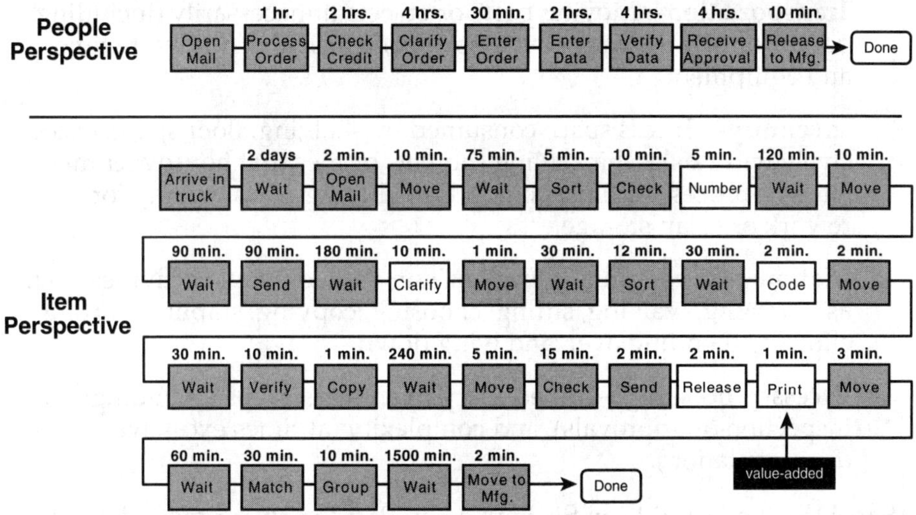

FIGURE V-1: Value Stream Perspectives

Waste Categories

Originally the Toyota version of Lean Manufacturing listed seven different categories of waste. Currently, given the western culture, an eighth waste category was added for under-utilized (or non-utilized) people (often classified as a management waste). An acronym to help with recalling the different wastes is DOWNTIME, which stands for:

- **Defects**—Repairs and rework needed to get something to function properly, which slows the process flow and reduces first-pass throughput yield.
- **Overproduction**—Producing parts ahead of schedule (before a customer requests or needs it), while other items in the process wait; working on unneeded parts (the wrong parts at the wrong time); evidenced by producing too much or too early of parts or finished goods.
- **Waiting**—People unnecessarily waiting, stalled in the workflow due to either shared equipment, unbalanced work activities (operations), decisions, approvals or inspections.
- **Non-utilized people** (a grammar stretch)—Under-utilized resources, lack of empowerment, or too many workers in the process causing inefficient operations.

- **Transportation**—Moving parts or objects unnecessarily (including papers or files); excess travel distance, which also can waste time and equipment.
- **Inventory**—Excess space consumed by shelving, floor space, excessively wide aisles, bins, filing cabinets or files that house accumulated in-process or finished goods, including parts waiting for rework or scrap storage.
- **Motion**—Excess, unnecessary, or non-valued people activities such as searching, walking, sitting, choosing, copying, stapling, sorting, climbing, bending over, and lying down.
- **Excess-processing**—Unnecessary operations, steps (including inspection or approvals), and complexity (at times excessive documentation).

(**See Also** "Lean and Lean Six Sigma," in Part I, **p. 29** for more detail.)

Value Stream Analysis Templates

The output of a Value Stream analysis can be displayed in multiple formats. Some of the more common templates include a matrix or a bar chart.

Summary Value Add Matrix

A summary Value Add matrix typically provides a high-level picture as to the number of process steps containing value and non-value activities and a categorization of what type of waste exists. Figure V-2 provides an illustration of a summary matrix that features three parts:

- Checkmarks to indicate the categorization (Value Add and the different types of waste)
- Total number of checkmarks per category
- Percent of the total for each checkmark

Detailed Value Add Matrix

The detailed Value Add matrix typically contains two parts—one for the current state and the other for the action plan to reach a lean state. Figure V-3 shows both parts as two separate sections for illustrative purposes; ideally, these two sections are part of a single matrix. If this template is created as a spreadsheet (such as Excel), it can easily sum the value-add and non-value-add times and calculate the analysis data.

Process Step	1	2	3	4	5	6	7	8	9	10	Total	%Total
Time (Hours)	12	10	1	10	20	6	10	1	10	20	100	100%
Value-Added			✔					✔			2	2%
Non-value-added												
Inventory (WIP, Backlog)									✔		10	10%
Delay (Operator)												
Duplication						✔					6	6%
Movement	✔				✔					✔	52	52%
Defects (Customer)		✔		✔			✔				30	30%
Rework (internal)												
Checking												
Lost Opportunities												
Total											100	100%

FIGURE V-2: Summary Value Add Matrix

Current State

	Process Step	Value Contribution (in Time)			Category of NonVA
		Value Add	Non VA Bus.	Non-VA	(Type of Waste)
1					
2					
3					
4					

Action Plan & Leaned Process

Improvement Action Taken	NEW Process Step	New Value Contribution (Time)			Comments
		Value Add	Non-VA Bus.	Non-VA	
	1				
	2				

Tally of Value Stream Analysis

Summary	Current State	Improvement
Total Steps	30	12
Total Cycle Time	25	16
# Value-Add Steps:	3	4
% Value-Add Steps:	10%	33%
Value-Add Time:	6	7
% Value-Add Time:	24%	44%
Total NVA Time:	19	9

FIGURE V-3: Detailed Value Add Matrix Template

A B C D E F G H I J K L M N O P Q R S T U V W X Y Z

Time-based Value Add Analysis

The time-based Value Add analysis uses a bar chart structure to depict a high-level view of the time allocation between Value Add and Non-Value Add activities. This is a good communication tool to focus the team and management on leaning the process to be more efficient. It provides the current state snapshot, the targeted lean process, and the generally unattainable process containing 100% value add activities. Figure V-4 shows an example of a time-base Value Add analysis for receiving a customer order.

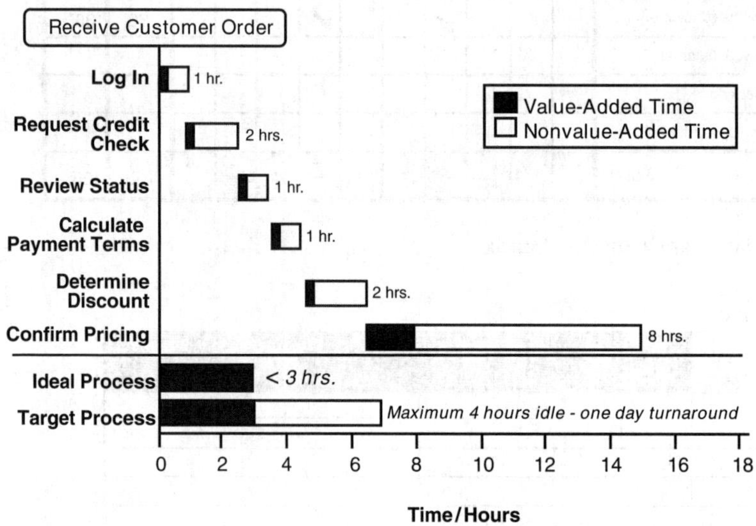

FIGURE V-4: Time-Based Value Add Analysis for a Customer Order

How to Use the Tool or Technique

A Value Stream analysis can be conducted by an individual or with a team, using the following procedure.

Step 1. Create or collect a detailed Process map of the process of interest.

a. Modify or ensure that the diagram maps both what happens to the item and what the process players do.

b. Calculate and record the cycle time (average and range—minimum and maximum) for each step.

(**See Also** "Process Map (or Flowchart)—7QC Tool," **p. 522**)

Step 2. Develop a Value Add template.

a. Select which type of matrix or bar chart best suits the analysis (reference Figures V-2, V-3 and V-4).

b. Record the process steps in the Value Add document. Consider numbering the steps if the process contains more than ten.

Step 3. Categorize the activities.

a. Examining each step one at a time, decide if the activity is necessary to meet the end-user customer requirements and produce the final output.

b. Classify necessary activities as value add and unnecessary activities as non-value add.

c. Document the value add activities on the Value Add matrix or bar chart.

d. Analyzing each of the non-value add activities, separate them into those that meet business requirements from those representing pure waste. Document them in the respective columns on the Value Stream matrix.

Step 4. Identify the ideal flow.

a. Analyze each step that contains waste, prioritize the waste, and develop an action plan to eliminate it. Potential strategies include

i. Eliminate loop-backs or rework

ii. Reduce batch size

iii. Re-sequence steps

iv. Standardize work

v. Eliminate work in process (WIP) and inventory

vi. Balance workloads

vii. Segregate work by complexity, re-sequence, and rebalance

viii. Establish visual workplace

b. Record the action plan on the Value Add document.

c. Define the improved, lean process after the implementation of the action plan.

Step 5. Execute the action plan and apply the appropriate tools to achieve the ideal flow.

Hints and Tips

- Value Stream analysis is a good starting point to work at the "door-to-door" level within your organization. This keeps the project at manageable level of complexity but retains a sufficient "big-picture" perspective.
- Waste is really a symptom rather than a root cause of the problem. Waste points to problems within the system at both the process and value stream levels.
- Working faster is **not** the answer; working smarter and adding value is.
- Work is to be done right the first time, every time; hence, activities such as inspection and rework are non-value adds.

Supporting or Linked Tools

Supporting tools that might provide input when developing a Value Stream analysis include

- Brainstorming (**See Also** "Brainstorming Technique," **p. 168**)
- Process Map (**See Also** "Process Map (or Flowchart)—7QC Tool," **p. 522**)
- Standard Operating Procedure (SOP)

A completed Value Stream analysis provides input to tools such as

- Standard Operating Procedure (SOP)
- Cause-and-Effect diagram (**See Also** "Cause-and-Effect Diagram—7QC Tool," **p. 173**)

- Control plan (**See Also** "Matrix Diagrams—7M Tool," **p. 399**, for a brief discussion on control plans.)
- FMEA (**See Also** "Failure Modes and Effects Analysis (FMEA)," **p. 287**)
- Training Plan (**See Also** "Matrix Diagrams—7M Tool," **p. 399**, for a brief discussion on training plans.)

Figure V-5 illustrates the link between a Value Stream analysis and its related tools and techniques.

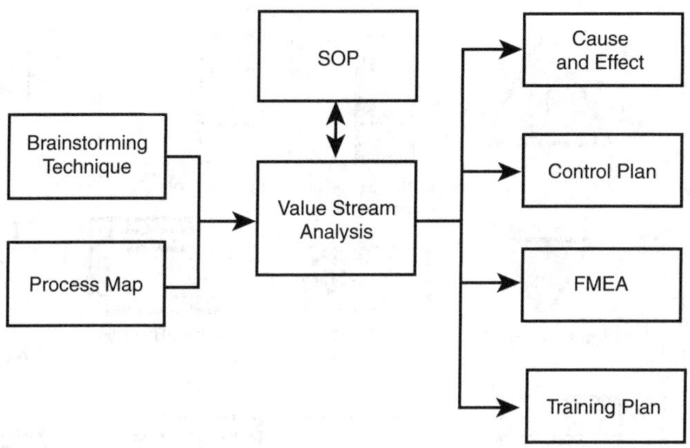

FIGURE V-5: Value Stream Analysis Tool Linkage

Variations

Value Stream Mapping (VSM) is a tool used understand and **visualize** the overall flow of material and information as a product or service makes its way through an end-to-end process. It focuses on the big picture in order to improve the whole system, not just sub-optimizing individual processes (holistic). It identifies sources of waste as opposed to just identifying waste, similar to the Value Stream analysis, but depicts it in a flowchart-like format. VSM provides a common language to describe processes and forms the basis of an implementation plan by helping you design how the whole door-to-door flow should operate. As part of VSM, process areas should assess their effective use of direct labor and overhead.

Similar to Process mapping, the VSM technique usually includes diagramming both the current state and improved state. It typically records the metrics including cycle times, down times, in-process inventory, number of workers, working time (less breaks), work-in-process (WIP), and scrap rate to mention a few. The Value Stream map format displays the process flow a bit differently than a traditional Process map. It tends to include more icons to highlight the potential areas of waste. Figure V-6 provides some of the common VSM icons, and Figure V-7 illustrates a simple Value Stream map.

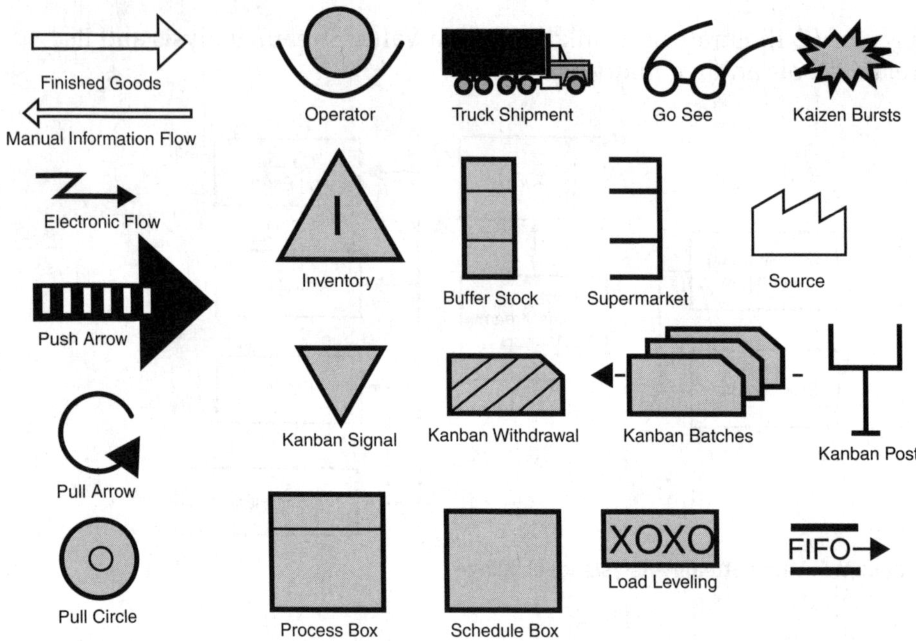

FIGURE V-6: Common Value Stream Map Icons

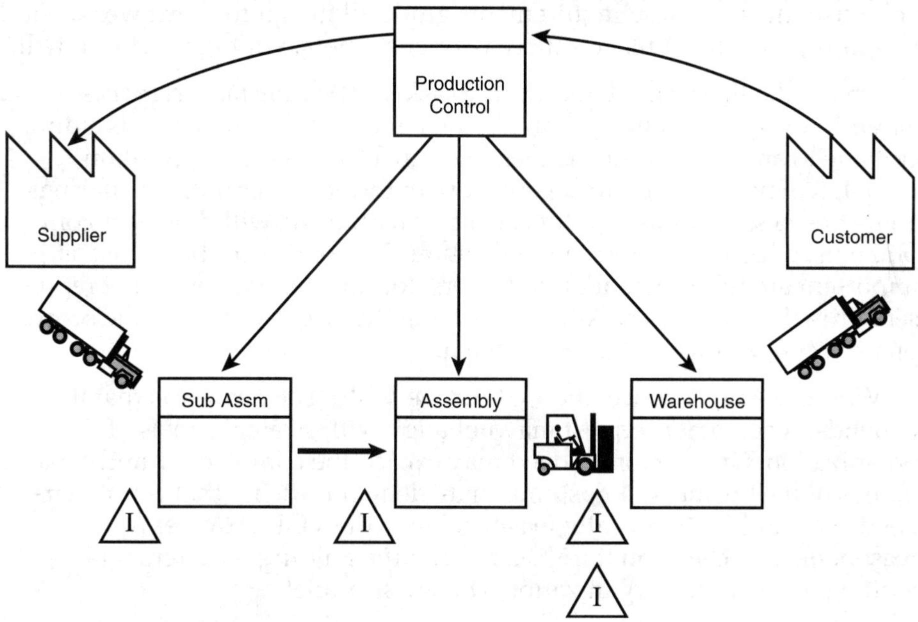

FIGURE V-7: Simple Value Stream Map

Voice of Customer Gathering Techniques

What Question(s) Does the Tool or Technique Answer?

What is the best way to capture customer requirements?

VOC gathering helps you to

- Understand what is important to customers and non-customers

When Best to Use the Tool or Technique

Before developing a strategy or action plan or before starting any work, first understand what the customer requirements are.

Brief Description

The Voice of the Customer (VOC) gathering techniques represent a suite of different approaches to capture their requirements. It also applies to gathering *Voice of the Business* (VOB) requirements. This technique aims to collect what is important to both the internal and external customers and non-customers, as these requirements guide all the go-forward work. For if your organization fails to satisfy your customers, your competition will.

One of the first critical questions to ask is, "*Who are the specific customers?*" Why that seems like an obvious question, it sometimes is a difficult one to answer. If a business engages in a two-tiered distribution model, it may not have direct contact with the end-user of their offerings. Some businesses only keep direct contact with those with issues or compliments, which may represent only about 20% of its user base. Just as important are those potential customers, for it is the prospects that represent growth opportunity. After identifying the target audience, the second phase focuses on what matters to them.

Who takes precedence, the VOC or the VOB? The answer is that it depends. A customer request may not align with strategic goals of the organization. Or a customer need may exceed the competency and capabilities of the business. A customer may demand pricing that is not market driven and/or financially feasible. Thus, the VOB provides a reasonableness check on the VOC to keep the guiding requirements within context of the organization's business model.

The VOC and VOB may represent contrasting views of a similar topic. For example, a customer may define quality according to an items' fitness for use, while the VOB may focus on the conformance to a specification. The end user may describe cost as the purchase price, operating costs, maintenance costs, downtime, and depreciation—the total cost of ownership (TCO). Alternatively, the VOB views cost through production/manufacturing lenses, taking into account the cost of raw materials, labor, overhead and operating costs. Responsibility for service from a customer's perspective lasts the entire useful life of the product, whereas the VOB may focus primarily on the warranty period. Spare parts are viewed as a necessary evil in a customer's eyes, while the VOB sees a profitable business opportunity.

Gathering Techniques

There are a range of gathering techniques that represent a varying degree of cost, reliability, and detail. Key considerations when selecting the optimal VOC gathering technique includes the available budget for the initiative, the required level of granularity and opportunity for follow-up, and the customers' profile and demographics.

With respect to the budget required to gather input, it needs to cover three dimensions—the costs associated with developing the approach, collecting and analyzing the data. The development cost may require the development of a survey tool or interview guide that could range in level of specificity and complexity. Consider whether the scenario requires customers to experience a prototype of a new offering, or if the required data already existed in a database, development work could be minimal.

Collecting the needed information ties to the level of specificity. Determine if generic input suffices, or if "what if" scenarios best capture the depth of thinking required. *Passive data collection* provides the lowest level of reliability and may fail to represent the entire cross-section of the customer base. Passive data typically is stored in a variety of functional databases including customer service and support, customer administration and financing, and customer relationship management (CRM)/sales. Other sources of useful data may be buried in other the company records including customer letters, complaints, and customer phone logs.

Market research falls into two categories—secondary and primary. *Secondary customer research* sources data from customer surrogates such as industry experts, partners, competitors, benchmarks, and the customer-facing value chain resources. This gathering technique can be passive or active, using either trade journals, reviews and editorials, or direct conversations with the opinion leaders. The data collected is more reliable than passive techniques but is less than primary customer research. Similar to the game of telephone, which tests the accuracy of a whispered message passed from one person to the next as compared with the original phrase, secondary research operates on similar principles. The more removed from the desired source of information, the less accurate the message will be. Sometimes talking directly with customers presents obstacles (expense and accessibility), but little else can substitute for it.

Primary customer research goes directly to the source. This approach covers existing customers, lost customers, and non-customers. Loyal customers kindly take the time to give you feedback, both positive and negative. Retaining current customers is crucial, given that it costs more to win a new customer than it does to keep those active customers. However, lost customers can be a gold-mine for improvement opportunities ideas; the challenges is to secure some of their time. Primary research techniques include face-to-face interviews or focus groups, observation, and surveys conducted in different media.

Typical market research spans many objectives, including a planning guide to determine the basic economic trends and how they may impact the offering. It seeks purchase patterns and key influencers that drive change—change in real and disposable income, in consumers' tastes and values, and in distribution patterns. Market research tries to identify new emerging market segments and other potential opportunities matching current offerings, as well as unfulfilled new, unique, and difficult (NUD) requirements. It also may attempt to solve current offerings' issues such as sluggish sales volume, weighing the price-to-value ratio relative to competitive offerings. (**See Also** the "NUD versus ECO," **p. 376,** section under "KJ Analysis," **p. 375**)

A B C D E F G H I J K L M N O P Q R S T U V W X Y Z

Note

Think Before Doing—Sampling requires planning and forethought. Take care to envelope a representative mix of the customer-base. Take time to understand the profile and demographics of a market to identify the various segments. The respondent mix should reflect that of the market and cover the target audience identified in the project charter or SIPOC. (**See Also** “Sampling,” **p. 618**; “SIPOC, (Supplier-Input-Process-Output-Customer)” **p. 663**; and “SMART Problem and Goal Statements for a Project Charter,” **p. 665**)

Trade-offs

Each gathering technique poses a set of pros and cons. Existing data such as industry studies or government publications (census data, for example) have the advantage of being relatively inexpensive and easy to find. In contrast, their disadvantages may be that they lack sufficient enough specificity or detail to satisfy the research objectives. Surveys (in various forms) are used to collect original data. They have the advantage of being designed to directly address the research objectives. Their disadvantages are not only do they cost more, but also they are more time consuming to design, collect, and analyze data.

Evaluate each during the planning process to make the appropriate trade-offs. Figure V-8 provides a quick summary of the major VOC gathering techniques and respective strengths and weaknesses.

Increasing data detail & effort required ↑ … ↓ Decreasing data detail & effort required

Method	+	-	Considerations
Focus groups	Interaction among participants	Influential participants may sway other's opinions	Skilled facilitator required
Interview	Direct & immediate feedback	Time consuming	Need to align schedules
Phone survey	Direct feedback	Selective response rate (customer can always hang up)	Bias inherent in data from selective response
Web survey	Ease of use for customer	If voluntary need to limit number of questions	Need to focus number and quantity of questions
Customer service data	Typically easily queried from a database	If voluntary may not capture all customer opinions	Data integrity

FIGURE V-8: VOC Selection Matrix

The active Focus Groups represent the most ergs of effort and gather the detail, while at the other extreme passive gathering technique of mining a customer service database is relatively easy to execute but yields less detail.

The budget and time required to develop and collect a VOC/VOB plan tend to enjoy much focus and scrutiny. Often the planning, time, resources, and budget to properly analyze the collected data is often overlooked or rushed in the process. However, this final step is when all the benefits are reaped. Taking the time upfront to plan this phase helps keep the end in mind and may inform how to design an interview guide and develop the overall implementation plan. A rule of thumb is that the time it takes to collect the data, usually requires double the amount of time to analyze the verbatim data.

Interview Guide

The Interview Guide should be built to aid in collecting input from either secondary or primary research. A data collection aid for passive sources includes a data collection sheet. An interview guide can take on many forms—a tool to guide someone through a one-on-one or phone interview or a questionnaire deployed electronically or hard copy. (**See Also** "Data Collection Matrix," **p. 248**)

Arriving at the optimal number of questions entails a mix of both art and science. The tool should comprise enough questions to reveal the answers to a few critical questions, perhaps allowing for multiple ways of posing the question to validate understanding and ensure consistency of related answers. In addition, judicious consideration of the respondents' time should inform as to the appropriate number of questions.

The interview guide is a formal set of questions, statements, and stimulus materials designed to elicit responses that will accomplish the objectives of the research project. The questionnaire responses measure individual attitudes, behavior, and feelings toward topic(s). It provides consistent framework for a collection of responses to support analysis and comparison of quantitative and qualitative data. Specific research objectives and questionnaire design should be driven from the business objectives and the marketing questions that fulfill them.

The guide can pose two basic types of questions—open-ended and closed.

Open-ended Questions

Open-ended questions accept any type of responses provided by the respondent. It allows them to answer in their own words. Within this category, free responses can be selected. Their advantages include receiving unanticipated responses. The input represents the real views of the respondent. This approach is useful when the list of alternatives under consideration is too long to present each option to the respondents. Their disadvantages include the potential difficulty interpreting what the respondents meant or intended by their answers. It can create interviewer/recorder bias and is less suited for self-administered questionnaires. More weight may be given to more articulate respondents. It can be costly to tabulate, code, and analyze responses.

Probing questions are another form of open-ended questions. Their advantages include the fact that this form elicits additional information. They often obtain more complete responses. However, their disadvantages are that they are expensive and can be difficult to interpret what the respondent meant. These questions may create interviewer/recorder bias. They are less suited for self administered questionnaires and may cause more weight to be applied to more articulate and verbose respondents.

Three broad categories describe this type of questioning—association, construction, and completion techniques. *Association techniques* present respondents with a series of words, sounds, images, or other stimuli and are asked to respond to each with the first word that comes to mind. *Construction techniques* ask respondents to view a stimulus and create a story or draw a picture to explain the stimulus. *Completion techniques* provide respondents with an incomplete sentence and asks them to complete it in any manner they choose.

At times a *projective-technique* is used, wherein the respondent projects into the future. This technique primarily focuses more on potential solutions, rather than simply requirements gathering. The advantages of this approach again is that it solicits more informative, probative information than closed-ended questions. It provides useful information in exploratory stages of a research process in which insights, ideas, and hypotheses are sought. The disadvantages of projecting is that it requires highly trained interviewers and analysts to interpret the data. Typically it involves a small sample size, which often cannot be projected onto the entire population.

Closed-ended Questions

Closed–ended questions limit the responses to those provided to the respondent. The respondent makes a selection among a set of options. Technically, the term "closed-end" characterizes both the questions and the responses for this technique. There are seven main subtypes for this approach, which depend on the type or responses collected. They are

- **Dichotomous**—Bipolar and unipolar.
- **Ranking**—Involving critical, ratio, and interval.
- **Checklist**—Both categorical and hierarchical.

- **Multiple choice**—Respondent chooses one response from a list of options.
- **Multiple response**—Respondent may give more than one response from a list of options.
- **Numerical.**
- **Rating Scales** (Likert or Semantic Differential).

Dichotomous questions allow for only two possible answers, such as, "*Do you own or lease your car?*" The most common variety are yes/no responses. Thus the name, dichotomous, meaning opposite pairs of descriptive words or phrases representing the two extremes. They are easy to ask and easy to answer, but they do not provide much information. Use this type of questioning format to collect classification data about the respondent and consider following it with an open-ended question to obtain explanatory comments if necessary. The advantages are that they can be good lead-in to questions that require more detail and that they are quick and easy to administer and answer. There is less chance of interviewer bias, and the responses are easy to analyze. The disadvantages are that they provide no detailed information and they can be difficult to word to obtain appropriate responses.

Ranking questions compare within a list of items, typically in order of preference or importance. The order of items relative to one another is the key, not the differences between values. An example is, "Please rank the following reasons from least important (1) to most important (7) to you." There are two kind of ranking systems—interval and ratio. *Interval ranking* uses a constant scale that lacks a natural zero value. The differences make sense, but ratios do not (that is, 40–30 and 30–20 decibels, but 40/20 decibels is not twice as loud. This covers topics such as loudness, brightness, saltiness, and temperature. *Ratio rankings* are ordered on a constant scale with a natural zero, for example height, weight, age, and length. The advantages are that they yield information quickly, and they are easy to ask, tabulate, and analyze. It usually is a familiar activity for the respondent, or it is easy to explain to them. The disadvantages include the number of responses should be limited to five or seven, and it assumes the respondent has information about all items in the question.

The question format using *checklists* asks the respondent to check one (or more) response categories. This approach is useful for factual answers (that is, demographics). Its advantages cover easy to ask and answer, easy to tabulate and analyze, and provides specific answers for the respondents. The disadvantages are that it assumes the researcher knows all relevant alternative responses and often requires a long list that can be boring for the respondent.

A B C D E F G H I J K L M N O P Q R S T U V W X Y Z

The *multiple choice* question format lists a number of answers and requires respondents to select the one answer that best approximates their own. The advantages include the fact that it overcomes many of the problems of open-ended questions, assures respondents will answer on the same dimension, and is less demanding than open-ended to administer. Moreover, this format is less expensive to administer and process than open-ended and is easy to edit, tabulate, and analyze. However, the disadvantages are that the responses must be mutually exclusive and all inclusive, and the order of responses can create response bias.

The *multiple response* format is similar to the multiple choice, except the respondent is allowed to make no choice or more than one choice (instead of just one). Its advantages are that it overcomes limitations of open-ended and multiple choice questions and is useful when research seeks responses; support induction data to theories. It is less demanding and expensive than open-ended to administer and process, and the analysis of structure in response clusters can offer insights. The disadvantages are that the responses must be inclusive of all possible responses, the order of responses can create response bias, and it requires advance multivariate analysis to yield insight into new theories and models.

When using *rating scales,* the respondents are given a range of categories, a continuum, or scale in which to express their opinions. There are a wide variety of scaling techniques. The most common are comparative, non-comparative, uni-polar, and bipolar.

Comparative scales involve the direct comparison of stimulus objects. The comparative scale data must be interpreted in relative terms and employ paired comparison, rank order, or constant sum scales. *Paired Comparison* is a technique wherein the respondents are asked to choose which item rates higher, according to a predetermined criterion. The *Rank Ordered* technique asks the respondents to rank products according to some predetermined criterion. The *Constant Sum* technique request the respondents to allocate a predetermined number of rating points among several items, according to some criterion. The *Q-Sort* technique involves the respondents ranking a group of items into sets according to some criterion, and it discriminates among a large group of items in a relatively short time. The advantages of these comparative techniques include that they are easily understood and applied, only small differences can be detected between stimulus objects, and they use the same known reference points for all respondents. These techniques tend to reduce bias introduced by one judgment on another. Conversely, the disadvantage is their inability to generalize beyond the stimulus objects scaled.

Non-comparative scales place each object in its own independent scale from the others in the stimulus set. The respondents evaluate only one object at a time, and the data are generally assumed to be interval or ratio-scaled. The non-comparative scale categories are 1) continuous rating scales, 2) graphic rating scale, and an 3) itemized rating scales, using a finite number of choices with numbers or descriptors for each choice.

The *Likert rating scale* is an ordinal measurement scale. It is named for Rensis Likert, who invented the scale in 1932, using a series of statements followed by five response alternatives reflecting a person's attitude—typically *strongly agree, agree, no opinion, disagree* or *strongly disagree.* Each category encompasses five equally weighted, bi-directional choices that represent both positive and negative options. The positive responses yield a higher score than the negative ones. A similar, but more specific scale is the *Thurstone* scale, which uses equal-appearing intervals and so provides interval level data and thus more powerful statistical procedures, such as 7, 9, 11, or 13 intervals reflecting unfavourable, neutral, and favourable (respectively) to determine where the respondent's attitude falls. Another option is the *Semantic Differential* scale using several criteria each with its own a bipolar scale to capture the respondent's attitude toward something, usually used to evaluate the strengths and weaknesses of a product or company. The *bipolar scale* involves opposite adjectives such as excellent-poor, good-bad, complex-simple, and typically have a numeric scale within these two extremes (-3, -2, -1, 0, 1, 2, 3). The *Stapel Scale* is a uni-polar similar to Semantic Differential Scale, except only one word or phrase is used instead of using two dichotomous descriptive words or phrases as choices, and each choice is numbered.

There are three types of scales—ordinal, interval and ratio. The *ordinal* uses numbers to indicate the response relative to some characteristic. The scale establishes relative magnitude (if a response has more or less of a characteristic), but not how much more or less. In addition to the counting operation allowable for nominal scale data, ordinal scales permit the use of statistics based on centiles, (that is, percentile, quartile, median). The *interval* scale employs numerically equal distances on the scale represent equal values in the characteristic being measured. It permits comparison of the differences between objects. The location of the zero point is not fixed, and both the zero point and the units of measurement are arbitrary. Any positive linear transformation of the form $y = a + bx$ will preserve the properties of the scale. Statistical techniques that may be used include all of those that can be applied to nominal and ordinal data and in addition the arithmetic mean, standard deviation, and other statistics commonly used in marketing research. The ***ratio*** scale possesses all the properties of the nominal, ordinal, and interval scales. It has an absolute zero point. It is meaningful to compute ratios of scale values. Only proportionate transformations of the form $y = bx$, where b is a positive constant, are allowed. All statistical techniques can be applied to ratio data. (**See Also** "Statistical Tools," **p. 684**)

The advantages of using a rating scale are that they can measure intensity of feelings toward issues, and the numerical values can be assigned to each point on the continuum, and statistical routines can be run. They are easy and efficient to ask, answer, and tabulate. The disadvantages are that the scale intervals may not provide clear distinction for respondents. The respondents may have difficulty interpreting questions, or may both agree and disagree depending on their interpretation, thereby loosing interest or getting frustrated by the question.

Format Considerations

The questionnaire design involves several decisions. Itemized scales can be balanced or unbalanced. In general, the scale should be balanced and bipolar to obtain objective data and minimize compliance bias. Although there is no single, optimal number of scale categories, the rule of thumb is to employ between five and seven (sometimes as many as nine) categories. Using and odd versus even number of categories needs to be decided. If a neutral or indifferent response is possible from at least some of the respondents, an odd number of categories should be used. In situations where the respondents can be expected not to have an opinion, the accuracy of the data may be improved by a non-forced scale and/or the inclusion of a "don't know/not applicable" response. Place the description of each scale response category near the response categories.

The wording of questions is critical. The questionnaire is the interface between researcher and respondent. There are no hard and fast rules in determining question wording, but the guidelines include **simple** language, **familiar** vocabulary, the **shorter** the questions the better, and the more **specific** and precise the wording the better. For example, *What kind of car do you own?* versus *What brand and model car do you own?* Newspapers aim to write to a fourth-grade level of reader to avoid misunderstanding and misinterpretation, so should a questionnaire. If heeded, the respondent can process the intent of the question faster, with less ambiguity, therein creating a less disruptive, more pleasant experience.

The strength of using closed-ended questions is that they are easy to complete for the respondents, simple to code and analyze, have reduced interpreter bias, are free of grading bias, and are more valid and reliable responses. Conversely, the development of such a tool should take time and effort to properly design a closed-ended questionnaire. Researchers need to create carefully worded questions and exhaustive lists of possible responses to ensure that the responses are representative. Care should be given that the questions are not leading, that the respondents answer thoughtfully, not mechanically, and that the number of questions are kept to the necessary minimum. The number of questions must be managed to the necessary few, and it takes time to word them succinctly and capture the exact essence of the question's intent.

> **Note**
>
> **Attribute Measurement System Analysis**—Prior to executing the questionnaire, ensure that appropriate rigor is present in conducting a measurement system analysis. Validate that the purpose of each question is clearly defined, but also that the wording and phrasing is understood by third parties (surrogate or sample respondent set) as intended. Moreover, ensure that each question's response is actionable—have the end in mind. Heed caution if asking questions that probe into uncontrollable items, as they may set improper expectations and clutter the field of actionable responses. Whatever time you can spend with prospective customers (whether blind or direct interaction) is precious, so use the time wisely. (**See Also** "Measurement System Analysis (MSA)," **p. 412**)

How to Use the Tool or Technique

VOC Data Collection Process

The collecting of VOC and VOB requirements generally involves a team of people. The following guidelines cover a six-step process:

Step 1. Identify the target audience and determine the objectives—what information is required?

Step 2. Develop a data collection plan.

a. Consider starting with a passive approach first, collect data from existing information, and then fill the information gaps afterward using either secondary or primary research.

b. Identify the various sources of data and best methods to collect it.

c. Define the profile and demographics of the target audience.

d. Document the data gather plan (who, what, when, where, why, how).

e. Develop a budget and timeline that incorporates developing the collection tools (if necessary), gathering, and analyzing the data.

Step 3. Develop the data gathering tool(s).

a. Develop the appropriate collection tool depending on the approach—passive, secondary, or primary research.

b. Refer to the following Interview Guide procedure, if appropriate.

Step 4. Execute the data collection plan.

Step 5. Analyze the data to generate a key list of customer requirements **using their language.**

a. Consolidate and organize findings into a useable format.

b. Consider using a KJ Analysis technique and an Affinity diagram if appropriate. (**See Also** "KJ Analysis," **p. 375** and "Affinity Diagram—7M Tool," **p. 436**)

Step 6. Translate the customer language into the actionable, measurable critical-to-quality (CTQ) items for the organization. (**See Also** "Critical to Quality (CTQ), **p. 242**)

a. Develop the appropriate level of detail, such that the necessary number of CTQs per requirement are developed and delegated to the accountable process players.

b. Set specifications (or metrics) for CTQs.

Interview Guide Development Process

An interview guide is the primary data collection vehicle for primary and some forms of secondary research. It can be built individually or with a team. However, ensure that the appropriate subject matter experts are consulted if they're not an active part of the development process.

The procedure to develop an interview guide is as follows:

Step 1. Build the analysis plan.

a. Refine research objectives within the context of the larger VOC Data Gathering Plan Review business and research objectives and determine the specific data needs.

b. Brainstorm measures to accomplish objectives, to avoid saying at the end of the research initiative, "*If only we had asked…*"

c. Reduce the list to its essential items.

d. Develop a draft analysis plan to answer the question, "*What will be done with the data once it is collected?*"

Step 2. Determine the primary research process.

a. Decide on the best research for the specific business situation. Consider the following options:

   i. Face-to-face interviews (Monad or Dyad)

   ii. Face-to-face Focus Groups

   iii. Direct or Indirect Observation Studies

   iv. Web-based Focus Groups and Bulletin Board Discussions

   v. Web-based Surveys

   vi. E-mail Surveys

   vii. Telephone Interviews

   viii. Telephone Surveys

   ix. Mail Surveys

   x. Delphi Studies

   xi. Omnibus Polls

   xii. Lab or Field Study

Step 3. Evaluate question content.

a. Are they aligned with business and research objectives?

b. Evaluate potential stimulus materials and research questions to determine whether or not they will elicit meaningful response data.

c. Consider the following criteria:

   i. Does the respondent understand the question?

   ii. Does the respondent have necessary information to respond?

   iii. Will respondent provide the right information?

   iv. Will stimulus or questions bias the responses?

Step 4. Determine the question format and structure.

a. Decide to use open-ended and/or closed questions.

b. Determine the structure—format and sequence to facilitate the interviewer or the respondent (if self-administered) through the various topics. Consider the following structure sequence:

i. **Introduction/Purpose of Survey**

ii. **Opening/Lead-in Question**—Should be easy to answer, not too personal, and may not contribute any useful data.

iii. **Qualifying Questions**—To ensure that the respondent qualifies as part of the target audience.

iv. **Warm-up Questions**—To break the ice and get the respondent comfortable.

v. **Specific Questions**—To capture the main objective of the research.

vii. **Demographics**—The data to categorize the respondent into groups or segments, which may be used later when applying stratification to the data to reveal certain patterns. (**See Also** "Stratification," **p. 697**)

c. Design the layout. The appearance should be clear and uncluttered. If using open-ended questions, leave enough space for recording answers. If using "skip patterns" for potentially inapplicable questions for some respondents, the instructions should be crystal clear.

Step 5. Pre-test and revise questions to arrive at the final draft.

a. Pre-test or pilot the questions to ensure that what is intended to be communicated is what is understood by the respondent. Test both the question wording and the analysis technique to ensure the respondent's answers are actionable.

Step 6. Execute.

Hints and Tips

- Consciously try to use precise, unambiguous language in an interview guide.
- When developing an interview guide, consider several cautions in crafting the actual wording:
    - Avoid double-barreled (two) questions, such as, "*Do you consider Bran cereal sweet and tasty?*"
    - Avoid leading questions, for example, "*Don't you agree that...?*" or "*Wouldn't you say that...?*"
    - Avoid loaded questions including, "*How do you like the performance of the car?*" with the option of selecting *Extremely well (), Very Well (), Pretty well (), Not too Well ().*
    - Avoid words with an emotional tone, such as *oil monopolies* and *luxury items.*
    - Avoid using labels for people (that is, right- or left-wing, conservative, liberal). They often are ambiguous (different meanings for different people).
    - Avoid questions that assume knowledge or rely on memory.
    - Avoid asking too many positive or negative questions.

Supporting or Linked Tools

Supporting tools that might provide input when gathering VOC include

- Benchmarking (**See Also** "Benchmarking," **p. 160**)
- Written reports
- Brainstorming (**See Also** "Brainstorming Technique," **p. 168**)

A completed VOC data provides input to tools such as

- Brainstorming (**See Also** "Brainstorming Technique," **p. 168**)
- Affinity diagrams (**See Also** "Affinity Diagram—7M Tool," **p. 136**)
- KJ Analysis (**See Also** "KJ Analysis," **p. 375**)
- CTQs (**See Also** "Critical To Quality (CTQ)," **p. 242**)

- Cause-and-Effect diagram (**See Also** "Cause-and-Effect Diagram," **p. 173**)
- QFD (**See Also** "Quality Function Deployment (QFD)," **p. 543**)
- Solution Generation (**See Also** "Brainstorming Technique," **p. 168**)

Figure V-9 illustrates the link between VOC data and its related tools and techniques.

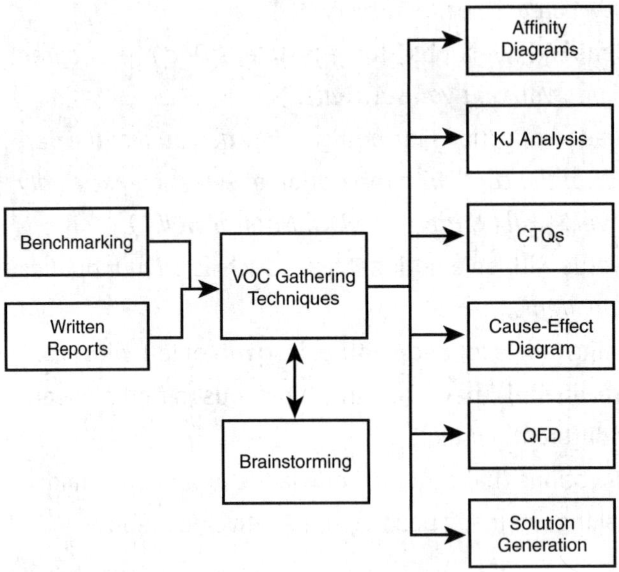

FIGURE V-9: VOC Gathering Techniques Tool Linkage

W

Work Breakdown Structure (WBS)

What Question(s) Does the Tool or Technique Answer?

What are the project deliverables, and what activities will produce them?

A WBS helps you to

- Communicate the project deliverables and define the respective activities
- Ensure the deliverables' subcomponents are identified as necessary and sufficient for project completion
- Provide a baseline by which to measure project performance and control
- Begin to estimate required time for every activity by understanding the work required

When Best to Use the Tool or Technique

At the beginning of the project, during its planning phase, use a Work Breakdown Structure to illustrate the detail project scope. It takes the milestones and deliverables from the project charter and begins to break them down and organize them into their associated activities.

Brief Description

The Work Breakdown Structure is an important project management tool that describes the project deliverables and their descriptions and is structured as a hierarchical Tree diagram. It defines the total scope of a project, its milestones, deliverables, and activities, wherein nouns depict the milestones and deliverables, and verb-noun combination define the activities and/or tasks. The tool further develops the deliverables from the project charter, maps them to the key milestones, and defines the activities and tasks needed to produce those project outputs. Hence, the tool helps to visualize the project deliverable-task linkage, and can be expanded to include assigning of the individual accountable to produce that deliverable. (**See Also** "Tree Diagram," **p. 712**)

A WBS can identify a project's key *milestones,* the critical reference points in a project that delineate a major event. Milestones are used to monitor the project's progress. A milestone has zero duration on a project schedule and represents zero effort or work. However, preparing for a milestone can involve significant work, as they serve as a checkpoint to help to manage and control a project. For example, if a project improvement team were following the Lean Six Sigma DMAIC methodology (Define-Measure-Analyze-Improve-Control), each of those five steps or phase-gates represent a milestone.

The basic building blocks of a WBS are referred to as *work packages.* These work packages may be sub-projects and contain one or more tasks. They list the most important tasks in the project, as tasks require time and resources (people, equipment, facilities, materials, and sometimes project funding). A WBS can help to ascertain which tasks depend on the completion of others, and find the relationships between tasks. A WBS also can identify points of completion that can be seen as milestones. The tool also can define the duration for each activity. However, the WBS does **not** show sequence of these activities and does **not** show dependencies between activities. (**See Also** "Activity Network Diagram (AND)—7M Tool," **p. 127**, "Critical Path Method (CPM)," **p. 242**, "Gantt Chart," **p. 317**, and "PERT (Program Evaluation and Review Technique) Chart," **p. 453**, for a discussion on project activity sequencing and interdependencies.)

Drill down into as many levels of detail as necessary, however the convention is a minimum of four levels. For example, each supporting task can be broken down further into either any in-process deliverables produced by a given task indicating any component parts to the higher-level project deliverable **or** the tool needed to support that activity. If the latter approach is utilized, it documents the project's tool-task-deliverable combination needed to manage the project performance. Figure W-1 illustrates the basic hierarchical structure of a WBS.

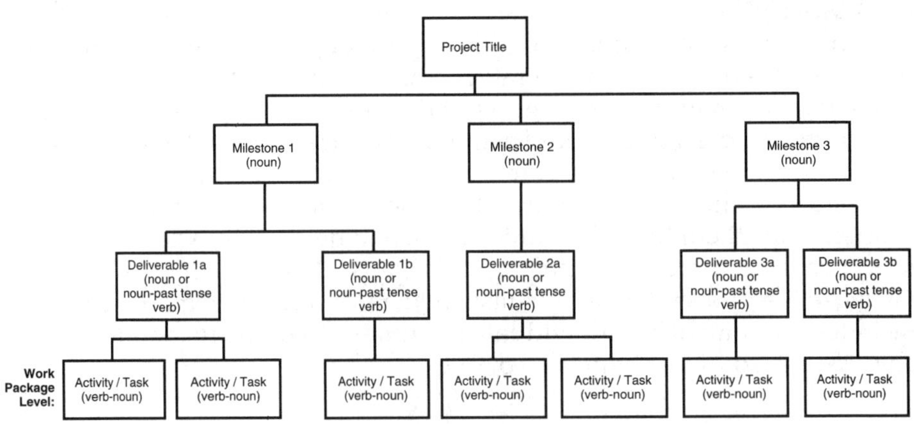

FIGURE W-1: WBS Basic Hierarchical Structure

Complex projects may require that the *manageable components* of the project be coded for easy referencing. In addition, the WBS can apply to organizational units, aligning the work packages to them, and being referred to as Organizational Breakdown Structures (OBS).

How to Use the Tool or Technique

A WBS can be developed individually or with a team. If created individually, the project team should be invited to edit and modify the WBS to ensure accuracy and gain ownership from the key individuals accountable to produce the project deliverables.

Develop a WBS by executing the following procedure:

Step 1. Identify the project milestones and deliverables defined in the project charter.

Step 2. Using a hierarchical Tree diagram structure, record the project title as the tip of the Tree and record the milestones as the first level of the Tree diagram.

a. Decide whether to build the WBS diagram from top-down (vertically) or left-to-right (horizontally).

b. Draw a connecting line between the milestones and the project title to indicate the relationship.

Step 3. Align each deliverable with the appropriate milestone and document as the next detail layer in the hierarchy. This represents the second level.

a. Draw a connecting line between the deliverables and the milestone to indicate the linkage.

Step 4. Determine the activities needed to produce each output and document as the third level of branches in the Tree.

a. Draw a connecting line between the activities and the deliverables to indicate the relationship.

Step 5. Determine if any subordinate tasks are required to support the activities and document them as the fourth level in the Tree hierarchy.

a. Draw a connecting line between the tasks and the activities to indicate the relationship.

b. *Note*: this fourth level of the Tree represents the *work packages*.

A B C D E F G H I J K L M N O P Q R S T U V W X Y Z

Step 6. *Optional*: Identify the one person accountable for each of the deliverables, activities, and tasks, and record his/her name in each corresponding box on the Tree diagram.

Step 7. *Optional*: Ask the one person accountable for a given activity or task to determine the amount of time required to complete that work and record it in the corresponding box on the Tree diagram.

Step 8. *Optional*: Ask the one person accountable for a given activity or task to document either any in-process deliverables produced by a given task indicating any component parts to the higher-level project deliverable or required tool to complete the task.

Supporting or Linked Tools

Supporting tools that might provide input when developing a WBS include

- Project Charter(**See Also** "SMART Problem nd Goal Statements for a Project Charter," **p. 665**)
- SIPOC

A completed WBS provides input to tools such as

- Activity Network Diagram (**See Also** "Activity Network Diagram (AND)—7M Tool," **p. 127**)
- Critical Path Method (CPM) (**See Also** "Critical Path Method (CPM)," **p. 242**)
- Gantt chart (**See Also** "Gantt Chart," **p. 317**)
- PERT chart (**See Also** "PERT (Program Evaluation and Review Tecnique) Chart," **p. 453**)
- RACI (**See Also** "RACI (Responsible, Accountable, Consulted, Informed) Matrix," **p. 554**)

Figure W-2 illustrates the link between a WBS and its related tools and techniques.

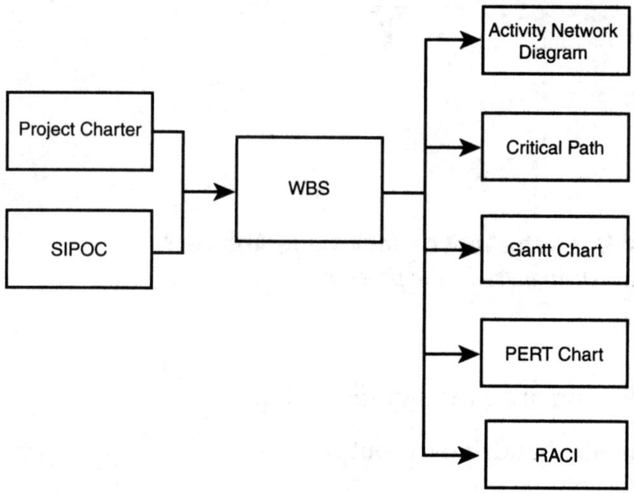

FIGURE W-2: WBS Tool Linkage

Variations

Cause-and-Effect matrix (**See Also** "Cause and Effect Diagram," **p. 173**)

A B C D E F G H I J K L M N O P Q R S T U V W X Y Z

Y

Y = f (X)

What Question(s) Does the Tool or Technique Answer?

What are the critical parameters in the process?

Y=f (X) helps you to

- Identify and summarize the significant inputs
- Process elements that drive the output

When Best to Use the Tool or Technique

Hypothesize the components of the Y = f (X) equation early in an improvement project, starting with the desired outputs, and then validate the critical variables that impact the output throughout the remainder of the project.

Brief Description

The equation, Y = f (X), reads "*Y is a function of X*," where Y is the output or final product of your process, and the *Xs* describe the critical inputs and the process elements that influence the output. In other words, it defines that the output depends on its key process and inputs variables, given that the process transforms the inputs to produce the output.

The Y = f (X) concept focuses improvement (or problem-solving) teams on identifying, optimizing and controlling the vital few inputs and process variables to successfully achieve the desired output level. The equation could involve several subordinate outputs, perhaps as leading indicators of the overall "big Y." For example, if growth were identified as the big Y, the improvement team may examine leading indicators, such as customer loyalty, customer satisfaction, and customer complaints (as little Ys). Each subordinate Y may flow down into its own Y= f (X) relationship, wherein some of the critical variables for one also may affect another little Y, as illustrated in Figure Y-1. Another example is for *Days Sales Outstanding*, wherein it could be determined as a function of invoice accuracy, payment terms, and the customers' processes.

This equation summarizes and the *Process map* illustrates the important input-process-output (IPO) model. Hence, Y=f(X) links to the Process map tool, whereby the big Y represents the overall process outputs, the little Ys represent the in-process outputs, and the Xs represent the process steps and the overall and in-process inputs.

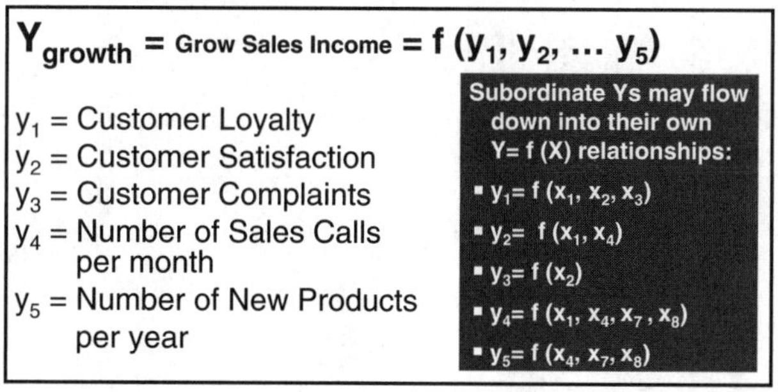

FIGURE Y-1: Y=f(: X) Example

Early in an improvement project, the team speculates as to the critical variables affecting the output. They begin to develop hypotheses as to the potential causes currently dampening the output—a theory of cause-and-effect. Next the team tries to validate its hypothesis. Depending on the type of data available, validation may occur through statistical means—either Hypothesis testing, regression analysis, and/or Design of Experiment. Regression analysis statistically verifies that this equation adequately models the results. Pilots testing potential improvement solutions and simulations (using historical data to model toe process) also may assist in verifying cause-and-effect, but the results yield less confidence. (**See Also** "Hypothesis Testing," **p. 335**, "Regression Analysis," **p. 571**, and "Design of Experiment (DOE)," **p. 250**)

A B C D E F G H I J K L M N O P Q R S T U V W X Y Z

PART III

Best Practices Articles

The Anatomy of Quality Loss in a Product

The Anatomy of Variations in Product Performance

Benchmarking—Avoid Arrogance and Lethargy

Building Strength via Communities of Practice and Project Management

Complex Organizational Change Through Discovery-based Learning

Lean Six Sigma for Fast Track Commercialization High Risk-High Reward, Rapid Commercialization: PROCEED WITH CAUTION!

Listening to the Customer First-Hand; Engineers Too

The Practice of Designing Relationships

A Process for Product Development

Selecting Project Portfolios using Monte Carlo Simulation and Optimization

These 10 articles are available in downloadable PDF format from www.prenhallprofessional.com/title/0132300214

PART II

Best Practices Articles

PART IV
Appendixes

A

Statistical Distribution Tables

TABLE A-1: t Distribution Critical Values
(t_{α} for a one-tail test; divide alpha by 2 for two-tailed tests – $t_{\alpha/2}$)

d.f.	$t_{0.100}$	$t_{0.050}$	$t_{0.025}$	$t_{0.010}$	$t_{0.005}$	d.f.
1	3.078	6.314	12.706	31.821	63.657	1
2	1.886	2.920	4.303	6.965	9.925	2
3	1.638	2.353	3.182	4.541	5.841	3
4	1.533	2.132	2.776	3.747	4.604	4
5	1.476	2.015	2.571	3.365	4.032	5
6	1.440	1.943	2.447	3.143	3.707	6
7	1.415	1.895	2.365	2.998	3.499	7
8	1.397	1.860	2.306	2.896	3.355	8
9	1.383	1.833	2.262	2.821	3.250	9
10	1.372	1.812	2.228	2.764	3.169	10
11	1.363	1.796	2.201	2.718	3.106	11
12	1.356	1.782	2.179	2.681	3.055	12
13	1.350	1.771	2.160	2.650	3.012	13
14	1.345	1.761	2.145	2.624	2.977	14
15	1.341	1.753	2.131	2.602	2.947	15
16	1.337	1.746	2.120	2.583	2.921	16
17	1.333	1.740	2.110	2.567	2.898	17

continues

TABLE A-1: Continued

d.f.	$t_{0.100}$	$t_{0.050}$	$t_{0.025}$	$t_{0.010}$	$t_{0.005}$	d.f.
18	1.330	1.734	2.101	2.552	2.878	18
19	1.328	1.729	2.093	2.539	2.861	19
20	1.325	1.725	2.086	2.528	2.845	20
21	1.323	1.721	2.080	2.518	2.831	21
22	1.321	1.717	2.074	2.508	2.819	22
23	1.319	1.714	2.069	2.500	2.807	23
24	1.318	1.711	2.064	2.492	2.797	24
25	1.316	1.708	2.060	2.485	2.787	25
26	1.315	1.706	2.056	2.479	2.779	26
27	1.314	1.703	2.052	2.473	2.771	27
28	1.313	1.701	2.048	2.467	2.763	28
29	1.311	1.699	2.045	2.462	2.756	29
infinity	1.282	1.645	1.960	2.326	2.576	infinity

Table A-2: Chi-Square ($\chi 2$) Distribution Critical Values

	Left-tail Test (1 - α)			Right-tail Test (α)			
d.f.	$\chi^2_{0.99}$	$\chi^2_{0.95}$	$\chi^2_{0.90}$	$\chi^2_{0.10}$	$\chi^2_{0.05}$	$\chi^2_{0.01}$	d.f.
1	0.00016	0.0039	0.0158	2.71	3.84	6.63	1
2	0.0201	0.1026	0.2107	4.61	5.99	9.21	2
3	0.115	0.352	0.584	6.25	7.81	11.34	3
4	0.297	0.711	1.064	7.78	9.49	13.28	4
5	0.554	1.15	1.61	9.24	11.07	15.09	5
6	0.872	1.64	2.20	10.64	12.59	16.81	6
7	1.24	2.17	2.83	12.02	14.07	18.48	7
8	1.65	2.73	3.49	13.36	15.51	20.09	8
9	2.09	3.33	4.17	14.68	16.92	21.67	9
10	2.56	3.94	4.87	15.99	18.31	23.21	10
11	3.05	4.57	5.58	17.28	19.68	24.73	11
12	3.57	5.23	6.30	18.55	21.03	26.22	12
13	4.11	5.89	7.04	19.81	22.36	27.69	13
14	4.66	6.57	7.79	21.06	23.68	29.14	14
15	5.23	7.26	8.55	22.31	25.00	30.58	15
16	5.81	7.96	9.31	23.54	26.30	32.00	16
18	7.01	9.39	10.86	25.99	28.87	34.81	18
20	8.26	10.85	12.44	28.41	31.41	37.57	20
24	10.86	13.85	15.66	33.20	36.42	42.98	24
30	14.95	18.49	20.60	40.26	43.77	50.89	30
40	22.16	26.51	29.05	51.81	55.76	63.69	40
60	37.48	43.19	46.46	74.40	79.08	88.38	60
120	86.92	95.70	100.62	140.23	146.57	158.95	120

TABLE A-3: F Distribution Critical Values (ANOVA) for $\alpha = 0.05$

d.f. in	Degrees of Freedom (d.f.) in Numerator for $\alpha = 0.05$											
Denominator	1	2	3	4	5	6	7	8	9	10	12	15
1	161.4	199.5	215.7	224.6	230.2	234.0	236.8	238.9	240.5	241.9	243.9	245.9
2	18.51	19.0	19.16	19.25	19.30	19.33	19.35	19.37	19.38	19.40	19.41	19.43
3	10.13	9.55	9.28	9.12	9.01	8.94	8.89	8.85	8.81	8.79	8.74	8.70
4	7.71	6.94	6.59	6.39	6.26	6.16	6.09	6.04	6.00	5.96	5.91	5.86
5	6.61	5.79	5.41	5.19	5.05	4.95	4.88	4.82	4.77	4.74	4.68	4.62
6	5.99	5.14	4.76	4.53	4.39	4.28	4.21	4.15	4.10	4.06	4.00	3.94
7	5.59	4.74	4.35	4.12	3.97	3.87	3.79	3.73	3.68	3.64	3.57	3.51
8	5.32	4.46	4.07	3.84	3.69	3.58	3.50	3.44	3.39	3.35	3.28	3.22
9	5.12	4.26	3.86	3.63	3.48	3.37	3.29	3.23	3.18	3.14	3.07	3.01
10	4.96	4.10	3.71	3.48	3.33	3.22	3.14	3.07	3.02	2.98	2.91	2.85
11	4.84	3.98	3.59	3.36	3.20	3.09	3.01	2.95	2.90	2.85	2.79	2.72
12	4.75	3.89	3.49	3.26	3.11	3.00	2.91	2.85	2.80	2.75	2.69	2.62
13	4.67	3.81	3.41	3.18	3.03	2.92	2.83	2.77	2.71	2.67	2.60	2.53
14	4.60	3.74	3.34	3.11	2.96	2.85	2.76	2.70	2.65	2.60	2.53	2.46
15	4.54	3.68	3.29	3.06	2.90	2.79	2.71	2.64	2.59	2.54	2.48	2.40

d.f. in Denominator	Degrees of Freedom (d.f.) in Numerator for $\alpha = 0.05$					
	20	30	40	50	60	Infinity
20	2.12	2.04	1.99	1.96	1.95	1.84
30	1.93	1.84	1.79	1.76	1.74	1.62
40	1.84	1.74	1.69	1.66	1.64	1.51
50	1.78	1.69	1.63	1.60	1.58	1.44
60	1.75	1.65	1.59	1.56	1.53	1.39
Infinity	1.57	1.46	1.39	1.35	1.32	1.00

TABLE A-4: F Distribution Critical Values (ANOVA) for $\alpha = 0.025$

d.f. in	Degrees of Freedom (d.f.) in Numerator for $\alpha = 0.025$											
Denominator	1	2	3	4	5	6	7	8	9	10	12	15
1	647.8	799.5	864.2	899.6	921.8	937.1	948.2	956.7	963.3	968.6	976.7	984.9
2	38.51	39.00	39.17	39.25	39.30	39.33	39.36	39.37	39.39	39.40	39.41	39.43
3	17.44	16.04	15.44	15.10	14.88	14.73	14.62	14.54	14.47	14.42	14.34	14.25
4	12.22	10.65	9.98	9.60	9.36	9.20	9.07	8.98	8.90	8.84	8.75	8.66
5	10.01	8.43	7.76	7.39	7.15	6.98	6.85	6.76	6.68	6.62	6.52	6.43
6	8.81	7.26	6.60	6.23	5.99	5.82	5.70	5.60	5.52	5.46	5.37	5.27
7	8.07	6.54	5.89	5.52	5.29	5.12	4.99	4.90	4.82	4.76	4.67	4.57
8	7.57	6.06	5.42	5.05	4.82	4.65	4.53	4.43	4.36	4.30	4.20	4.10
9	7.21	5.71	5.08	4.72	4.48	4.32	4.20	4.10	4.03	3.96	3.87	3.77
10	6.94	5.46	4.83	4.47	4.24	4.07	3.95	3.85	3.78	3.72	3.62	3.52
11	6.72	5.26	4.63	4.28	4.04	3.88	3.76	3.66	3.59	3.53	3.43	3.33
12	6.55	5.10	4.47	4.12	3.89	3.73	3.61	3.51	3.44	3.37	3.28	3.18
13	6.41	4.97	4.35	4.00	3.77	3.60	3.48	3.39	3.31	3.25	3.15	3.05
14	6.30	4.86	4.24	3.89	3.66	3.50	3.38	3.29	3.21	3.15	3.05	2.95
15	6.20	4.77	4.15	3.80	3.58	3.41	3.29	3.20	3.12	3.06	2.96	2.86

d.f. in	Degrees of Freedom (d.f.) in Numerator for $\alpha = 0.025$					
Denominator	**20**	**30**	**40**	**50**	**60**	**Infinity**
20	2.46	2.35	2.29	2.25	2.22	2.09
30	2.20	2.07	2.01	1.97	1.94	1.79
40	2.07	1.94	1.88	1.83	1.80	1.64
50	1.99	1.87	1.80	1.76	1.72	1.55
60	1.94	1.82	1.74	1.70	1.67	1.48
Infinity	1.71	1.57	1.48	1.43	1.39	1.00

TIP

The following table is used to calculate area or probability for a **left-tail test** of the normal curve at the different Z values. For the area under the curve to the right of the Z-critical value, subtract the table value from 1.

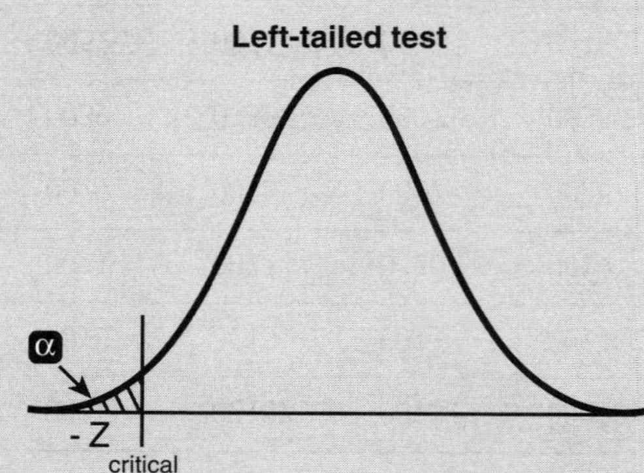

TABLE A-5: Standard Normal Table (Z values)

Z	X.X0	X.X1	X.X2	X.X3	X.X4	X.X5	X.X6	X.X7	X.X8	X.X9
-3.0	.00135	.00131	.00126	.00122	.00118	.00114	.00111	.00107	.00104	.00010
-2.9	.0019	.0018	.0017	.0017	.0016	.0016	.0015	.0015	.0014	.0014
-2.8	.0026	.0025	.0024	.0023	.0023	.0022	.0021	.0021	.0020	.0019
-2.7	.0035	.0034	.0033	.0032	.0031	.0030	.0029	.0028	.0027	.0026
-2.6	.0047	.0045	.0044	.0043	.0041	.0040	.0039	.0038	.0037	.0036
-2.5	.0062	.0060	.0059	.0057	.0055	.0054	.0052	.0051	.0049	.0048
-2.4	.0082	.0080	.0078	.0075	.0073	.0071	.0069	.0068	.0066	.0064
-2.3	.0107	.0104	.0102	.0099	.0096	.0094	.0091	.0089	.0087	.0084
-2.2	.0139	.0136	.0132	.0129	.0125	.0122	.0119	.0116	.0113	.0110
-2.1	.0179	.0174	.0170	.0166	.0162	.0158	.0154	.0150	.0146	.0143
-2.0	.0228	.0222	.0217	.0212	.0207	.0202	.0197	.0192	.0188	.0183
-1.9	.0287	.0281	.0274	.0268	.0262	.0256	.0250	.0244	.0239	.0233
-1.8	.0359	.0351	.0344	.0336	.0329	.0322	.0314	.0307	.0301	.0294
-1.7	.0446	.0436	.0427	.0418	.0409	.0401	.0392	.0384	.0375	.0367
-1.6	.0548	.0537	.0526	.0516	.0505	.0495	.0485	.0475	.0465	.0455

Z	X.X0	X.X1	X.X2	X.X3	X.X4	X.X5	X.X6	X.X7	X.X8	X.X9
-1.5	.0668	.0655	.0643	.0630	.0618	.0606	.0594	.0582	.0571	.0559
-1.4	.0808	.0793	.0778	.0764	.0749	.0735	.0721	.0708	.0694	.0681
-1.3	.0968	.0951	.0934	.0918	.0901	.0885	.0869	.0853	.0838	.0823
-1.2	.1151	.1131	.1112	.1093	.1075	.1057	.1038	.1020	.1003	.0985
-1.1	.1357	.1335	.1314	.1292	.1271	.1251	.1230	.1210	.1190	.1170
-1.0	.1587	.1562	.1539	.1515	.1492	.1469	.1446	.1423	.1401	.1379
-0.9	.1841	.1814	.1788	.1762	.1736	.1711	.1685	.1660	.1635	.1611
-0.8	.2119	.2090	.2061	.2033	.2005	.1977	.1949	.1922	.1894	.1867
-0.7	.2420	.2389	.2358	.2327	.2297	.2266	.2236	.2207	.2177	.2148
-0.6	.2743	.2709	.2676	.2643	.2611	.2578	.2546	.2514	.2483	.2451
-0.5	.3085	.3050	.3015	.2981	.2946	.2912	.2877	.2843	.2810	.2776
-0.4	.3446	.3409	.3372	.3336	.3300	.3264	.3228	.3192	.3156	.3121
-0.3	.3821	.3783	.3745	.3707	.3669	.3632	.3594	.3557	.3520	.3483
-0.2	.4207	.4168	.4129	.4090	.4052	.4013	.3974	.3936	.3897	.3859
-0.1	.4602	.4562	.4522	.4483	.4443	.4404	.4364	.4325	.4286	.4247
0.0	.5000	.4960	.4920	.4880	.4840	.4801	.4761	.4721	.4681	.4641

Table A-6: Standard Normal Table (Z values)

TIP

The following table is used to calculate area or probability for a **right-tail test** of the normal curve at the different Z values. For the area under the curve to the left of the Z-critical value, subtract the table value from 1.

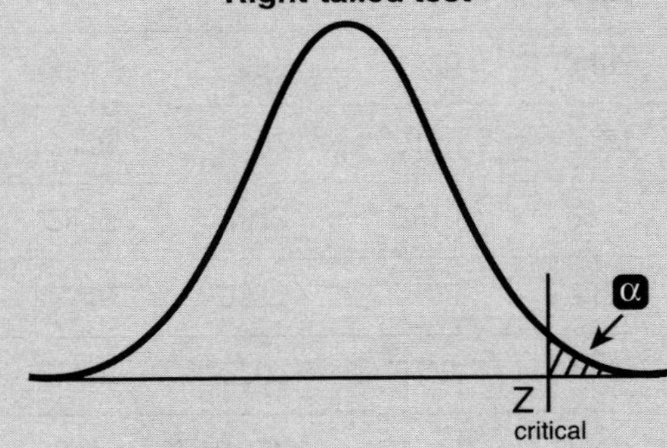

Z	X.X0	X.X1	X.X2	X.X3	X.X4	X.X5	X.X6	X.X7	X.X8	X.X9
0.0	.5000	.5040	.5080	.5120	.5160	.5199	.5239	.5279	.5319	.5359
0.1	.5398	.5438	.5437	.5517	.5557	.5596	.5636	.5675	.5714	.5753
0.2	.5793	.5832	.5871	.5910	.5948	.5987	.6026	.6064	.6103	.6141
0.3	.6179	.6217	.6255	.6293	.6331	.6368	.6406	.6443	.6480	.6517
0.4	.6554	.6591	.6628	.6664	.6700	.6736	.6772	.6808	.6844	.6879
0.5	.6915	.6950	.6985	.7019	.7054	.7088	.7123	.7157	.7190	.7224

Z	X.X0	X.X1	X.X2	X.X3	X.X4	X.X5	X.X6	X.X7	X.X8	X.X9
0.6	.7257	.7291	.7324	.7357	.7389	.7422	.7454	.7486	.7517	.7549
0.7	.7580	.7611	.7642	.7673	.7704	.7734	.7764	.7794	.7823	.7852
0.8	.7881	.7910	.7939	.7967	.7995	.8023	.8051	.8079	.8106	.8133
0.9	.8159	.8186	.8212	.8238	.8264	.8289	.8315	.8340	.8365	.8389
1.0	.8413	.8438	.8461	.8485	.8508	.8531	.8554	.8577	.8599	.8621
1.1	.8643	.8665	.8686	.8708	.8729	.8749	.8770	.8790	.8810	.8830
1.2	.8849	.8869	.8888	.8907	.8925	.8944	.8962	.8990	.8997	.9015
1.3	.9032	.9049	.9066	.9082	.9099	.9115	.9131	.9147	.9162	.9177
1.4	.9192	.9207	.9222	.9236	.9251	.9265	.9279	.9292	.9306	.9319
1.5	.9332	.9345	.9357	.9370	.9382	.9394	.9406	.9418	.9429	.9441
1.6	.9452	.9463	.9474	.9484	.9495	.9505	.9515	.9525	.9535	.9545
1.7	.9554	.9564	.9573	.9582	.9591	.9599	.9608	.9616	.9625	.9633
1.8	.9641	.9649	.9658	.9664	.9671	.9678	.9686	.9693	.9699	.9706
1.9	.9713	.9719	.9726	.9732	.9738	.9744	.9750	.9756	.9761	.9767

continues

TABLE A-6: Continued

Z	X.X0	X.X1	X.X2	X.X3	X.X4	X.X5	X.X6	X.X7	X.X8	X.X9
2.0	.9773	.9778	.9783	.9788	.9793	.9798	.9803	.9808	.9812	.9817
2.1	.9821	.9826	.9830	.9834	.9838	.9842	.9846	.9850	.9854	.9857
2.2	.9861	.9864	.9868	.9871	.9875	.9878	.9881	.9884	.9887	.9890
2.3	.9893	.9896	.9898	.9901	.9904	.9906	.9909	.9911	.9913	.9916
2.4	.9918	.9920	.9922	.9925	.9927	.9929	.9931	.9932	.9934	.9936
2.5	.9938	.9940	.9941	.9943	.9945	.9946	.9948	.9949	.9951	.9952
2.6	.9953	.9955	.9956	.9957	.9959	.9960	.9961	.9962	.9963	.9964
2.7	.9965	.9966	.9967	.9968	.9969	.9970	.9971	.9972	.9973	.9974
2.8	.9974	.9975	.9976	.9977	.9977	.9978	.9979	.9979	.9980	.9981
2.9	.9981	.9982	.9983	.9983	.9983	.9984	.9985	.9985	.9986	.9986
3.0	.99865	.99869	.99874	.99878	.99882	.99886	.99889	.99893	.99896	.99900

B
Glossary

Accuracy: The precision with which a target is reached. Accuracy is measured as the difference between the outcome and a target or standard value.

Affinity Diagram: A tool used to gather and group ideas; usually depicted as a Tree diagram.

Alias: A synonym for "confounded"; a common DOE term.

Alternative hypothesis (Ha): A statement of what is desired to be concluded through sample data. The alternative statement is dependent on the Null hypothesis.

ANOM: (Analysis of Mean) An analytical process that quantifies the mean response for each individual Control Factor level. ANOM can be performed on data that is in regular engineering units or data that has been transformed into some form of signal-to-noise ratio or other data transform. Main effects and interaction plots are created from ANOM data.

ANOVA: (Analysis of Variance) An analytical process that decomposes the contribution each individual control factor has on the overall experimental response. The ANOVA process is also capable of accounting for the contribution of interactive effects between control factors and experimental error in the response provided enough degrees of freedom are established in the experimental array. The value of Epsilon Squared (% contribution to overall CFR variation) is calculated using data from ANOVA.

Array: An arithmetically derived matrix or table of rows and columns that is used to impose an order for efficient experimentation. The rows contain the individual experiments. The columns contain the experimental factors and their individual levels or set points.

ASQ: American Society for Quality, an associate recognized for its work in promoting quality and six sigma; www.asq.org/.

Attribute (or discrete) data: Has clear boundaries between values. There are several types of discrete data—count (for example, number of defects), nominal with more than two categories (e.g. people's names, colors), binary with a rank order (yes/no, good/better/best), and binary without rank order (male/female, heads/tails).

Balanced design: A classic DOE term that describes when the test design involves an equal number of runs for each combination of each setting (or level) for each factor.

Benchmarking: The process of comparative analysis between two or more concepts, components, subassemblies, subsystems, products or processes. The goal of Benchmarking is to qualitatively and quantitatively identify a superior subject within the competing choices. Often the benchmark is used as a standard to meet or surpass. Benchmarks are used in building Houses of Quality, Concept Generation, and the Pugh Concept Selection Process.

Best Practice: A preferred and repeatable action or set of actions completed to fulfill a specific requirement or set of requirements. Often used during planning phases, including within a product development process.

Beta (β): The Greek letter, β, is used to represent the slope of a best fit line. It indicates the linear relationship between the signal factor(s) (Critical Adjustment Parameters) and the measured Critical Functional Response in a dynamic robustness optimization experiment.

Bias: The difference between a standard value and the average of multiple measurements taken of different items.

Black Belt: A job title or role indicating that the person has been certified as having mastered the Six Sigma DMAIC (Define-Measure-Analyze-Improve-Control) content and demonstrated expertise in leading one or more projects. Title usually designates the team leader of a Six Sigma project and is often a coach of Green Belts.

Blocking: A technique used in classical DOE to remove the effects of unwanted, assignable cause noise or variability from the experimental response so that only the effects from the control factors are present in the response data. Blocking is a data purification process used to help assure the integrity of the experimental data used in constructing a statistically significant math model.

Capability Growth Index (CGI): The calculated percentage between 0% and 100% that a group of System, Subsystem, or Subassembly CFRs have attained in getting their Cp Indices to equal a value of 2 (indicating how well their Critical Functional Responses have attained Six Sigma performance during Product Development). The CPI for critical functions is a metric often found on an Executive Gate Review scorecard.

Capability Index: Cp and Cpk Indices that calculate the ratio of the Voice of the Customer versus the Voice of the Product or Process. Cp is a measure of capability based on short-term or small samples of data—usually what is available during Product development. Cpk is a measure of long-term or large samples of data that include not only variation about the mean, but also the shifting of the mean itself—usually available during steady state production.

Checklist: A simple list of action items, steps, or elements needed to complete a task. Each item is "checked off" as it is completed.

Classical Design of Experiments (DOE): Experimental methods employed to construct math models relating a dependent variable (the measured Critical Functional Response) to the set points of any number of independent variables (the experimental control factors). DOE is used sequentially to build knowledge of fundamental functional relationships (Ideal/Transfer Functions) between various factors and a response variable.

Coefficient of Determination (r2): Determines how well the scatter plot's best fit line fits the data; sometimes called R-square. See Regression Analysis.

Commercialization: A business process that harnesses the resources of a company in the endeavor of conceiving, developing, designing, optimizing, certifying design and process capability, producing, selling, distributing, and servicing a product.

Compensation: The use of feedforward or feedback control mechanisms to intervene when certain noise effects are present in a product or process. The use of compensation is only done when insensitivity to noise cannot be attained through robustness optimization.

Component: A single part in a subassembly, subsystem or system. An example would be a stamped metal part prior to having anything assembled to it.

Component Requirements Document: The document that contains all the requirements for a given component. They are often converted into a Quality Plan that is given to the production supply chain to set the targets and constrain the variation allowed in the incoming components.

Confidence Interval: The probability range within which the population data (parameters) exists. This region can be described two ways, either a single limit (an upper or lower) or a two-sided limit (both an upper and lower). Outside these limits is defined as the "rejection region," wherein the "test variable" is said to not be part of the population.

Confidence Level: Determined by the formula one minus alpha (or 1- α), to connote a value with a degree of statistical significance or certainty.

Confounded: A classic DOE term describing when the effects of two or more factors are indistinguishable from one another (also called "alias").

Conjoint Analysis: A structured technique that reveals the underlying customer preferences that guide purchasing decisions, based on trade-off decisions among combined alternatives (or bundles).

Contingency table: A matrix used to organize categorical data in preparation for analysis such as a Chi-Square hypothesis test. The columns and rows represent two different ways to analyze the same set of data and are used to calculate degrees of freedom (i.e. df = [(# rows–1) x (# columns–1)]).

Continuous (or variable) data: A type of data that has no boundaries between values, hence the number can be divided by two and still make sense. Continuous data in non-counting intervals and ratios such as time, height, weight, length, and temperature.

Control Charts: A tool set used to monitor and control a process for variation over time. Varies with the type of data it monitors.

Control Factor: The factors or parameters (CFP or CTF Spec.) in a design or process that the engineer can control and specify to define the optimum combination of set points for satisfying the Voice of the Customer. As defined by DOE, it is an input (X) of a process having an effect on a response and whose value can be easily selected.

Correlation: A metric that measures the linear relationship between two process variables. Correlation describes the X and Y relationship with a single number (the Pearson's Correlation Coefficient (r)), whereas regression summarizes the relationship with a line—the regression line. See Regression Analysis.

Critical Adjustment Parameter (CAP): A specific type of CFP that controls the mean of a CFR. They are identified using sequential DOE and engineering analysis. They are the input parameters for Response Surface Methods for the optimization of mean performance optimization after Robust Design is completed. They enable Cpk to be set equal to Cp, thus enabling entitlement to be approached if not attained.

Critical Functional Parameter (CFP): An input variable (usually an engineered additivity grouping) at the subassembly or subsystem level that controls the mean or variation of a CFR.

Critical Functional Response (CFR): A measured scalar or vector (complete, fundamental, continuous engineering variable) output variable that is critical to the fulfillment of a critical (highly important) customer requirement. Some refer to these critical customer requirements as CTQs. A metric often found on an Executive Gate Review scorecard.

Critical-to-Function Specification (CTF): A dimension, surface or bulk characteristic (typically a scalar) that is critical to a component's contribution to a subassembly, subsystem, or system-level CFR.

Critical Parameter Management (CPM): The process that develops critical requirements and measures critical functional responses to design, optimize, and certify the capability of a product and its supporting network of manufacturing and service processes.

Critical Path: The sequence of tasks in a project that take the greatest amount of time for completion.

Critical Value: For hypothesis testing, the critical value demarcates the point at which the Null hypothesis (HO) can be rejected or not. It is determined by the given type of hypothesis test, its distribution, the confidence level (or significance value), and often the degrees of freedom. If the test statistic (or calculated statistic, based on the data) falls within the "non-rejection" region, then the Null hypothesis is not rejected, as shown in Figure G-1.

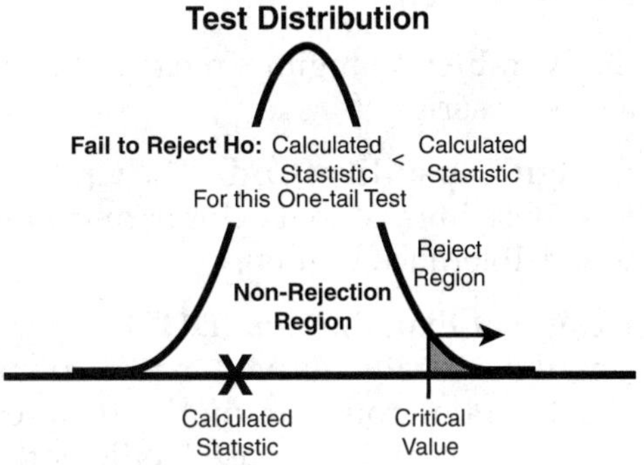

FIGURE G-1: Critical Value Illustration

Criticality: A measurable requirement or functional response that is highly important to a customer. All requirements are important, but only a few are truly critical.

Cross-functional Team: A group of people representing multiple functional disciplines and possessing a wide variety of technical and experiential background and skills working together. Particularly applicable in the product commercialization process. See Multi-functional teams.

CTQ: Critical to Quality translates the customer's generic requirements into specific, actionable, and measurable definition that describes how a process player's work helps to fulfill those requirements. Often displayed in either a tree or matrix format.

Dashboard: A summary and reporting tool of critical success metrics and information about a process and/or product performance. Usually viewed as more complex than scorecards and depicts the critical parameters necessary to run the business.

Degrees of Freedom (df): A statistical term that represents the amount of freedom (or "float") the data has to "represent" the population. The number of measurements that are independently available to estimate a population parameter. As data is collected, degrees of freedom (df) are earned. As statistical information is calculated (or a parameter is estimated), degrees of freedom are "spent" on describing the population from the sample data. The mean is a calculated statistic, it uses up one degree of freedom, resulting in (n-1) degrees of freedom, where "n" represents the sample size. If a population average is represented by four numbers, there is freedom for the first three numbers to be whatever they want to be, but the fourth number must be "dedicated" (as a calculation) to achieve the same population average. Hence, that fourth number does not have the freedom to be whatever it wants to be; so a four-number sample (n=4) has three degrees of freedom to describe the average (e.g. df = n–1; 4–1 = 3).

Deliverable: A tangible, measurable output completed as an outcome of a task or series of tasks.

Design Capability (Cpd): The Cp Index for a design's Critical Functional Response in ratio to its Upper and Lower Specification Limits (VOC-based Tolerance Limits).

Design of Experiments (DOE): A process for generating data that utilizes a mathematically derived matrix to methodically gather and evaluate the effect of numerous parameters on a response variable. Designed experiments, when properly used, efficiently produce useful data for model building or engineering optimization activities. It examines the interactions of multiple variables and is a preferred approach to One Factor At a Time (OFAT) experimentation.

Deterioration Noise Factor: A source of variability that results in some form of physical deterioration or degradation of a product or process. This is also referred to as an "inner noise" because it refers to variation inside the controllable factor levels.

DFSS: A Six Sigma concept used by the engineering technical community to design and develop a product. The acronym represents Design-For-Six-Sigma.

Discrete (or attribute) data: A type of data that has clear, distinct boundaries between values. There are several types of discrete data—count (for example, number of defects), nominal with more than two categories (people's names, colors), binary with a rank order (yes/no, good/better/best), and binary without rank order (male/female, heads/tails).

DMADV: A 5-step Six Sigma method used primarily to re-design a broken process, as well as to solve problems and/or improve processes or products of defects. The acronym stands for Define-Measure-Analyze-Design-Validate.

DMAIC: A 5-step Six Sigma method used to solve problems and/or improve processes or products of defects. The acronym stands for Define-Measure-Analyze-Improve-Control.

DMEDI : A 5-step method combining classic Six Sigma and Lean concepts to re-design a broken process, as well as to solve problems and/or improve processes or products of defects. The acronym stands for Define-Measure-Explore-Develop-Implement.

DPMO: Defects per Million Opportunities measurement, a calculation to count failures within a process and corresponds with a sigma level used in Six Sigma.

Economic Coefficient: The economic coefficient is used in the Quality Loss Function. It represents the proportionality constant in the loss function of the average dollars lost (A_0) due to a customer reaction to off-target performance and the square of the deviation from the target response (Δ_0^2). This is typically, but not exclusively, calculated when approximately 50% of the customers are motivated to take some course of economic action due to poor performance (but not necessarily out-right functional failure). This is often referred to as the LD 50 point in the literature.

ECV: Expected Commercial Value; a financial metric often found on an Executive Gate Review scorecard.

Energy Flow Map: A representation of an engineering system that shows the paths of energy divided into productive and non-productive work. Analogous to a free body diagram from an energy perspective. This accounts for the law of conservation of energy and is used in preparation to math modeling and design of experiments.

Energy Transformation: The physical process a design or product system uses to convert some form of input energy into various other forms of energy that ultimately produce a measurable response. The measurable response may itself be a form of energy or the consequence of energy transformations, which have taken place within the design.

Engineering Metrics: A scalar or vector that is usually called a CTF Spec., CFP, CAP, or a CFR. They are greatly preferred as opposed to Quality Metrics (yield, defects, and so on) in DFSS.

Engineering Process: A set of disciplined, planned, and interrelated activities that are employed by engineers to conceive, develop, design, optimize, and certify the capability of a new product or process design.

Environmental Noise Factors: Sources of variability that are due to effects that are external to the design or product, also referred to as "outer noise." They can also be sources of variability that one neighboring subsystem imposes on another neighboring subsystem or component. Examples include vibration, heat, contamination, misuse, overloading, and so on.

Evolutionary Operation (EVOP): A type of DOE that describes a sequence of experimental designs wherein the cumulative learning (often small and incremental) helps to improve the prediction of future treatments to arrive at a better response.

Experiment: An evaluation or series of evaluations that explore, define, quantify, and build data that can be used to model or predict functional performance in a component, subassembly, subsystem or product. Experiments can be used to build fundamental knowledge for scientific research or they can be used to design and optimize product or process performance in the engineering context of a specific commercialization process.

Experimental Efficiency: This is a process-related activity that is facilitated by intelligent application of engineering knowledge and the proper use of designed experimental techniques. Examples include the use of fractional factorial arrays, control factors that are engineered for additivity and compounded noise factors.

Experimental Error: The variability present in experimental data that is caused by meter error and drift, human inconsistency in taking data, random variability taking place in numerous noise factors not included in the noise array and control factors that have not been included in the inner array. In the Taguchi approach, variability in the data due to interactive effects is often but not always included as experimental error.

Experimental Factors: Independent parameters that are studied in an orthogonal array experiment. Robust Design classifies experimental factors as either control factors or noise factors.

Experimental Space: The combination of the entire control factor, noise factor and signal factor (CAP) levels that produce the range of measured response values in an experiment.

5Ms and P: Potential root cause memory trigger acronym used primarily for despersion problems, and represents Machines, Methods, Materials, Measurements, Mother Nature, and People.

4Ps: Potential root cause memory trigger acronym used primarily for services problems and represents Policies, Procedures, People, and Plant.

F-Ratio: The ratio formed in the ANOVA process by dividing the mean square of each experimental factor effect by the MS of the error variance. This is the ratio of variation occurring between each of the experimental factors in comparison to the variation occurring within all the experimental factors being evaluated in the entire experiment. It is a form of signal-to-noise ratio in a statistical sense. The noise in this case is random experimental error—not variability due to the assignable cause noise factors in the Taguchi noise array.

Feedback Control System: A method of compensating for the variability in a process or product by sampling output response and sending a feedback signal that changes a Critical Adjustment Parameter to put the mean of the response back on its intended target.

FMEA: Failure Modes & Effects Analysis. A risk analysis technique that identifies and ranks the potential failure modes of a design or process and then prioritizes improvement actions.

Fractional Factorial Design: A DOE design describing fewer combinations than a full design is tested to reduce the number of experimental runs needed, but at the cost of examining fewer interactions among the factors.

Full Factorial Design: A DOE design wherein all the possible combinations of all levels of all factors are tested; none are omitted. Two- and three-level orthogonal arrays that include every possible combination between the experimental factors. Full factorial experimental designs use degrees of freedom to account for all the main effects and all interactions between factors included in the experimental array. Basically all of the interactions beyond two way interactions are likely to be of negligible consequence, thus there is little need to use large arrays to rigorously evaluate such rare and unlikely 3-way interactions and above.

Fundamental: The property of a Critical Functional Response that expresses the basic or elemental physical activity that is ultimately responsible for delivering customer satisfaction. A response is fundamental if it does not mix mechanisms together and is uninfluenced by factors outside of the component, subassembly, subsystem, and system design or production process being optimized.

Gantt Chart: A horizontal bar chart used for project planning and control that lists the necessary project activities as row headings against horizontal lines showing the dates and duration of each activity.

Gate: A short period of time during a process when the team reviews and reacts to the results against requirements from the previous Phase and proactively plans for the smooth execution of the next Phase.

Gate Reviews: Meetings with the project team and sponsors to inspect completed deliverables. Focus is on the results from specific tools and best practices and manages the associated risks and problems. They also make sure the team has everything they need to apply the tools and best practices for the next Phase with discipline and rigor. A Gate Review's time should be 20% reactive and 80% proactive.

Goal Post Mentality: A philosophy about quality that accepts anything within the tolerance band (USL-LSL) as equally good and anything that falls outside of the tolerance band as equally bad. See Soccer, Hockey, LaCrosse, and Football rulebooks.

Goal Statement: Identifies the critical parameters (including timeframe) for a targeted improvement. (Use SMART technique to ensure completeness.)

GOSPA: Goals, Objectives, Strategies, Plans & Actions planning methodology.

Grand Total Sum of Squares: The value obtained when squaring the response of each experimental run from a matrix experiment and then adding the squared terms together.

Green Belt: A job title or role indicating that the person has been certified as having demonstrated an understanding of the basic Six Sigma DMAIC (Define-Measure-Analyze-Improve-Control) concepts. This role may support a Black Belt on a Six Sigma project, or in some companies work on a small-scale project directly related to their respective job.

Histogram: A graphical display of the frequency distribution of a set of data. Histograms display the shape, dispersion, and central tendency of the distribution of a data set.

House of Quality: An input—output relationship matrix used in the process of Quality Function Deployment.

Hypothesis Testing: A statistical evaluation that checks the validity of a statement to a specified degree of certainty. These tests are done using well-known and quantified statistical distributions.

IDEA: A 4-step Six Sigma method used by strategic marketing to define, develop, manage and refresh a portfolio of offerings (products and services). The acronym represents Identify-Define-Evaluate-Activate.

Ideal/Transfer Function: Fundamental functional relationships between various engineering control factors and a measured critical functional response variable. The math model of Y = f(x) that represents the customer focused response that would be measured if there were no noise or only random noise acting on the design or process.

Inbound Marketing: Marketing activities that are focused on providing deliverables for "internal consumption," as opposed to deliverables intended for the marketplace.

Independent Effect: The nature of an experimental factor's effect on the measured response when it is acting independently with respect to any other experimental factor. When all control factors are producing independent effects, the design is said to be exhibiting an additive response.

Inference: Drawing some form of conclusion about a measurable functional response based on representative or sample experimental data. Sample size, uncertainty, and the laws of probability play a major role in making inferences.

Inner Array: An orthogonal matrix that is used for the control factors in a designed experiment and that is crossed with some form of outer noise array during Robust Design.

Inspection: The process of examining a component, subassembly, subsystem, or product for off-target performance, variability, and defects either during product development or manufacturing. The focus is typically on whether the item under inspection is within the allowable tolerances or not. Like all processes, inspection itself is subject to variability, and out-of-spec parts or functions may pass inspection inadvertently.

Interaction: The dependence of one experimental DOE factor on the level set point of another experimental factor for its contribution to the measured response. There are two types of interaction: synergistic (mild to moderate and useful in its effect) and anti-synergistic (strong and disruptive in its effect).

Interaction Graph: A plot of the interactive relationship between two experimental factors as they affect a measured response. The ordinate (vertical axis) represents the response being measured, and the abscissa (horizontal axis) represents one of the two factors being evaluated. The average response value for the various combinations of the two experimental factors is plotted. The points representing the second factor's low level are connected by a line. Similarly, the points representing the second factor's next higher level are connected by a line.

IRR: Internal Rate of Return (%IRR); a financial metric often found on an Executive Gate Review scorecard.

KJ Analysis: Stands for "Jiro Kawakita," a Japanese anthropologist who treated attributes (or language) data in a similar manner as variables data by grouping and prioritizing it. A KJ diagram (similar to an Affinity diagram) focuses on the unique and different output, linking the critical customer priorities to the project team's understanding and consensus.

Kurtosis: A term that refers to a graphical distribution peakedness; it describes the amount that the curve's middle is flattened or spiked.

Lagging Indicator: An indicator that follows the occurrence of something; hence used to determine the performance of an occurrence or an event. By tracking lagging indicators, one reacts to the results. For example, the high and low temperature, precipitation, and humidity of a given day.

Leading Indicator: An indicator that precedes the occurrence of something; hence used to "signal" the upcoming occurrence of an event. By tracking leading indicators, one can prepare or anticipate the subsequent event and be proactive. For example, barometric pressure and doplar radar of surrounding region gives indication of ensuing weather.

Lean Six Sigma: Modified Six Sigma approach to emphasize improving speed of a process by making "lean" its "non-value-add" steps. Started in a manufacturing environment, its concepts have expanded to multiple industries and applications. Its common metrics include zero wait time, zero inventory, line balancing, cutting batch sizes to improve flow through, and reducing overall process time.

Least Squares method: A regression analysis technique that selects the minimum sum of the squared deviations about the regression line. The approach uses the smallest sum of all the squared residual values calculated for each data point to determine the regression line.

Level: The DOE settings or value for a given factor. The set point at which a control factor, signal factor (CAP), or noise factor is placed during a designed experiment.

Life Cycle Cost: The costs associated with making, supporting, and servicing a product or process over its intended life.

Linear Combination: This term has a general mathematical definition and a specific mathematical definition associated with the dynamic robustness case. In general, a linear combination is the simple summation of terms. In the dynamic case it is the specific summation of the product of the signal level and its corresponding response.

Linear Graph: A graphical aid used to assign experimental factors to specific columns when evaluating or avoiding specific interactions.

Linear Regression: A quantitative model building tool that relates one or more independent variables (Xs) to a single dependent variable (Y). See "Regression Analysis."

Linearity: 1) A term that describes a measurement device's inherent bias across its operating range, which is defined by its measurement scale. For example, a device may introduce variability at either extreme of its scale—bias relative to a true value. 2) The relationship between a dependent variable (the response) and an independent variable (such as the signal or control factor) that is graphically expressed as a straight line. Linearity is typically a topic within the dynamic cases of the robustness process and in Linear Regression analysis.

LMAD: A 4-step Six Sigma method used by marketing to manage the ongoing operations of a portfolio of launched offerings (products and services) across the value chain. The acronym represents Launch-Manage-Adapt-Discontinue.

Loss to Society: The economic loss that society incurs when a product's functional performance deviates from its targeted value. The loss is often due to economic action taken by the consumer reacting to poor product performance but can also be due to the effects that spread out through society when products fail to perform as expected. For example, a new car breaks down in a busy intersection due to a transmission defect, and 14 people are 15 minutes late for work (cascading loss to many points in society).

Lower Specification Limit: The lowest functional performance set point that a design or component can attain before functional performance is considered unacceptable.

Main Effect: The effect of a single DOE factor that is independent of any other factors. The contribution an experimental factor makes to the measured response independent of experimental error and interactive effects. The sum of the half effects for a factor is equal to the main effect.

Manufacturing Process Capability (Cpm): The ratio of the manufacturing tolerances to the measured performance of the manufacturing process.

Matrix: An array of experimental set points that is derived mathematically. The matrix is composed of rows (containing experimental runs) and columns (containing experimental factors).

Matrix Experiment: A series of evaluations that are conducted under the constraint of a matrix.

Mean: The average value of a sample of data that is typically gathered in a matrix experiment.

Mean Square Deviation (MSD): A mathematical calculation that quantifies the average variation a response has with respect to a target value.

Mean Square Error: A mathematical calculation that quantifies the variance within a set of data.

Measured Response: The quality characteristic that is a direct measure of functional performance.

Measurement Error: The variability in a data set that is due to poorly calibrated meters and transducers, human error in reading and recording data and normal, random effects that exist in any measurement system used to quantify data.

Meter: A measurement device usually connected to some sort of transducer. The meter supplies a numerical value to quantify functional performance.

Monte Carlo Simulation: A computer simulation technique that uses sampling from a random number sequence to simulate characteristics or events or outcomes with multiple possible values.

MSA: Measurement System Analysis tool to understand the level of reproducibility and repeatability.

MTBF: Mean Time Between Failure measurement of the lapsed time from one failure to the next.

Multi-disciplined Team: A group of people possessing a wide variety of technical and experiential background and skills working together. Particularly applicable in the product commercialization process. See Cross-functional teams.

Noise: Any source of variability. Typically noise is either external to the product (such as environmental effects), a function of unit to unit variability due to manufacturing, or it may be associated with the effects of deterioration. In this context, noise is an assignable, non-random cause of variation.

Noise Factor: An input (X) to a process having an effect on a response but is not easily managed as defined in terms of DOE.

Noise Directionality: A distinct upward or downward trend in the measured response depending on the level at which the noises are set. Noise factor set points can be compounded depending upon the directional effect on the response.

Noise Experiment: An experiment designed to evaluate the strength and directionality of noise factors on a product or process response.

Noise Factor: Any factor that promotes variability in a product or process.

Normal Distribution: The symmetric distribution of data about an average point. The normal distribution takes on the form of a bell shaped curve. It is a graphic illustration of how randomly selected data points from a product or process response will mostly fall close to the average response with fewer and fewer data points falling farther and farther away from the mean. The normal distribution can also be expressed as a mathematical function and is often called a Gaussian Distribution.

NPV: Net Present Value; a financial metric often found on an Executive Gate Review scorecard.

NUD: Acronym representing New, Unique and Difficult customer requirements, that if fulfilled will delight them and help to outpace competition.

Null Hypothesis (H_0): A statement of what typically is desired to be disproved through sample data.

Off-Line Quality Control: The processes included in pre-production commercialization activities. The processes of concept design, parameter design, and tolerance design make up the elements of off-line quality control. It is often viewed as the area where quality is designed into the product or process.

One Factor at a Time (OFAT) Experiment: An experimental technique that examines one factor at a time, determines the best operational set point, locks in on that factor level, and then moves on to repeat the process for the remaining factors. This technique is widely practiced in scientific circles but lacks the circumspection and discipline provided by full and fractional factorial designed experimentation. Sometimes one factor at a time experiments are used to build knowledge prior to the design of a formal factorial experiment.

On-line Quality Control: The processes included in the production phase of commercialization. The processes of statistical process control (loss function based and traditional), inspection and evolutionary operation (EVOP) are examples of on-line quality control.

Operating Income: Calculated as gross profit minus operating expenses. A financial metric often found on an Executive Gate Review scorecard.

Operational Definition: Defines what good looks like, often used for attribute data, and describes a standard criteria to measure against. Needs to be concise and easily understood by the full range of its users. It might include prose (or text) descriptions, pictures, photographs, models, or samples.

Operational Marketing: Pertains to marketing's activities in support of launching and managing an offering (product and/or service) or set of offerings across the value chain.

Optimize: Finding and setting control factor levels at the point where their mean, standard deviation, or S/N (signal-to-noise) ratios are at the desired or maximum value. Optimized performance means the control factors are set such that the design is least sensitive to the effects of noise and the mean is adjusted to be right on the desired target.

Orthogonal: The property of an array or matrix that gives it balance and the capability of producing data that allow for the independent quantification of independent or interactive factor effects.

Orthogonal Array: A balanced matrix that is used to lay out an experimental plan that avoids any confounding of the main effects or any interactions; wherein each factor level appears the same number of times. A full factorial is described as balanced and orthogonal because the design represents an equal number of runs for each level and combinations of all the factors.

Outbound Marketing: Marketing activities that are focused on providing deliverables for the customers as opposed to deliverables intended for "internal consumption."

Outer Array: The orthogonal array used in dynamic robust design that contains the noise factors and signal factors. Each treatment combination of the control factors specified in the inner array is repeated using each of the treatment combinations specified by the outer array.

p-Value: In hypothesis testing, the p-value represents the probability of the Null hypothesis (H_0) being true, based on chance alone. It is a "goodness of fit" statistic. Typically the Null hypothesis states that the factors of interest are equal, or from the same population. If the p-value is large (>0.05 for a 95% confidence level), then there is insufficient evidence to reject the Null hypothesis of equality; therefore, fail to reject the Null hypothesis; therefore, presume they are the same. If the calculated p-value < 0.05, for a 95% confidence test and the test statistic is beyond the critical value that determines the hypothesis is true, then there is sufficient evidence to reject the Null hypothesis; therefore, they are different. A small p-value indicates that it is very unlikely the items are the same (by chance alone), and the distance from the means is far apart. Figure G-2 illustrates a small p-value, where the Null hypothesis that Sample B is from the same population as Sample A is rejected (i.e. H_0 mean of A = mean of B).

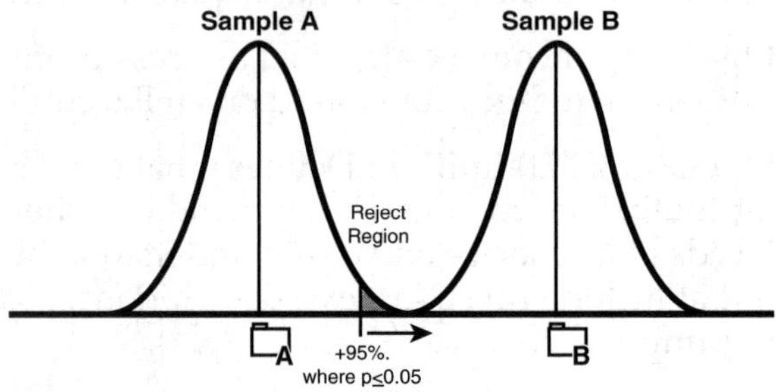

FIGURE G-2: Statistically Small p-value (< 0.05) Illustration

Parameter: 1) A population characteristic; and 2) A factor used in the design, optimization, and certification of capability processes. Experimental parameters are CFRs, CFPs, CTF Specs, and Noise Factors.

Parameter Design: The process employed to optimize the levels of control factors against the effect of noise factors. Signal factors (dynamic cases) or tuning factors (NTB Cases) are used in the two-step optimization process to adjust the performance onto a specific target during parameter (robust) design.

Parameter Optimization Experiment: This is the main experiment in parameter design that is used to find the optimum level for the control factors. Usually this experiment is done using some form of dynamic crossed array design.

Part Worth: The simplest "part" or utility that defines the value for each of the attribute levels in a Conjoint Analysis.

PERT Chart: Diagram that displays the dependency relationships that exist between tasks and helps to discern the variation in expected task time.

PEST: Potential root cause memory trigger acronym used primarily for external environment problems and represents Political, Economic, Social, and Technological.

PFQT: An acronym that stands for Productivity (how many), Financial (how much money), Quality (how well), and Time (how fast).

Phase: A period of time that is designed to conduct work to produce specific results that meet the requirements for a given project wherein specific tools and best practices are used.

Phase/Gate Product Development Process: A series of time periods that are rationally divided into Phases for the development of new Products and Processes. Gates are checkpoints at the end of each Phase to review progress, assess risks, and plan for efficiency in future Phase performance.

Population Parameter or Statistic: A statistic such as the mean or standard deviation that is calculated with all the possible values that make up the entire population of data in an experiment. Samples are just a portion of a population.

Porter's 5 Forces Analysis: A business analysis model developed by Michael Porter to determine an industry's competitive nature by evaluating five forces of rivalry among current players, barriers to entry, threat of substitutes, buyer (bargaining) power, and supplier bargaining power.

Power: Defines making a correct test decision—correctly rejecting the Null hypothesis because the Null is false. Power represents the ability to detect a difference when one exists. It is determined by the formula one minus beta (or $1-\beta$). Often the default is set at 80%.

Precision: Defines the variation in repeated measurements of the same item. There are two major ways to measure precision—repeatability and reproducibility.

Probability: The likelihood or chance that an event or response will occur out of some number (n) of possible opportunities.

Problem Statement or Opportunity for Improvement: A clear, concise definition of what is wrong with a current process or product/services offering. Should be aligned with company strategies and/or annual plan. (Use SMART technique to ensure completeness.)

Process: A set sequence of steps to make something or do something.

Process Capability Analysis: Quantifies the ability of a process to produce output that meets customer requirements. Various capability metrics include DPMO (Defects Per Million Opportunities); Cp, Cpk (Potential Process Capability, Short-term); Pp, Ppk (Process Capability, Long-term); and Rolled Throughput Yield (RTY).

Process Map: A type of flow chart depicting the steps in a process, its inputs/outputs, and often identification of responsibility for each step and the key measures.

Product Commercialization: The act of gathering customer needs, defining requirements, conceiving product concepts, selecting the best concept, designing, optimizing, and certifying the capability of the superior product for production, delivery, sales and service.

Product Development: The continuum of tasks, from In-bound Marketing to Technology Development to certified Technology being transferred into Product Design to the final step of the certified Product Design being transferred into Production.

Project Cycle-Time: The time that elapses from the beginning to the end of a project.

Project Management: The methods of planning, designing, managing, and completing Projects. Project Management designs and controls the micro timing of tasks and actions (underwritten by tools and best practices) within each of the phases of a product development process.

Pugh Process: A structured concept selection process used by multidisciplinary teams to converge on superior concepts. The process uses a matrix consisting of criteria based on the Voice of the Customer and its relationship to specific, candidate design concepts. The evaluations are made by comparing the new concepts to a benchmark called the datum. The process uses the classification metrics of "same as the datum," "better than the datum," or "worse than the datum." Several iterations are employed wherever increasing superiority is developed by combining the best features of highly ranked concepts until a superior concept emerges and becomes the new benchmark.

QFD: Quality Function Deployment. A process for translating the Voice of the Customer into Technical Requirements at the Product Level. QFD, as part of the Critical Parameter Management Process, uses a series of matrices called Houses of Quality to translate and link system requirements to subsystem requirements which, in turn, are translated and linked to subassembly requirements which, in turn, are translated and linked to component requirements which, in turn, are translated and linked to manufacturing process requirements.

Quality: The degree or grade of excellence [*The American Heritage Dictionary*, 2nd College Edition; Houghton Mifflin Co. (1985)]. In a product development context, it is a product with superior features that performs on target with low variability throughout its intended life. In an economic context, it is the absence or minimization of costs associated with the purchase and use of a product or process.

Quality Characteristic: A measured response that relates to a general or specific requirement that can be an attribute or a continuous variable. The quantifiable measure of performance that directly affects the customer's satisfaction. Often in DFSS these have to be converted to an engineering scalar or vector.

Quality Engineering: Most often referred to as Taguchi's approach to offline quality control (concept, parameter, and tolerance design) and online quality control.

Quality Function Deployment (QFD): A disciplined process for obtaining, translating, and deploying the Voice of the Customer into the various Phases of Technology Development and the ensuing commercialization of products or processes during Product Design. See QFD.

Quality Loss Cost: The costs associated with the loss to customers and society when a product or process performs off the targeted response.

Quality Loss Function: The relationship between the dollars lost by a customer due to off-target product performance and the measured deviation of the product from its intended performance. Usually described by the quadratic loss function.

Quality Metrics: A category of metrics used to evaluate the level of quality, for example, defects, time-to-failure, yield, go-no go. See Quality Characteristic.

Quality Plan: The document that is used to communicate specifications to the production supply chain. Often the Component House of Quality is converted into a Quality Plan.

RACI Matrix: A 2-dimensional table that lists tasks or deliverables as the row headings and roles (or people's names) as the column headings. The cell data contains the responsibility, by task, by role (or person) R = Responsible; A = Accountable; C = Consulted; I = Informed.

Random Error: The non-systematic variability that is present in experimental data due to random effects occurring outside of the control factor main effects. The residual variation in a data set that was induced by unsuppressed noise factors and error due to human or meter error.

Randomization: The technique employed to remove or suppress the effect of systematic or biased order (a form of assignable cause variation) in running designed experiments. Randomizing is especially important when applying classic DOE in the construction of math models. It helps assure the data is as random as possible.

Real Win Worth (RWW): Technique to analyze market potential, competitive position, and financial return. The composite value is often used as a summary metric on an Executive Gate Review scorecard.

Regression: A statistical tool that measures the strength of relationship between one or more independent variables with a dependent variable. It builds upon the correlation concepts to develop an empirical, data-based model. Correlation describes the X and Y relationship with a single number (the Pearson's Correlation Coefficient (r)), whereas regression summarizes the relationship with a line—the regression line.

Relational Database: The set of requirements and fulfilling data that is developed and documented during Critical Parameter Management. They link many-to-many relationships up and down throughout the hierarchy of the System being developed.

Reliability: The measure of robustness over time. The length of time a product or process performs as intended.

Repeat Measurement: The taking of data points where the multiple measured responses are taken without changing any of the experimental set points. Repeat measurements provide an estimate of measurement error only.

Repeatability: Variation of repeated measurements of the same item, by the same person, with the same measuring device.

Replicate: The taking of data in which the design or process set points have all been changed since the previous readings were taken. Often a replicate is taken for the first experimental run, then again at the middle and end of an experiment (for a total of three replicates of the first experimental run). Replicate measurements provide an estimate of total experimental error.

Replication: Repeating all DOE treatment combinations more than once (not repeating a measurement of a single treatment combination). Provides a better estimate of random error.

Reproducibility: 1) The variation in the averages from repeated measurements made by different people (or machines) on the same item. 2) The ability of a design to perform as targeted throughout the entire development, design and production phases of commercialization. Verification tests provide the data on reproducibility in light of the noise imposed on the design.

Requirement: Criteria that must be fulfilled; something wanted or needed.

Residual: A term that describes the difference between each data point and the best fit line (or residual line) used to model the data set (also called error).

Resolution Level: The amount of confounding found in a DOE design, either main effects with interactions or interactions with interactions, based on the number of factors and levels in a design, wherein the higher the resolution, the less confounding exists. Selecting the appropriate resolution level depends upon the tolerable level of uncertainty about the experimentation balanced against both time and economic constraints on the number of experiments (runs) that can be conducted.

Response: A measurable output (Y) of a process as defined by DOE. The measured value taken during an experimental run. Also called the quality characteristic. In DFSS we prefer to focus on Critical Functional Responses (CFR)s.

Risk Mitigation: A planning process to identify, prevent, remove, or reduce risk if it occurs and define actions to limit the severity/impact of a risk, should it occur.

Robust Design: A process within the domain of Quality Engineering for making a product or process insensitive to the effects of variability without actually removing the sources of variability. Synonymous with Parameter Design.

ROI: Return on Investment calculated as the annual benefit divided by the investment amount. A financial metric often found on an Executive Gate Review scorecard.

Run Order: When a DOE test is conducted in a random order to eliminate bias.

Sample: A select, usually random set of data points that are taken out of a greater population of data.

Sample Size: The measure of how many samples have been taken from a larger population. Sample size has a notable effect on the validity of making a statistical inference.

Sample Statistic: A statistic such as the mean or standard deviation that is calculated using a sample from the values that make up the entire population of data.

Saturated Experiment: The complete loading of an orthogonal array with experimental factors. There is the possibility of confounding main effects with potential interactions between the experimental factors within a saturated experiment.

Scalar: A continuous engineering variable that is measured by its magnitude alone (no directional component exists for a scalar).

Scaling Factor: A Critical Adjustment Parameter that is known to have a strong effect on the mean response and a weak effect on the standard deviation. The scaling factor often has the additional property of possessing a proportional relationship to both the standard deviation and the mean.

Scaling Penalty: The inflation of the standard deviation in a response as the mean is adjusted using a CAP.

Scorecards: A set of summary data evaluating the tool-task-deliverable output to fulfill requirements and used to predict outcomes or evaluate performance of a process or product when making decisions and managing risk. Should be used in conjunction with a dashboard.

Screening Experiment: Typically a small, limited experiment that is used to determine which factors are important to the response of a product or process. Screening experiments are used to build knowledge prior to the main modeling experiments in sequential DOE methodology.

Sensitivity: The change in a CFR based upon unit changes in a CFP, a CAP, or a CTF Spec. Also a measure of the magnitude (steepness) of the slope between the measured response and the signal factor (CAP) levels in a dynamic robustness experiment.

Sigma (σ): The standard deviation (technically a measure of the population standard deviation).

Signal-to-Noise Ratio: A ratio or value formed by transforming the response data from a robust design experiment using a logarithm to help make the data more additive. Classically, signal to noise is an expression relating the useful part of the response to the non-useful variation in the response.

Signal Factor: A Critical Adjustment Parameter that is known to be capable of adjusting the average output response of the design in a linear manner. Signal factors are used in dynamic robustness cases as well as in Response Surface Methods.

Significance level: The pre-defined level of certainty required to "test" such that the observed effect (a test statistic or "test factor") actually caused the response or is different from the known distribution, and not by chance alone. Denoted by the term alpha α.

SIPOC: A summary tool to communicate the Suppliers, Inputs, Process, Outputs, and Customers of a process.

Six Sigma (6s): A disciplined approach to enterprise-wide quality improvement and variation reduction. Technically, it is the denominator of the Capability (Cp) Index.

Slack Time: The amount of time a task can be delayed without delaying a project (often denoted as S).

SLEPT: Potential root cause memory trigger acronym used primarily for external environment problems; represents Social, Legal, Economic, Political, and Technological.

Slope: The term quantifies the steepness of a linear relationship between two factors.

Smaller-the-Better (STB): A static case where the smaller the measured response is, the better the quality of the product or process.

SMART Problem and Goal Statement: An acronym specifying that a project problem and goal statement need to be Specific, Measurable, Achievable (but aggressive), Relevant (to the project team and business), and Time-bounded.

SPC: Represents Statistical Process Control and is a statistical method used to monitor process variation relative to its control limits.

Specification: A specific quantifiable set point that typically has a nominal or target value and a tolerance of acceptable values associated with it. This is what results when a team tries to use tools and best practices to fulfill a requirement. Requirements are hopefully completely fulfilled by a final design specification—but many times they are not.

Spider Chart: Also called a spider diagram, radar chart, web, or star chart. A graph displaying multiple dimensions as spokes on a wheel and connects them from both the centerpoint and the circumference. Used in techniques such as the Porter's 5 Forces to diagram an industry segment's competitive characteristics. See Porter's 5 Forces.

Splines: A statistical technique that uses a mix of different lines and curves to generate a better fitting model. See Regression Analysis.

Sponsor: A role indicating that the person has ultimate accountability for a Six Sigma project, its direction and funding. This person is a senior manager whose job is highly dependent on the outcome of the project—often the Functional Process Owner of the process on which the project is focused. The project team reports to the Sponsor for this project. The Sponsor conducts Gate Reviews of the project and serves as the project's liaison to the company's goals and missions.

Stability: Describes a measurement system that lacks change in bias or precision over time when repeatedly measuring the same items. Hence, a stable measurement system is said not to drift.

Standard Deviation: A measure of the variability in a set of data. It is calculated by taking the square root of the variance. Standard deviations are not additive, while the variances are.

Standard Normal Distribution: Represents a conventional scale for a normally distributed bell-shaped curve that has a central tendency of zero and a standard deviation of one unit, wherein the units are called sigma (σ).

Standard Order: When a DOE test is conducted such that its treatment combinations are in order.

Static Robustness Case: One of the two major types of experimental robustness cases to study a product or process response as related to specific design parameters. The static case has no pre-designed signal factor associated with the response. Thus the response target is considered fixed or static. Control factors and noise factors are used to find local optimum set points for static robustness performance.

STEEPLE: Potential root cause memory trigger acronym used primarily for external environment problems and represents Social/demographic, Technological, Economic, Environmental (natural), Political, Legal, and Ethical.

Strategic Marketing: Pertains to marketing's activities in support of the definition, development, management, and refreshing of a portfolio of offerings (products or services) at an enterprise, business unit, or division level.

Subassembly: Any two or more components that can be assembled into a functioning assembly.

Subassembly Requirements Document: A document that contains the requirement, both critical and all others, for a subassembly.

Subsystem: A group of individual components and subassemblies that perform a specific function within the total product system. Systems consist of two or more subsystems.

Subsystem Requirements Document: A document that contains the requirement, both critical and all others, for a subsystem.

Sum of Squares: A calculation technique used in the ANOVA process to help quantify the effects of the experimental factors and the mean square error (if replicates have been taken).

Sum of Squares Due to the Mean: The calculation of the sum of squares to quantify the overall mean effect due to the experimental factors being examined.

Supply Chain: The network of suppliers that provide raw materials, components, subassemblies, subsystems, software, or complete systems to your company.

Surrogate Noise Factor: Time, position, and location are not actual noise factors but stand in nicely as surrogate "sources of noise" in experiments that do not have clearly defined physical noises. These are typically used in process robustness optimization cases.

Synergistic Interaction: A mild to moderate form of interactivity between control factors. Synergistic interactions display monotonic but non-parallel relationships between two control factors when their levels are changed. They are typically not disruptive to robust design experiments.

System: An integrated group of subsystems, subassemblies, and components that make up a functioning unit that harmonizes the mass, energy, and information flows and transformations of the elements to provide an overall product output that fulfills the customer requirements.

System Integration: The construction and evaluation of the system from its subsystems, subassemblies, and components.

System Requirements Document: A document that contains the requirements, both critical and all others, for a system.

Tactical Marketing: Pertains to marketing's activities in support of the design and development of an offering (product or service) to prepare it for commercialization or launch.

Taguchi, Genichi: The originator of the well-known system of Quality Engineering. Dr. Taguchi is an engineer, former university professor, author, and global quality consultant.

Target: The ideal point of performance known to provide the ultimate in customer satisfaction. Often called the nominal set point or the ideal performance specification.

Task: A specific piece of work (or definable activity) required to be performed as a duty.

Technology Development: The building of new or leveraged technology (R&D) in preparation for transfer of certified technology into Product Design programs.

Technology Transfer: The handing over of certified (robust and tunable) technology and data acquisition systems to the design organization.

Test Statistic: A calculated number that summarizes sample data and is used to analyze a hypothesis. Examples of a statistic include average (or mean), variance, and proportion.

Tool: An instrument or device that aids the completion of a task.

Total Sum of Squares: The part of the ANOVA process that calculates the summation of all the square roots. It decomposes into the sum of squares from individual experimental factor effects and the sum of squares from experimental error.

Transfer Function: Fundamental functional relationships between various engineering control factors and a measured critical functional response variable. The math model of Y = f(x) that represents the customer focused response that would be measured if there were no noise or only random noise acting on the design or process. Sometimes they are called "transfer functions" because they help model how energy, mass, and logic and control signals are transferred across system, subsystem, and subassembly boundaries.

Treatment Combination: A single experimental run or the combination of control factors and levels from an orthogonal array, as described by the DOE technique.

TRIZ: A systematic innovation process originally developed in Russia to generate ideas that are ingenious and viable.

Two-Step Optimization Process: The process of first finding the optimum control factor set points to minimize sensitivity to noise and then adjusting the mean response onto the customer-focused target.

Type I Error: A type of error used in hypothesis testing that arises when incorrectly rejecting the null hypothesis, although it is actually true. Thus, based on the test statistic, the final conclusion rejects the Null hypothesis, but in truth it should be accepted. Type I error equates to the alpha (α) or significance level, whereby the generally accepted default is 5%.

Type II Error: A type of error used in hypothesis testing that occurs when the test decision incorrectly "accepts" the null hypothesis. Based on the test statistic, the final decision fails to reject the Null when it is actually false. Type II error also is called "beta"(β), and the default is typically set at 20%.

UAPL: A 4-step Six Sigma method used by tactical marketing to ready an offering (products and services) for commercialization. The acronym represents Understand-Analyze-Plan-Launch. This process is designed to be conducted in parallel with the technical community developing a product or service.

UMC: Unit manufacturing cost. The cost associated with making a product or process.

Unit-to-Unit Variability: Variation in a product or process due to noises in the production and assembly process.

Upper Specification Limit: The largest functional performance set point that a design or component is allowed before functional performance is considered unacceptable.

Value Chain: For a launched offer, the go-to-market value chain describes those functions that add value (according to the customer or client) to an already launched offering (product or service). For example, sales, value-added resellers, other distribution channels, consultants (delivering incremental software and services), service (break-fix, maintenance, parts), training, customer support call center, and marketing support materials and advertising.

Variable (or Continuous) Data: A type of data that has no boundaries between values, hence the number can be divided by two and still make sense. Continuous data in non-counting intervals and ratios such as time, height, weight, length, and temperature.

Variance: The mean squared deviation of the measured response values from their average value.

Variation: Changes in parameter values due to systematic or random effects. Variation is the root cause of poor quality and the monetary losses associated with it.

Vector: An engineering measure that has both magnitude and directionality associated with it. They are highly valued metrics for Critical Parameter Management.

Verification: The process of validating the results from a model or a designed experiment.

Voice of the Business (VOB): The internal requirements of the business in the words of senior or executive management. The VOB is used throughout the product commercialization process and must be balanced with customer requirements.

Voice of the Customer (VOC): The wants and needs of the customer in their own words. The VOC is used throughout the product commercialization process to keep the requirements and the designs that fulfill them focused on the needs of the customer.

Voice of the Process: The measured functional output of a manufacturing process.

Voice of the Product: The measured functional output of a design element at any level within the engineered system.

WCBF: Worldwide Conventions and Business Forums, an organization dedicated to furthering Six Sigma through conferences and meetings; www.wcbf.com.

Work Breakdown Structure (WBS): The process of dividing a project into manageable tasks and sequencing them to ensure a logical flow between tasks.

Workflow Charts: Diagrams that depict the flow of work in a process.

Y = f(x): A mathematical equation read as "Y equals f of x," which means the result (output) measures (represented by Y) are a function of the process (input) measures (represented by x). Used to understand how the inputs affect the outputs.

Yield: The percentage of the total number of units produced that are acceptable; the "percent good."

C
References

- _______. *A Guide to the Project Management Body of Knowledge*. Pennsylvania: Project Management Institute, Inc., 2000.
- Breyfogle, Forrest W. III. *Implementing Six Sigma, 2nd Edition*. New Jersey: John Wiley & Sons, Inc., 2003.
- Creveling, C.M., L. Hambleton, and Burke McCarthy. *Six Sigma for Marketing Processes, An Overview for Marketing Executives, Leaders, and Managers.* New Jersey: Prentice-Hall, 2006.
- Creveling, C.M., J.L. Slutsky, and D. Antis, Jr. *Design for Six Sigma in Technology and Product Development*. New Jersey: Prentice-Hall, 2003.
- Eckes, George. *Six Sigma for Everyone.* Hoboken, New Jersey: John Wiley & Sons, Inc., 2003.
- Gale, B. T., *Managing Customer Value*, ISBN: 0-02-911045-9.
- Garvin, D. A., *Managing Quality*, ISBN: 0-02-911380-6.
- George, Michael L.. *Lean Six Sigma, Combining Six Sigma Quality with Lean Speed*. New York: McGraw-Hill, 2002.
- George, Mike, Dave Rowlands, and Bill Kastle. *What is Lean Six Sigma?* New York: McGraw-Hill, 2004.
- George, Michael L, David Rowlands, Mark Price, and John Maxey. *The Lean Six Sigma Pocket Toolbook*. New York: McGraw-Hill, 2005.
- Ishikawa, Kaoru. *Guide to Quality Control*. New York: Quality Resources, 1982.
- Kemp, Steven M. and Sid Kemp. *Business Statistics Demystified*. New York: McGraw-Hill, 2004. ISBN 0-07-144024-0.

- Mann, Darrell. *Hands-On Systematic Innovation for Business and Management.* Devon, UK: Lazarus Press, 2004.
- Pande, Pete and Larry Holpp. *What is Six Sigma?* New York: McGraw-Hill, 2002.
- Pande, Peter S., Robert P. Neuman, and Roland R. Cavanagh. *The Six Sigma Way Team Fieldbook.* New York: McGraw-Hill, 2002.
- Pyzdek, Thomas. *The Six Sigma Handbook.* New York: McGraw-Hill, 2003.
- Tague, Nancy R. *The Quality Toolbox, 2nd Edition.* Wisconsin: ASQ Quality Press, 2005.
- Sleeper. Andrew. *Design for Six Sigma Statistics.* New York: McGraw-Hill, 2006.
- Urban, Glen L. and John R. Hauser. *Design and Marketing of New Products.* New Jersey: Prentice-Hall, 1980.
- Wenger, Etienne, Richard McDermott, and William M. Snyder. *Cultivating Communities of Practice.* Massachusetts: Harvard Business School Publishing, 2002. ISBN 1-57851-330-8.
- Wilson, Paul F., Larry D. Dell, and Gaylord F. Anderson. *Root Cause Analysis, A Tool for Total Quality Management.* Wisconsin: ASQ Quality Press, 1993.
- Wortman, Bill, W.R. Richardson, G. Gee, M. Williams, T. Pearson, F. Bensley, J.P. Patel, J. DeSimone, and D.R. Carlson. *Villanova University Comprehensive Six Sigma Black Belt Handbook.* Indiana: Quality Council of Indiana, 2006.

Internet Sites

- www.aia-consulting.com; see Dana W. Clarke's *Innovation Magic* newsletter.
- www.aitriz.org
- www.isixsigma.com
- www.patientsafety.gov
- www.pmi.org
- www.triz-journal.com

Index

Page numbers preceded by PDF: indicate topics found online in downloadable PDF format at www.prenhallprofessional.com/title/0132300214

C

D

E

F

G

H

I

J–K

L

M

N

Q

R

S

T

U

V

W

X–Z